VECTOR CALCULUS

VECTOR CALCULUS

Thomas H. Barr
Rhodes College

PRENTICE HALL
Upper Saddle River, NJ 07458

Library of Congress Cataloging-in-Publication Data

Barr, Thomas H.
Vector calculus/Thomas H. Barr
p. cm.
Includes bibliographical references (p. -) and index.
ISBN 0-13-400037-4
1. Vector analysis. I. Title.
QA433.B35 1997
515'.63–dc20 96-27591
CIP

Acquisitions Editor: George Lobell
Editorial Assistant: Gale Epps
Editorial Director: Tim Bozik
Editor-in-Chief: Jerome Grant
Assistant Vice President of Production and Manufacturing: David W. Riccardi
Editorial/Production Supervision: Robert C. Walters
Managing Editor: Linda Mihatov Behrens
Executive Managing Editor: Kathleen Schiaparelli
Manufacturing Buyer: Alan Fischer
Manufacturing Manager: Trudy Pisciotti
Marketing Manager: John Tweeddale
Marketing Assistant: Diane Penha
Creative Director: Paula Maylahn
Art Director: Jayne Conte
Cover Photo: Ferald Murphy's "Watch," Dallas Museum of Art

Simon & Schuster / A Viacom Company
Upper Saddle River, NJ 07458

Printed in the United States of America
10 9 8 7 6 5 4 3 2 1

ISBN 0-13-400037-4

PRENTICE-HALL INTERNATIONAL (UK) LIMITED, LONDON
PRENTICE-HALL OF AUSTRALIA PTY. LIMITED, SYDNEY
PRENTICE-HALL OF CANADA, INC., TORONTO
PRENTICE-HALL HISPANOAMERICANA, S.A., MEXICO
PRENTICE-HALL OF INDIA PRIVATE LIMITED, NEW DELHI
PRENTICE-HALL OF JAPAN, INC., TOKYO
SIMON & SCHUSTER ASIA PTE. LTD., SINGAPORE
EDITORA PRENTICE-HALL DO BRASIL, LTDA., RIO DE JANEIRO

For Kathryn, Rebecca, Elizabeth,
and my parents Harold and Sara

Contents

Preface

Vector calculus is a natural extension of differential and integral calculus for functions of a single variable. In this subject, the variables and values of functions are vectors instead of real numbers, as they are in single-variable calculus. The subject is a natural outgrowth of attempts in the nineteenth century to express and answer questions arising in mechanics, electromagnetic theory, and fluid dynamics. In the present day, it is a mature mathematical area, and one whose domain of applications has spread to such areas as chemistry, engineering, biology, and economics. It is a rich subject, worth the attention and study of students in any of these disciplines.

This book is an introduction to vector calculus; it is written for students who have taken an introductory calculus course that includes concepts, techniques, and applications of differentiation and integration, and also some discussion of Taylor's Theorem. The approach is to present the subject in an intuitive fashion, stressing geometric notions and their algebraic and analytical counterparts. Part of this approach is to introduce and use some of the rudiments of linear algebra; this makes vector calculus conceptually and formally parallel with single-variable calculus. There is a plethora of exercises, spanning a spectrum from "drill" exercises (necessary in any introduction, even though computers and calculators can do amazing things) to conceptual questions, to somewhat challenging multi-step problems that may set an ambitious student on any of several paths of inquiry.

Graphical software and graphing calculators can help greatly in visualizing curves and surfaces in three dimensional space. By no means are such electronic tools essential tools for this subject, but used to any extent, they can make it more stimulating than it might be when hand-drawn pictures are the only images possible. In many cases, they can aid greatly in solving problems, settling conjectures, and raising fascinating questions. Some examples in this text are presented using Mathematica graphics, and some of the exercises will be expedited by using this or a similar program. In addition, there are exercises in which a computer algebra system or calculator that can do symbolic integration may be helpful. It is very satisfying, for example, to set up a triple integral as an iterated integral and then turn the antidifferentiation over to a machine to produce a final answer. In doing such an exercise in this way, the student can concentrate more on the essential points of the exercise: visualizing and describing the region of integration, reflecting on the theorem that connects triple integrals with iterated integrals, and interpreting the final answer.

This text contains more than ample material for a one-semester course at the second year college level. For example, the instructor could make a fairly complete three semester-hour course by covering chapters 1, 2, and 3 (skipping 3.8), and sections 4.1 through 4.5. In a four-hour course, it may be possible to cover most of the book. At a slightly more leisurely pace, it would be possible to cover the whole book in a two-quarter sequence.

At Rhodes College, students who take the semester-long third calculus course using this text are either sophomores who have taken the year-long single-variable calculus sequence or first-year students who have scored 4 or 5 on the BC Advanced Placement Test. The subsequent mathematics course is usually differential equations or linear algebra.

A book of this sort is the product of thousands of hours of labor, and also of the input of hundreds of people. Many of these people have been students, both in classes I have taught myself and classes taught by others who have used this manuscript in earlier versions. It is not possible for me to name them all here. There are several people, though, to whom I am profoundly indebted and who must be named. My colleague Ken Williams has provided me countless hours of conversation about mathematics, teaching, and vector calculus in particular; it was his encouragement that eventually led to my embarking on this project. Ferebee Tunno, Mike Rosolino, and Stephen Baumert, all Rhodes students, have read the text, worked out solutions to exercises, and offered many wonderful suggestions; to them also I owe much. Andrew Simoson at King College has contributed greatly, both by classroom testing the manuscript and producing solutions to exercises. Janet Andersen, Tim Pennings, and John Stoughton and their students at Hope College have offered helpful criticisms, corrections, and moral support. Etta Coughlan TEX-ed most of the manuscript, and for her diligence, good-humor, and patience I am greatly indebted. My colleague Brian Stuart and students Scott Wells and John Schafer have provided tremendous technical assistance in helping to educate me about computer graphics, typesetting, and combining these elements. For the insight, honesty, and kindness of the reviewers of the manuscript, I am grateful; I have taken many of those comments and suggestions to heart. George Lobell and Jerome Grant at Prentice Hall have shepherded this project along; it has been my pleasure to work with and get to know them. Bob Walters, the editorial and production supervisor, has taught me a great deal about fine points of typesetting. Barbara Zeiders, the copy editor, has helped greatly to improve the manuscript and to educate me in many matters of grammar, punctuation, usage, and typesetting. Tom Kiffe at Texas A & M has provided invaluable assistance with TEXand related software. My wife Kathryn and daughters Rebecca and Elizabeth have contributed greatly by giving up the time and attention that I have diverted from them to write the words and draw the pictures in the following pages. And the countless words of encouragement from family and friends have helped me maintain a reasonable work pace.

Readers who spot misprints or errors will garner my gratitude by communicating them to me either by regular mail (Mathematics and Computer Science Department, Rhodes College, 2000 N. Parkway, Memphis, TN 38112) or electronic mail (`barr@mathcs.rhodes.edu`).

Thomas H. Barr

Chapter 1

Coordinate and Vector Geometry

Elementary calculus presupposes a general understanding of numbers and their geometric representation on the number line. In a similar way, multivariate calculus relies on an understanding of such notions as ordered pairs and ordered triples of numbers and their geometric representation in the Cartesian plane and coordinatized three-dimensional space. Thus it is the purpose of Chapter 1 to help the reader form a literal picture of these ideas that is adequate to prepare for "doing" calculus on multivariate functions.

Much of what we discuss in this book will focus on what happens in three-dimensional space, the geometric model for the physical world about us. For that reason, and also because most students of this subject will already have studied plane analytic geometry (i.e., points, distance, lines, conic sections, and the formulas and algebraic methods by which we talk about them), it is appropriate to begin by discussing analytic geometry in three-dimensional space. The analogies with the planar theory will, in many cases, be quite direct: we will discuss graphs of simple equations in three variables and briefly examine the three-dimensional analogs of such objects as circles, ellipses, and hyperbolas.

The physical notions of displacement, velocity, acceleration, and force all have the common features of "direction" and "magnitude." These two features can be captured geometrically by representing the physical quantities as directed line segments (i.e., as **vectors**). We will see that defined in this way, vectors can also be represented in terms of components, and consequently in terms of coordinates. It is this connection between vectors and coordinate geometry, along with algebraic properties of vectors, that makes them an essential part of the rest of this book. The basic idea is that vectors play the role in this subject that numbers played in elementary calculus. In Section 1.4 through the end of Chapter 1, we see how geometry can be done with vectors, and we explore many formal properties and applications of vectors.

1.1 Rectangular Coordinates and Distance

Points in space can be located by three **coordinates**, that is three numbers written in a particular order. This can be done by first establishing a **coordinate system** of three mutually perpendicular number lines whose origins coincide at a single point, called the **origin** of the coordinate system. It is conventional to orient these **axes** and label them x, y, and z, so that as the fingers on one's right hand rotate from the positive x-axis to the positive y-axis through the right angle between them, the thumb points in the positive z direction. See Figure 1.1.1. For obvious reasons, such a system is called

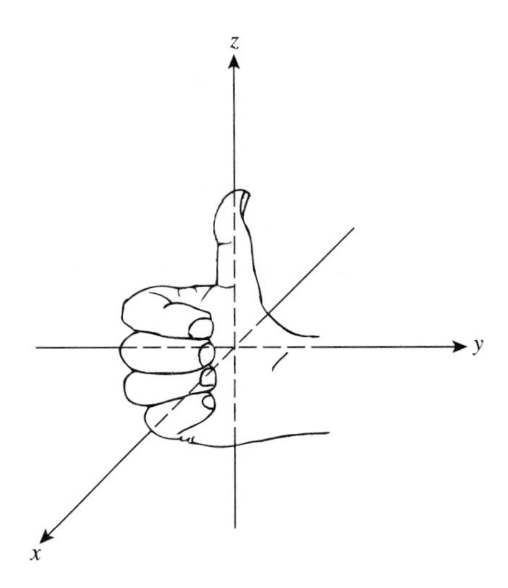

Figure 1.1.1 A right-handed coordinate system

right-handed. It is also conventional to draw two-dimensional representations of this system as if the positive x-axis were protruding toward the viewer perpendicular to the plane of the paper.

Given a point P in space, draw a perpendicular line from it to the x-axis. The coordinate a of the point of intersection with the x-axis is called the x-**coordinate** of P. See Figure 1.1.2. Next, construct a line from P

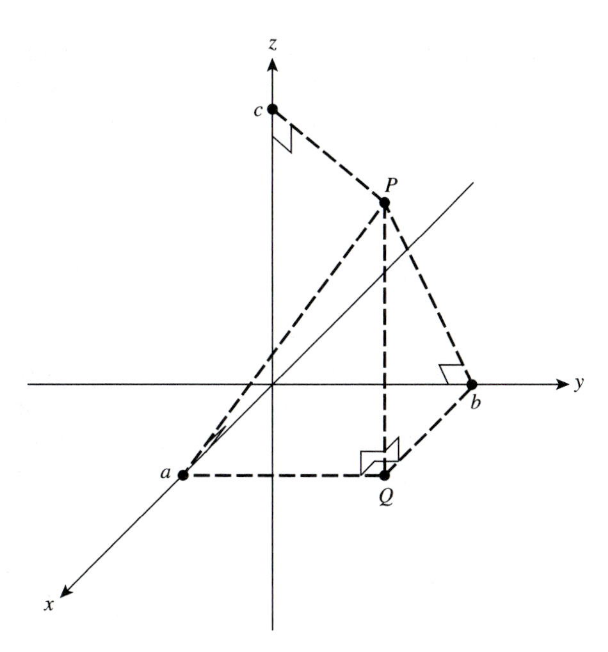

Figure 1.1.2 A point and its coordinates

perpendicular to the y-axis; the coordinate b of the intersection is the **y-coordinate** of P. The **z-coordinate** is obtained in a similar fashion. This procedure returns an ordered triple (a, b, c) of numbers for any given point P.

Conversely, given an ordered triple (a, b, c), this process can be reversed to produce a point in space whose coordinates are (a, b, c). Through the point on the x-axis with coordinate a draw a line parallel to the y-axis, and through the point on the y-axis with coordinate b, draw a line parallel to the x-axis. Call the intersection of these lines Q. See Figure 1.1.2 again. Through Q draw a line parallel to the z-axis, and find the point P such that the segment from P to the point on the z-axis with coordinate c is perpendicular to the z-axis. Then P has coordinates (a, b, c).

To summarize, there is a one-to-one correspondence between points in space and ordered triples of real numbers. It is this observation that makes applications meaningful, and it also gives us perfect freedom to refer to "the point (a, b, c)" instead of saying "the point with the coordinates (a, b, c)."

Example 1.1.1 Plot the points $(3, 0, 0)$, $(0, -2, 0)$, $(-4, 3, 0)$, and $(2, 2, 4)$ on one set of coordinate axes.

Solution The first two points lie on the x- and y-axes, respectively, and

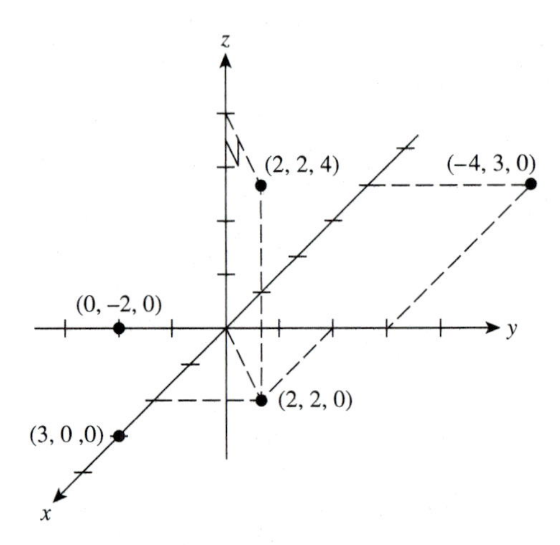

Figure 1.1.3 Plotting points in $\mathbb{R}^3$

they are easy to locate in Figure 1.1.3. The third point lies in the xy-plane, and it is readily drawn in the figure by sketching a dashed line parallel to the y-axis through the point on the x-axis with coordinate -3 and a second dashed line parallel to the x-axis through the point on the y-axis with coordinate 3. Where the two intersect is the desired point $(-3,\ 3,\ 0)$. To indicate clearly where $(2,\ 2,\ 4)$ lies, first sketch $(2,\ 2,\ 0)$. Then sketch the vertical line through $(2,\ 2,\ 0)$ that is parallel to the z-axis; the point $(2,\ 2,\ 4)$ will lie on this line. In order to locate $(2,\ 2,\ 4)$ with the proper perspective, lightly draw the segment from $(0,\ 0,\ 0)$ to $(2,\ 2,\ 0)$ and then draw the parallel segment that starts at $(0,\ 0,\ 4)$. Where it intersects the vertical line is the point $(2,\ 2,\ 4)$. ♦

We write $\mathbb{R}^3$ for the set $\{(x,\ y,\ z)|x,\ y,\ z \in \mathbb{R}\}$, where $\mathbb{R}$ denotes the real numbers, and we *think* of $\mathbb{R}^3$ geometrically because of the direct correspondence between ordered triples and points in space. Indeed, we may refer to three-dimensional space simply as "$\mathbb{R}^3$."

The three planes in $\mathbb{R}^3$ determined by the three possible pairs of coordinate axes are called the **coordinate planes**. These planes subdivide $\mathbb{R}^3$ into eight (count them!) different parts called **octants**. There is no generally-agreed-upon numbering of them, but the one consisting of points with nonnegative coordinates is called the **first octant**.

The coordinate planes and the ones parallel to them are all easily described by equations. For instance, the yz-plane consists of all points (x, y, z) with zero x-coordinate. That is, a point $(x,\ y,\ z)$ lies in the yz-plane if and only if $x = 0$. Thus the *equation* $x = 0$ completely describes the yz-plane. We may wish to be more formal and less wordy by writing $\{(x, y, z)|x = 0\}$ for the yz-plane.

As another example, consider the equation $y = -3$. It describes the collection of points (x, y, z) with y coordinate -3. Thus it represents the plane parallel to the xz-plane which passes through the point $(0, -3, 0)$.

Example 1.1.2 Describe and sketch the set of points in $\mathbb{R}^3$ that satisfy the equation $x + y = 2$.

Solution If the space under consideration were the xy-plane then the graph of the equation would be the line with x- and y-intercepts 2. Indeed, this observation reveals that the set of points in $\mathbb{R}^3$ that satisfy this equation intersects the plane $z = 0$ in this line. See Figure 1.1.4. Since any point above

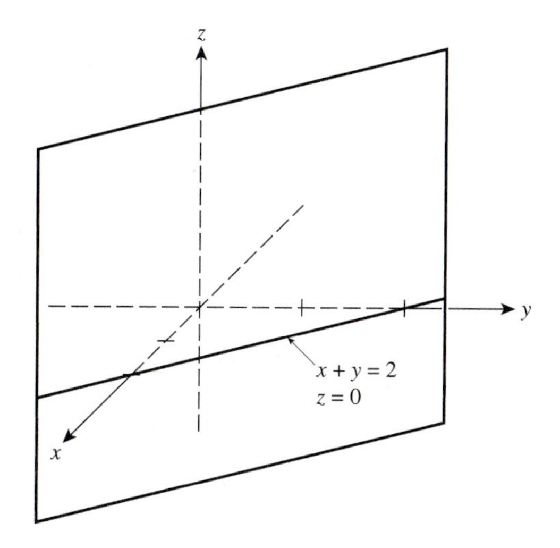

Figure 1.1.4 The plane $x + y = 2$ in $\mathbb{R}^3$

or below the line has the same x- and y-coordinates as the corresponding point on the line, such a point also lies in the set described by the equation. The set of points in $\mathbb{R}^3$ lying above and below the line is simply the plane parallel to the z-axis that passes through the line. Any point that is off this plane has x- and y-coordinates that do not satisfy the equation, so there are no other points in the graph of the equation except those in the plane. ♦

In Example 1.1.2, to make a sketch that is somewhat realistic, I have drawn the portions of the coordinate axes hidden by the surface as dashed lines. This convention of dashing lines that are hidden by surfaces is a simple and effective aid to sketching many three-dimensional figures. Also, I am only able to suggest where the plane is by drawing a rectangular portion of it (which actually appears as a parallelogram because of the perspective). Finally, notice that I have left a little gap in the right-hand edge of this rectangle where it passes behind the y-axis. This is another simple and subtle trick that can be very helpful whenever depth is to be suggested. While initially your own attempts at sketching may involve several erasures (*always* use pencil!), effort and practice in the beginning will help you to cultivate your ability to visualize and draw.

Given two points $P_1 = (a_1, a_2, a_3)$ and $P_2 = (b_1, b_2, b_3)$ in $\mathbb{R}^3$, we want to define the distance between the points in a way that is consistent with the way it is calculated in the plane. As shown in Figure 1.1.5, drop segments

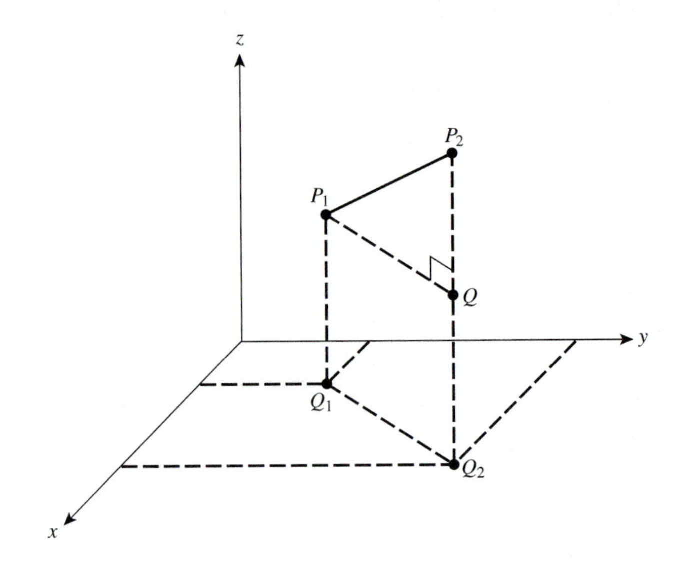

Figure 1.1.5 Deriving the formula for the distance between P_1 and P_2

from P_1 and P_2 perpendicular to the xy-plane, and let Q_1 and Q_2 be the corresponding points in the xy-plane. Construct a segment from P_1 parallel to Q_1Q_2 and label its intersection with P_2Q_2 by Q. Then ΔP_1QP_2 is a right triangle with hypotenuse P_1P_2. By the Pythagorean Theorem,

$$|P_1P_2|^2 = |P_1Q|^2 + |QP_2|^2. \tag{1.1.1}$$

(Here the vertical bars indicate the length of the segment determined by the two points between the bars.) Also note that $|P_1Q| = |Q_1Q_2|$. Since $Q_1 = (a_1, a_2, 0)$ and $Q_2 = (b_1, b_2, 0)$, by the usual distance formula in the plane we have $|Q_1Q_2|^2 = (b_1 - a_1)^2 + (b_2 - a_2)^2$. Since Q has the same z-coordinate as P_1, we see that $|P_2Q| = |b_3 - a_3|$, so substituting into (1.1.1), we obtain

$$|P_1P_2| = \sqrt{(b_1 - a_1)^2 + (b_2 - a_2)^2 + (b_3 - a_3)^2}. \tag{1.1.2}$$

This serves as our definition of the **distance** between two points in $\mathbb{R}^3$.

Example 1.1.3 A pedestrian in a large city walks one block east, two blocks south, and then enters a building and rides the elevator to the 42nd floor. If city blocks are 1000 ft long and each story in this building is 10 ft high, how far is she from her starting point?

Solution By establishing a coordinate system with its origin at her starting point and x-, y-, and z-axes pointing east, north, and straight up, respectively, we see that the question is, "What is the distance between the points (0, 0, 0) and (1000, −2000, 420)?" Using the distance formula (1.1.2), we calculate that she is

$$\sqrt{1000^2 + (-2000)^2 + 420^2} = \sqrt{5176400} \approx 2275 \text{ ft}$$

from her starting point. ♦

The set of all points in space that are a fixed distance from some given point is a **sphere**[1]. If the fixed distance is r, say, and the given point is (a, b, c), then a point (x, y, z) lies on the sphere if and only if the distance between $(x,\ y,\ z)$ and (a, b, c) equals r. That is, (x, y, z) is on the sphere if and only if x, y, and z satisfy the equation

$$(x - a)^2 + (y - b)^2 + (z - c)^2 = r^2.$$

Example 1.1.4 Find the equation for the sphere centered at the point (5, −2, 6) that is tangent to the xz-plane.

Solution Clearly, we need to determine the sphere's radius. By making a sketch (see Figure 1.1.6) we are easily convinced that the sphere is tangent to the xz-plane at the point $(5, 0, 6)$ and thus that its radius must be 2. Thus the equation of the sphere is

$$(x - 5)^2 + (y + 2)^2 + (z - 6)^2 = 4$$

[1]In this book we strive to make distinctions among objects of different dimensions. Thus a *sphere* is a two-dimensional object, the "skin" of a three-dimensional, solid *ball*. This is somewhat at variance with ordinary usage, where often the term *sphere* is used to refer to both a round solid object and to its surface. Similarly, we will at times make a distinction between a *circle* and the *disk* that it bounds.

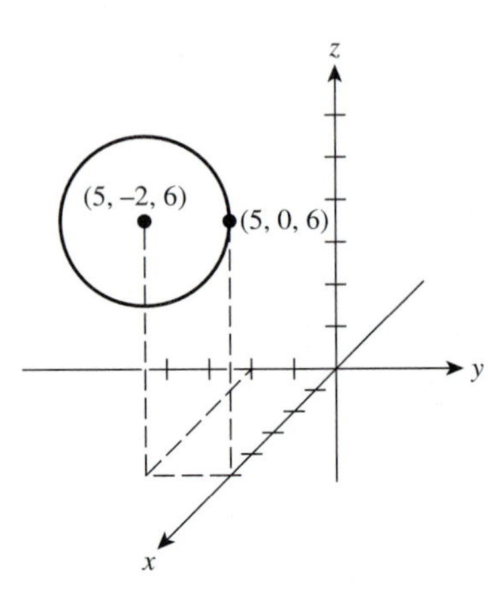

Figure 1.1.6 The sphere centered at (5, −2, 6) that is tangent to the xz-plane

or, in expanded form,

$$x^2 + y^2 + z^2 - 10x + 4y - 12z + 61 = 0. \quad \blacklozenge$$

Example 1.1.5 Find the largest sphere centered at $(3, 9, 11)$ that is contained in the first octant.

Solution The radius of the sphere can be no more that the shortest of the distances from (3, 9, 11) to the coordinate planes. Since that distance is 3 (why?), the largest such sphere has radius 3 and its equation is therefore

$$(x - 3)^2 + (y - 9)^2 + (z - 11)^2 = 9. \quad \blacklozenge$$

Inequalities involving x, y, and z often describe *regions* of space. We illustrate this in the next example.

Example 1.1.6 Describe the sets of points that satisfy the following inequalities: **(a)** $y > 4$, **(b)** $x^2 + y^2 + z^2 - x - 2y + 3z + \frac{5}{2} \leq 0$, **(c)** $0 \leq x \leq 1$, $0 \leq y \leq 2$, $0 \leq z \leq 3$.

Solution

(a) The points in (x, y, z) for which $y > 4$ is simply all points in $\mathbb{R}^3$ to the right of the plane $y = 4$.

(b) By completing the square, we can write the inequality as

$$\left(x-\frac{1}{2}\right)^2+(y-1)^2+\left(z+\frac{3}{2}\right)^2\le 1.$$

In words, this says that the distance between (x, y, z) and $\left(\frac{1}{2}, 1, -\frac{3}{2}\right)$ is less than or equal to 1. Thus the inequality describes the **ball** of radius 1 centered at $(\frac{1}{2}, 1, -\frac{3}{2})$, that is, the sphere of radius 1 centered at this point along with all the points inside it.

(c) For any point (x, y, z) that satisfies all three inequalities, the first inequality $0 \le x \le 1$ requires that the point lie between or on the planes $x = 0$ and $x = 1$. Similarly, the second inequality requires the point to lie between or on the planes $y = 0$ and $y = 2$, and the third requires it to lie between or on $z = 0$ and $z = 3$. Thus the set of points forms a rectangular solid with faces formed by these six planes. ♦

The **midpoint** of a segment joining two points $P_1 = (a_1, a_2, a_3)$ and $P_2 = (b_1, b_2, b_3)$ is the point $P = (\overline{x}, \overline{y}, \overline{z})$ on the segment that is equidistant from P_1 and P_2. Referring to Figure 1.1.7, you will see that the z coordinate of P is the same as that of R and, by similar triangles, R is the midpoint of P_2Q. Thus the z-coordinate of R is the z coordinate of Q, plus half the difference between b_3 and a_3. That is, $\overline{z} = a_3 + \frac{1}{2}(b_3 - a_3) = \frac{1}{2}(a_3 + b_3)$.

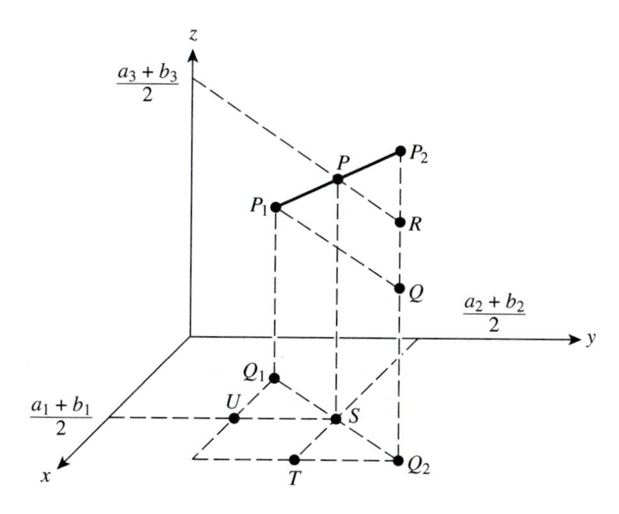

Figure 1.1.7 Finding the coordinates of the point P midway between P_1 and P_2

The x- and y-coordinates of P are the same as those of S, and these are, respectively, the x-coordinate of U and the y-coordinate of T. Reasoning as for R, we have

$$\overline{x} = a_1 + \frac{1}{2}(b_1 - a_1) = \frac{1}{2}(a_1 + b_1)$$

and

$$\overline{y} = a_2 + \frac{1}{2}(b_2 - a_2) = \frac{1}{2}(a_2 + b_2).$$

Thus the midpoint is

$$\left(\frac{a_1 + b_1}{2}, \frac{a_2 + b_2}{2}, \frac{a_3 + b_3}{2}\right).$$

Example 1.1.7 Determine whether the midpoint of the segment joining (−8, 11, 13) to (2, 7, 5) lies inside the ball of radius 4 centered at (2, 5, 6).

Solution The midpoint is simply

$$\left(\frac{-8+2}{2}, \frac{11+7}{2}, \frac{13+5}{2}\right) = (-3, 9, 9)\,.$$

Since the distance from (−3, 9, 9) to (2, 5, 6) is $\sqrt{25+16+9} = \sqrt{50} = 5\sqrt{2}$ and this is larger than 4, the midpoint is outside the ball. ♦

Exercises 1.1

1. Sketch the following points on a single set of coordinate axes.

   (a) $(0, 0, -\frac{3}{2})$ **(b)** (2, 3, 0)

   (c) (−1, −1, −1) **(d)** (4, 0, 1)

2. The points (0, 0, 0) and (3, 2, 1) are to be opposite vertices of a rectangular box with faces parallel to the coordinate planes. Find the remaining vertices and sketch the box.

3. Same as 2 with (−1, −1, 1) and (4, 3, 5).

4. Find the distances between the following pairs of points and determine the midpoints of the segments determined by the pairs.
   (a) (0, −11, 0), (4, 0, 13)
   (b) (1/3, 1/4, 1/5), (−8/3, 1/2, −7/5)
   (c) (3.7, 1.2, 5.6), (7.3, −1.2, 6.5)
   (d) $(\pi, 3\pi, \pi/2)$, $(-\pi/6, 2\pi/3, -\pi/2)$
   (e) $(1-\sqrt{5}, \sqrt{2}, -1)$, $(1+\sqrt{5}, -\sqrt{2}, 2)$

In Exercises 5 through 11, describe in words and sketch the sets of points that satisfy the given equation.

5. $z = -3$

6. $x = 1$

7. $2x + z = 2$

8. $x^2 + y^2 + z^2 - 12y = 36$

9. $x^2 + y^2 + z^2 + 4x - 10y - 2z + 26 = 0$

10. $x^2 + y^2 = 1$

11. $(y - 1)^2 + (z + 3)^2 = 0$

In Exercises 12 through 16, describe and sketch the sets of points that satisfy the given inequality or inequalities.

12. $y < -2$

13. $-1 \leq z \leq 1$

14. $0 \leq x \leq 1$, $3 \leq y \leq 4$, $1 \leq z \leq 2$

15. $(x - 1)^2 + (y + 1)^2 + (z + 3)^2 > 81$

16. $x^2 + y^2 + z^2 \leq 4$, $z \geq 1$

17. The Great American Pyramid is a basketball arena in Memphis whose exterior is in the shape of a pyramid with a square base covering approximately $531,000$ ft^2 and a height of approximately 321 ft. About how far is it from one corner of the pyramid to its top?

18. What is the radius of the smallest ball centered at the point $(3, -8, 6)$ that contains the origin?

19. Find the equation for the largest sphere that can be inscribed in the cubical region given by $1 \leq x \leq 3$, $-2 \leq y \leq 0$, $6 \leq z \leq 8$.

20. Find the largest sphere centered at (0, 5, 3) that will fit inside the rectangular box given by the inequalities $-1 \leq x \leq 2$, $0 \leq y \leq 6$, $1 \leq z \leq 7$.

21. How many points with integer coordinates lie a distance no more than 2 from the origin?

22. In words (Do not try to obtain a general formula!), write a reasonable definition for the distance from a point to a line. Use your definition to find the distance from the point (6, 7, −6) to the x-, y-, and z-axes.

23. Geometrically, what is the set of points that are a fixed distance from a given line? Write an equation for the set of points that have distance 5 from the y-axis.

24. What is a reasonable way to define the distance from a point to a plane? (As in Exercise 22, state your definition in words, and do not attempt to find a general formula.) Use your definition to find the distance from the point $(-5, 2, 8)$ to the xy-, xz-, and yz-planes.

25. The set of points in $\mathbb{R}^3$ that are equidistant from two given points $P_1 = (a_1, a_2, a_3)$ and $P_2 = (b_1, b_2, b_3)$ is a plane. Show that the equation for this plane is

$$(b_1 - a_1)(x - \overline{x}) + (b_2 - a_2)(y - \overline{y}) + (b_3 - a_3)(z - \overline{z}) = 0,$$

where $(\overline{x}, \overline{y}, \overline{z})$ is the midpoint of the segment P_1P_2.

26. Using the result in Exercise 25, find the equation for the plane consisting of points equidistant from the following pairs of points:
(a) $(0, 0, 0)$ and $(-2, 7, 3)$
(b) $(3, -3, 2)$ and $(2, 1, 4)$

27. Suppose that the midpoint and one endpoint of a segment are known. Find the other endpoint in terms of the two known points.

28. Use Exercises 25 and 27 to find the equation for the plane that is tangent to the sphere $x^2 + y^2 + z^2 = 9$ at the point $(1, -1, \sqrt{7})$.

29. Use Exercises 25 and 27 to show that if $(\overline{x}, \overline{y}, \overline{z})$ is a point on the sphere $x^2 + y^2 + z^2 = r^2$, then the equation for the plane tangent to the sphere at $(\overline{x}, \overline{y}, \overline{z})$ is

$$\overline{x}x + \overline{y}y + \overline{z}z = r^2.$$

1.2 Surfaces and Equations

The objective in this section is for us to become familiar with geometrical representation of equations that involve three variables. To do this we discuss both some general methods that can aid in sketching the graphs of equations and some specific graphs that will serve as ready examples later.

Graphs of Functions of Two Variables

Definition 1.2.1 A **function** of two real variables x and y is a rule, f, that associates with each choice of x and y a single real number $f(x, y)$ called the **value of** f **at** (x, y). The set of ordered pairs (x, y) for which the rule is specified is the **domain** of the function.

Example 1.2.1 The function f that associates with x and y the sum of the squares of x and y can be expressed by

$$f(x, y) = x^2 + y^2.$$

The value of this function, for instance, at $(2, -5)$ is

$$f(2, -5) = 2^2 + (-5)^2 = 29.$$

Its domain consists of all ordered pairs $(x, y) \in \mathbb{R}^2$. ♦

Example 1.2.2 The function f defined by $f(x, y) = (\ln y)/x$ has values such as $f(-3, 2) = -(\ln 2)/3$ and $f(4, e) = (\ln e)/4 = \frac{1}{4}$. Its domain consists of ordered pairs (x, y) such that $x \neq 0$ and $y > 0$. In set notation, we write the domain as $\{(x, y) \in \mathbb{R}^2 | x \neq 0, y > 0\}$. ♦

The adage about one picture being worth a thousand words holds true when we try to explain the qualitative and global behavior of a function of two variables: a graph of a function conveys a great deal of information about it.

Definition 1.2.2 The **graph** of a function f of two variables is the set of all points (x, y, z) in $\mathbb{R}^3$ such that $z = f(x, y)$ and (x, y) is in the domain of f.

Example 1.2.3 Given a constant c, the function $f(x, y) = c$ has as domain all ordered pairs (x, y). Its graph is then all points (x, y, z) such that $z = c$. Evidently, this is the plane parallel to the xy-plane that cuts the z-axis at c. ♦

Example 1.2.4 Draw the graph of $f(x, y) = \sin x$.

Solution No matter what value we assign to y, $f(x, y)$ is still $\sin x$. Geometrically, for each number c, the plane $y = c$ cuts the graph of f in a curve exactly like the curve $z = \sin x$ in the xz-plane. See Figure 1.2.1. Indeed, the graph consists of the union of these curves. This surface is like a landscape of infinitely many ridges parallel to the y-axis. ♦

In general, if f is a function of two variables in which one of the variables does not appear, then the graph is the union of copies of a single curve that lies in either the xz-, yz-, or xy-plane. The graph of such a function is usually fairly easy to visualize and sketch, as Example 1.2.4 indicates. The

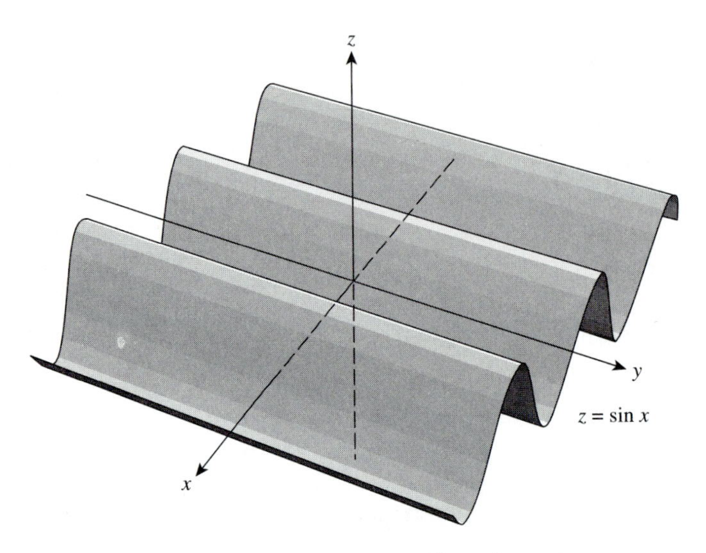

Figure 1.2.1 Graph of $f(x, y) = \sin x$

graph of a function of two variables in which one of the variables does not appear is called a **cylinder**.

Often, the graph of a function is not easily visualized or sketched by hand. There are, however, many pieces of software available that have fairly sophisticated three-dimensional graphing capabilities. Most work in the same basic way. The user specifies the function f to be graphed and a range of (x, y) values for which the plotting is to be done, typically a rectangle in the xy-plane. The software then automatically subdivides the rectangle into a grid of several small subrectangles, calculates the value of f at the corners of the rectangles, plots the points, and then connects with straight lines the points on the graph corresponding to adjacent corners of each subrectangle. This gives a "wire-frame" approximation to the actual surface. Then the software projects the straight-line approximating image onto a viewing plane that it usually selects automatically. The resulting picture on the computer screen is this projection, a jumble of overlapping quadrilaterals that may or may not suggest the shape of the original surface. In Figure 1.2.2(a), for instance, we see the result of this process applied to the function $f(x, y) = x^2 - y^2$.

The program may add realism to the two dimensional scene by automatically "hiding" portions of some of these quadrilaterals that correspond to portions of the actual surface that are obscured when viewed from the viewing plane. This is illustrated in Figure 1.2.2(b).

Example 1.2.5 The graph of $f(x, y) = 3(x - x^3)(1/4 - y^2)$ for $-2 \le x \le 2$ and $-1.5 \le y \le 1.25$ is shown in Figure 1.2.3 from two different points of

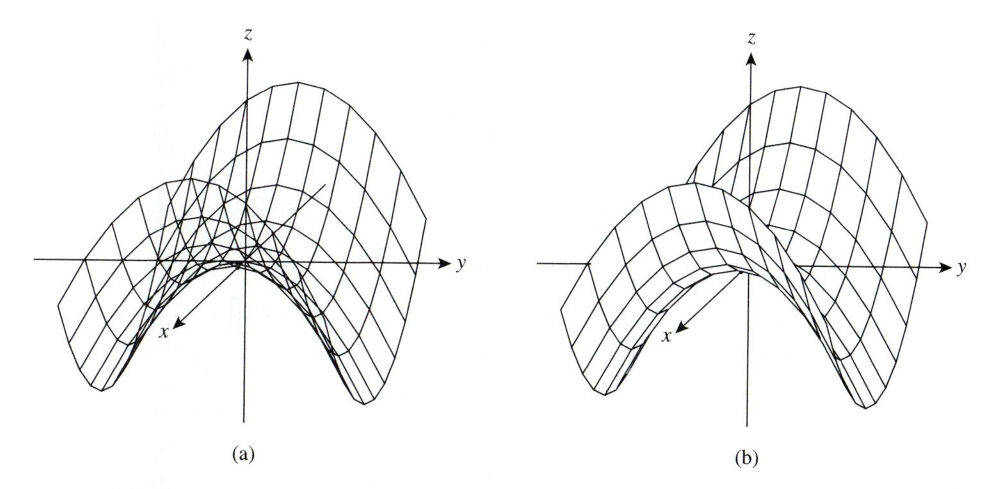

Figure 1.2.2 Wire-frame and hidden-line renderings of the graph of $f(x, y) = x^2 - y^2$

view. The first view is from (20, 4, 4), a point that is about five times as far out along the x-axis as it is along the y- or z-axis. The second view is from (10, 20, 5). Notice that the software has not drawn the coordinate axes but has instead drawn a box around the portion of the graph displayed. This automatically drawn box helps to guide the eye and keeps the figure from being cluttered.

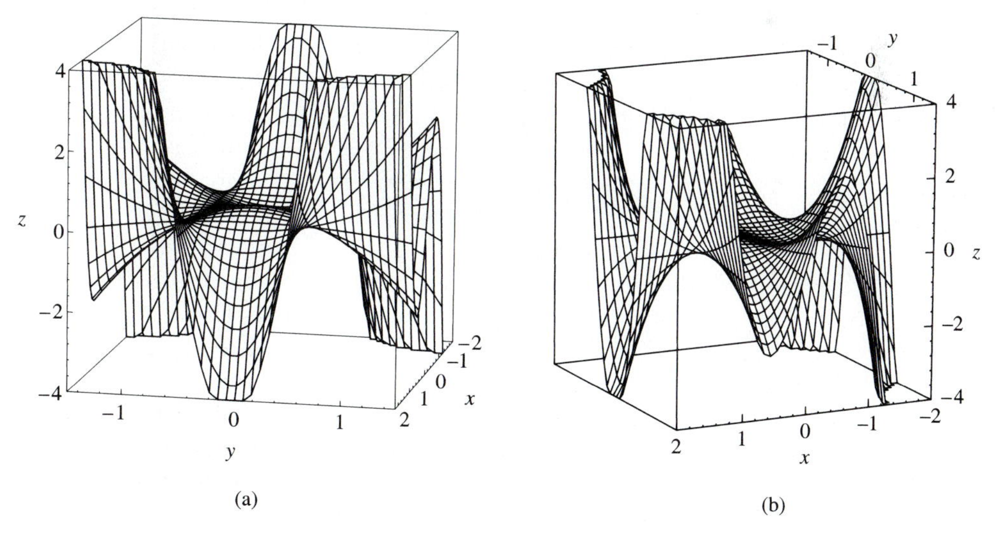

Figure 1.2.3 Two views of the graph of $f(x, y) = 3(x - x^3)(1/4 - y^2)$

Can you determine roughly the coordinates of the points on the surface where the "peaks," "passes," and "valleys," in this "landscape" occur? ♦

Example 1.2.6 In modeling the concentration of a pollutant leaking from a ship at sea, we might make the qualitatively reasonable assumption that the concentration of pollutant falls off in inverse proportion to the square of the distance from the ship. What is the shape of the graph of the pollutant concentration function?

Solution If we put the ship at the origin of an xy-coordinate system then the distance of a point (x, y) from the ship is $\sqrt{x^2 + y^2}$. Our assumption about the pollutant means that its concentration is proportional to

$$f(x, y) = \frac{1}{(\sqrt{x^2 + y^2})^2} = \frac{1}{x^2 + y^2}.$$

The graph of this function is shown in Figure 1.2.4. Notice that the values of f get arbitrarily large for (x, y) near $(0, 0)$. ♦

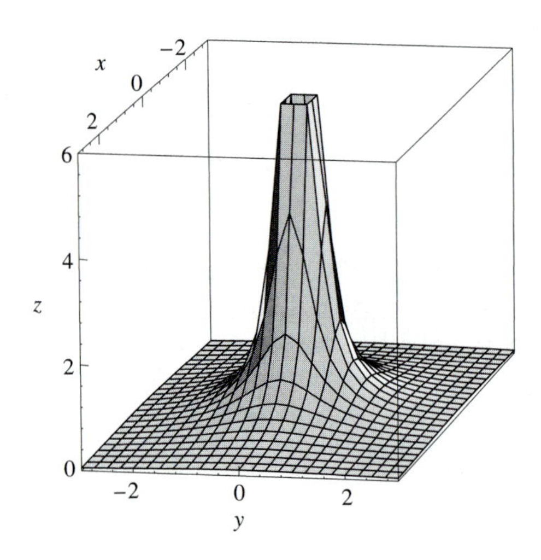

Figure 1.2.4 Graph of $f(x, y) = \dfrac{1}{x^2 + y^2}$

The graphs of functions of two variables are examples of what we will call **surfaces**. More generally, a set of points (x, y, z) that satisfy an equation that relates all three variables is often a surface. For instance, the set of points (x, y, z) satisfying

$$x^2 + (y - 1)^2 + (z + 3)^2 = 25$$

is a sphere of radius 5 centered at (0, 1, −3). It is not, though, the graph of a function of two variables (why?). The equation can be written as $F(x, y, z) = 0$ where F is the function of three variables defined by

$$F(x, y, z) = x^2 + (y - 1)^2 + (z + 3)^2 - 25.$$

Generally, an equation in x, y, and z can be written in the form

$$F(x, y, z) = 0$$

for some function F, and often the points satisfying such an equation form a surface in $\mathbb{R}^3$.

There is a class of equations in three variables whose graphs are readily analyzed with a little knowledge of conic sections. These also provide a traditional repertoire of examples to illustrate general concepts of differentiation and integration. A few of these will be important in our discussion of local extrema in Chapter 3.

Quadric Surfaces

The **general second-degree equation in three variables** is

$$Ax^2 + By^2 + Cz^2 + Dxy + Exz + Fyz + Gx + Hy + Iz + J = 0, \quad (1.2.1)$$

where A, B, C, D, E, and F are not all zero, and the graphs of these equations for various choices of the coefficients are the **quadric surfaces**. These are directly analogous to the conic sections, which are the graphs in the plane of the general second-degree equation in *two* variables.

Example 1.2.7 Consider the case where $A = 1/a^2$, $B = 1/b^2$, $C = 1/c^2$, $D = E = F = G = H = I = 0$, and $J = -1$ in Equation 1.2.1, where a, b, and c are all nonzero. Then the equation has the form

$$\frac{x^2}{a^2} + \frac{y^2}{b^2} + \frac{z^2}{c^2} = 1. \quad (1.2.2)$$

To analyze the graph of this equation, consider "slicing" it with planes parallel to the xy-plane. For a constant k, let $z = k$ in Equation 1.2.2. Then

$$\frac{x^2}{a^2} + \frac{y^2}{b^2} + \left(\frac{k}{c}\right)^2 = 1 \quad (1.2.3)$$

or

$$\frac{x^2}{a^2(1 - (k/c)^2)} + \frac{y^2}{b^2(1 - (k/c)^2)} = 1,$$

which is an ellipse for each value of k between $-c$ and c. When k equals c or $-c$, the graph of (1.2.3) is just the point $(0, 0 \pm c)$. By the same reasoning, the cross sections parallel to the xz-plane and the yz-plane are also ellipses. Those in the coordinate planes have the equations

$$\frac{x^2}{a^2} + \frac{y^2}{b^2} = 1, \quad \frac{x^2}{a^2} + \frac{z^2}{c^2} = 1, \quad \text{and} \quad \frac{y^2}{b^2} + \frac{z^2}{c^2} = 1,$$

which, when sketched, provide a reasonable picture of the **ellipsoid**. See Figure 1.2.5. ♦

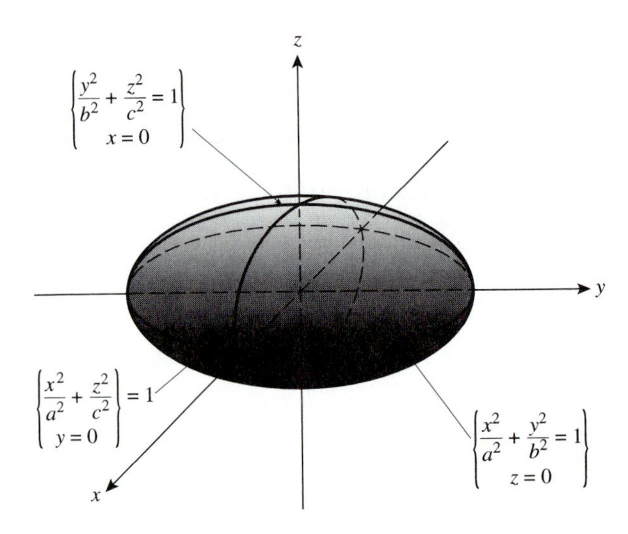

Figure 1.2.5 The ellipsoid $\dfrac{x^2}{a^2} + \dfrac{y^2}{b^2} + \dfrac{z^2}{c^2} = 1$

Example 1.2.8 Consider the case where C, D, E, F, G, H, and J are zero, $I = -1$, and A and B are positive in (1.2.1). Then the equation can be written

$$z = Ax^2 + By^2. \tag{1.2.4}$$

The same technique as we used in the preceding example reveals the general shape. The cross sections in the planes $z = k$ for $k > 0$ are the ellipses

$$\frac{x^2}{k/A} + \frac{y^2}{k/B} = 1.$$

These are larger for larger k and they shrink to the point $(0, 0, 0)$ when k decreases toward 0. When $k < 0$ there is no intersection, so the surface lies on and above the xy-plane. The cross sections in the planes $y = k$ are the parabolas

$$z = Ax^2 + Bk^2,$$

and similarly, the cross sections in the planes $x = k$ are the parabolas

$$z = By^2 + Ak^2.$$

The cross sections in the xz- and yz-planes have the equations $z = Ax^2$ and $z = By^2$. Sketching these along with one of the elliptical horizontal cross sections gives a picture of the **elliptic paraboloid**, which is shown in Figure 1.2.6. ♦

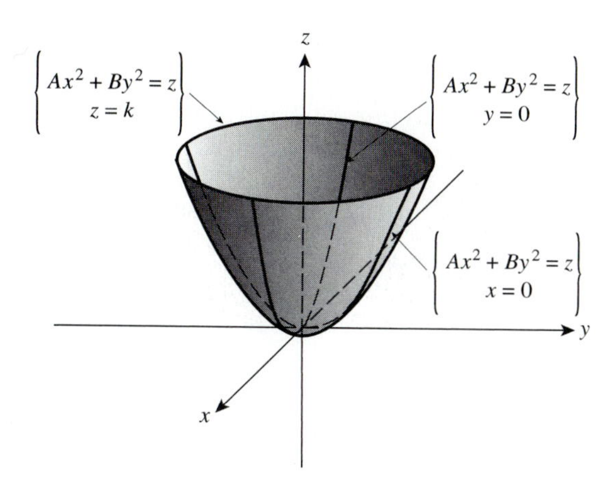

Figure 1.2.6 The elliptic paraboloid $z = Ax^2 + By^2$, where $A > 0$ and $B > 0$

Example 1.2.9 When, in the preceding example, A and B have opposite signs, we obtain a **hyperbolic paraboloid**. For definiteness, suppose that $A < 0$ and $B > 0$. Then the cross sections in the planes $x = k$ are the parabolas

$$z = By^2 + Ak^2, \tag{1.2.5}$$

which open upward. The cross sections in the planes $y = k$ are the parabolas

$$z = Ax^2 + Bk^2, \tag{1.2.6}$$

which open downward. By sketching two cross sections from (1.2.5) in the planes $x = \pm k$, the cross section $z = By^2$ in the plane $x = 0$, and two other parabolic cross sections in the planes $y = \pm k$, we obtain a sketch as in

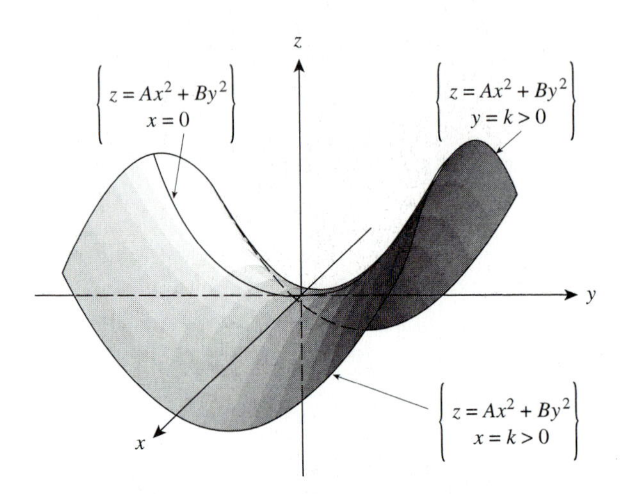

Figure 1.2.7 The hyperbolic paraboloid $z = Ax^2 + By^2$, where $A < 0$ and $B > 0$

Figure 1.2.7. Note also that for $k > 0$, the cross sections in the planes $z = k$ are the *hyperbolas*

$$\frac{y^2}{k/B} - \frac{x^2}{k/(-A)} = 1.$$

(Remember that $A < 0$ and $B > 0$.) Similarly for $k < 0$, the cross section in the plane $z = k$ is the hyperbola

$$\frac{x^2}{k/A} - \frac{y^2}{-k/B} = 1.$$

This explains why this surface has the name *hyperbolic* paraboloid. ♦

Example 1.2.10 The graph of the equation

$$z^2 = \frac{x^2}{a^2} + \frac{y^2}{b^2} \tag{1.2.7}$$

is an **elliptic cone** with its vertex at the origin and axis of symmetry the z-axis. See Figure 1.2.8. To see this, first consider the cross sections in the planes $z = k$, where $k \neq 0$. These are the ellipses

$$\frac{x^2}{k^2a^2} + \frac{y^2}{k^2b^2} = 1.$$

The cross section in the plane $x = 0$ is the graph of

$$z^2 = \frac{y^2}{b^2}.$$

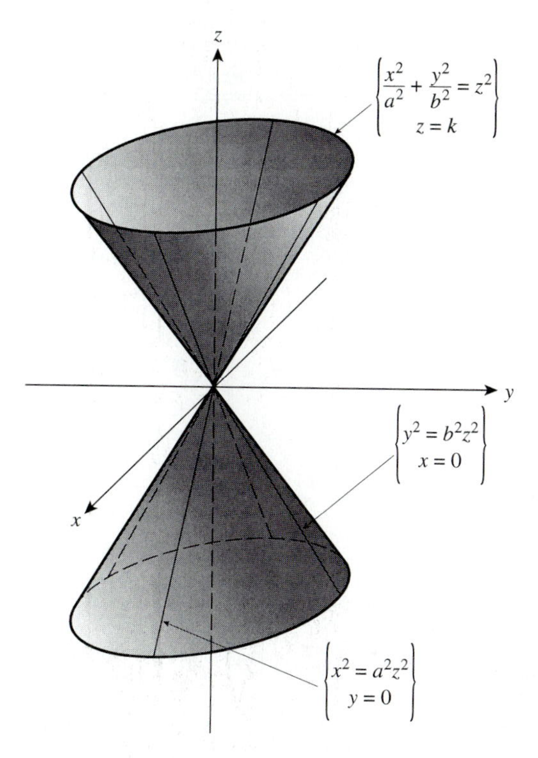

Figure 1.2.8 An elliptic cone, graph of the equation $z^2 = x^2/a^2 + y^2/b^2$

Since this equation can be written as

$$(bz - y)(bz + y) = 0,$$

we see that either $y = bz$ or $y = -bz$, a pair of lines in the yz-plane that intersect in the origin. Similarly, the cross section in the plane $y = 0$ is the pair of intersecting lines $x = az$ and $x = -az$. By taking the square root of both sides of (1.2.7), we obtain

$$z = \pm\sqrt{\frac{x^2}{a^2} + \frac{y^2}{b^2}}. \tag{1.2.8}$$

Clearly the graph of $z = \sqrt{x^2/a^2 + y^2/b^2}$ is the upper half of the cone, and the minus sign gives the lower half. Note that the surface is not smooth at the point (0, 0, 0). Later, this feature will serve as an example of a function of two variables that is not differentiable, directly analogous to the example in elementary calculus where $f(x) = |x|$ is shown to be continuous but not differentiable at 0. ♦

The graph of	is obtained from the graph of	by
$F(y, x, z) = 0$	$F(x, y, z) = 0$	reflection in the plane $x = y$
$F(x, z, y) = 0$	$F(x, y, z) = 0$	reflection in the plane $y = z$
$F(z, y, x) = 0$	$F(x, y, z) = 0$	reflection in the plane $x = z$
$F(y, z, x) = 0$	$F(x, y, z) = 0$	reflection in the plane $x = y$ followed by reflection in the plane $y = z$.
$F(x - a, y - b, z - c) = 0$	$F(x, y, z) = 0$	translation of each point a units in the x-direction, b units in the y-direction, and c units in the z direction.
$F\left(\frac{x}{a}, \frac{y}{b}, \frac{z}{c}\right) = 0$	$F(x, y, z) = 0$	multiplying the x- y- and z-coordinates of each point by a, b, and c respectively.

Table 1.2.1 Transformations of variables in equations and their geometric effects

Example 1.2.11 Identify the graphs of

$$3z^2 - 4x^2 - 5y^2 = 0 \quad \text{and} \quad 3x^2 - 4y^2 - 5z^2 = 0.$$

Solution By writing the first equation as

$$3z^2 \;=\; 4x^2 + 5y^2$$

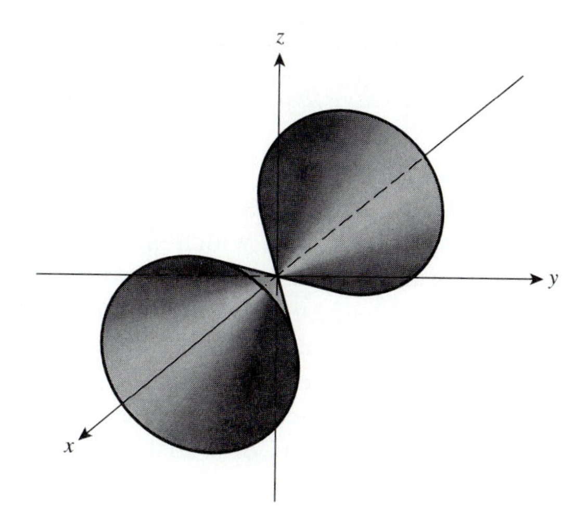

Figure 1.2.9 The elliptic cone $3x^2 - 4y^2 - 5z^2 = 0$

$$z^2 = \frac{x^2}{(\sqrt{3}/2)^2} + \frac{y^2}{(\sqrt{5}/3)^2}$$

we see this is of the form (1.2.7), so its graph is an elliptic cone with axis the z-axis and elliptical cross sections in planes parallel to the xy-plane.

The second equation is essentially identical to the first except that the variables play different roles. For instance, the role of x in the second equation is that of z in the first. This means that the graph of the second equation is also an elliptical cone, but with axis the x-axis and elliptical cross sections in the planes parallel to the yz-plane. See Figure 1.2.9. Table 1.2.1 summarizes the geometric effect of changing the variables in these and similar ways. ♦

Example 1.2.12 Identify the graph of

$$z - 2x^2 - 4x - y^2 + 10y = 29. \tag{1.2.9}$$

Solution Completing the square in the variables whose squares appear in the equation will simplify its appearance:

$$z - 2(x^2 + 2x + 1 - 1) - (y^2 - 10y + 25 - 25) = 29$$

$$z - 2(x + 1)^2 - (y - 5)^2 + 27 = 29$$

$$z - 2 = 2(x + 1)^2 + (y - 5)^2.$$

With $z' = z - 2$, $x' = x + 1$, and $y' = y - 5$, this can be written as

$$z' = 2(x')^2 + (y')^2,$$

the equation of an elliptic paraboloid with axis the z'-axis, vertex at (0, 0, 0) in $x'y'z'$-coordinates, and elliptical cross sections in planes parallel to the $x'y'$-plane. But the z'-axis is the line in which the planes $x' = 0$ and $y' = 0$ intersect. These planes are also given by $x = -1$ and $y = 5$. When $x' = 0$, $y' = 0$, and $z' = 0$, we have $x = -1$, $y = 5$, and $z = 2$, so the vertex is at the point with xyz-coordinates $(-1, 5, 2)$. In other words, the graph of (1.2.9) is the graph of $z = 2x^2 + y^2$ translated -1 unit in the x-direction, 5 units in the y-direction, and 2 units in the z-direction. See Table 1.2.1 for a general description of the geometry of this sort of transformation. ♦

Exercises 1.2

Sketch the graphs of the following functions by hand.

1. $f(x, y) = -2$

2. $g(x, y) = \cos y$

3. $h(x, y) = 1 - x$

4. $F(x, y) = 2x^2 + 2y^2 - 1$

5. $G(x, y) = \sqrt{x^2 + y^2/4}$

6. $H(x, y) = \sqrt{1 - x^2 - y^2}$

7. Match each of the following functions with its graph in Figure 1.2.10.

(a) $A(x, y) = \dfrac{1}{1 + x^2 + y^2}$

(b) $B(x, y) = \dfrac{x^2 - y^2}{x^2 + y^2}$

(c) $C(x, y) = \sin^2 x + \cos^2 y$

(d) $D(x, y) = e^{-(x+1)^2-(y+1)^2} + \frac{1}{2}e^{-(x-1)^2-(y-1)^2}$

(e) $E(x, y) = (x^2 - y^2)^2$

(f) $F(x, y) = 1 - x - y$

8. Find a function whose graph reasonably approximates the shape of the dome on top of the U.S. Capitol building.

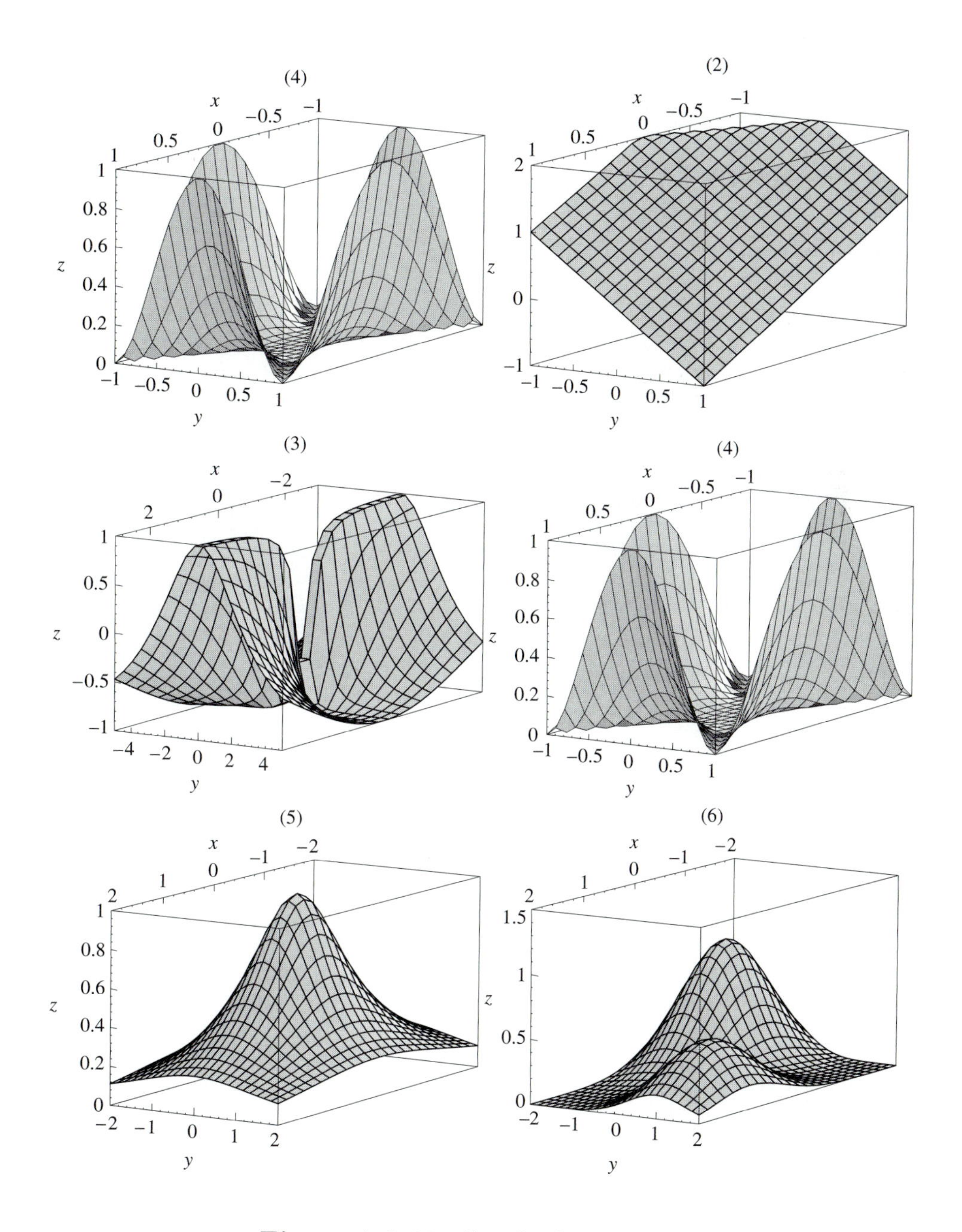

Figure 1.2.10 Graphs for Exercise 7

9. Find a function whose graph reasonably approximates the shape of a conical funnel sitting upside down.

10. Find a function whose graph approximates the shape of the double-gabled roof shown in Figure 1.2.11(a). (*Hint*: Try a two-line formula involving $|x|$ and $|y|$.)

11. Find a function whose graph approximates the shape of the pyramid shown in Figure 1.2.11(b).

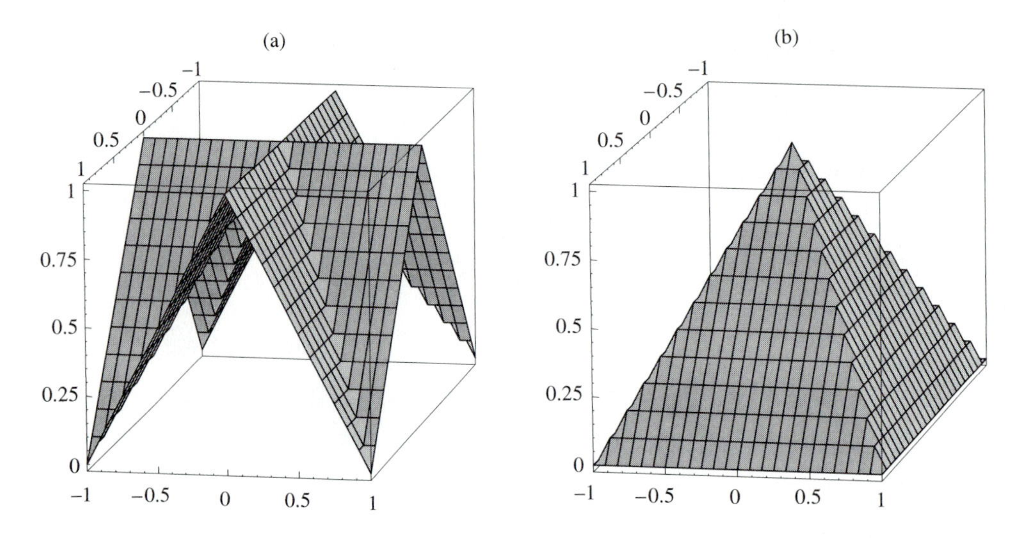

Figure 1.2.11 Figures for Exercises 10 and 11

12. Determine reasonable values for a, b, and c so that the ellipsoid

$$\frac{x^2}{a^2} + \frac{y^2}{b^2} + \frac{z^2}{c^2} = 1$$

bears a reasonable resemblance to the surface of

a a rugby ball

b a cucumber

c the earth, an **oblate spheroid** where the distance from the center to the north or south pole is about 6350 km and the distance from the center to any point on the equator is about 6370 km.

Identify and sketch the quadric surfaces.

13. $x^2 + y^2 + z^2 - 2x + 4y + 10z + 21 = 0$

14. $x^2 - y^2 + 16z^2 = 0$

15. $x^2 - 2y^2 + z = 0$

16. $x^2 + y^2 + z - 1 = 0$

17. $y = x^2 + z^2$

18. $y^2 = x^2 + z^2$

19. $9x^2 + 4y^2 + 36z^2 - 18x + 16y = 0$

20. $z = x^2 - 4y^2 + 8y - 3$

21. Analyze the equation $-\frac{x^2}{a^2} - \frac{y^2}{b^2} + \frac{z^2}{c^2} = 1$ and sketch its graph. This is a **hyperboloid of two sheets.**

22. Analyze the equation $\frac{x^2}{a^2} + \frac{y^2}{b^2} - \frac{z^2}{c^2} = 1$ and sketch its graph. This is a **hyperboloid of one sheet.**

Use Exercises 21 and 22 to help identify and sketch the graphs of the following equations.

23. $4x^2 - y^2 + 9z^2 = 36$

24. $-x^2 - 4y^2 + 4z^2 = 4$

25. $x^2 + y^2 - 2z^2 = -5$

26. $4x^2 - 5y^2 - 6z^2 = 120$

Describe and sketch the relationship between the graphs of the following pairs of equations.

27. $z = x^2 + y^2$ and $z = -2x^2 - 2y^2$

28. $z = \sqrt{x^2 + y^2}$ and $z = 2 - \sqrt{x^2 + y^2}$

29. $z = e^x$ and $z = \frac{1}{4}e^x - 2$

30. $z - x^2 - y^2 = 0$ and $y - z^2 - x^2 = 0$

31. $x^2 + 4y^2 + 25z^2 = 100$ and $25x^2 + y^2 + 4(z - 1)^2 = 100$

32. $z = e^{-x^2-y^2}$ and $z = 2 - e^{-25x^2-25y^2}$

1.3 Cylindrical and Spherical Coordinates

Many physical problems involve geometry with symmetry about a line or a point. For instance, if you are concerned with the flow of a fluid through a cylindrical pipe, it is natural to model the molecules of fluid as points with position and velocity specified with reference to the axis of the pipe. If you are concerned with tracking and predicting the position of a spacecraft from some location on the earth, it is natural to represent its position and velocity in terms of the spacecraft's distance from the center of the earth and the angles between the position vector and two chosen reference directions. These two illustrations are instances in which cylindrical and spherical coordinates would be useful.

Cylindrical Coordinates

In three-dimensional space, establish a directed line and a point O on it that will be the **origin**. Call this line the w-axis. In the plane through O that is perpendicular to this line, establish a ray with its initial point at O. Call this ray the **polar axis**. See Figure 1.3.1. Now, given a point P off

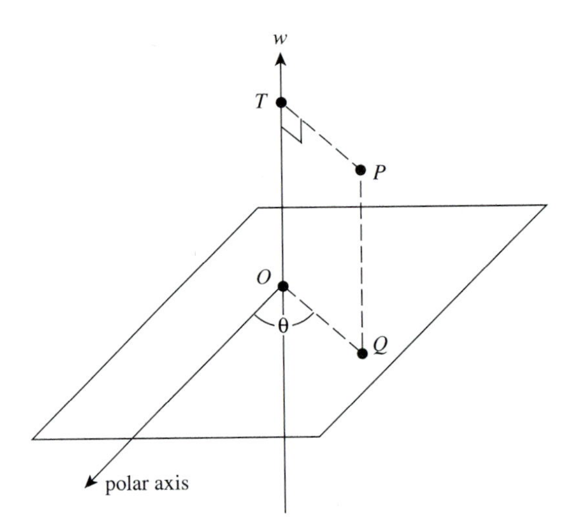

Figure 1.3.1 Cylindrical coordinates of a point P

the w-axis, construct the line through P parallel to the w-axis, and let Q be its intersection with the plane through O perpendicular to the w-axis. Also construct the line through P and the w-axis that is perpendicular to the w-axis, and let T be its point of intersection with the w-axis. The w-

coordinate of P is the coordinate of T on the w-axis. The angle θ, measured counterclockwise as viewed from the positive w-axis, which the segment OQ forms with the polar axis is the θ-coordinate of P. Finally, the length of the segment OQ is the r-coordinate of P. This procedure assigns to each point P in space at least one ordered triple $(r,\ \theta,\ w)$ called cylindrical coordinates of P.

Now superimpose on the polar- and w-axes the xyz rectangular coordinate system so that the origins, w- and z-axes, and positive x-axis and polar axis coincide. Given a point P, construct the points Q, R, S, and T as shown in Figure 1.3.2. Clearly, the z and w coordinates of P are the same.

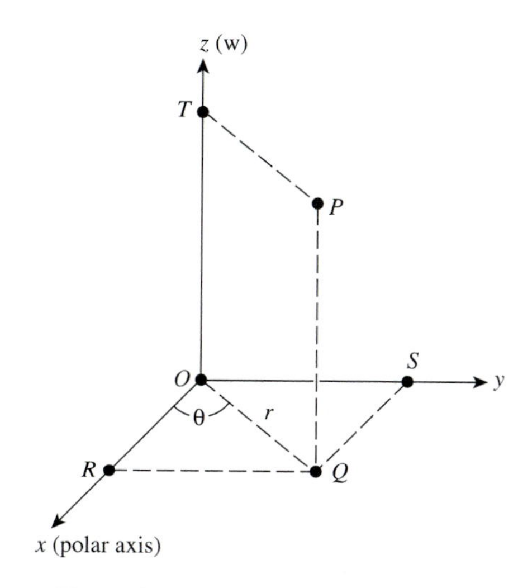

Figure 1.3.2 Transformation from cylindrical to rectangular coordinates

Since $r = |OQ|$ and OR is a leg of the right triangle $\triangle ORQ$, from the figure it is clear that the x-coordinate of R is $r\cos\theta$. Also, OS is a leg of the right triangle $\triangle OSQ$, so the y-coordinate of S is $r\sin\theta$. To summarize, given the cylindrical coordinates $(r,\ \theta,\ w)$ of a point, the rectangular coordinates are

$$x = r\cos\theta \qquad (1.3.1)$$

$$y = r\sin\theta \qquad (1.3.2)$$

$$z = w. \qquad (1.3.3)$$

Conversely, given rectangular coordinates $(x,\ y,\ z)$, we see that the cylindrical coordinates satisfy

$$r^2 = x^2 + y^2 \qquad (1.3.4)$$

$$\tan\theta = \frac{y}{x},\ x \neq 0 \tag{1.3.5}$$
$$w = z. \tag{1.3.6}$$

The first of these follows by squaring (1.3.1) and (1.3.2) and adding; the second comes from dividing (1.3.2) by (1.3.1); the third is obvious.

Example 1.3.1 The point with cylindrical coordinates $(7, 3\pi/4, -2)$ has rectangular coordinates

$$\begin{aligned} x &= 7\cos\left(\frac{3\pi}{4}\right) = \frac{-7}{\sqrt{2}} \\ y &= 7\sin\left(\frac{3\pi}{4}\right) = \frac{7}{\sqrt{2}} \\ z &= -2. \quad \blacklozenge \end{aligned}$$

Example 1.3.2 Find cylindrical coordinates for the point with rectangular coordinates $(-2,\ -2\sqrt{3},\ 3)$.

Solution The w-coordinate, by (1.3.3), is $w = 3$. Also, r and θ satisfy

$$r^2 = (-2)^2 + (-2\sqrt{3})^2 = 4 + 12 = 16$$

and

$$\tan\theta = \frac{-2\sqrt{3}}{-2} = \sqrt{3}.$$

From Figure 1.3.3 we see that $r = 4$ and $\theta = 4\pi/3$ satisfy these equations.

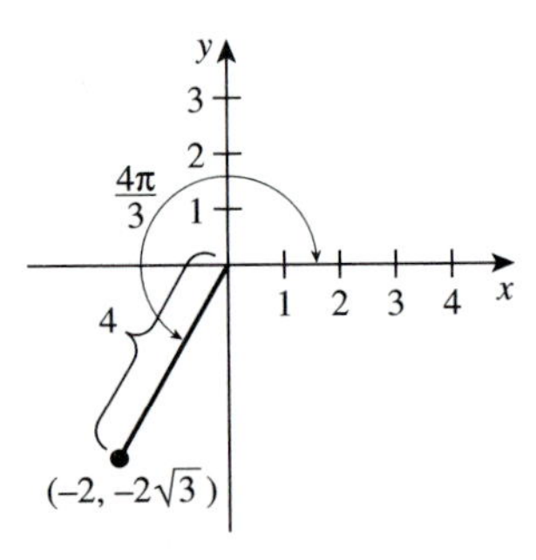

Figure 1.3.3 One possible choice of r and θ

If we had selected $r = -4$ as the solution to $r^2 = 16$, then $\theta = \pi/3$. $\blacklozenge$

Example 1.3.2 shows that a point may have many different cylindrical coordinates. From now on, however, unless we explicitly relax it, we make the following restrictions: $r \geq 0$ and $0 \leq \theta < 2\pi$. This reduces the nonuniqueness of cylindrical coordinates but does not quite eliminate it. See the exercises.

Example 1.3.3 The points with coordinates satisfying $r = 10$ lie on a circular cylinder of radius 10 whose axis is the w-axis. In rectangular coordinates the equation is $\sqrt{x^2 + y^2} = 10$. See Figure 1.3.4. ♦

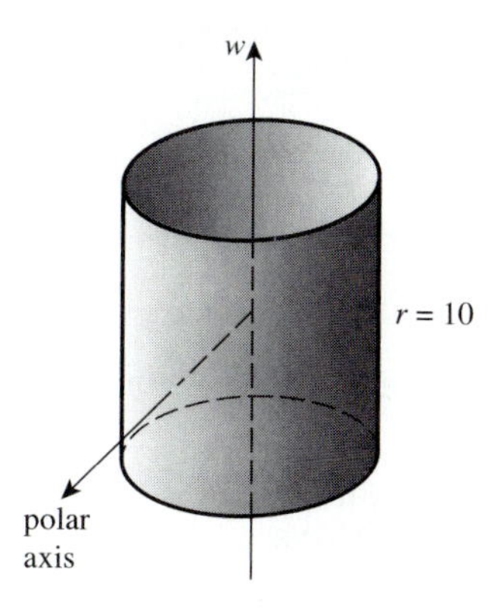

Figure 1.3.4 The graph of the cylindrical coordinate equation $r = 10$

Example 1.3.4 The graph of the equation $\theta = 5\pi/6$ is the half-plane, with the w-axis as its edge, that forms an angle of $5\pi/6$ with the polar axis. Since $\tan 5\pi/6 = -1/\sqrt{3}$, by (1.3.5), this set is given by $\{(x, y, z) | x \leq 0,\ y = -x/\sqrt{3}\}$ in rectangular coordinates. ♦

Example 1.3.5 What is the equation for the circular cone $z = \sqrt{x^2 + y^2}$ in cylindrical coordinates? Simply substitute the expressions for x, y, and z from (1.3.1):

$$w = \sqrt{(r \cos \theta)^2 + (r \sin \theta)^2}.$$

This simplifies to

$$w = r. \quad ♦$$

Example 1.3.6 Some simple equations in cylindrical coordinates do not correspond to simple equations in rectangular coordinates. For instance, the equation $r = \theta$, where we allow θ to range over $[0, \infty)$, has as its graph a

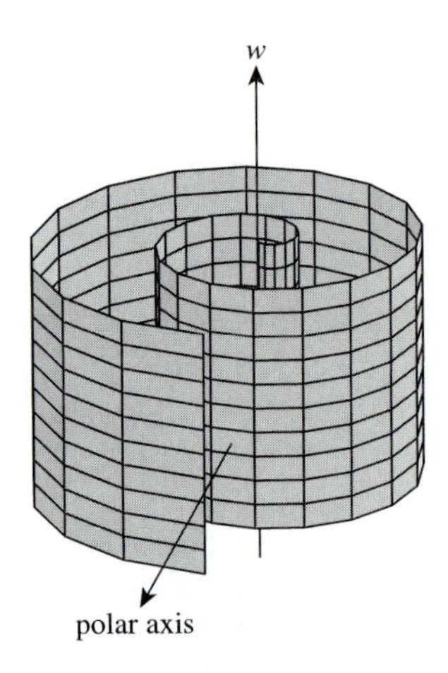

Figure 1.3.5 The graph of $r = \theta$ for $0 \leq \theta \leq 4\pi$ and $-3 \leq w \leq 3$

spiral sheet that resembles a very loosely wound, infinitely wide roll of paper. See Figure 1.3.5.

This figure can be produced by having the computer plot the points (x, y, z) generated by the equations

$$\begin{aligned} x &= \theta \cos \theta \\ y &= \theta \sin \theta \\ z &= w \end{aligned}$$

as θ ranges over the interval $[0, 4\pi]$ and w ranges over $[-3, 3]$. These equations, of course, are obtained by substituting θ for r in (1.3.1), (1.3.2), and (1.3.3). ♦

Spherical Coordinates

To locate a point P in space, repeat the construction shown in Figure 1.3.1. Let $\rho = |OP|$, let θ be the angle between the polar axis and the segment OQ, and let ϕ be the angle $\angle TOP$. By restricting $0 \leq \theta < 2\pi$ and $0 \leq \phi \leq \pi$, we have essentially unique **spherical coordinates** (ρ, θ, ϕ) for the point P. (Where is there a lack of uniqueness in this scheme?) As for cylindrical coordinates, superimpose, as shown in Figure 1.3.6, a rectangular coordinate system on this one to obtain transformations between the systems.

From Figure 1.3.6, the distance $|OQ|$ is $\rho \sin \phi$, and since OQ forms an angle of θ with the positive x-axis, the x-coordinate R of P is $\rho \sin \phi \cos \theta$.

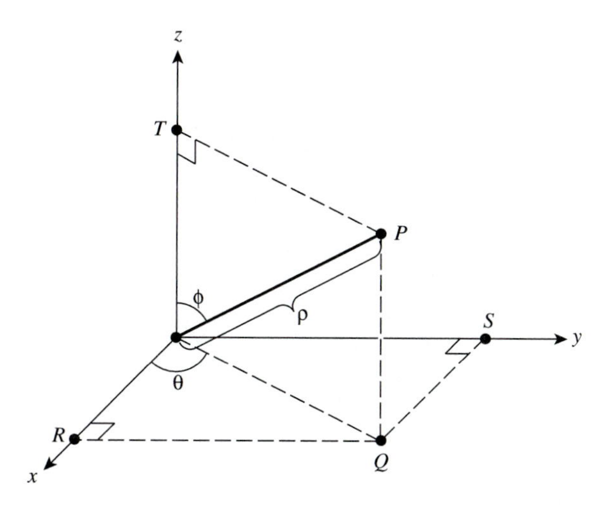

Figure 1.3.6 Transformation from spherical to rectangular coordinates

Similarly, the y-coordinate S of P is $\rho \sin \phi \sin \theta$. Finally, the z-coordinate T is $\rho \cos \phi$. To summarize, given spherical coordinates (ρ, θ, ϕ), the corresponding rectangular coordinates (x, y, z) are

$$x = \rho \sin \phi \cos \theta \tag{1.3.7}$$

$$y = \rho \sin \phi \sin \theta \tag{1.3.8}$$

$$z = \rho \cos \phi. \tag{1.3.9}$$

By squaring and adding these three equations, we obtain $x^2 + y^2 + z^2 = \rho^2$. Since $\rho \geq 0$,

$$\rho = \sqrt{x^2 + y^2 + z^2}. \tag{1.3.10}$$

Solving (1.3.9) for $\cos \phi$ and using (1.3.10), we obtain

$$\cos \phi = \frac{z}{\rho} = \frac{z}{\sqrt{x^2 + y^2 + z^2}}.$$

Finally, dividing (1.3.8) by (1.3.7), we obtain

$$\tan \theta = \frac{y}{x}, \quad x \leq 0.$$

So, given rectangular coordinates (x, y, z), the spherical coordinates can be found from

$$\rho = \sqrt{x^2 + y^2 + z^2} \tag{1.3.11}$$

$$\tan\theta = \frac{y}{x}, \quad x \neq 0 \tag{1.3.12}$$

$$\cos\phi = \frac{z}{\sqrt{x^2+y^2+z^2}}. \tag{1.3.13}$$

Example 1.3.7 The graph of the equation $\phi = \pi/4$ is a (half) cone that opens along the positive z-axis. This can be seen by substituting $\pi/4$ for ϕ in (1.3.13):

$$\frac{z}{\sqrt{x^2+y^2+z^2}} = \frac{1}{\sqrt{2}}.$$

Upon solving for z in terms of x and y, we obtain

$$z = \sqrt{x^2+y^2},$$

which is a familiar equation. ♦

Example 1.3.8 The sphere $x^2+y^2+z^2 = 16$ is succinctly written as $\rho = 4$ simply by using (1.3.11). ♦

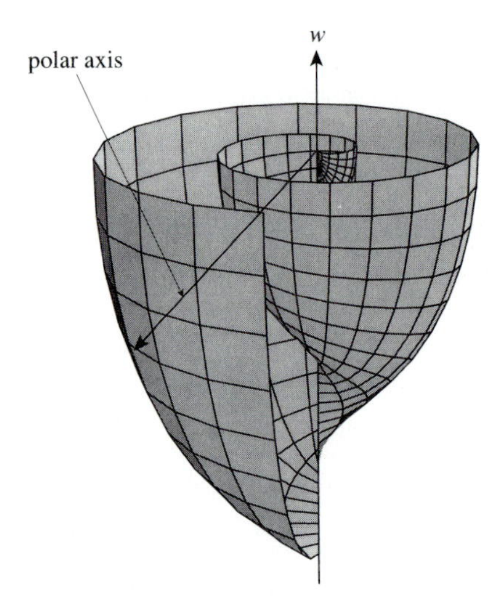

Figure 1.3.7 The graph of the spherical coordinate equation $\rho = \theta$

Example 1.3.9 The sphere $x^2 + y^2 + (z-1)^2 = 1$ becomes, in spherical coordinates,

$$\rho^2 \sin^2\phi\cos^2\theta + \rho^2\sin^2\phi\sin^2\theta + \rho^2\cos^2\phi - 2\rho\cos\phi + 1 = 1$$

when we substitute for x, y, and z the expressions in (1.3.7), (1.3.8), and (1.3.9). Simplifying, we obtain $\rho^2 - 2\rho\cos\phi = 0$. Finally, canceling ρ, we find that the equation for this sphere reduces to $\rho = 2\cos\phi$. ♦

Example 1.3.10 Graphs of many simple equations in spherical coordinates are difficult to sketch by hand or visualize without significant effort. Computer graphics packages can aid with visualization. It is at first hard to believe that the graph of an equation as simple as $\rho = \theta$ would be as complex and beautiful as that shown in Figure 1.3.7. The figure shows only the portion of the graph corresponding to $0 \le \theta \le 4\pi$ and $\pi/2 \le \phi \le \pi$; the "top" half (the portion corresponding to $0 \le \phi \le \pi/2$) has been "cut" off to permit a view of the inside of this infinite "snail shell." ♦

Exercises 1.3

1. Plot the points with the following cylindrical coordinates:

(a) $(3,\ \pi/4,\ 2)$

(b) $(1,\ 0,\ -4)$

(c) $(2,\ 2\pi/3,\ 1)$

(d) $(1,\ 7\pi/4,\ 0)$

2. Plot the points with the following spherical coordinates:

(a) $(1,\ \pi/3,\ \pi/2)$

(b) $(6,\ 3\pi/4,\ \pi/4)$

(c) $(1/2,\ 3\pi/2,\ 3\pi/4)$

(d) $(6,\ 11\pi/12,\ 0)$

3. Find the rectangular coordinates of the points in Exercise 1.

4. Find the rectangular coordinates of the points in Exercise 2.

5. Find the cylindrical and spherical coordinates of the points with the following rectangular coordinates:

(a) $(1,\ 1,\ 1)$

(b) $(-1,\ 1,\ -1)$

(c) $(-\sqrt{2},\ \sqrt{6},\ 2\sqrt{2})$

(d) $(0,\ -\pi\sqrt{3}/2,\ \pi/2)$

6. Write the following Cartesian equations in cylindrical coordinates and sketch the graph:

(a) $x^2 + y^2 + (z-1)^2 = 1$

(b) $z = 2(x^2 + y^2)$

(c) $x = 1$

(d) $z = -3$

(e) $x^2 + y^2 + z^2 = 100$

7. Sketch the graphs of the following equations in cylindrical coordinates:

(a) $r = 5$

(b) $r = 2\cos\theta$

(c) $\theta = 5\pi/6$

(d) $z = \theta$

(e) $z = 2r$

(f) $z = r^2$

(g) $z = r^2 \cos 2\theta$

8. Convert the following spherical coordinate equations to Cartesian coordinates and sketch the graph:

(a) $\rho = 18$

(b) $\phi = \pi/6$

(c) $\theta = \pi/4$

(d) $\rho = 4\sec\phi$

(e) $\rho = 5\cos\phi$

9. Describe the points specified by the following sets of cylindrical coordinates:

(a) $\{(r,\, \theta,\, w) | 1 \le w \le 2\}$

(b) $\{(r,\, \theta,\, w) | 3 \le r \le 5\}$

(c) $\{(r,\, \theta,\, w) | \pi/4 \le \theta \le \pi/2,\ 1 \le r \le 2\}$

(d) $\{(r,\, \theta,\, w) | 1 \le r \le 2,\ 0 \le \theta \le \pi/4,\ -1 \le w \le 1\}$

10. Describe the points specified by the following sets of spherical coordinates:

(a) $\{(\rho,\, \theta,\, \phi) | \rho = 2,\ \theta = \pi/4\}$

(b) $\{(\rho,\, \theta,\, \phi) | \phi = \pi/3,\ \rho \le 1\}$

(c) $\{(\rho,\, \theta,\, \phi) | 2 \le \rho \le 3,\ 0 \le \theta \le \pi/4,\ \pi/6 \le \phi \le \pi/3\}$

(d) The set of points bounded by the cones $\phi = \pi/6$, $\phi = \pi/3$ and the spheres $\rho = 1$, $\rho = 2$.

11. Which points in space do not have unique cylindrical coordinates? Which points in space do not have unique spherical coordinates?

12. Given a sphere with center at the origin, the **great circle** through two points P_1 and P_2 on the sphere is the circle with center at the origin that passes through P_1 and P_2. The **great circle distance** between P_1 and P_2 is the length of the arc $\overset{\frown}{P_1P_2}$ on the great circle. Find the great circle distance between the following pairs of points (given in spherical coordinates):

(a) $(3, \pi/4, \pi/6)$, $(3, \pi/4, 3\pi/4)$

(b) $(2, \pi/3, \pi/2)$, $(2, 7\pi/6, \pi/2)$

(c) $(1, \pi/4, 0)$, $(1, \pi/2, \pi/4)$

13. Given a cylinder of radius r and two points P_1 and P_2 on it, the **geodesic** on the cylinder connecting P_1 to P_2 is the curve on the cylinder of least length joining P_1 to P_2. Show that the length of the geodesic on the cylinder with axis the w-axis and radius r that joins two points (r, θ_1, w_1) and (r, θ_2, w_2) is

$$\sqrt{r^2(\theta_2 - \theta_1)^2 + (w_2 - w_1)^2}.$$

14. The **celestial sphere** is the imaginary sphere with infinite radius which is centered at the center of the earth and on which all astronomical objects appear to move. We can also visualize it as a sphere of finite radius where the earth, placed at its center O, has zero radius. The **north celestial pole** N is the point on the celestial sphere in which the ray from earth's center through its north pole intersects the celestial sphere. The **south celestial pole** S is defined similarly. The **celestial equator** is the great circle on the celestial sphere obtained by projecting earth's equator onto it from earth's center.

Very distant objects (stars and galaxies, objects outside the solar system) appear to move along circles of constant latitude on the celestial sphere, but the motion of the sun and planets is more complicated. Because the axis of earth's rotation is inclined about 23° from a perpendicular to the plane in which it revolves about the sun, the sun appears to move (more or less) along a great circle (called the **ecliptic**) that is inclined about 23° with the celestial equator and that intersects the equator in two points, the **vernal equinox** and the **autumnal equinox**. The vernal equinox lies in the constellation Aries; for this reason it is often symbolized by the capital Greek letter Υ (upsilon), which serves as an effigy of a ram's head.

A point P on the celestial sphere is located as follows: Construct the great circle through P and N, and let Q be its intersection with the celestial equator. The **declination** of P is the angle $\delta = \angle POQ$ measured in degrees as positive if P is in the northern hemisphere and negative if in the southern hemisphere. The **right ascension** of P is the angle $\alpha = \angle \Upsilon OQ$ measured in hours, minutes, and seconds clockwise. (One hour of angle measure is 1/24 of the angle measure of an entire circle.)

(a) Sketch the celestial sphere, labeling the celestial poles and equator, the ecliptic, and the vernal equinox. For a point P on the celestial sphere, sketch the right ascension and declination.

(b) Suppose that an observer stands on earth with the vernal equinox straight overhead. Can she see a star whose declination is $+48°$ and whose right ascension is $4\,\text{hr}\,23'\,28''$?

(c) By putting a "standard" coordinate system in the celestial sphere (origin at O, x-axis piercing Υ, z-axis piercing N) write equations that give spherical coordinates θ and ϕ in terms of right ascension α and declination δ.

1.4 Vectors in Three-Dimensional Space

The notion of a vector is a convenient way of expressing mathematically how we experience certain aspects of physical reality. For instance, when you play with a couple of magnets, note how when the north poles are brought closer together, greater effort must be exerted on one magnet in order to keep them closer together. In a very informal way, we can represent the direction of the hands' effort with an arrow, and more effort by a longer arrow. As a second example, consider asking a clerk in the grocery store where you can find the canned tomatoes. If they are on the aisle where you stand, he or she may make a small pointing gesture and say, "End of this aisle. . . ." On the other hand, if they are on the other side of the building, the clerk may make a big pointing gesture with his or her whole arm and say, "Aisle 14 – B." The position of the tomatoes relative to your present position *is*, in a sense merely a hand pointing from where you stand to where they sit, simply an arrow. These two examples show that such ordinary notions as **force** and **position** have two common aspects: magnitude and direction.

We begin by defining a vector in the terms of ordinary geometry.

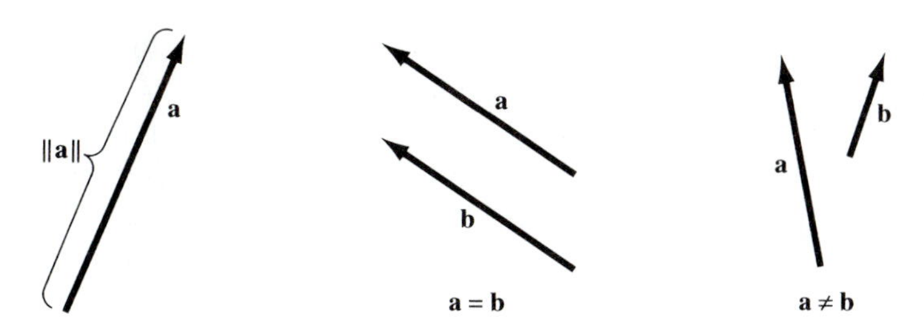

Figure 1.4.1 Magnitude, equality, and inequality of vectors

Definition 1.4.1 A **vector** is a line segment with a direction. We draw an arrow on one end (the **head**; the end without an arrow is called the **tail**) to indicate the direction, and we will typically symbolize a vector with a boldface letter such as **a**. The **magnitude** of a vector **a** is the length of the segment; it is denoted $||\mathbf{a}||$. See the first part of Figure 1.4.1. Two vectors are **equal** if and only if they have the same magnitude and direction. See the second and third parts of Figure 1.4.1. The **zero vector**, denoted **0**, has magnitude 0 and no direction. A **unit vector** is a vector with magnitude 1. If a vector goes from one point P to another point Q, then we write for it $\vec{PQ}$.

Example 1.4.1 With $P = (1, 1, 1)$ and $Q = (3, 5, 6)$ the vector $\vec{PQ}$ has magnitude the distance from P to Q:

$$||\vec{PQ}|| = \sqrt{(3-1)^2 + (5-1)^2 + (6-1)^2} = 3\sqrt{5}.$$

With $R = (0, 0, 0)$ and $S = (2, 4, 5)$, you can readily verify by way of a sketch that $\vec{RS}$ is parallel to $\vec{PQ}$, that the two point in the same direction, and that $||\vec{RS}|| = ||\vec{PQ}||$. By our definition, then, $\vec{RS} = \vec{PQ}$. ♦

By defining equality of two vectors in the way given in Definition 1.4.1, we are regarding as the same every pair of parallel segments with the same length and orientation. Thus a given vector remains unchanged if it is translated to various locations in space.

Definition 1.4.2 The **sum** of two vectors **a** and **b**, denoted $\mathbf{a} + \mathbf{b}$, is the vector obtained by first translating the tail of **b** so that it lies at the head of **a** and then drawing the vector with tail at the tail of **a** and head at the head of the translated **b**. See Figure 1.4.2.

By placing copies of the vectors **a** and **b** tail to tail, and in the same figure forming $\mathbf{a}+\mathbf{b}$ and (the potentially different sum) $\mathbf{b}+\mathbf{a}$, we see that the two

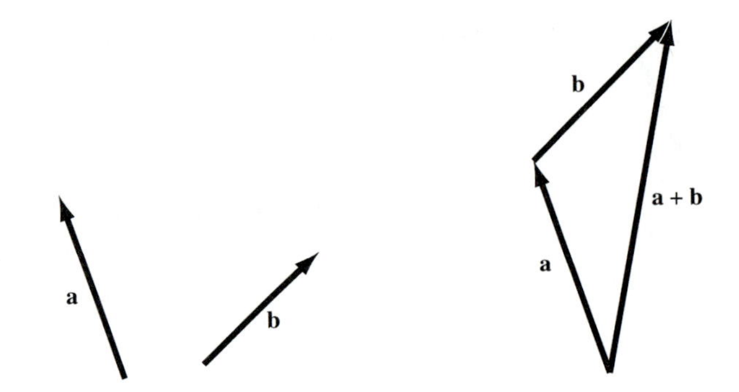

Figure 1.4.2 The sum $\mathbf{a} + \mathbf{b}$ of vectors $\mathbf{a}$ and $\mathbf{b}$

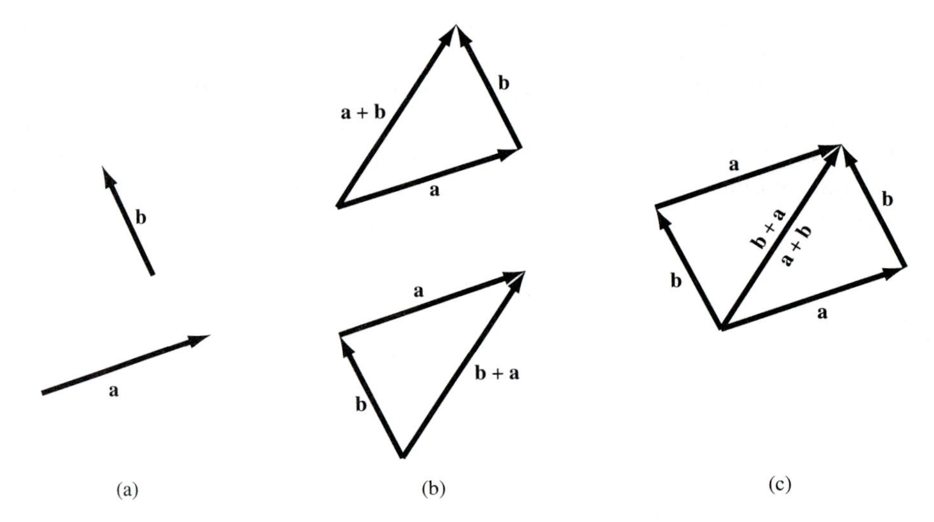

Figure 1.4.3 The commutativity of vector addition is equivalent to the Parallelogram Law.

copies of $\mathbf{a}$ and $\mathbf{b}$ form opposite sides of a parallelogram. See Figure 1.4.3. Since the diagonal of this parallelogram is both $\mathbf{a} + \mathbf{b}$ and $\mathbf{b} + \mathbf{a}$, this shows that

$$\mathbf{a} + \mathbf{b} = \mathbf{b} + \mathbf{a};$$

vector addition is **commutative**. For obvious reasons, this is sometimes referred to as the **Parallelogram Law**.

Definition 1.4.3 Multiplication of a vector $\mathbf{a}$ by a real number (i.e., a **scalar**) c produces a new vector $c\,\mathbf{a}$ obtained in the following way: $c\,\mathbf{a}$ is the vector of magnitude $|c|\,||\mathbf{a}||$ with the same direction as $\mathbf{a}$ if c is positive and the opposite direction to $\mathbf{a}$ if c is negative. See Figure 1.4.4. If $c \neq 0$, then

$\mathbf{a}/c$ represents $(1/c)\,\mathbf{a}$. The **difference** of two vectors is given by $\mathbf{a}-\mathbf{b} = \mathbf{a}+(-1)\,\mathbf{b}$.

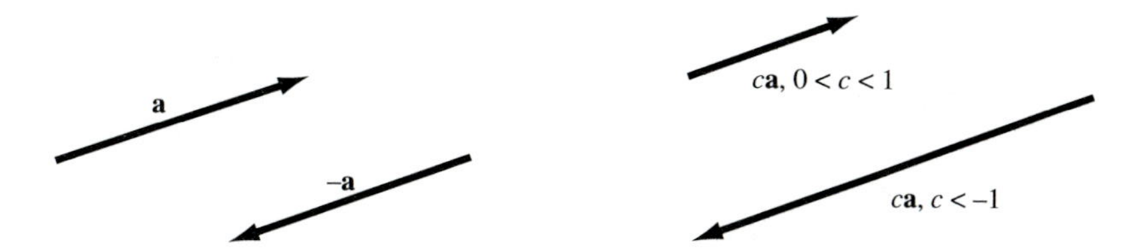

Figure 1.4.4 Multiplication of vector **a** by a scalar c

You should verify that if **a** and **b** are placed tail to tail, then $\mathbf{b}-\mathbf{a}$ is the vector that points from the head of **a** to the head of **b**.

There are two other important properties of vector arithmetic that can be shown geometrically. First, vector addition is **associative**. That is, given three vectors **a**, **b**, and **c**, the sums $(\mathbf{a}+\mathbf{b})+\mathbf{c}$ and $\mathbf{a}+(\mathbf{b}+\mathbf{c})$ are equal. To see this, place **a**, **b**, and **c** tail to head as shown in Figure 1.4.5 so that the points A, B, C, and D satisfy $\mathbf{a}=\vec{AB}$, $\mathbf{b}=\vec{BC}$, and $\mathbf{c}=\vec{CD}$. Then

$$(\mathbf{a}+\mathbf{b})+\mathbf{c} = \vec{AC}+\vec{CD} = \vec{AD}$$

and

$$\mathbf{a}+(\mathbf{b}+\mathbf{c}) = \vec{AB}+\vec{BD} = \vec{AD}.$$

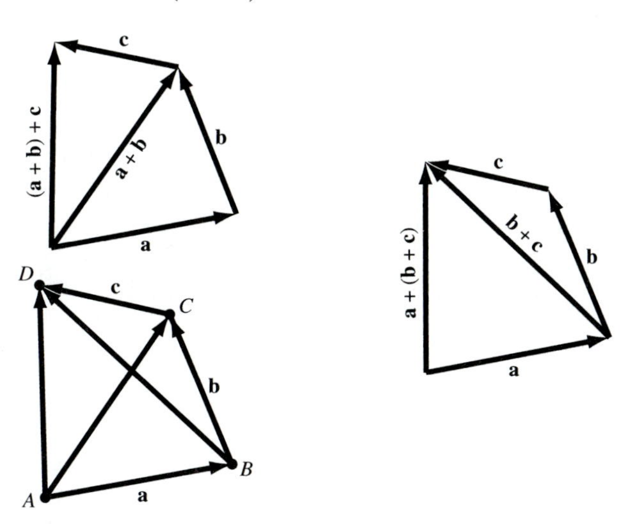

Figure 1.4.5 Vector addition is associative.

Thus

$$\mathbf{a} + (\mathbf{b} + \mathbf{c}) = (\mathbf{a} + \mathbf{b}) + \mathbf{c}. \tag{1.4.1}$$

(Keep in mind that this works whether or not A, B, C, and D all lie in the same plane!)

Second, scalar multiplication **distributes** over vector addition. That is, given a scalar k and two vectors $\mathbf{a}$ and $\mathbf{b}$, the vectors $k\mathbf{a} + k\,\mathbf{b}$ and $k(\mathbf{a} + \mathbf{b})$ are the same. To see this, put the tail of $\mathbf{b}$ on the head of $\mathbf{a}$ and let A, B, and C satisfy $\mathbf{a} = \vec{AB}$ and $\mathbf{b} = \vec{BC}$. See Figure 1.4.6. Assuming for the

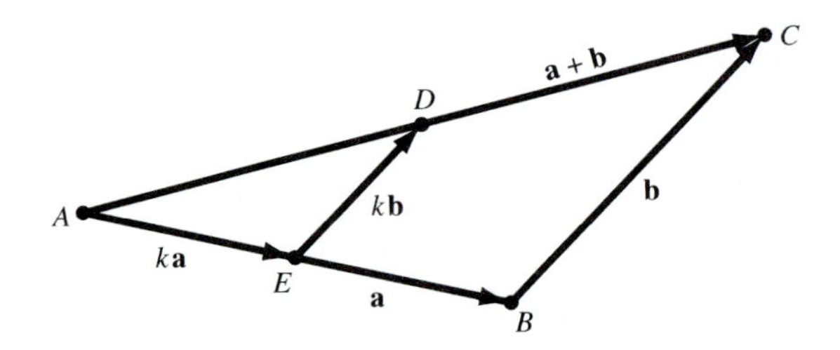

Figure 1.4.6 Proving that scalar multiplication distributes over vector addition

moment that $k > 0$, let $\vec{AE} = k\,\mathbf{a}$ and $\vec{ED} = k\,\mathbf{b}$. Then $\vec{AD} = k\,\mathbf{a} + k\,\mathbf{b}$. On the other hand, since $\triangle AED$ is similar to $\triangle ABC$, the vectors $\vec{AD}$ and $\vec{AC}$ are parallel. Indeed, $\vec{AD} = k\,\vec{AC}$; that is,

$$k\,\mathbf{a} + k\,\mathbf{b} = k(\mathbf{a} + \mathbf{b}). \tag{1.4.2}$$

Example 1.4.2 Let A, B, C, and D be four points in space such that no three are collinear. Show that the segments joining the midpoints of the segments AB, BC, CD, and DA form a parallelogram.

Solution From Figure 1.4.7, we can see it is sufficient to show that the quadrilateral joining the midpoints M, N, P, and Q has parallel and congruent sides. Equivalently, it suffices to show that the vector $\vec{MN}$ equals the vector $\vec{QP}$. Note that

$$\vec{MN} = \vec{MB} + \vec{BN} = \frac{1}{2}\vec{AB} + \frac{1}{2}\vec{BC} = \frac{1}{2}(\vec{AB} + \vec{BC}) \tag{1.4.3}$$

and

$$\vec{QP} = \vec{QD} + \vec{DP} = \frac{1}{2}\vec{AD} + \frac{1}{2}\vec{DC} = \frac{1}{2}(\vec{DC} + \vec{AD}). \tag{1.4.4}$$

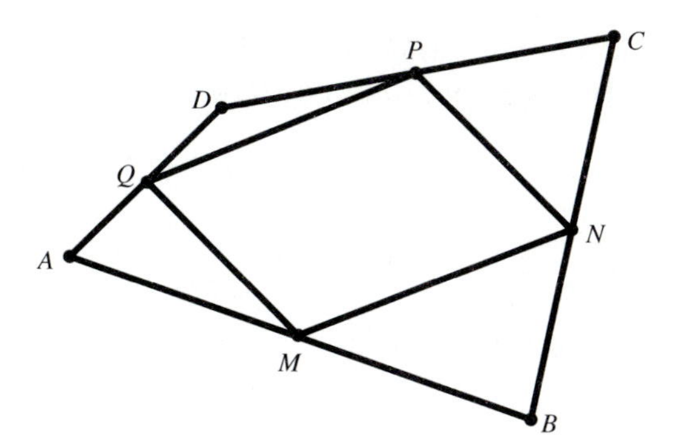

Figure 1.4.7 The midpoints of four joined segments in space determine a parallelogram.

When we subtract $\vec{QP}$ from $\vec{MN}$, formulas (1.4.3) and (1.4.4) give

$$\begin{aligned}\vec{MN} - \vec{QP} &= \frac{1}{2}(\vec{AB} + \vec{BC}) - \frac{1}{2}(\vec{DC} + \vec{AD}) \\ &= \frac{1}{2}(\vec{AB} + \vec{BC} + \vec{DA} + \vec{CD}) = \frac{1}{2}\mathbf{0} = \mathbf{0},\end{aligned}$$

which is equivalent to $\vec{MN} = \vec{QP}$. Therefore, the points M, N, P, and Q are the vertices of a parallelogram. ♦

The following definition simply gives a term to apply to two parallel vectors. Because the notion of parallelness generalizes in important ways, we begin with this restricted definition of linear dependence.

Definition 1.4.4 Two vectors $\mathbf{a}$ and $\mathbf{b}$ are **linearly dependent** if one can be written as a scalar multiple of the other. Otherwise, $\mathbf{a}$ and $\mathbf{b}$ are **linearly independent**.

Linear dependence can be phrased in a way that turns out to be more useful than the definition in some cases.

Theorem 1.4.1 *Two vectors* $\mathbf{a}$ *and* $\mathbf{b}$ *are linearly dependent if and only if there are scalars* c_1 *and* c_2 *not both zero such that*

$$c_1\mathbf{a} + c_2\,\mathbf{b} = \mathbf{0}. \tag{1.4.5}$$

Proof Suppose that $\mathbf{a}$ and $\mathbf{b}$ are linearly dependent. If $\mathbf{b} = \mathbf{0}$, then

$$0 \cdot \mathbf{a} + c_2 \cdot \mathbf{b} = \mathbf{0}$$

for any choice of c_2. On the other hand, if $\mathbf{b} \neq \mathbf{0}$ then, since $\mathbf{a}$ and $\mathbf{b}$ are linearly dependent, $\mathbf{b} = k\,\mathbf{a}$ for some nonzero scalar k. This can be written as

$$-k\mathbf{a} + \mathbf{b} = \mathbf{0}$$

($c_1 = -k$ and $c_2 = 1$).

Conversely, suppose that (1.4.5) holds for some constants c_1 and c_2 not both zero. If $c_2 \neq 0$, then

$$\mathbf{b} = \frac{-c_1}{c_2}\,\mathbf{a},$$

so $\mathbf{a}$ and $\mathbf{b}$ are linearly dependent. If $c_2 = 0$, then $c_1 \neq 0$ and (1.4.5) reduces to $c_1\,\mathbf{a} = \mathbf{0}$, which yields $\mathbf{a} = \mathbf{0} = 0 \cdot \mathbf{b}$. In either case, one vector is a scalar multiple of the other. ♦

The contrapositive of this theorem is also very useful.

Corollary 1.4.1 *Two vectors are linearly independent if and only if*

$$c_1\,\mathbf{a} + c_2\,\mathbf{b} = \mathbf{0}$$

implies that $c_1 = c_2 = 0$.

Example 1.4.3 Show that the diagonals of a parallelogram bisect one another.

Solution From Figure 1.4.8, it is sufficient to show that $\vec{AE} = \vec{EC}$ and

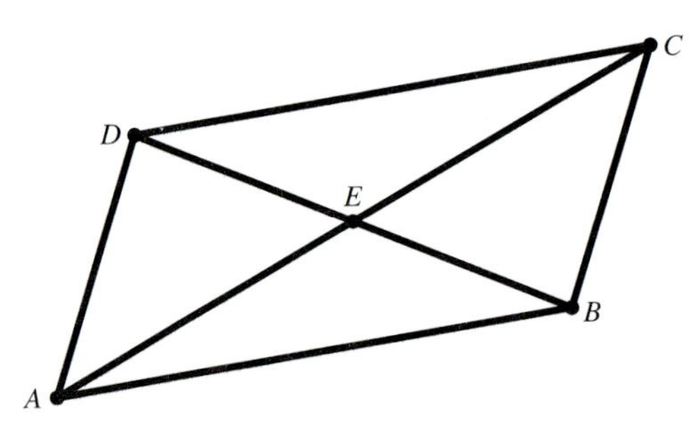

Figure 1.4.8 The diagonals of a parallelogram bisect one another

$\vec{DE} = \vec{EB}$. Since $ABCD$ is a parallelogram, $\vec{AB} = \vec{DC}$. Also note that

$$\vec{AB} = \vec{AE} + \vec{EB}$$

and

$$\vec{DC} = \vec{DE} + \vec{EC}.$$

Subtracting $\vec{DC}$ from $\vec{AB}$ gives us $\mathbf{0} = \vec{AE} - \vec{EC} + \vec{EB} - \vec{DE}$ or, rewriting slightly,

$$\mathbf{0} = (\vec{AE} - \vec{EC}) + (\vec{EB} - \vec{DE}). \tag{1.4.6}$$

But $\vec{AE}$ and $\vec{AC}$ are linearly dependent; $\vec{EC}$ and $\vec{AC}$ are also. It then follows that $\vec{AE} - \vec{EC} = c_1\,\vec{AC}$ for some constant c_1. Similarly, $\vec{EB} - \vec{DE} = c_2\,\vec{DB}$ for some c_2. This gives

$$\mathbf{0} = c_1\,\vec{AC} + c_2\,\vec{DB}.$$

Since $\vec{AC}$ and $\vec{DB}$ are linearly independent, $c_1 = c_2 = \mathbf{0}$. We conclude that $\vec{AE} - \vec{EC} = \mathbf{0}$ and $\vec{EB} - \vec{DE} = \mathbf{0}$. ♦

We now introduce the ideas that make vectors more than mere conceptual aids. By representing them in terms of their components, we can do computations with vectors.

Establish a rectangular coordinate system in three-dimensional space. Any vector **a** can be put into **standard position** by translating it so that its tail lies at the origin. We call the coordinates $(a_1,\, a_2,\, a_3)$ of the head of **a** in standard position the **components** of **a**. See Figure 1.4.9. This means

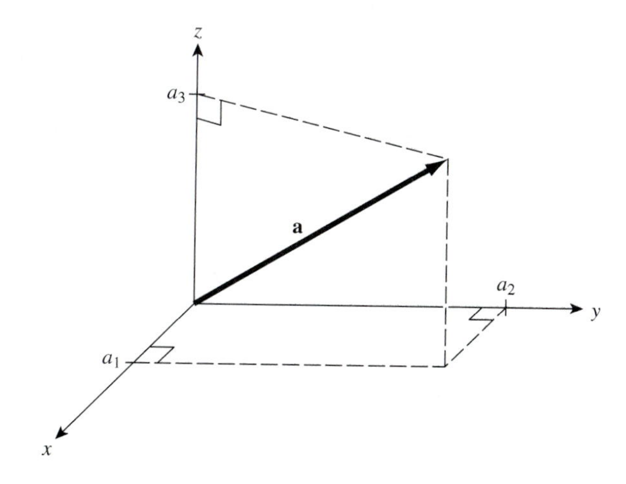

Figure 1.4.9 The components of a vector **a**

that two vectors are equal whenever they have the same components.

We denote by **i** the vector with components (1, 0, 0), by **j** the vector with components (0, 1, 0) and **k** the vector with components (0, 0, 1). (When

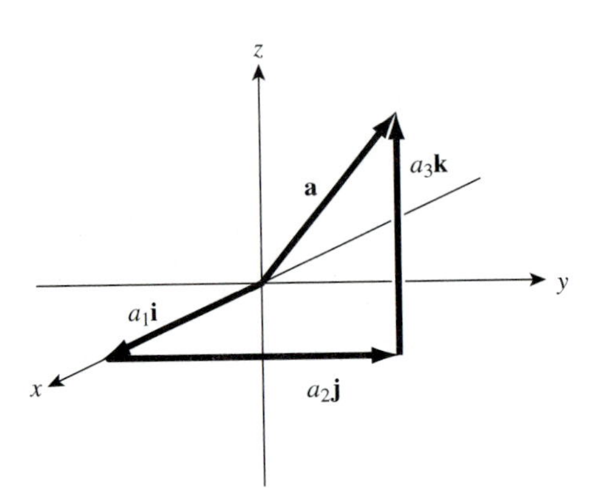

Figure 1.4.10 A vector **a** in terms of **i**, **j**, and **k**

writing these vectors by hand, we usually indicate that they are unit vectors by drawing a "hat" or circumflex atop the letter: $\hat{\imath}$, $\hat{\jmath}$, $\hat{k}$.) If **a** is any vector with components (a_1, a_2, a_3) then, by Figure 1.4.10,

$$\mathbf{a} = a_1\mathbf{i} + a_2\mathbf{j} + a_3\mathbf{k}.$$

Sometimes we will abuse our notation and write (a_1, a_2, a_3) instead. Also, we will frequently use the convention that if a boldface letter represents a vector, the same letter in normal typeface with subscripts represents the components of the vector. For instance, if **b** is a vector, then $\mathbf{b} = (b_1, b_2, b_3) = b_1\,\mathbf{i}+b_2\,\mathbf{j}+b_3\,\mathbf{k}$; if **x** is a vector, then $\mathbf{x} = (x_1, x_2, x_3) = x_1\,\mathbf{i}+x_2\,\mathbf{j}+x_3\,\mathbf{k}$. Since we've made this identification of ordered triples and vectors, we will often speak of points in $\mathbb{R}^3$ and vectors interchangeably. For instance, $(-3, 5, 7)$ has two meanings: the point $(-3, 5, 7)$ and the vector $-3\,\mathbf{i} + 5\,\mathbf{j} + 7\,\mathbf{k}$.

Representing vectors with components, we can now obtain formulas for addition, subtraction, and scalar multiplication of vectors in terms of components. If $\mathbf{a} = a_1\,\mathbf{i} + a_2\,\mathbf{j} + a_3\,\mathbf{k}$ and $\mathbf{b} = b_1\,\mathbf{i} + b_2\,\mathbf{j} + b_3\,\mathbf{k}$, then placing **a** in standard position and the tail of **b** to the head of **a**, we see from Figure 1.4.11 that the resultant vector $\mathbf{a} + \mathbf{b}$ has components $(a_1 + b_1, a_2 + b_2, a_3 + b_3)$. That is,

$$(a_1\,\mathbf{i} + a_2\,\mathbf{j} + a_3\,\mathbf{k}) + (b_1\,\mathbf{i} + b_2\,\mathbf{j} + b_3\,\mathbf{k}) = (a_1 + b_1)\,\mathbf{i} + (a_2 + b_2)\,\mathbf{j} + (a_3 + b_3)\,\mathbf{k}. \tag{1.4.7}$$

Also, scalar multiplication is defined so that $c\,\mathbf{a}$ has magnitude $|c|\,\|\mathbf{a}\|$ and the direction of **a** if $c > 0$ and the opposite direction to **a** if $c < 0$. By examining the two pairs of similar right triangles in Figure 1.4.12, we see that the x-component of $c\,\mathbf{a}$ is ca_1, the y-component is ca_2, and the z-component

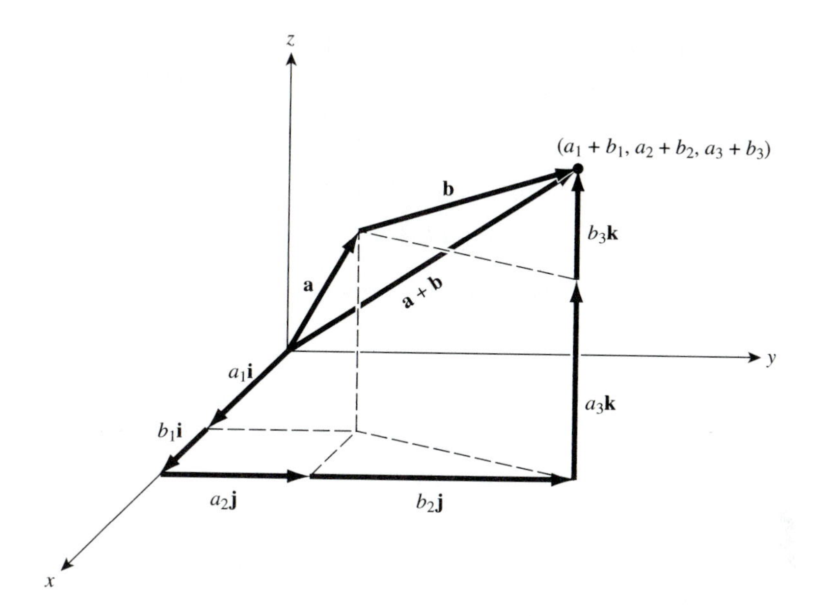

Figure 1.4.11 Component-wise addition of vectors

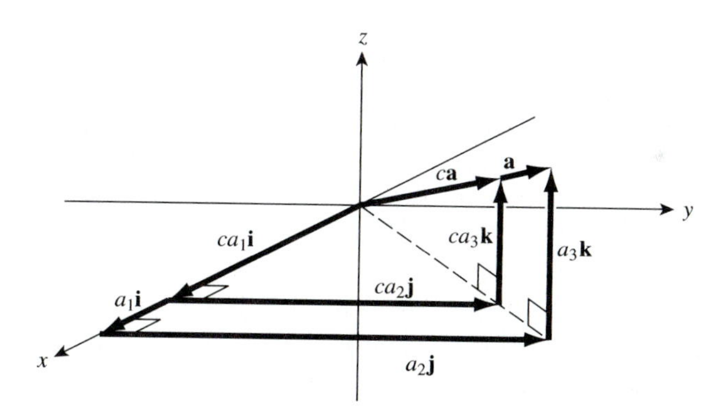

Figure 1.4.12 The components of $c\mathbf{a}$ are ca_1, ca_2, and ca_3.

is ca_3. Thus

$$c(a_1\,\mathbf{i} + a_2\,\mathbf{j} + a_3\,\mathbf{k}) = (ca_1)\,\mathbf{i} + (ca_2)\,\mathbf{j} + (ca_3)\,\mathbf{k}. \tag{1.4.8}$$

Finally, the magnitude of $\mathbf{a} = a_1\,\mathbf{i} + a_2\,\mathbf{j} + a_3\,\mathbf{k}$ is

$$||\mathbf{a}|| = \sqrt{a_1^2 + a_2^2 + a_3^2}, \tag{1.4.9}$$

by the Pythagorean Theorem.

Example 1.4.4 Let $\mathbf{a} = 10\,\mathbf{i} - \mathbf{j} + 14\,\mathbf{k}$, $\mathbf{b} = 6\,\mathbf{i} + 7\,\mathbf{j} - 8\,\mathbf{k}$, and $k = \frac{1}{2}$. Calculate $\mathbf{a} + \mathbf{b}$, $k\,\mathbf{b}$, $||\mathbf{a}||$, and a unit vector in the direction of $\mathbf{a}$.

Solution By (1.4.7),

$$\begin{aligned} \mathbf{a} + \mathbf{b} &= (10 + 6)\,\mathbf{i} + (-1 + 7)\,\mathbf{j} + (14 - 8)\,\mathbf{k} \\ &= 16\,\mathbf{i} + 6\,\mathbf{j} + 6\,\mathbf{k}. \end{aligned}$$

Also, by (1.4.8),

$$\begin{aligned} k\mathbf{b} &= \frac{1}{2} \cdot 6\,\mathbf{i} + \frac{1}{2} \cdot 7\,\mathbf{j} + \frac{1}{2} \cdot (-8)\,\mathbf{k} \\ &= 3\,\mathbf{i} + \frac{7}{2}\,\mathbf{j} - 4\,\mathbf{k}. \end{aligned}$$

By (1.4.9),

$$||\mathbf{a}|| = \sqrt{10^2 + (-1)^2 + 14^2} = \sqrt{297}.$$

Finally, a unit vector in the direction of $\mathbf{a}$ is

$$\begin{aligned} \frac{\mathbf{a}}{||\mathbf{a}||} = \frac{1}{||\mathbf{a}||}\,\mathbf{a} &= \frac{1}{\sqrt{297}}(10\,\mathbf{i} - \mathbf{j} + 14\,\mathbf{k}) \\ &= \frac{10}{\sqrt{297}}\,\mathbf{i} - \frac{1}{\sqrt{297}}\,\mathbf{j} + \frac{14}{\sqrt{297}}\,\mathbf{k}. \quad \blacklozenge \end{aligned}$$

Example 1.4.5 Find the vector with tail at the point $(-4,\ 2,\ 7)$ and head at $(2,\ 3,\ 10)$.

Solution We are really asking, find the vector $\mathbf{c}$ that points from the head of $\mathbf{a} = -4\,\mathbf{i} + 2\,\mathbf{j} + 7\,\mathbf{k}$ to the head of $\mathbf{b} = 2\,\mathbf{i} + 3\,\mathbf{j} + 10\,\mathbf{k}$ when $\mathbf{a}$ and $\mathbf{b}$ are in standard position. This is just

$$\mathbf{c} = \mathbf{b} - \mathbf{a} = (2\,\mathbf{i} + 3\,\mathbf{j} + 10\,\mathbf{k}) - (-4\,\mathbf{i} + 2\,\mathbf{j} + 7\,\mathbf{k}) = 6\,\mathbf{i} + 1\,\mathbf{j} + 3\,\mathbf{k}. \quad \blacklozenge$$

Example 1.4.6 Let $\mathbf{a}$ be a unit vector with tail at the origin that forms angles of $45°$ and $82°$ with the positive x- and y-axes, respectively. Find the components of $\mathbf{a}$.

Solution Let $\mathbf{a} = (a_1,\ a_2,\ a_3)$. The component a_1 of the vector is simply the x-coordinate of the vector's head when it is in standard position. In this position, the vector forms the hypotenuse of a right triangle with vertices $(0,\ 0,\ 0)$, $(a_1,\ a_2,\ a_3)$, and $(a_1,\ 0,\ 0)$. Since $||\mathbf{a}|| = 1$ and the angle between the x-axis and the vector is $45°$, it follows that $a_1 = 1 \cdot \cos 45° = 1/\sqrt{2} = 0.07071$. Similarly, $a_2 = 1 \cdot \cos 82° = 0.1392$. Finally, since $a_1^2 + a_2^2 + a_3^2 = 1$,

$$a_3 = \sqrt{1 - a_1^2 - a_2^2} = \sqrt{1 - 0.5194} = 0.6933.$$

So $\mathbf{a} = 0.7071\mathbf{i} + 0.1392\mathbf{j} + 0.6933\mathbf{k}$. ♦

Example 1.4.7 A hiker walks 2 miles southeast on flat ground, then turns east and goes 1 mile along an upward incline of $10°$ with the horizontal. Then she turns north and goes down an incline of $5°$ for 3 miles. Where is she relative to the starting point of this trip?

Solution We can readily use vectors to answer this question. Place the origin of a coordinate system so that the x-axis points east, the y-axis north, and the z-axis vertically. Her position after the first leg of the hike is represented by a vector of length 2 that is parallel to $\mathbf{i} - \mathbf{j}$. A unit vector in this direction is

$$\frac{\mathbf{i} - \mathbf{j}}{||\mathbf{i} - \mathbf{j}||} = \frac{1}{\sqrt{2}}(\mathbf{i} - \mathbf{j}).$$

So the first leg is represented by

$$\mathbf{a} = \frac{2}{\sqrt{2}}(\mathbf{i} - \mathbf{j}) = 1.4142\,\mathbf{i} - 1.4142\,\mathbf{j}.$$

The second leg is a vector of length 1 parallel to $x\mathbf{i} + z\mathbf{k}$, where x and z are positive numbers such that $\cos(10°) = x/1$ and $\sin(10°) = z/1$. See Figure 1.4.13. Thus $x = 0.9849$ and $z = 0.1736$, so the second leg of the hike

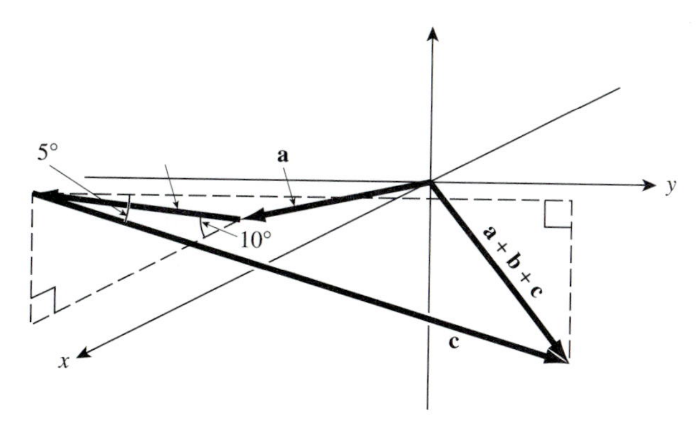

Figure 1.4.13 Three legs of a hike

is represented by

$$\mathbf{b} = 0.9849\,\mathbf{i} + 0.1736\,\mathbf{k}.$$

Finally, the third leg of the hike is a vector of length 3 that is parallel to $y\,\mathbf{j} - z\mathbf{k}$, where y and z satisfy $y/3 = \cos(5°)$ and $z/3 = \sin(5°)$. Again, consult Figure 1.4.13. Then $y = 3\cos(5°) = 2.9886$ and $z = 3\sin(5°) = 0.2615$, so the third part of the hike is given by

$$\mathbf{c} = 2.9886\,\mathbf{j} - 0.2615\,\mathbf{k}.$$

The hiker's position at the end of this walk is

$$\begin{aligned}
&\mathbf{a} + \mathbf{b} + \mathbf{c} \\
&\quad = (1.4142\,\mathbf{i} - 1.4142\,\mathbf{j}) + (0.9849\,\mathbf{i} + 0.1736\,\mathbf{k}) + (2.9886\,\mathbf{j} - 0.2615\,\mathbf{k}) \\
&\quad = 2.399\,\mathbf{i} + 1.5744\,\mathbf{j} - 0.0879\,\mathbf{k}
\end{aligned}$$

Thus she is 2.399 miles east of, 1.5744 miles north of, and 464 ft below the point at which she started. ♦

Example 1.4.8 Kay, a college student, goes to a party being held in a large square room 100 ft by 100 ft. There is a coordinate grid on the floor marked in feet. She has been standing at the position (80, 80) talking to an acquaintance when she notices that her best friend Gigi is standing at position (10, 20) and simultaneously that her worst enemy Dee is standing at position (60, 10). Kay wants to get to her friend but in such a way that she avoids getting too close to Dee. Assuming that Kay is repulsed by her enemy in inverse proportion to the distance between them and that she is attracted to her friend in the same way, in what direction will she start walking in order to get to her friend? See Figure 1.4.14.

Solution The basic idea here is that the forces of attraction and repulsion will combine in an additive fashion and that Kay will go in the direction of the sum of these two force vectors.

The attracting force $\mathbf{A}$ will be $c/||\vec{KG}||$ times a unit vector parallel to $\vec{KG}$, where c is some constant. Now $\vec{KG} = (10 - 80)\,\mathbf{i} + (20 - 80)\,\mathbf{j} = -70\,\mathbf{i} - 60\,\mathbf{j}$, so

$$\begin{aligned}
\mathbf{A} &= \frac{c}{||\vec{KG}||}\frac{\vec{KG}}{||\vec{KG}||} = \frac{c}{||\vec{KG}||^2}\vec{KG} \\
&= \frac{c}{70^2 + 60^2}(-70\,\mathbf{i} - 60\,\mathbf{j}) \\
&= c(-0.00824\,\mathbf{i} - 0.00706\,\mathbf{j}).
\end{aligned}$$

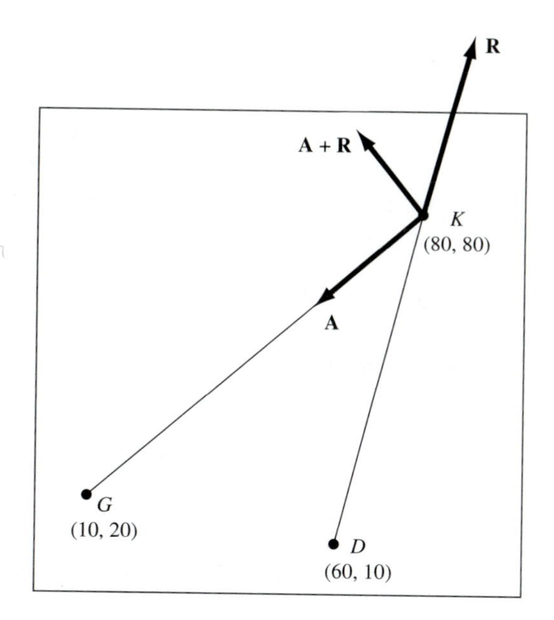

Figure 1.4.14 Attracting and repulsive forces between friends and enemies

Similarly, the repulsive force **R** will be $c/||\vec{DK}||$ times a unit vector parallel to $\vec{DK}$. Since $\vec{DK} = (80 - 60)\,\mathbf{i} + (80 - 10)\,\mathbf{j} = 20\,\mathbf{i} + 70\,\mathbf{j}$, we see that

$$\begin{aligned}\mathbf{R} &= \frac{c}{||\vec{DK}||}\frac{\vec{DK}}{||\vec{DK}||} \\ &= \frac{c}{20^2 + 70^2}(20\,\mathbf{i} + 70\,\mathbf{j}) \\ &= c(0.00377\,\mathbf{i} + 0.0132\,\mathbf{j}).\end{aligned}$$

Thus Kay will start walking in the direction of

$$\begin{aligned}\mathbf{A} + \mathbf{R} &= c(-0.00824\,\mathbf{i} - 0.00706\,\mathbf{j}) + c(0.00377\,\mathbf{i} + 0.0132\,\mathbf{j}) \\ &= c(-0.00447\,\mathbf{i} + 0.00615\,\mathbf{j}).\end{aligned}$$

Do you think that the path Kay follows under these conditions will cause her to bump into the wall? ♦

Exercises 1.4

1. Let $\mathbf{a} = 5\,\mathbf{i} - \mathbf{j}$, $\mathbf{b} = 7\,\mathbf{i} - 3\,\mathbf{j} + 2\,\mathbf{k}$, $\mathbf{c} = -4\,\mathbf{i} + 9\,\mathbf{j} - 8\,\mathbf{k}$, and $\mathbf{d} = -3\,\mathbf{i} - 11\,\mathbf{j} + 7\,\mathbf{k}$. Calculate the following or indicate that the expression is not defined.

   (a) $\mathbf{a} + \mathbf{b} + \mathbf{c} + \mathbf{d}$ (b) $2\,\mathbf{a} + 3\,\mathbf{b}$
   (c) $\mathbf{b}/\mathbf{d}$ (d) $\mathbf{c} - 4\,\mathbf{a}$
   (e) $\mathbf{a}\,\mathbf{d}$ (f) $\|\mathbf{c} - \mathbf{b}\|$
   (g) $\mathbf{a}/\|\mathbf{a}\|$ (h) $\|\mathbf{a}\|\mathbf{b} - \|\mathbf{b}\|\mathbf{a}$

2. Give an example of each of the following (there may be only one!).
   (a) a vector that points in the direction opposite $\mathbf{a} = \mathbf{i} + 2\,\mathbf{j} + 3\,\mathbf{k}$
   (b) a vector with tail at the point $(-8, 2, 7)$ and head at the point $(10, 12, 17)$
   (c) a unit vector parallel to $\mathbf{a} = 4\,\mathbf{i} + 2\,\mathbf{j} - \mathbf{k}$
   (d) a vector of magnitude 13 parallel to $\mathbf{a} = 5\,\mathbf{i} + 2\,\mathbf{k}$
   (e) a unit vector that in standard position forms an angle of $60°$ with both the x- and y-axes
   (f) two vectors that in standard position form a right angle between them

3. A vector has its tail at the origin, has magnitude 3, and forms angles of $68°$ and $53°$ with the positive x- and y-axes, respectively. Find the components of the vector.

4. Show that a vector $\mathbf{a}$ satisfies $\mathbf{a} = \mathbf{0}$ if and only if $\|\mathbf{a}\| = 0$.

5. Show that $\|k\mathbf{a}\| = |k|\,\|\mathbf{a}\|$ for a vector $\mathbf{a}$ and a scalar k.

6. In the following equations, solve for the unknown(s).
   (a) $t\,\mathbf{i} + 3\,\mathbf{j} + \mathbf{k} = (4t + 3)\,\mathbf{i} + 3\,\mathbf{j} + \mathbf{k}$
   (b) $s\,\mathbf{i} - 2t\,\mathbf{j} = (1 - t)\,\mathbf{i} + (s + 2)\,\mathbf{j}$
   (c) $s\,\mathbf{i} + t\,\mathbf{j} + u\,\mathbf{k} = 5(\mathbf{j} + \mathbf{k})$
   (d) $s(\mathbf{i} + \mathbf{j} + \mathbf{k}) = 4t\,\mathbf{i} + 3u\,\mathbf{j} + 8\,\mathbf{k}$

7. Determine whether the following pairs of vectors are linearly independent.
   (a) $\mathbf{a} = \mathbf{i} + 5\mathbf{j}$, $\mathbf{b} = 2\,\mathbf{i} - 6\,\mathbf{j} + \mathbf{k}$
   (b) $\mathbf{a} = (2, 3, -1)$, $\mathbf{b} = (-4, -6, 2)$

(c) $\mathbf{a} = \mathbf{i} + \mathbf{k}$, $\mathbf{b} = \mathbf{j}$

(d) $\mathbf{a} = \mathbf{0}$, $\mathbf{b} = (6,\ 12,\ 12)$

8. Suppose that a boat charts a course at follows: go 20 miles northeast, turn 30° to port and go 30 miles, turn 105° to starboard and go 10 miles, turn 90° to port and go 20 miles. What is the final destination in relation to the starting point?

9. A hiker walks 4 miles north up an incline of 8°, $\frac{1}{2}$ mile east up an incline of 12°, and 1 mile northeast on level ground. What is his position relative to his starting point?

10. A swimmer is attempting to cross from the west to the east side of a 100-meter-wide river that flows south at 3 km/hr. At a particular moment, he is 40 m from the west bank swimming in the direction of a tree on the east bank that is 30 m upstream from his current position. If he is swimming at 4 km/hr, in what direction is he actually moving?

11. In Example 1.4.8, suppose that instead of an enemy, Kay's friend Bea is standing at position (60, 10). Assuming that Kay is attracted toward Bea in the same way that she is attracted toward Gigi, in what direction will Kay initially walk? To which friend will she actually go?

12. Suppose that an object with position vector $\mathbf{x}$ experiences an attractive force toward another object with fixed position vector $\mathbf{a}$ that is inversely proportional to the square of the distance between the objects. Also, suppose that the first object is repelled from an object at position $\mathbf{b}$ in the same way. Find a formula for the resultant force experienced by the object at position $\mathbf{x}$.

13. Use vectors to show that the triangle joining the midpoints of an isosceles triangle is isosceles.

14. A **median** of a triangle is a segment joining a vertex to the midpoint of the opposite side. Using vectors, show that the point in which two medians intersect cuts them both into two segments such that the lengths of the subsegments are in the ratio 2 to 1. See Figure 1.4.15. (*Suggestion*: The vectors $\vec{AF}$ and $\vec{FD}$ are linearly dependent, so $\vec{AF} = k_1\,\vec{FD}$ for some scalar k_1. Similarly, $\vec{BF} = k_2\,\vec{FE}$ for some other constant k_2. Verify that

$$\begin{aligned} \vec{AF} &= k_1\left(\vec{FE} + \vec{EC} + \vec{CD}\right) \\ &= k_1\left(\frac{1}{k_2}\vec{BF} + \frac{1}{2}\vec{AC} + \frac{1}{2}\vec{CB}\right) \\ &= k_1\left(\frac{1}{k_2}\vec{BF} + \frac{1}{2}\vec{AB}\right) \end{aligned}$$

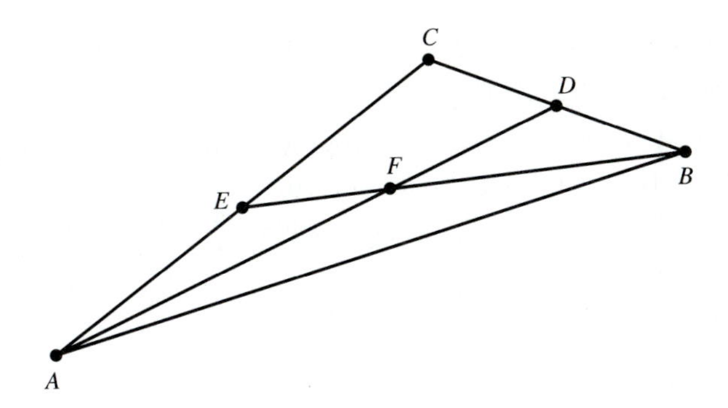

Figure 1.4.15 Intersecting medians of a triangle

$$= \quad k_1\left(\frac{1}{k_2}\vec{BF} + \frac{1}{2}\left(\vec{AF} + \vec{FB}\right)\right).$$

Use this to show that

$$\left(1 - \frac{k_1}{2}\right)\vec{AF} + k_1\left(\frac{1}{2} - \frac{1}{k_2}\right)\vec{BF} = \mathbf{0}$$

and use linear independence.)

15. Find the angles that the vector with tail at (0, 0, 0) and head at (3, 7, −4) forms with the x-, y-, and z-axes.

16. Find the vector that points from (0, 5, 10) to (−4, 11, 1). Then find its magnitude and the angles it forms with the positive $x-$, $y-$, and $z-$axes.

17. Show that if a vector $\mathbf{a}$ forms angles α, β, and γ with the coordinate axes, then $\cos^2\alpha + \cos^2\beta + \cos^2\gamma = 1$. (*Hint*: Show that the components of the vector are $||\mathbf{a}||\cos\alpha$, $||\mathbf{a}||\cos\beta$, and $||\mathbf{a}||\cos\gamma$.) The cosines of α, β, and γ are called the *direction cosines* of $\mathbf{a}$.

18. A vector forms an angle of $\pi/3$ with the positive x-axis, $\pi/3$ with the positive y-axis, and has magnitude 6. Find all of the direction cosines.

19. The **center of mass** of a collection of n objects with masses m_1, m_2, m_3, …, m_n and respective position vectors $\mathbf{a}_1$, $\mathbf{a}_2$, $\mathbf{a}_3$, …, $\mathbf{a}_n$ is the point with position vector

$$\mathbf{a} = \frac{\sum_{i=1}^{n} m_i\,\mathbf{a}_i}{\sum_{i=1}^{n} m_i}.$$

Calculate the center of mass of the objects with the given masses and positions.

(a) $m_1 = 1$, $m_2 = 3/2$, $m_3 = 5$, $m_4 = 1/2$; $\mathbf{a}_1 = (0, 2, 0)$, $\mathbf{a}_2 = (-7, 2, 0)$, $\mathbf{a}_3 = (-1, -1, -1)$, $\mathbf{a}_4 = (8, 4, -2)$.

(b) $m_1 = m_2 = m_3 = m_4 = m_5 = 6$; $\mathbf{a}_1 = (1/\sqrt{2}, 1/\sqrt{2}, 0)$, $\mathbf{a}_2 = (-1/\sqrt{2}, 1/\sqrt{2}, 0)$, $\mathbf{a}_3 = (-1/\sqrt{2}, -1/\sqrt{2}, 0)$, $\mathbf{a}_4 = (1/\sqrt{2}, -1/\sqrt{2}, 0)$, $\mathbf{a}_5 = (0, 0, 1)$.

20. Three objects are to be placed at the positions $(1, 0, 0)$, $(-\sqrt{3}/2, 1/2, 0)$, and $(1/\sqrt{2}, -1/\sqrt{2}, 0)$ so that the center of mass is at the origin. Find the masses that accomplish this.

21. A vector $\mathbf{a}$ is a **linear combination** of vectors $\mathbf{a}_1, \mathbf{a}_2, \mathbf{a}_3, \ldots, \mathbf{a}_n$ if there are constants $k_1, k_2, k_3, \ldots, k_n$ such that

$$\mathbf{a} = k_1\,\mathbf{a}_1 + k_2\,\mathbf{a}_2 + k_3\,\mathbf{a}_3 + \cdots + k_n\,\mathbf{a}_n.$$

Show that $\mathbf{a} = (4/3, 1/3, 10/3)$ is a linear combination of $\mathbf{a}_1 = (1, 4, 5)$ and $\mathbf{a}_2 = (-2, 7, 0)$.

22. Show that $\mathbf{a} = (3, -2, 1)$ is not a linear combination of $\mathbf{a}_1 = (1, 4, 1)$ and $\mathbf{a}_2 = (6, 2, 2)$.

23. Show that any vector $\mathbf{x} = (x_1, x_2, x_3)$ is a linear combination of $\mathbf{a}_1 = (1, 0, 1)$, $\mathbf{a}_2 = (2, -1, 3)$, and $\mathbf{a}_3 = (1, 1, 4)$.

24. A set of vectors $\mathbf{a}_1, \mathbf{a}_2, \mathbf{a}_3, \ldots, \mathbf{a}_n$ is **linearly dependent** if at least one of them can be written as a linear combination of the others. Show that $\mathbf{a}_1 = (4, 1, 9)$, $\mathbf{a}_2 = (1, -1, 1)$, and $\mathbf{a}_3 = (7, -2, 12)$ is a linearly dependent set.

1.5 The Dot Product, Projection, and Work

In this section we extend our ability to do geometry with the analytical tool of the dot product. Besides this we will explore a few geometric and physical applications of this tool.

The general idea is to design a "product" of two vectors that measures to what extent they point in the same direction. The product should be positive if the vectors point in the same general direction, zero if they are perpendicular, and negative if they point in generally opposite directions.

Definition 1.5.1 The **angle** between two nonzero vectors $\mathbf{a}$ and $\mathbf{b}$ in $\mathbb{R}^3$ (or $\mathbb{R}^2$) is the smaller of the angles subtended by the vectors when they are placed tail to tail. The vectors are **orthogonal** (or **perpendicular**) if the angle between them is a right angle.

Definition 1.5.2 The **dot product** of two vectors $\mathbf{a}$ and $\mathbf{b}$ in $\mathbb{R}^2$ or $\mathbb{R}^3$ is the scalar given by

$$\mathbf{a} \cdot \mathbf{b} = ||\mathbf{a}||\,||\mathbf{b}||\cos\theta,$$

where θ is the angle between $\mathbf{a}$ and $\mathbf{b}$. We agree that $\mathbf{a} \cdot \mathbf{0} = 0$.

Example 1.5.1 Find the angle between $\mathbf{a} = 2\,\mathbf{i}$ and $\mathbf{b} = \frac{1}{2}\,\mathbf{i} + (\sqrt{3}/2)\,\mathbf{j}$. Calculate $\mathbf{a} \cdot \mathbf{b}$.

Solution By sketching $\mathbf{a}$ and $\mathbf{b}$ (see Figure 1.5.1) we see that the angle θ

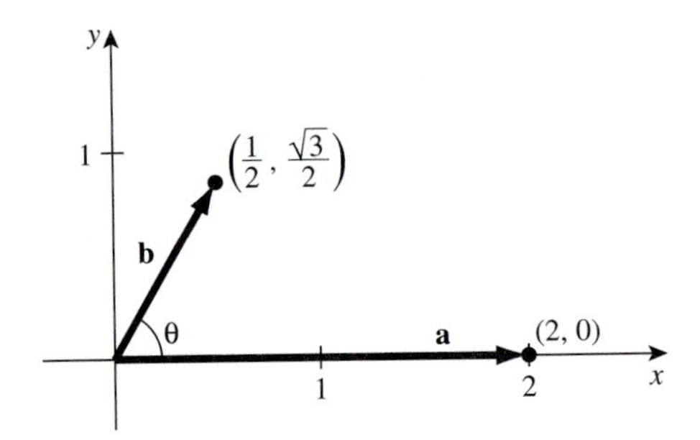

Figure 1.5.1 Finding the angle between $\mathbf{a}$ and $\mathbf{b}$

between $\mathbf{a}$ and $\mathbf{b}$ satisfies

$$\tan\theta = \frac{\sqrt{3}/2}{1/2} = \sqrt{3}.$$

Thus $\theta = \pi/3$. By the definition of dot product then,

$$\begin{aligned} \mathbf{a} \cdot \mathbf{b} &= ||2\,\mathbf{i}|| \left|\left| \frac{1}{2}\mathbf{i} + \frac{\sqrt{3}}{2}\mathbf{j} \right|\right| \cos\left(\frac{\pi}{3}\right) \\ &= 2 \cdot 1 \cdot \frac{1}{2} = 1. \quad \blacklozenge \end{aligned}$$

If $\mathbf{a}$ and $\mathbf{b}$ are given in terms of components then there is another way to calculate $\mathbf{a} \cdot \mathbf{b}$.

Theorem 1.5.1 *If* $\mathbf{a} = a_1\,\mathbf{i} + a_2\,\mathbf{j} + a_3\,\mathbf{k}$ *and* $\mathbf{b} = b_1\,\mathbf{i} + b_2\,\mathbf{j} + b_3\,\mathbf{k}$, *then*

$$\mathbf{a} \cdot \mathbf{b} = a_1 b_1 + a_2 b_2 + a_3 b_3. \tag{1.5.1}$$

Proof The proof is just an application of the familiar law of cosines. Drawing $\mathbf{a}$ and $\mathbf{b}$ tail to tail, we see in Figure 1.5.2 that the vectors $\mathbf{a}$, $\mathbf{b} - \mathbf{a}$, and

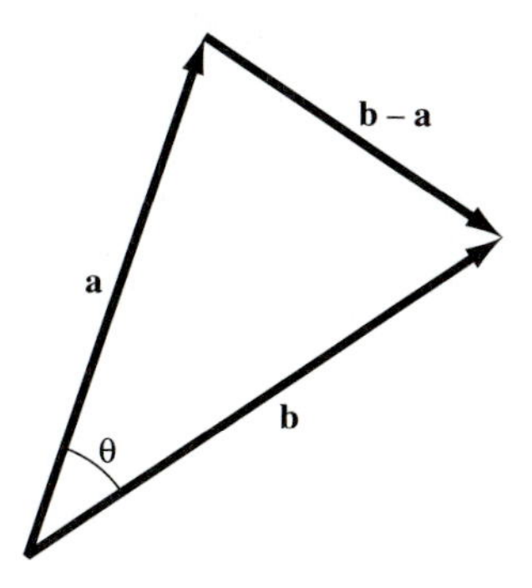

Figure 1.5.2 Applying the Law of Cosines to obtain a new formula for the dot product $\mathbf{a} \cdot \mathbf{b}$

$\mathbf{b}$ form the sides of a triangle. By the law of cosines,

$$||\mathbf{b} - \mathbf{a}||^2 = ||\mathbf{a}||^2 + ||\mathbf{b}||^2 - 2||\mathbf{a}||\,||\mathbf{b}||\cos\theta, \tag{1.5.2}$$

where θ is the angle between $\mathbf{a}$ and $\mathbf{b}$. Since the third term on the right is just -2 times $\mathbf{a} \cdot \mathbf{b}$, we solve (1.5.2) for $\mathbf{a} \cdot \mathbf{b}$ and then write everything else in terms of the components:

$$\begin{aligned}
\mathbf{a} \cdot \mathbf{b} &= \frac{1}{2}\left(||\mathbf{a}||^2 + ||\mathbf{b}||^2 - ||\mathbf{b} - \mathbf{a}||^2\right) \\
&= \frac{1}{2}\Big[\left(a_1^2 + a_2^2 + a_3^2\right) + \left(b_1^2 + b_2^2 + b_3^2\right) \\
&\qquad - \left((b_1 - a_1)^2 + (b_2 - a_2)^2 + (b_3 - a_3)^2\right)\Big] \\
&= \frac{1}{2}(a_1^2 + a_2^2 + a_3^2 + b_1^2 + b_2^2 + b_3^2 - b_1^2 + 2b_1 a_1 - a_1^2 - b_2^2 \\
&\qquad + 2b_2 a_2 - a_2^2 - b_3^2 + 2b_3 a_3 - a_3^2) \\
&= \frac{1}{2} \cdot 2\,(a_1 b_1 + a_2 b_2 + a_3 b_3) \\
&= a_1 b_1 + a_2 b_2 + a_3 b_3. \quad \blacklozenge
\end{aligned}$$

Example 1.5.2 If $\mathbf{a} = 2\,\mathbf{i} - \mathbf{j} + 3\,\mathbf{k}$ and $\mathbf{b} = 6\,\mathbf{i} + 7\,\mathbf{j} + \mathbf{k}$ then

$$\mathbf{a} \cdot \mathbf{b} = 2 \cdot 6 + (-1) \cdot 7 + 3 \cdot 1 = 8. \quad \blacklozenge$$

The dot product has many of the same properties as ordinary multiplication. By writing vectors in terms of their components, we can prove the following theorem.

Theorem 1.5.2 *For any vectors* $\mathbf{a}$, $\mathbf{b}$, *and* $\mathbf{c}$, *and any scalar* p,

1. $\mathbf{a} \cdot \mathbf{b} = \mathbf{b} \cdot \mathbf{a}$.

2. $(p\mathbf{a}) \cdot \mathbf{b} = \mathbf{a} \cdot (p\mathbf{b}) = p(\mathbf{a} \cdot \mathbf{b})$.

3. $\mathbf{a} \cdot (\mathbf{b} + \mathbf{c}) = \mathbf{a} \cdot \mathbf{b} + \mathbf{a} \cdot \mathbf{c}$.

Proof We prove property 3 and leave the other parts as exercises.

3. Let $\mathbf{a} = a_1\,\mathbf{i} + a_2\,\mathbf{j} + a_3\,\mathbf{k}$, $\mathbf{b} = b_1\,\mathbf{i} + b_2\,\mathbf{j} + b_3\,\mathbf{k}$, and $\mathbf{c} = c_1\,\mathbf{i} + c_2\,\mathbf{j} + c_3\,\mathbf{k}$. Then, by the definition of $\mathbf{b} + \mathbf{c}$ and Theorem 1.5.1,

$$\mathbf{a} \cdot (\mathbf{b} + \mathbf{c}) = a_1(b_1 + c_1) + a_2(b_2 + c_2) + a_3(b_3 + c_3). \tag{1.5.3}$$

By the distributive property of ordinary multiplication and commutativity of addition the right side of (1.5.3) can be written as

$$\begin{aligned} & a_1b_1 + a_1c_1 + a_2b_2 + a_2c_2 + a_3b_3 + a_3c_3 \\ = \;& (a_1b_1 + a_2b_2 + a_3b_3) + (a_1c_1 + a_2c_2 + a_3c_3). \end{aligned}$$

This last expression is, by Theorem 1.5.1, the sum of the dot products $\mathbf{a} \cdot \mathbf{b}$ and $\mathbf{a} \cdot \mathbf{c}$. Thus

$$\mathbf{a} \cdot (\mathbf{b} + \mathbf{c}) = \mathbf{a} \cdot \mathbf{b} + \mathbf{a} \cdot \mathbf{c}. \quad \blacklozenge$$

Two other properties that follow directly from the definitions are given in the following theorem. The proof is an exercise.

Theorem 1.5.3

1. $||\mathbf{a}|| = \sqrt{\mathbf{a} \cdot \mathbf{a}}$ *for any vector* $\mathbf{a}$.

2. *Nonzero vectors* $\mathbf{a}$ *and* $\mathbf{b}$ *are orthogonal if and only if* $\mathbf{a} \cdot \mathbf{b} = 0$.

Example 1.5.3 Find the angle between the two vectors in Example 1.5.2.

Solution By the definition of the dot products and Theorem 1.5.1, we have

$$\cos\theta = \frac{\mathbf{a}\cdot\mathbf{b}}{||\mathbf{a}||||\mathbf{b}||} = \frac{8}{\sqrt{14}\sqrt{86}} \approx 0.2306.$$

Thus $\theta \approx 1.338\,\text{rad} \approx 76.7°$. ♦

Example 1.5.4 Using vectors, prove that if A and B are endpoints of a diameter of a circle and C is any other point on the circle, then $\triangle ABC$ is a right triangle.

Solution Figure 1.5.3 summarizes the situation. It will suffice to show that

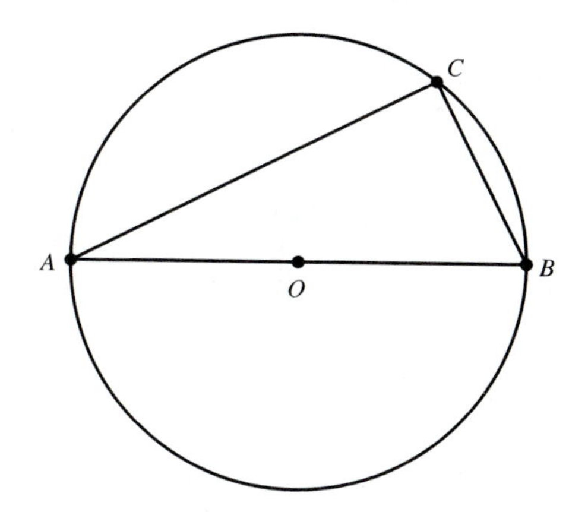

Figure 1.5.3 A well-known theorem of geometry proved using vectors

$\vec{CA}$ and $\vec{CB}$ are orthogonal vectors. Note, first, that

$$\vec{CA} = \vec{CO} + \vec{OA}$$

and second, that

$$\vec{CB} = \vec{CO} + \vec{OB} = \vec{CO} - \vec{BO} = \vec{CO} - \vec{OA}.$$

Then

$$\begin{aligned}
\vec{CA}\cdot\vec{CB} &= (\vec{CO}+\vec{OA})\cdot(\vec{CO}-\vec{OA}) \\
&= (\vec{CO}+\vec{OA})\cdot\vec{CO} - (\vec{CO}+\vec{OA})\cdot\vec{OA} \\
&= \vec{CO}\cdot\vec{CO} + \vec{OA}\cdot\vec{CO} - \vec{CO}\cdot\vec{OA} - \vec{OA}\cdot\vec{OA} \\
&= ||\vec{CO}||^2 - ||\vec{OA}||^2.
\end{aligned}$$

Here we have used Theorems 1.5.2 and 1.5.3 for these steps. Since A and C are both on the circle with center O, $||\vec{CO}|| = ||\vec{OA}||$. Thus $\vec{CA} \cdot \vec{CB} = 0$, and by Theorem 1.5.3 the vectors $\vec{CA}$ and $\vec{CB}$ are orthogonal; $\triangle ABC$ is a right triangle. ♦

Vectors represent physical quantities such as displacement, velocity, and force. What is a physical interpretation of the dot product? We answer this question in some examples.

Example 1.5.5 It is noon on the first day of autumn at the equator. A 50-ft pole leans 10° from the vertical, casting a shadow on the ground. How long is the shadow?

Solution At noon on the first day of autumn, the sun's rays are perpendicular to the ground at the equator. Then one way of solving this is just by simple trigonometry. See Figure 1.5.4. We may also view the problem in

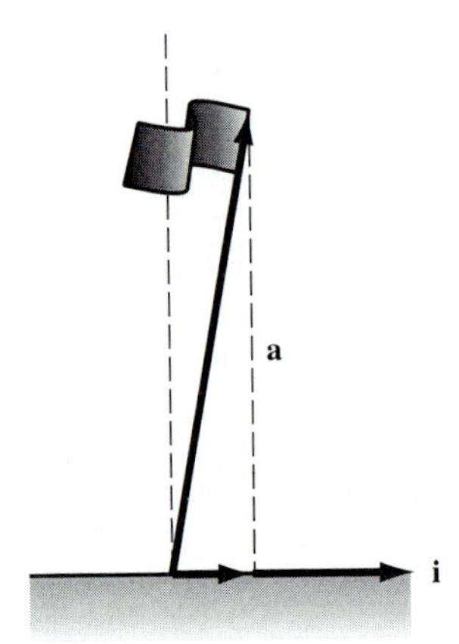

Figure 1.5.4 At noon on the first day of autumn, a leaning flagpole at the equator casts a shadow that is how long?

terms of vectors. Place the pole's base at the origin of a coordinate system with the x-axis parallel to the ground as shown. The length of the shadow is $50 \cos 80° \approx 8.6824$, or

$$50 \cdot 1 \cdot \cos 80° = ||\mathbf{a}||||\mathbf{i}|| \cos 80° = \mathbf{a} \cdot \mathbf{i},$$

where $\mathbf{a}$ is the position vector of the top of the pole. The position vector of the tip of the shadow is then $8.6824\,\mathbf{i}$. We have *projected* the vector $\mathbf{a}$ onto $\mathbf{i}$. ♦

Definition 1.5.3 Given vectors $\mathbf{a}$ and $\mathbf{b}$, the **projection** of $\mathbf{a}$ onto $\mathbf{b}$ is the vector $\mathrm{proj}_{\mathbf{b}}(\mathbf{a})$ obtained by placing $\mathbf{a}$ and $\mathbf{b}$ tail to tail, dropping a perpendicular from the head of **a** to the line on which **b** lies, and drawing the vector from the tail of **b** to the point of intersection of the line and the perpendicular. The **component of a in the direction of b** is the scalar obtained by multiplying the magnitude of the projection $\text{comp}_{\mathbf{b}}(\mathbf{a})$ of **a** onto **b** by $+1$ if the projection has the same direction as **b** and by -1 if it has the opposite direction.

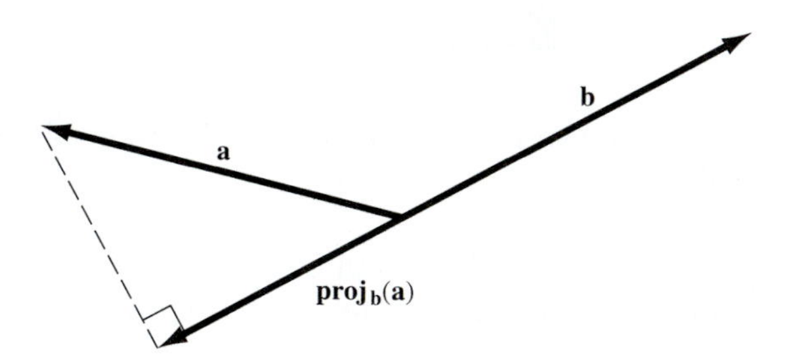

Figure 1.5.5 The projection of **a** onto **b**

A look at Figure 1.5.5 reveals that the component of **a** in the direction of **b** is

$$\text{comp}_{\mathbf{b}}(\mathbf{a}) = ||\mathbf{a}|| \cos\theta = \frac{||\mathbf{a}||\,||\mathbf{b}|| \cos\theta}{||\mathbf{b}||} = \frac{\mathbf{a}\cdot\mathbf{b}}{||\mathbf{b}||}, \tag{1.5.4}$$

where θ is the angle between **a** and **b**. It follows from Definition 1.5.3 that the projection vector is equal to the component of **a** in the direction of **b** multiplied by a unit vector in the direction of **b**. Thus the projection of **a** onto **b** is

$$\text{proj}_{\mathbf{b}}(\mathbf{a}) = \left(\frac{\mathbf{a}\cdot\mathbf{b}}{||\mathbf{b}||}\right)\frac{\mathbf{b}}{||\mathbf{b}||} = \left(\frac{\mathbf{a}\cdot\mathbf{b}}{||\mathbf{b}||^2}\right)\mathbf{b}. \tag{1.5.5}$$

Example 1.5.6 Find $\text{comp}_{\mathbf{b}}(\mathbf{a})$ and $\text{proj}_{\mathbf{b}}(\mathbf{a})$ when $\mathbf{a} = 2\,\mathbf{i}+\mathbf{j}-\mathbf{k}$ and $\mathbf{b} = 4\,\mathbf{i}+\mathbf{j}-5\,\mathbf{k}$.

Solution We have

$$\text{comp}_{\mathbf{b}}(\mathbf{a}) = \frac{\mathbf{a}\cdot\mathbf{b}}{||\mathbf{b}||} = \frac{8+1+5}{\sqrt{16+1+25}} = \frac{14}{\sqrt{42}}$$

and

$$\begin{aligned}\text{proj}_{\mathbf{b}}(\mathbf{a}) &= \text{comp}_{\mathbf{b}}(\mathbf{a})\cdot\frac{\mathbf{b}}{||\mathbf{b}||}\\ &= \frac{14}{\sqrt{42}}\frac{1}{\sqrt{42}}(4\,\mathbf{i}+\mathbf{j}-5\,\mathbf{k})\\ &= \frac{1}{3}(4\,\mathbf{i}+\mathbf{j}-5\,\mathbf{k}). \quad \blacklozenge\end{aligned}$$

As another application of the dot product, we consider the notion of **work**. In physics, the work done by a *constant* force F acting upon an object that undergoes a displacement d along a straight line is Fd, where both F and d have associated with them signs to indicate their orientations. For instance, in Figure 1.5.6, since F and d have opposite orientation (signs),

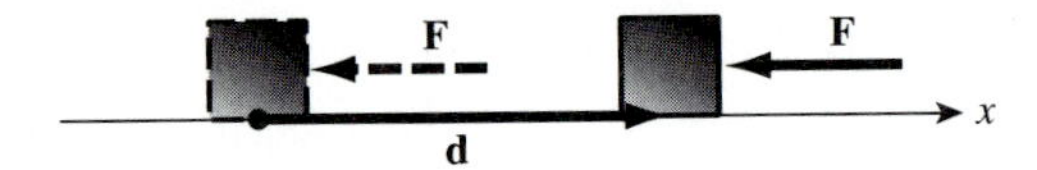

Figure 1.5.6 The work done by a force F on an object that undergoes a displacement d is Fd.

the work done by F is negative. This sign convention allows us to keep track of whether the applied force promotes the motion (positive work) or opposes it (negative work).

Also, if a constant vector force $\mathbf{F}$ acts on an object that undergoes a displacement $\mathbf{d}$, the **work** done by $\mathbf{F}$ is the component of $\mathbf{F}$ in the direction of $\mathbf{d}$ multiplied by the magnitude of $\mathbf{d}$. So we have

$$W = \text{comp}_{\mathbf{d}}(\mathbf{F})||\mathbf{d}|| = \frac{\mathbf{F}\cdot\mathbf{d}}{||\mathbf{d}||}||\mathbf{d}|| = \mathbf{F}\cdot\mathbf{d}. \tag{1.5.6}$$

Example 1.5.7 Find the work done by the gravitational force that acts on a 100-kg mass that slides 10 m down a frictionless incline forming an angle of 30° with the horizontal.

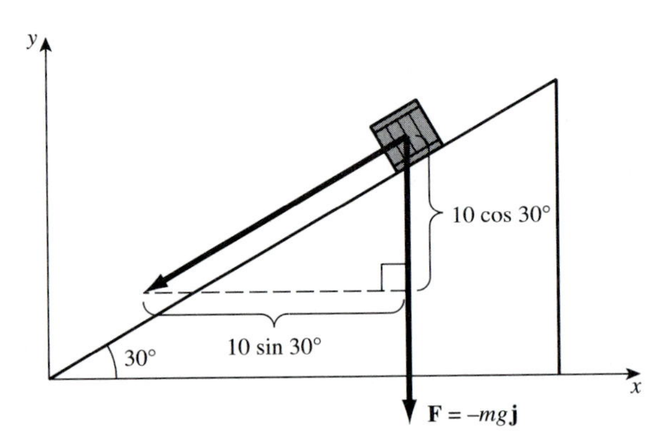

Figure 1.5.7 Finding the work done by gravity on an object that moves along and incline

Solution If we set up a coordinate system as shown in Figure 1.5.7, then $\mathbf{F} = -mg\mathbf{j}$, where m is the mass of the object and g is the gravitational constant (measured in the appropriate units; e.g., if m is in kilograms, then g is 9.8 kg·m/s^2). Since $\mathbf{d}$ points downward and to the left, both of its components will be negative. However, the absolute values of the x and y components of $\mathbf{d}$ are, respectively, $10\cos 30° = 10 \cdot \sqrt{3}/2 = 5\sqrt{3}$ and $10 \sin 30° = 10 \cdot \frac{1}{2} = 5$. Thus $\mathbf{d} = -5\sqrt{3}\,\mathbf{i} - 5\,\mathbf{j}$ and the work done by the gravitational force is

$$\mathbf{F} \cdot \mathbf{d} = -100 \cdot 9.8\,\mathbf{j} \cdot (-5\sqrt{3}\,\mathbf{i} - 5\,\mathbf{j}) = 4900 \text{ N} \cdot \text{m}. \quad \blacklozenge$$

Exercises 1.5

1. Calculate the dot products $\mathbf{a} \cdot \mathbf{b}$ in the following:

(a) $\mathbf{a} = 2\,\mathbf{i} + \mathbf{j}$; $\mathbf{b} = 3\,\mathbf{i} + 3\,\mathbf{j}$

(b) $\mathbf{a} = 4\,\mathbf{i} + \mathbf{k}$; $\mathbf{b} = 2\,\mathbf{j} + 7\,\mathbf{k}$

(c) $\mathbf{a} = 6\,\mathbf{i} + 4\,\mathbf{j} - 2\,\mathbf{k}$; $\mathbf{b} = -\mathbf{i} - \mathbf{j} + \mathbf{k}$

(d) $\mathbf{a} = -\dfrac{\mathbf{i}}{\sqrt{3}} + \dfrac{\mathbf{j}}{\sqrt{3}}$; $\mathbf{b} = \dfrac{2\,\mathbf{i}}{\sqrt{29}} - \dfrac{5\,\mathbf{k}}{\sqrt{29}}$

2. Find the angle between the following pairs of vectors.

(a) $\mathbf{i} + 2\,\mathbf{j} - \mathbf{k}$; $3\,\mathbf{i} + 6\,\mathbf{j} - 3\,\mathbf{k}$

(b) $\dfrac{\mathbf{i}}{2} - \dfrac{\mathbf{k}}{3}$; $-3\,\mathbf{i} + 2\,\mathbf{j}$

(c) $6\,\mathbf{i} + 5\,\mathbf{j} + 4\,\mathbf{k}$; $\mathbf{i} - \mathbf{j} + \mathbf{k}$

(d) $8\,\mathbf{i} + 5\,\mathbf{j} + \mathbf{k}$; $-3\,\mathbf{i} + 5\,\mathbf{j} - \mathbf{k}$

3. For each expression, determine whether or not it makes sense. Explain in either case.

(a) $||\mathbf{a}||\mathbf{b} - ||\mathbf{b}||\mathbf{a}$ (b) $\mathbf{a} \cdot (\mathbf{b} \cdot \mathbf{c})$

(c) $(\mathbf{a} - \mathbf{b}) \cdot \mathbf{c}/2$ (d) $\dfrac{\mathbf{a} \cdot \mathbf{b}}{\mathbf{c}}$

(e) $||\mathbf{a}||^2\mathbf{b} - (\mathbf{a} \cdot \mathbf{b})^2\mathbf{a}$ (f) $\dfrac{\mathbf{b}}{\mathbf{a} \cdot \mathbf{c}}$

4. In the following, calculate $\text{comp}_{\mathbf{b}}(\mathbf{a})$ and $\text{proj}_{\mathbf{b}}(\mathbf{a})$.

(a) $\mathbf{a} = 4\,\mathbf{i} + 6\,\mathbf{j} - 5\,\mathbf{k}$; $\mathbf{b} = -\mathbf{i} - \mathbf{j} - \mathbf{k}$

(b) $\mathbf{a} = \dfrac{1}{\sqrt{2}}\mathbf{i} - \sqrt{2}\,\mathbf{j} + \dfrac{1}{\sqrt{2}}\mathbf{k}$; $\mathbf{b} = \mathbf{i} + \mathbf{j} + \mathbf{k}$

(c) $\mathbf{a} = 4\,\mathbf{i} + 8\,\mathbf{j} + 16\,\mathbf{k}$; $\mathbf{b} = 3\,\mathbf{i} + 9\,\mathbf{j} + 27\,\mathbf{k}$

5. Give an example of:

(a) a vector that is orthogonal to $\mathbf{i} + \mathbf{j} - \mathbf{k}$.

(b) a vector that is orthogonal to both $\mathbf{i} + \mathbf{j}$ and $\mathbf{i} - \mathbf{j}$.

(c) a vector whose projection onto $\mathbf{k}$ is the zero vector.

(d) two vectors that form an angle of $\pi/4$ with $\mathbf{j}$.

6. Find the value of t such that $\mathbf{a} = t\,\mathbf{i} + \mathbf{j} + 7\,\mathbf{k}$ is orthogonal to $\mathbf{b} = 2\,\mathbf{i} - 2\,\mathbf{j} + \mathbf{k}$.

7. A triangle in $\mathbb{R}^3$ has vertices $(6, 11, 3)$, $(1, 5, -7)$, and $(2, 0, 2)$. Find the angles at each vertex.

8. A triangle in $\mathbb{R}^2$ has vertices $(4, 2)$, $(-3, 7)$, and $(-1, -3)$. Find the angles at each vertex.

9. Find the value of t such that the projection of $\mathbf{a} = \mathbf{i} + t\,\mathbf{j} + \mathbf{k}$ onto $\mathbf{b} = 9\,\mathbf{i} + \mathbf{j} + \mathbf{k}$ satisfies

$$\mathrm{proj}_{\mathbf{b}}(\mathbf{a}) = \mathbf{b}.$$

10. Prove Theorem 1.5.2(1).

11. Prove Theorem 1.5.2(2).

12. Prove Theorem 1.5.3.

In Exercises 13 through 18, use only the properties in Theorems 1.5.2 and 1.5.3 to make the proofs.

13. Prove that $(\mathbf{a} \cdot \mathbf{b})\mathbf{c} - (\mathbf{b} \cdot \mathbf{c})\mathbf{a}$ is orthogonal to $\mathbf{b}$ for any three vectors $\mathbf{a}$, $\mathbf{b}$, and $\mathbf{c}$.

14. Prove that for any two vectors $\mathbf{a}$ and $\mathbf{b}$,

$$||\mathbf{a}||^2 + ||\mathbf{b}||^2 = \frac{1}{2}\left(||\mathbf{a} + \mathbf{b}||^2 + ||\mathbf{a} - \mathbf{b}||^2\right).$$

15. Prove that for any two vectors $\mathbf{a}$ and $\mathbf{b}$,

$$\mathbf{a} \cdot \mathbf{b} = \frac{1}{4}\left(||\mathbf{a} + \mathbf{b}||^2 - ||\mathbf{a} - \mathbf{b}||^2\right).$$

16. Prove that if $||\mathbf{a} + \mathbf{b}|| = ||\mathbf{a} - \mathbf{b}||$, then $\mathbf{a}$ is orthogonal to $\mathbf{b}$.

17. Prove that if $||\mathbf{a} + \mathbf{b}|| > ||\mathbf{a} - \mathbf{b}||$, then the angle between $\mathbf{a}$ and $\mathbf{b}$ is less than $\pi/2$.

18. Prove, using vectors, that the diagonals of a rhombus are perpendicular.

19. A block of metal with mass 1 kg is dragged up a frictionless incline of 36° for a distance of 5 m. How much work is done by the gravitational force during this motion?

20. In Exercise 19, some force is necessary to cause the block to move up the incline. Assuming that the block's velocity is constant during this trip, how much work is done by the force that acts along the incline to move the block?

21. A methane molecule has one carbon and four hydrogen atoms arranged so that the hydrogen atoms are at vertices of a regular tetrahedron and the carbon is at the center. The angle between each pair of carbon-hydrogen bonds is the same. Find this angle. (*Suggestion*: Assume that the distance from C to each H is 1; put the C atom at the origin, find the spherical coordinates of the H atoms, convert to rectangular coordinates, and examine the dot products of pairs of vectors from C to various H's.)

22. Given two points P_1 and P_2 with cylindrical coordinates (r_1, θ_1, z_1) and (r_2, θ_2, z_2), show that

$$\vec{OP_1} \cdot \vec{OP_2} = r_1 r_2 \cos(\theta_1 - \theta_2) + z_1 z_2.$$

Use this to find a formula for the angle between these two vectors.

23. Given two points P_1 and P_2 with spherical coordinates $(\rho_1, \theta_1, \phi_1)$ and $(\rho_2, \theta_2, \phi_2)$, show that

$$\begin{aligned} \vec{OP_1} \cdot \vec{OP_2} &= \rho_1 \rho_2 \Bigg(\cos(\phi_1 - \phi_2) \cos^2\left(\frac{\theta_1 - \theta_2}{2}\right) \\ &\qquad + \cos(\phi_1 + \phi_2) \sin^2\left(\frac{\theta_1 - \theta_2}{2}\right) \Bigg) \end{aligned}$$

(*Hint*: You will need several trigonometric identities.)

24. Use the result in Exercise 23 to find a formula for the great circle distance between two points on a sphere of radius R with coordinates (R, θ_1, ϕ_1) and (R, θ_2, ϕ_2). (See Exercise 12 in Section 1.3.)

25. The **prime meridian** is the great circle arc that passes through the earth's north pole, south pole, and Greenwich, England. Place the origin of a coordinate system at the center of the earth so that the positive z-axis passes through the north pole and the positive x-axis passes through the intersection of the prime meridian with the equator.

We define the **longitude** of a point on the earth's surface with spherical coordinates (R, θ, ϕ) to be the angle given by

$$\Theta = 360^\circ - \theta,$$

and the **latitude** of such a point is

$$\Phi = 90^\circ - \phi$$

(where both θ and ϕ are measured in degrees). Find a formula for the great circle distance between one point with latitude and longitude Θ_1 and Φ_1 and a second with latitude and longitude Θ_2 and Φ_2. (Take the radius of the earth to be 6370 km.)

26. Use the formula in Exercise 25 to calculate the great circle distance between

(a) Washington, DC (39° north latitude, 77° west longitude), and Moscow, Russia (56° north latitude, 283° west longitude)

(b) Blountville, Tennessee (36° north latitude, 82° west longitude), and Seattle, Washington (47° north latitude, 123° west longitude)

(c) two of your favorite spots on the earth

1.6 The Cross Product and Determinants

The Cross Product

To begin this section we take a sort of "bare-hands" approach to solving the following problem: Given two nonzero, nonparallel vectors, find a third vector that is perpendicular to both of them. In Section 1.7 and at other times, we will need to be able to find such a vector. The result of our work will be a motivation for the definition of the cross product.

Given vectors $\mathbf{a} = (a_1, a_2, a_3)$ and $\mathbf{b} = (b_1, b_2, b_3)$, we want to obtain a vector $\mathbf{x} = (x, y, z)$ that is perpendicular to both. The vector $\mathbf{x}$ must be nonzero and satisfy $\mathbf{x} \cdot \mathbf{a} = 0$ and $\mathbf{x} \cdot \mathbf{b} = 0$:

$$a_1 x + a_2 y + a_3 z = 0 \quad (1.6.1)$$

$$b_1 x + b_2 y + b_3 z = 0. \quad (1.6.2)$$

Assume for the moment that a_1 and b_1 are both nonzero. Multiplying (1.6.1) by b_1, multiplying (1.6.2) by a_1, and subtracting the resulting equations, we obtain

$$D_3 y + D_2 z = 0,$$

where $D_3 = a_2b_1 - a_1b_2$ and $D_2 = a_3b_1 - a_1b_3$. If $D_3 \neq 0$, then

$$y = \frac{-D_2z}{D_3}.$$

Since there are only two equations in the three unknowns, we are free to choose a value for one of the variables. So let $z = -D_3$, which makes $y = D_2$. Substituting these values of y and z into (1.6.1), we obtain

$$x = \frac{1}{a_1}(a_3D_3 - a_2D_2) = a_2b_3 - a_3b_2.$$

Thus, under the assumptions $a_1 \neq 0$, $b_1 \neq 0$, and $a_3b_1 - a_1b_3 \neq 0$, the vector

$$(a_2b_3 - a_3b_2,\ D_2,\ -D_3) = (a_2b_3 - a_3b_2,\ a_3b_1 - a_1b_3,\ a_1b_2 - a_2b_1) \quad (1.6.3)$$

is perpendicular to $\mathbf{a}$ and $\mathbf{b}$. We shall show shortly that the vector in (1.6.3) is always perpendicular to given linearly independent vectors $\mathbf{a}$ and $\mathbf{b}$. This motivates the following definition.

Definition 1.6.1 The **cross product** of vectors $\mathbf{a} = (a_1, a_2, a_3)$ and $\mathbf{b} = (b_1, b_2, b_3)$ is the vector given by

$$\begin{aligned} \mathbf{a} \times \mathbf{b} &= (a_2b_3 - a_3b_2,\ a_3b_1 - a_1b_3,\ a_1b_2 - a_2b_1) \\ &= (a_2b_3 - a_3b_2)\,\mathbf{i} + (a_3b_1 - a_1b_3)\,\mathbf{j} + (a_1b_2 - a_2b_1)\,\mathbf{k}. \end{aligned}$$

Example 1.6.1 If $\mathbf{a} = (5, 0, 7)$ and $\mathbf{b} = (2, 4, -10)$, then

$$\mathbf{a}\times\mathbf{b} = (0(-10)-7\cdot4)\,\mathbf{i}+(7\cdot2-5\cdot(-10))\,\mathbf{j}+(5\cdot4-0\cdot2)\,\mathbf{k} = -28\,\mathbf{i}+64\,\mathbf{j}+20\,\mathbf{k}. \quad \blacklozenge$$

The following theorem states that the cross product not only possesses the orthogonality property we wanted but also others that relate to the angle between the two vectors we cross together.

Theorem 1.6.1 *For any two vectors* $\mathbf{a}$ *and* $\mathbf{b}$ *in* $\mathbb{R}^3$,

1. $\mathbf{a} \times \mathbf{b}$ *is orthogonal to* $\mathbf{a}$ *and* $\mathbf{b}$ *if the vectors are linearly independent.*

2. *If the vectors are both nonzero then* $||\mathbf{a} \times \mathbf{b}|| = ||\mathbf{a}||\,||\mathbf{b}||\sin\theta$, *where* θ *is the angle between* $\mathbf{a}$ *and* $\mathbf{b}$.

3. $\mathbf{a} \times \mathbf{b} = \mathbf{0}$ *if and only if* $\mathbf{a}$ *and* $\mathbf{b}$ *are linearly dependent.*

Proof 1. We just verify that $\mathbf{a}\cdot(\mathbf{a}\times\mathbf{b})$ is zero.

$$\begin{aligned}\mathbf{a}\cdot(\mathbf{a}\times\mathbf{b}) &= a_1(a_2b_3 - a_3b_2) + a_2(a_3b_1 - a_1b_3) + a_3(a_1b_2 - a_2b_1)\\ &= a_1a_2b_3 - a_1a_3b_2 + a_2a_3b_1 - a_2a_1b_3 + a_3a_1b_2 - a_3a_2b_1\\ &= 0,\end{aligned}$$

and similarly, $\mathbf{b}\cdot(\mathbf{a}\times\mathbf{b}) = 0$.
2. Starting with the right-hand side of the equation, we can write

$$\begin{aligned}||\mathbf{a}||||\mathbf{b}||\sin\theta &= ||\mathbf{a}||||\mathbf{b}||\sqrt{1-\cos^2\theta}\\ &= ||\mathbf{a}||||\mathbf{b}||\sqrt{1-\left(\frac{\mathbf{a}\cdot\mathbf{b}}{||\mathbf{a}||||\mathbf{b}||}\right)^2}\\ &= \sqrt{||\mathbf{a}||^2||\mathbf{b}||^2 - (\mathbf{a}\cdot\mathbf{b})^2}.\end{aligned} \tag{1.6.4}$$

By expanding the magnitudes and dot product under the radical in terms of components and simplifying, we can write this last expression as

$$\begin{aligned}&\Big(a_2^2b_3^2 + a_3^2b_2^2 + a_1^2b_3^2 + a_3^2b_1^2 + a_1^2b_2^2 + a_2^2b_1^2\\ &\quad -2a_2a_3b_2b_3 - 2a_1a_3b_1b_3 - 2a_1a_2b_1b_2\Big)^{1/2}\\ &= \left((a_3b_2 - a_2b_3)^2 + (a_1b_3 - a_3b_1)^2 + (a_1b_2 - a_2b_1)^2\right)^{1/2}\\ &= ||\mathbf{a}\times\mathbf{b}||.\end{aligned}$$

3. By 2, if $\mathbf{a}\times\mathbf{b} = \mathbf{0}$, then $||\mathbf{a}||\,||\mathbf{b}||\sin\theta = 0$. If $||\mathbf{a}|| = 0$, then $\mathbf{a} = \mathbf{0}$, so $\mathbf{a}$ and $\mathbf{b}$ are linearly dependent, and the same is true if $||\mathbf{b}|| = 0$. For $\mathbf{a}$ and $\mathbf{b}$ both nonzero, this means that $\sin\theta = 0$, so that $\theta = 0$ or π, in which case, again, $\mathbf{a}$ and $\mathbf{b}$ are linearly dependent.

Conversely, if $\mathbf{a}$ and $\mathbf{b}$ are nonzero vectors such that $\mathbf{a} = k\mathbf{b}$, then the angle between them is zero or π. Since $\sin 0 = \sin\pi = 0$, this says that $||\mathbf{a}\times\mathbf{b}|| = ||\mathbf{a}||\,||\mathbf{b}||\cdot 0 = 0$, and $\mathbf{a}\times\mathbf{b} = \mathbf{0}$. ♦

We can visualize the direction in which the cross product of two vectors points. Place $\mathbf{a}$ and $\mathbf{b}$ tail to tail, and imagine placing the wrist of your right hand on their tails with your extended fingers pointing in the direction of $\mathbf{a}$ so that as they bend, they curl toward $\mathbf{b}$ through the (smaller) angle between $\mathbf{a}$ and $\mathbf{b}$. Then the thumb points in the direction of $\mathbf{a}\times\mathbf{b}$. This is called the **right-hand rule**, for reasons that are evident. See Figure 1.6.1.

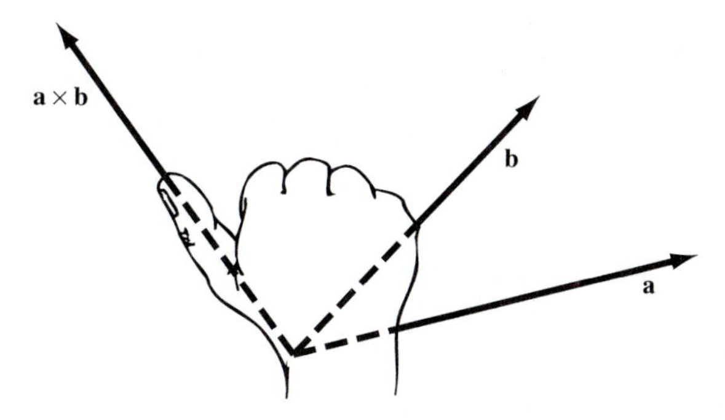

Figure 1.6.1 The right-hand rule for visualizing the direction in which $\mathbf{a} \times \mathbf{b}$ points

Determinants

We devised the cross product in order to obtain a vector that is perpendicular to two given vectors $\mathbf{a}$ and $\mathbf{b}$. The components of $\mathbf{a} \times \mathbf{b}$ are of special interest because they are two-by-two (2×2) **determinants**, quantities that have geometric significance. We conclude this section by examining two-by-two and three-by-three determinants and properties that make them useful and important in calculus of several variables.

We can "stack up" vectors from $\mathbb{R}^2$ or $\mathbb{R}^3$ to form a rectangular array of numbers called a **matrix**. For example, if $\mathbf{a} = (a_1, a_2)$ and $\mathbf{b} = (b_1, b_2)$ are vectors in $\mathbb{R}^2$, we write

$$\begin{bmatrix} a_1 & a_2 \\ b_1 & b_2 \end{bmatrix} \tag{1.6.5}$$

for the 2×2 matrix with $\mathbf{a}$ and $\mathbf{b}$ as its rows. Alternatively, we may write the matrix as

$$\begin{bmatrix} \mathbf{a} \\ \mathbf{b} \end{bmatrix}.$$

Conversely, given any 2×2 matrix of numbers such as in (1.6.5), we can extract the vectors that are its rows. Similarly, from three vectors $\mathbf{a} = (a_1, a_2, a_3)$ $\mathbf{b} = (b_1, b_2, b_3)$, and $\mathbf{c} = (c_1, c_2, c_3)$ in $\mathbb{R}^3$ we write

$$\begin{bmatrix} a_1 & a_2 & a_3 \\ b_1 & b_2 & b_3 \\ c_1 & c_2 & c_3 \end{bmatrix}$$

for the three-by-three (3×3) matrix with $\mathbf{a}$, $\mathbf{b}$, and $\mathbf{c}$ as its rows. We may also write the matrix as

$$\begin{bmatrix} \mathbf{a} \\ \mathbf{b} \\ \mathbf{c} \end{bmatrix}.$$

Definition 1.6.2 The **determinant** of a 2×2 matrix $\begin{bmatrix} a_1 & a_2 \\ b_1 & b_2 \end{bmatrix}$ is the number

$$\det\left(\begin{bmatrix} a_1 & a_2 \\ b_1 & b_2 \end{bmatrix}\right) = a_1 b_2 - a_2 b_1.$$

We will also frequently symbolize the determinant by $\begin{vmatrix} a_1 & a_2 \\ b_1 & b_2 \end{vmatrix}$.

A word of caution about the notation is in order. Do not confuse a matrix with its determinant! A matrix is an array of numbers, analogous to a vector, whereas a determinant is a single number.

The first property of 2×2 determinants we examine is that they can represent area.

Theorem 1.6.2 *The area of the parallelogram in* $\mathbb{R}^2$ *spanned by two vectors* $\mathbf{a} = (a_1, a_2)$ *and* $\mathbf{b} = (b_1, b_2)$ *is*

$$A = \left|\det\left(\begin{bmatrix} a_1 & a_2 \\ b_1 & b_2 \end{bmatrix}\right)\right|.$$

(Here the vertical bars represent absolute value.)

Proof Let θ be the angle between $\mathbf{a}$ and $\mathbf{b}$. See Figure 1.6.2. Then the

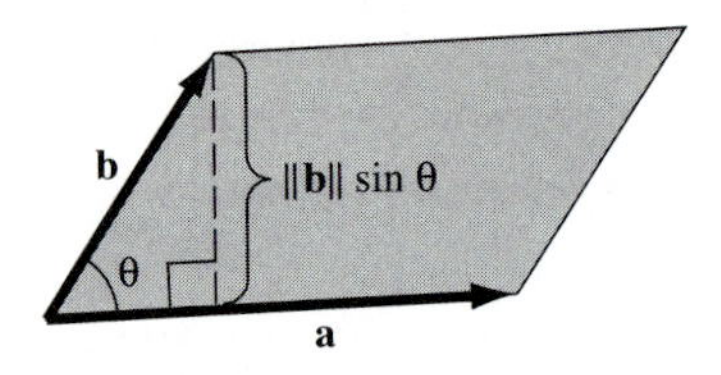

Figure 1.6.2 Finding the area of the parallelogram spanned by **a** and **b**

area of the parallelogram is the product of its base and altitude:

$$A = ||\mathbf{a}||\,||\mathbf{b}||\sin\theta \quad = \quad ||\mathbf{a}||\,||\mathbf{b}||\sqrt{1-\cos^2\theta}$$

$$= ||\mathbf{a}||\,||\mathbf{b}||\sqrt{1 - \left(\frac{\mathbf{a}\cdot\mathbf{b}}{||\mathbf{a}||\,||\mathbf{b}||}\right)^2}$$
$$= \sqrt{||\mathbf{a}||^2||\mathbf{b}||^2 - (\mathbf{a}\cdot\mathbf{b})^2}.$$

When expanded and simplified in terms of components, this last expression becomes (see the proof of Theorem 1.6.1)

$$\sqrt{(a_1b_2 - a_2b_1)^2} = |a_1b_2 - a_2b_1| = \left|\det\left(\begin{bmatrix} a_1 & a_2 \\ b_1 & b_2 \end{bmatrix}\right)\right|. \quad \blacklozenge$$

Example 1.6.2 Find the area of the parallelogram in $\mathbb{R}^2$ with vertices (2, 2), (5, 7), (4, 10), and (1, 5).

Solution The figure is indeed a parallelogram since the vectors $\mathbf{a} = (5 - 2, 7 - 2) = (3, 5)$ and $\mathbf{a}' = (4 - 1, 10 - 5) = (3, 5)$ are equal. With $\mathbf{b} = (1-2, 5-2) = (-1, 3)$, the parallelogram is spanned by $\mathbf{a}$ and $\mathbf{b}$, so its area, by Theorem 1.6.2, is

$$A = \left|\det\left(\begin{bmatrix} 3 & 5 \\ -1 & 3 \end{bmatrix}\right)\right| = |3\cdot 3 - (-1)\cdot 5| = 9 + 5 = 14. \quad \blacklozenge$$

We also have an area formula for parallelograms in $\mathbb{R}^3$.

Theorem 1.6.3 *The area A of a parallelogram spanned by vectors $\mathbf{a}$ and $\mathbf{b}$ in $\mathbb{R}^3$ is*

$$A = ||\mathbf{a}\times\mathbf{b}||.$$

Proof A three-dimensional analog of Figure 1.6.2 reveals that the area of the parallelogram is the product of the length of its base $||\mathbf{a}||$ and its altitude $||\mathbf{b}||\sin\theta$, where θ is the angle between $\mathbf{a}$ and $\mathbf{b}$. This is

$$||\mathbf{a}||\,||\mathbf{b}||\sin\theta = ||\mathbf{a}\times\mathbf{b}||,$$

by Theorem 1.6.1. $\blacklozenge$

Example 1.6.3 Find the area of the triangle with vertices $(5, 1, 2)$, $(-2, 1, 7)$, $(4, 1, 0)$.

Solution The area of the triangle is half the area of the parallelogram spanned by

$$\mathbf{a} = (5 - (-2),\ 1 - 1,\ 2 - 7) = (7,\ 0,\ -5)$$

and

$$\mathbf{b} = (4-(-2),\ 1-1,\ 0-7) = (6,\ 0,\ -7).$$

Thus

$$\begin{aligned} A &= \frac{1}{2}\|(7,\ 0,\ 5)\times(6,\ 0,\ -7)\| \\ &= \frac{1}{2}\|0\,\mathbf{i}+79\,\mathbf{j}+0\,\mathbf{k}\| = \frac{79}{2}. \quad \blacklozenge \end{aligned}$$

By Definition 1.6.1 and our definition of 2×2 determinants, we see that the cross product of two vectors **a** and **b** can be written as

$$\mathbf{a}\times\mathbf{b} = \begin{vmatrix} a_2 & a_3 \\ b_2 & b_3 \end{vmatrix}\mathbf{i} - \begin{vmatrix} a_1 & a_3 \\ b_1 & b_3 \end{vmatrix}\mathbf{j} + \begin{vmatrix} a_1 & a_2 \\ b_1 & b_2 \end{vmatrix}\mathbf{k}. \tag{1.6.6}$$

Taking the magnitude of both sides of (1.6.6), we obtain

$$\|\mathbf{a}\times\mathbf{b}\| = \sqrt{\begin{vmatrix} a_2 & a_3 \\ b_2 & b_3 \end{vmatrix}^2 + \begin{vmatrix} a_1 & a_3 \\ b_1 & b_3 \end{vmatrix}^2 + \begin{vmatrix} a_1 & a_2 \\ b_1 & b_2 \end{vmatrix}^2}. \tag{1.6.7}$$

Referring to Figure 1.6.3, we see that $\|\mathbf{a}\times\mathbf{b}\|$ is the magnitude of the vector whose components are the areas of the projections of the parallelogram onto the coordinate planes! This idea will appear later when we talk about surface integrals.

Definition 1.6.3 The **determinant** of a 3×3 matrix with rows $\mathbf{a} = (a_1,\ a_2,\ a_3)$, $\mathbf{b} = (b_1,\ b_2,\ b_3)$, and $\mathbf{c} = (c_1,\ c_2,\ c_3)$ is the number

$$\det\left(\begin{bmatrix} a_1 & a_2 & a_3 \\ b_1 & b_2 & b_3 \\ c_1 & c_2 & c_3 \end{bmatrix}\right) = a_1\det\left(\begin{bmatrix} b_2 & b_3 \\ c_2 & c_3 \end{bmatrix}\right) - a_2\det\left(\begin{bmatrix} b_1 & b_3 \\ c_1 & c_3 \end{bmatrix}\right) + a_3\det\left(\begin{bmatrix} b_1 & b_2 \\ c_1 & c_2 \end{bmatrix}\right).$$

We often symbolize the determinant by

$$\begin{vmatrix} a_1 & a_2 & a_3 \\ b_1 & b_2 & b_3 \\ c_1 & c_2 & c_3 \end{vmatrix} \quad \text{or} \quad \begin{vmatrix} \mathbf{a} \\ \mathbf{b} \\ \mathbf{c} \end{vmatrix}.$$

Analogous to Theorem 1.6.2, we have a formula for the volume of a parallelepiped spanned by three vectors in $\mathbb{R}^3$.

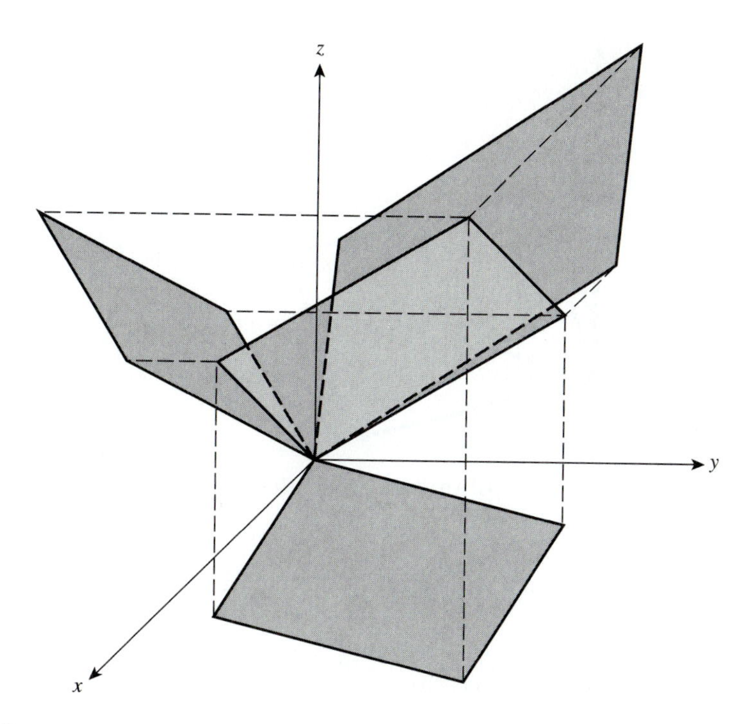

Figure 1.6.3 An analog to the Pythagorean Theorem. The square of the area of a parallelogram in $\mathbb{R}^3$ is the sum of the squares of the areas of its projections into the coordinate planes.

Theorem 1.6.4 *The volume of a parallelepiped spanned by vectors* $\mathbf{a} = (a_1, a_2, a_3)$, $\mathbf{b} = (b_1, b_2, b_3)$, *and* $\mathbf{c} = (c_1, c_2, c_3)$ *is*

$$V = \left| \det \left(\begin{bmatrix} a_1 & a_2 & a_3 \\ b_1 & b_2 & b_3 \\ c_1 & c_2 & c_3 \end{bmatrix} \right) \right| = |\mathbf{c} \cdot (\mathbf{a} \times \mathbf{b})|.$$

Proof The volume of the parallelepiped is the product of the area of its base with its altitude. By Figure 1.6.4, if ϕ is the angle between $\mathbf{c}$ and $\mathbf{a} \times \mathbf{b}$, then the altitude is $||\mathbf{c}||\,|\cos\phi|$. The area of the base, by Theorem 1.6.3, is $||\mathbf{a} \times \mathbf{b}||$. Thus the volume is

$$\begin{aligned} ||\mathbf{c}||\,||\mathbf{a} \times \mathbf{b}||\,|\cos\phi| &= |\,||\mathbf{c}||\,||\mathbf{a} \times \mathbf{b}||\cos\phi| \\ &= |\mathbf{c} \cdot (\mathbf{a} \times \mathbf{b})|. \end{aligned}$$

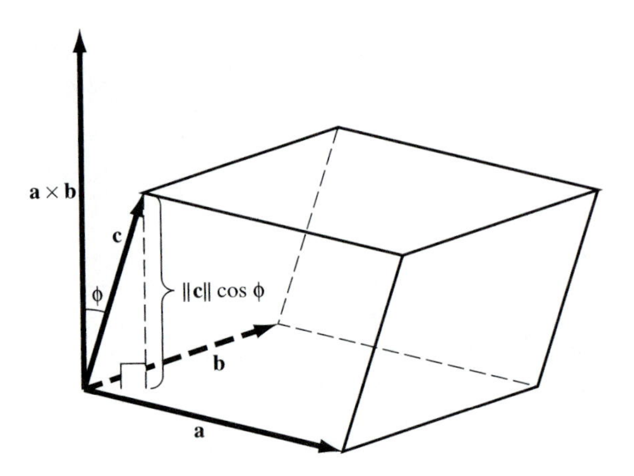

Figure 1.6.4 The volume of a parallelepiped spanned by **a**, **b**, and **c** is $|\mathbf{c}\cdot(\mathbf{a}\times\mathbf{b})|$.

By writing the dot product in component form and using (1.6.6), we have

$$\begin{aligned}\mathbf{c}\cdot(\mathbf{a}\times\mathbf{b}) &= c_1\begin{vmatrix} a_2 & a_3 \\ b_2 & b_3 \end{vmatrix} - c_2\begin{vmatrix} a_1 & a_3 \\ b_1 & b_3 \end{vmatrix} + c_3\begin{vmatrix} a_1 & a_2 \\ b_1 & b_2 \end{vmatrix} \\ &= \begin{vmatrix} c_1 & c_2 & c_3 \\ a_1 & a_2 & a_3 \\ b_1 & b_2 & b_3 \end{vmatrix},\end{aligned}$$

we see that the volume of the parallelepiped is

$$\left|\det\left(\begin{bmatrix} c_1 & c_2 & c_3 \\ a_1 & a_2 & a_3 \\ b_1 & b_2 & b_3 \end{bmatrix}\right)\right|. \quad \blacklozenge \tag{1.6.8}$$

The quantity $\mathbf{c}\cdot(\mathbf{a}\times\mathbf{b})$ is called the **scalar triple product**. It suggests a handy mnemonic device for remembering the definition of the cross product. If we imagine that the components of **c** are the *vectors* **i**, **j**, and **k**, then

$$\mathbf{c}\cdot(\mathbf{a}\times\mathbf{b}) = \mathbf{i}\begin{vmatrix} a_2 & a_3 \\ b_2 & b_3 \end{vmatrix} - \mathbf{j}\begin{vmatrix} a_1 & a_3 \\ b_1 & b_3 \end{vmatrix} + \mathbf{k}\begin{vmatrix} a_1 & a_2 \\ b_1 & b_2 \end{vmatrix} = \mathbf{a}\times\mathbf{b}.$$

That is, we can write

$$\mathbf{a}\times\mathbf{b} = \begin{vmatrix} \mathbf{i} & \mathbf{j} & \mathbf{k} \\ a_1 & a_2 & a_3 \\ b_1 & b_2 & b_3 \end{vmatrix}. \tag{1.6.9}$$

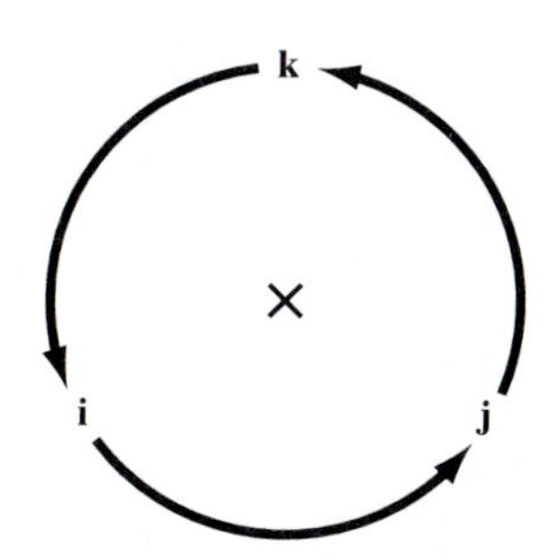

Figure 1.6.5 Mnemonic for remembering the cross product of the basis vectors

Another mnemonic provides a way of remembering the cross products of the basis vectors **i**, **j**, and **k**. If two vectors in Figure 1.6.5 are crossed in the order shown by the arrows, then the result is the next vector in the circle (e.g., $\mathbf{i} \times \mathbf{j} = \mathbf{k}$). If two of these vectors are crossed in the order opposite the arrows, the result is the negative of the next vector encountered by continuing about the circle in a clockwise direction (e.g., $\mathbf{k} \times \mathbf{j} = -\mathbf{i}$). (Verify this using the definition, and also check that it is consistent with the right-hand rule.)

Example 1.6.4 Find the volume of the parallelepiped that lies above the rectangle $2 \leq x \leq 4$, $-1 \leq y \leq 3$ in the xy-plane and between the planes $z = 5x - 8y$ and $z = 5x - 8y + 6$.

Solution The "bottom" of the parallelepiped has vertices obtained by substituting the ordered pairs $(2, -1)$, $(4, -1)$, $(4, 3)$, and $(2, 3)$ into the first plane equation. These vertices are $(2, -1, 18)$, $(4, -1, 28)$, $(4, 3, -4)$, and $(2, 3, -14)$. Thus the vectors $\mathbf{a} = (4 - 2, -1 - (-1), 28 - 18) = (2, 0, 10)$ and $\mathbf{b} = (2 - 2, 3 - (-1), -14 - 18) = (0, 4, -32)$ span the bottom of the parallelepiped. Similarly, by substituting $(2, -1)$ into the equation for the second plane, we see that $(2, -1, 24)$ is a point of the "top" of the parallelepiped right above $(2, -1, 18)$. Thus $\mathbf{c} = (2-2, -1-(-1), 24-18) = (0, 0, 6)$, along with **a** and **b** span the parallelepiped. By Theorem 1.6.4, the volume is $V = |\mathbf{c} \cdot (\mathbf{a} \times \mathbf{b})| = |(0, 0, 6) \cdot (-40, 64, 8)| = |6 \cdot 8| = 48$. ♦

Exercises 1.6

1. Calculate the cross products $\mathbf{a} \times \mathbf{b}$ where:

(a) $\mathbf{a} = \mathbf{i} + \mathbf{j}$, $\mathbf{b} = 2\mathbf{j} + 3\mathbf{k}$

(b) $\mathbf{a} = 3\mathbf{j} + \mathbf{k}$, $\mathbf{b} = 6\mathbf{j} - 4\mathbf{k}$

(c) $\mathbf{a} = \mathbf{i} + 2\mathbf{j} - 7\mathbf{k}$, $\mathbf{b} = 2\mathbf{i} + \mathbf{j} + 3\mathbf{k}$

(d) $\mathbf{a} = \mathbf{i}/2 + \mathbf{j}/3 - \mathbf{k}/5$; $\mathbf{b} = 15\mathbf{i} + 10\mathbf{j} - 6\mathbf{k}$

2. (a) Show by an example that, in general, $\mathbf{a} \times \mathbf{b} \neq \mathbf{b} \times \mathbf{a}$.

(b) Show by example that $\mathbf{a} \times (\mathbf{b} \times \mathbf{c}) \neq (\mathbf{a} \times \mathbf{b}) \times \mathbf{c}$.

3. Show that $(\mathbf{a} \cdot \mathbf{b})^2 + ||\mathbf{a} \times \mathbf{b}||^2 = ||\mathbf{a}||^2||\mathbf{b}||^2$.

4. Prove that $\mathbf{b} \times \mathbf{a} = -\mathbf{a} \times \mathbf{b}$ in two different ways.

5. Prove that $\mathbf{a} \times (\mathbf{b} + \mathbf{c}) = \mathbf{a} \times \mathbf{b} + \mathbf{a} \times \mathbf{c}$.

6. Show that, in general, $(\mathbf{a} \times \mathbf{b}) \times \mathbf{c} \neq \mathbf{a} \times (\mathbf{b} \times \mathbf{c})$.

7. Prove that $(\mathbf{a} - \mathbf{b}) \times (\mathbf{a} + \mathbf{b}) = 2\mathbf{a} \times \mathbf{b}$.

8. Verify that for any vector $\mathbf{a} = a_1\mathbf{i} + a_2\mathbf{j} + a_3\mathbf{k}$,

$$\begin{aligned} \mathbf{i} \times \mathbf{a} &= -a_3\mathbf{j} + a_2\mathbf{k} \\ \mathbf{j} \times \mathbf{a} &= a_3\mathbf{i} - a_1\mathbf{k} \\ \mathbf{k} \times \mathbf{a} &= -a_2\mathbf{i} + a_1\mathbf{j}. \end{aligned}$$

9. Show that for any vector $\mathbf{a} = a_1\mathbf{i} + a_2\mathbf{j} + a_3\mathbf{k}$,

$$\begin{aligned} \mathbf{k} \times (\mathbf{j} \times (\mathbf{i} \times \mathbf{a})) &= a_2\mathbf{j} \\ \mathbf{i} \times (\mathbf{k} \times (\mathbf{j} \times \mathbf{a})) &= a_3\mathbf{k} \\ \mathbf{j} \times (\mathbf{i} \times (\mathbf{k} \times \mathbf{a})) &= a_1\mathbf{i} \end{aligned}$$

(*Hint*: Exercise 8 and Figure 1.6.5.)

10. Prove that if $\mathbf{a} \times \mathbf{x} = \mathbf{0}$ for all $\mathbf{x} \in \mathbb{R}^3$, then $\mathbf{a} = \mathbf{0}$. (*Hint*: Exercise 9.)

11. Prove that $\mathbf{a} \times (\mathbf{b} \times \mathbf{c}) = (\mathbf{a} \cdot \mathbf{c})\mathbf{b} - (\mathbf{a} \cdot \mathbf{b})\mathbf{c}$. This identity means that $\mathbf{a} \times (\mathbf{b} \times \mathbf{c})$ is a linear combination of $\mathbf{b}$ and $\mathbf{c}$, and geometrically, that $\mathbf{a} \times (\mathbf{b} \times \mathbf{c})$ is parallel to a plane spanned by $\mathbf{b}$ and $\mathbf{c}$.

12. Prove that $\mathbf{a} \times (\mathbf{b} \times \mathbf{a}) = ||\mathbf{a}||^2\,\mathbf{b} - (\mathbf{a} \cdot \mathbf{b})\mathbf{a}$.

13. Given two nonzero vectors $\mathbf{a}$ and $\mathbf{b}$ in $\mathbb{R}^3$, show that $\mathbf{v}_1 = \mathbf{a}/||\mathbf{a}||$, $\mathbf{v}_2 = \mathbf{a} \times \mathbf{b}/||\mathbf{a} \times \mathbf{b}||$, and $\mathbf{v}_3 = \mathbf{a} \times (\mathbf{a} \times \mathbf{b})/(||\mathbf{a}||\,||\mathbf{a} \times \mathbf{b}||)$ are mutually orthogonal and that they form a right-handed system in that $\mathbf{v}_1 \times \mathbf{v}_2 = \mathbf{v}_3$, $\mathbf{v}_2 \times \mathbf{v}_3 = \mathbf{v}_1$, and $\mathbf{v}_3 \times \mathbf{v}_1 = \mathbf{v}_2$.

14. Calculate the following determinants.

(a) $\begin{vmatrix} 5 & 4 \\ 3 & -2 \end{vmatrix}$ (b) $\begin{vmatrix} 1/2 & 3/4 \\ 3/4 & 2/3 \end{vmatrix}$

(c) $\begin{vmatrix} 6 & 0 \\ 0 & 8 \end{vmatrix}$ (d) $\begin{vmatrix} 13 & 12 \\ -1 & 12 \end{vmatrix}$

(e) $\begin{vmatrix} 4 & 6 & 1 \\ 0 & 2 & 0 \\ 3 & 9 & -7 \end{vmatrix}$ (f) $\begin{vmatrix} -8 & 0 & 0 \\ 1 & 1 & 1 \\ 3 & -4 & 9 \end{vmatrix}$

(g) $\begin{vmatrix} 3 & 0 & 0 \\ 0 & -9 & 0 \\ 0 & 0 & 27 \end{vmatrix}$ (h) $\begin{vmatrix} 1 & 5 & 4 \\ -5 & 1 & 3 \\ -4 & -3 & 1 \end{vmatrix}$

15. Find the area of the parallelogram in the xy-plane spanned by $\mathbf{a} = 5\mathbf{i} + 2\mathbf{j}$ and $\mathbf{b} = \mathbf{i} + 7\mathbf{j}$.

16. Find the area of the parallelogram in the xy-plane with vertices $(-1, -1)$, $(2, 3)$, $(3, 1)$, and $(0, -3)$.

17. Find the area of the parallelogram in $\mathbb{R}^2$ bounded by the lines $x = 3$, $x = 9$, $y = x/3 - 4$, and $y = x/3 + 4$.

18. Find the area of the parallelogram in $\mathbb{R}^2$ bounded by the lines $y = 0$, $y = 1$, $x = -y$, and $x = -y + 1$.

19. Find the area of the triangle in $\mathbb{R}^2$ with vertices $(4, 5)$, $(-3, 7)$, and $(1, 1)$.

20. Find the area of the parallelogram in $\mathbb{R}^3$ spanned by $\mathbf{a} = 4\mathbf{i} + 2\mathbf{j} - \mathbf{k}$ and $\mathbf{b} = -\mathbf{i} + 10\mathbf{j} + 7\mathbf{k}$.

21. Find the area of the triangle with vertices $(0, 6, 0)$, $(-5, 7, 1)$, $(2, 1, 2)$.

22. (a) Show that the portion of the plane $z = 5x + 2y$ lying above the rectangle in the xy-plane with vertices $(1, 1, 0)$, $(3, 1, 0)$, $(3, 2, 0)$, and $(1, 2, 0)$ is a parallelogram.

(b) Find the areas of the projections of this parallelogram onto the yz- and xz-planes.

(c) Find the area of the parallelogram.

23. Prove that $\det\left(\begin{bmatrix} b_1 & b_2 \\ a_1 & a_2 \end{bmatrix}\right) = -\det\left(\begin{bmatrix} a_1 & a_2 \\ b_1 & b_2 \end{bmatrix}\right)$.

24. Prove that $\begin{vmatrix} a_1 & b_1 \\ a_2 & b_2 \end{vmatrix} = \begin{vmatrix} a_1 & a_2 \\ b_1 & b_2 \end{vmatrix}$.

25. Prove that

$$\begin{vmatrix} a_1 & a_2 \\ b_1 & b_2 \end{vmatrix} = \begin{vmatrix} a_1 & 0 \\ 0 & b_2 \end{vmatrix} + \begin{vmatrix} 0 & a_2 \\ b_1 & 0 \end{vmatrix} = \begin{vmatrix} a_1 & 0 \\ 0 & b_2 \end{vmatrix} - \begin{vmatrix} b_1 & 0 \\ 0 & a_2 \end{vmatrix}.$$

26. Prove that

$$\begin{vmatrix} ka_1 & ka_2 \\ b_1 & b_2 \end{vmatrix} = k\begin{vmatrix} a_1 & a_2 \\ b_1 & b_2 \end{vmatrix} = \begin{vmatrix} a_1 & a_2 \\ kb_1 & kb_2 \end{vmatrix}.$$

27. Show by direct calculation that for any four vectors **a**, **b**, **c**, and **d** in $\mathbb{R}^2$,

$$\begin{vmatrix} \mathbf{a}\cdot\mathbf{c} & \mathbf{a}\cdot\mathbf{d} \\ \mathbf{b}\cdot\mathbf{c} & \mathbf{b}\cdot\mathbf{d} \end{vmatrix} = \begin{vmatrix} \mathbf{a} \\ \mathbf{b} \end{vmatrix}\begin{vmatrix} \mathbf{c} \\ \mathbf{d} \end{vmatrix}.$$

28. (a) Use Exercises 23, 25, and 26 (or direct calculation) to prove that $\begin{vmatrix} \mathbf{b} \\ \mathbf{a} \\ \mathbf{c} \end{vmatrix} = \begin{vmatrix} \mathbf{c} \\ \mathbf{b} \\ \mathbf{a} \end{vmatrix} = \begin{vmatrix} \mathbf{a} \\ \mathbf{c} \\ \mathbf{b} \end{vmatrix} = -\begin{vmatrix} \mathbf{a} \\ \mathbf{b} \\ \mathbf{c} \end{vmatrix}$ for any three vectors **a**, **b**, and **c** in $\mathbb{R}^3$.

(b) Show that $\begin{vmatrix} \mathbf{b} \\ \mathbf{c} \\ \mathbf{a} \end{vmatrix} = \begin{vmatrix} \mathbf{c} \\ \mathbf{a} \\ \mathbf{b} \end{vmatrix} = \begin{vmatrix} \mathbf{a} \\ \mathbf{b} \\ \mathbf{c} \end{vmatrix}$ for any three vectors **a**, **b**, and **c** in $\mathbb{R}^3$.

29. Exercise 28 shows that we can calculate a 3×3 determinant by "expanding" along any row and remembering the pattern of signs $\begin{bmatrix} + & - & + \\ - & + & - \\ + & - & + \end{bmatrix}$ associated with the terms of the expansion. For instance,

$$\begin{vmatrix} a_1 & a_2 & a_3 \\ b_1 & b_2 & b_3 \\ c_1 & c_2 & c_3 \end{vmatrix} = -b_1\begin{vmatrix} a_2 & a_3 \\ c_2 & c_3 \end{vmatrix} + b_2\begin{vmatrix} a_1 & a_3 \\ c_1 & c_3 \end{vmatrix} - b_3\begin{vmatrix} a_1 & a_2 \\ c_1 & c_2 \end{vmatrix}.$$

Use this observation to provide a quick proof of the following.

(a) $\begin{vmatrix} a_1 & a_2 & a_3 \\ 0 & b_2 & b_3 \\ 0 & 0 & c_3 \end{vmatrix} = a_1b_2c_3$

(b) $\begin{vmatrix} 0 & 1 & 0 \\ 0 & 0 & 1 \\ 1 & 0 & 0 \end{vmatrix} = 1$

30. Use Exercises 23, 24, 25, and 26 (or direct calculation) to show that

$$\begin{vmatrix} a_1 & b_1 & c_1 \\ a_2 & b_2 & c_2 \\ a_3 & b_3 & c_3 \end{vmatrix} = \begin{vmatrix} a_1 & a_2 & a_3 \\ b_1 & b_2 & b_3 \\ c_1 & c_2 & c_3 \end{vmatrix}.$$

31. Use Exercise 26 to prove that

$$\begin{vmatrix} ka_1 & ka_2 & ka_3 \\ b_1 & b_2 & b_3 \\ c_1 & c_2 & c_3 \end{vmatrix} = k \begin{vmatrix} a_1 & a_2 & a_3 \\ b_1 & b_2 & b_3 \\ c_1 & c_2 & c_3 \end{vmatrix}.$$

32. Find the volume of the parallelepiped spanned by $\mathbf{a} = 4\,\mathbf{i} + 2\,\mathbf{j} - \mathbf{k}$, $\mathbf{b} = -\mathbf{i} + 10\,\mathbf{j} + 7\,\mathbf{k}$, and $\mathbf{c} = \mathbf{i} - \mathbf{k}$.

33. Suppose that $\mathbf{a}$ in Exercise 32 is replaced by $8\,\mathbf{i} + 4\,\mathbf{j} - 2\,\mathbf{k}$. What is the volume of the resulting parallelepiped? What property of determinants does this illustrate?

34. Suppose that $\mathbf{a}$, $\mathbf{b}$, and $\mathbf{c}$ in Exercise 32 are replaced by $8\,\mathbf{i} + 4\,\mathbf{j} - 2\,\mathbf{k}$, $-3\,\mathbf{i} + 30\,\mathbf{j} + 21\,\mathbf{k}$, and $\frac{1}{4}\,\mathbf{i} - \frac{1}{4}\,\mathbf{k}$. Without evaluating a determinant, find the volume of the resulting parallelepiped.

35. Find the volume of the parallelepiped that lies above the rectangle $0 \le x \le 5$, $0 \le y \le 6$ and between the planes $z = \frac{1}{2}x + y$ and $z = \dfrac{1}{2}x + y - 3$.

36. An $n \times n$ **determinant** can be defined as follows:

$$\begin{vmatrix} a_{11} & a_{12} & a_{13} & \cdots & a_{1n} \\ a_{21} & a_{22} & a_{23} & \cdots & a_{2n} \\ a_{31} & a_{32} & a_{33} & \cdots & a_{3n} \\ \vdots & \vdots & \vdots & \ddots & \vdots \\ a_{n1} & a_{n2} & a_{n3} & \cdots & a_{nn} \end{vmatrix}$$

$$= a_{11} \begin{vmatrix} a_{22} & a_{23} & \cdots & a_{2n} \\ a_{32} & a_{33} & \cdots & a_{3n} \\ \vdots & \vdots & \ddots & \vdots \\ a_{n2} & a_{n3} & \cdots & a_{nn} \end{vmatrix} - a_{12} \begin{vmatrix} a_{21} & a_{23} & \cdots & a_{2n} \\ a_{31} & a_{33} & \cdots & a_{3n} \\ \vdots & \vdots & \ddots & \vdots \\ a_{n1} & a_{n3} & \cdots & a_{nn} \end{vmatrix}$$

$$+ a_{13} \begin{vmatrix} a_{21} & a_{22} & \cdots & a_{2n} \\ a_{31} & a_{32} & \cdots & a_{3n} \\ \vdots & \vdots & \ddots & \vdots \\ a_{n1} & a_{n2} & \cdots & a_{nn} \end{vmatrix} + \cdots + (-1)^n a_{1n} \begin{vmatrix} a_{21} & a_{22} & \cdots & a_{2,n-1} \\ a_{31} & a_{32} & \cdots & a_{3,n-1} \\ \vdots & \vdots & \ddots & \vdots \\ a_{n1} & a_{n2} & \cdots & a_{n,n-1} \end{vmatrix}$$

where each $(n-1) \times (n-1)$ determinant is expanded using this same formula. Applying this procedure iteratively, we eventually reduce the $n \times n$ determinant to many 2×2 determinants which we compute by Definition 1.6.2. Use this formula to calculate the following determinants.

(a) $\begin{vmatrix} 1 & 2 & 3 & 4 \\ 4 & 3 & 2 & 1 \\ 0 & 1 & 0 & 1 \\ 1 & 1 & 1 & 0 \end{vmatrix}$ (b) $\begin{vmatrix} 8 & 0 & 0 & 0 & 0 \\ 0 & 3 & 0 & 0 & 0 \\ 2 & -10 & 1/2 & 0 & 0 \\ -4 & 1 & 1 & -1 & 0 \\ 1 & 1 & 9 & 4 & 1/3 \end{vmatrix}$

1.7 Planes and Lines in $\mathbb{R}^3$

Planes and lines are natural abstractions of everyday experience. For instance, we can idealize the vertical exterior walls and the roof of a house as several planes. The intersections of planes are lines, and this realization permits us to draw and make sense of pictures that represent such objects as houses. In order to treat the concepts of line and plane analytically, however, we must provide algebraic definitions for them. The purpose of this section is to provide such definitions and to examine some of the consequences.

Definition 1.7.1 Given a nonzero vector $\mathbf{n} = (A, B, C)$ and a point $\mathbf{x}_0 = (x_0, y_0, z_0)$, we define the **plane through $\mathbf{x}_0$ perpendicular to n** to be the set

$$\{\mathbf{x} \in \mathbb{R}^3 | (\mathbf{x} - \mathbf{x}_0) \cdot \mathbf{n} = 0\}.$$

The vector $\mathbf{n}$ is called a **normal** to the plane. See Figure 1.7.1.

When we write the equation for the plane in terms of components, we obtain

$$(x - x_0, y - y_0, z - z_0) \cdot (A, B, C) = 0,$$

which simplifies to

$$A(x - x_0) + B(y - y_0) + C(z - z_0) = 0 \tag{1.7.1}$$

or

$$Ax + By + Cz = D. \tag{1.7.2}$$

where $D = Ax_0 + By_0 + Cz_0$.

Example 1.7.1 Find the equation for the plane through the point $(4, 6, -5)$ that is perpendicular to the segment joining the points $(1, 2, 3)$ and $(5, 7, 0)$.

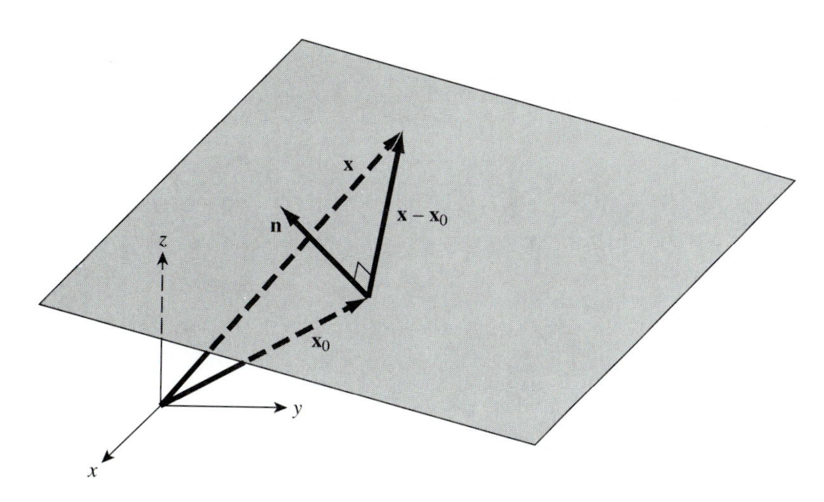

Figure 1.7.1 The plane through $\mathbf{x}_0$ perpendicular to $\mathbf{n}$ consists of all points $\mathbf{x}$ such that $(\mathbf{x} - \mathbf{x}_0) \cdot \mathbf{n} = 0$.

Solution A normal vector is $\mathbf{n} = (1 - 5,\ 2 - 7,\ 3 - 0) = (-4,\ -5,\ 3)$. By (1.7.1), the plane's equation is

$$-4(x - 4) - 5(y - 6) + 3(z + 5) = 0,$$

or, simplifying,

$$-4x + 5y + 3z + 61 = 0. \quad \blacklozenge$$

Example 1.7.2 Find a vector perpendicular to the plane with the equation $x + 2y + 3z - 6 = 0$. Sketch the plane.

Solution By (1.7.2), the coefficients on x, y, and z, respectively, provide components of a normal vector, so $\mathbf{n} = (1,\ 2,\ 3)$.

To sketch a portion of the plane, it will be sufficient to find the points where the coordinate axes intersect it (these are called the **intercepts** or **piercing points**) and simply join these with three line segments. By setting $x, y = 0$ in the equation for the plane, we find that $z = 2$, so $(0,\ 0,\ 2)$ is the piercing point on the z-axis. Similarly, $(0,\ 3,\ 0)$ and $(6,\ 0,\ 0)$ are the piercing points on the y- and x-axes, respectively. By connecting these points with line segments we can easily visualize the plane's orientation. In Figure 1.7.2 we see the portion of the plane in the first octant. $\blacklozenge$

We define a line in $\mathbb{R}^3$ as the collection of all points P such that the vector from a fixed point to P is parallel to a given vector. The following definition spells this out precisely.

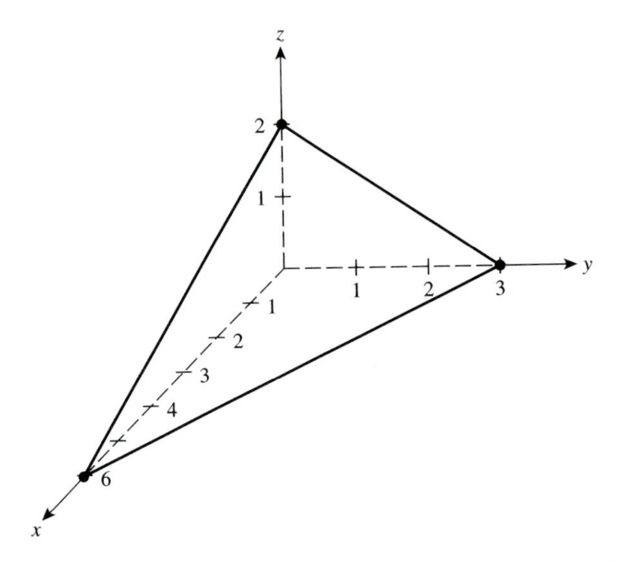

Figure 1.7.2 A sketch of the plane $x + 2y + 3z - 6 = 0$

Definition 1.7.2 Given a vector $\mathbf{m} = (a,\ b,\ c)$ and a point $\mathbf{x}_0 = (x_0,\ y_0,\ z_0)$, we define the **line through $\mathbf{x}_0$ parallel to $\mathbf{m}$** to be the set

$$\{\mathbf{x} \in \mathbb{R}^3 | \mathbf{x} = t\,\mathbf{m} + \mathbf{x}_0, t \in \mathbb{R}\}.$$

The vector $\mathbf{m}$ is called a **direction vector** for the line. We speak of the vector equation

$$\mathbf{x} = t\,\mathbf{m} + \mathbf{x}_0 \tag{1.7.3}$$

as a **parametrization** of the line, the **parameter** being t. See Figure 1.7.3.

With $\mathbf{x} = (x,\ y,\ z)$, we can equate components in (1.7.3) to obtain the **parametric equations** for the line:

$$\begin{aligned} x &= at + x_0 \\ y &= bt + y_0 \\ z &= ct + z_0, \end{aligned} \tag{1.7.4}$$

where we allow t to range over all of $\mathbb{R}$.

Example 1.7.3 Find the parametric equations for the line that passes through the points $(0,\ -10,\ 13)$ and $(1,\ 2,\ 1)$. Where does this line intersect the xy plane?

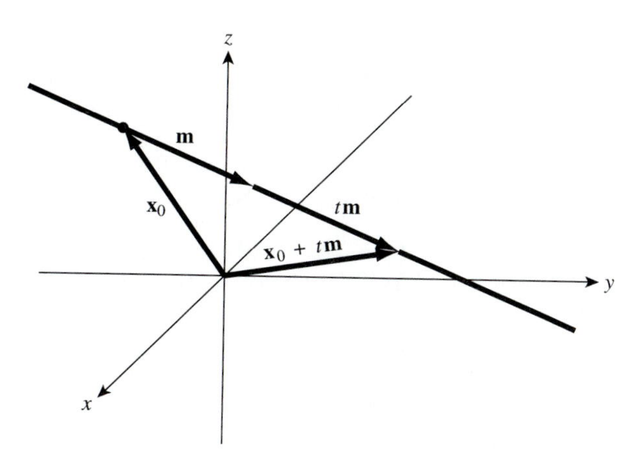

Figure 1.7.3 The line through $\mathbf{x}_0$ parallel to the vector $\mathbf{m}$

Solution A direction vector for the line is $\mathbf{m} = (0-1, -10-2, 13-1) = (-1, -12, 12)$. Thus the parametric equations are

$$\begin{aligned} x &= (-1)t + 0 \\ y &= (-12)t - 10 \\ z &= 12t + 13. \end{aligned}$$

To find where the line intersects the xy-plane, set the z-coordinate to zero in the third equation:

$$12t + 13 = 0.$$

Thus the intersection occurs when $t = -13/12$. Substituting this value of t into the x- and y-coordinates gives $x = 13/12$ and $y = -12(-13/12) - 10 = 3$. Thus the point of intersection is $(13/12, 3, 0)$. ♦

Example 1.7.4 Find the equation for the line that is perpendicular to the plane $7x + y - 10 = 0$ and that passes through the point $(1, 2, 1)$. Determine where the line intersects this plane.

Solution A normal to the plane is $\mathbf{n} = (7, 1, 0)$. This vector is also a direction vector for the line, so a parametrization of the line is $\mathbf{x} = (1 + 7t, 2 + t, 1)$. At the point of intersection, the x-, y-, and z-coordinates of $\mathbf{x}$ must satisfy the equation for the plane. So substituting these for x, y, and z in the equation for the plane,

$$7(1 + 7t) + (2 + t) - 10 = 0.$$

Solving for t gives $t = 1/50$, so the point of intersection is $(1 + 7 \cdot 1/50,\ 2 + 1/50,\ 1) = (57/50,\ 101/50,\ 1)$. ♦

Example 1.7.5 Find the equation for the plane determined by the three points $\mathbf{x}_1 = (0,\ 1,\ 5)$, $\mathbf{x}_2 = (4,\ 0,\ -1)$, and $\mathbf{x}_3 = (-2,\ 3,\ 0)$.

Solution The vectors $\mathbf{a} = \mathbf{x}_2 - \mathbf{x}_1 = (4,\ -1,\ -6)$ and $\mathbf{b} = \mathbf{x}_3 - \mathbf{x}_1 = (-2,\ 2,\ -5)$ for instance are parallel to the plane, so a normal is

$$\begin{aligned} &\mathbf{a} \times \mathbf{b} \\ &= ((-1)(-5) - (-6)2)\mathbf{i} + ((-6)(-2) - 4 \cdot (-5))\mathbf{j} + (4 \cdot 2 - (-1)(-2))\mathbf{k} \\ &= 17\,\mathbf{i} + 32\,\mathbf{j} + 6\,\mathbf{k}. \end{aligned}$$

Using $\mathbf{x}_1$ as a point on the plane, we obtain $23(x-0)+32(y-1)+6(z-5) = 0$ as the equation for the plane. ♦

Example 1.7.6 Find a formula for the distance from a point $\mathbf{x}_0$ in space to a given line $\mathbf{x} = t\,\mathbf{m} + \mathbf{p}_0$.

Solution From Figure 1.7.4 we see that the distance in question is the length of the vector $\mathbf{p} - \mathbf{x}_0$. This is the leg opposite the angle θ in a right

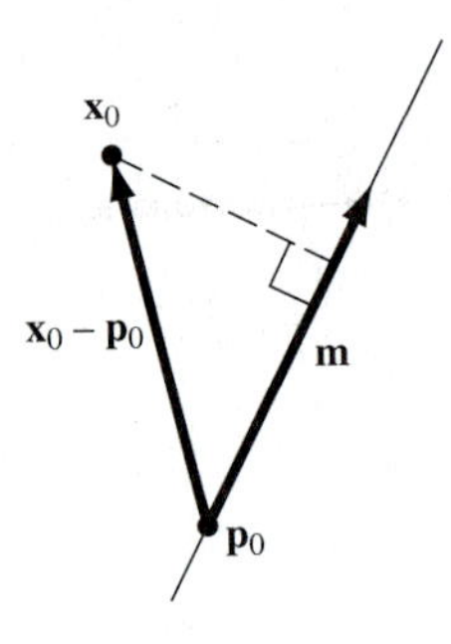

Figure 1.7.4 The distance from $\mathbf{x}_0$ to the line is $||(\mathbf{x}_0 - \mathbf{p}_0) \times \mathbf{m}||/||\mathbf{m}||$.

triangle with hypotenuse of length $\mathbf{x}_0 - \mathbf{p}_0$. So the distance is

$$||\mathbf{x}_0 - \mathbf{p}_0|| \sin\theta.$$

Since θ is the angle between $\mathbf{x}_0 - \mathbf{p}_0$ and $\mathbf{m}$, we write this as

$$\frac{||\mathbf{x}_0 - \mathbf{p}_0||\,||\mathbf{m}|| \sin\theta}{||\mathbf{m}||} = \frac{||(\mathbf{x}_0 - \mathbf{p}_0) \times \mathbf{m}||}{||\mathbf{m}||}. \tag{1.7.5}$$

This last step follows by Theorem 1.6.1. ♦

Example 1.7.7 In drawing, painting, and computer graphics, to represent three-dimensional objects realistically on a two-dimensional medium is often an important goal. One method of injecting realism is to make a **perspective projection** of the object onto a plane. We make a perspective projection as follows. Put the object on one side of the plane, and fix a projection point (the observer's eye) on the other side. From a point on the object, draw the line to the projection point. The intersection of this line with the plane is the *projection* of the point onto the plane, and the collection of all such projected points forms a picture in the plane. See Figure 1.7.5. We can readily describe this process using what we know about lines and

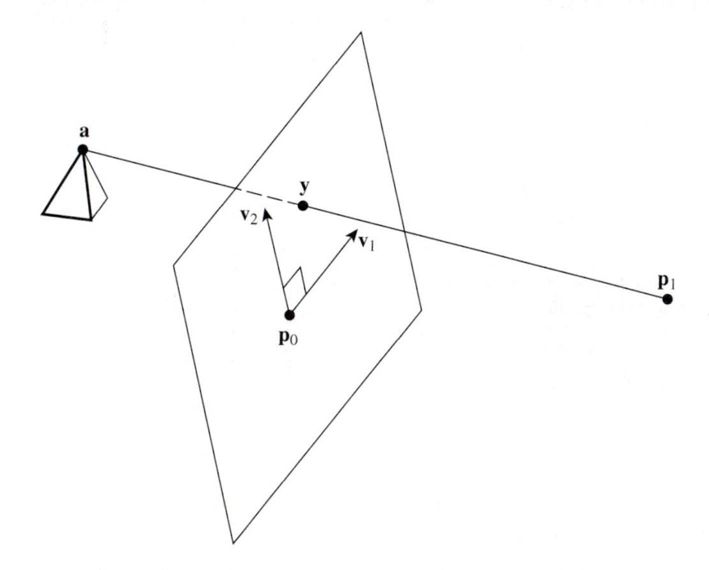

Figure 1.7.5 Perspective projection of a three-dimensional object (a tetrahedron) onto a plane

planes. Let $\mathbf{p}_0$ be the position vector of a point on the plane, and let $\mathbf{v}_1$ and $\mathbf{v}_2$ be two perpendicular vectors with $||\mathbf{v}_1|| = ||\mathbf{v}_2|| = 1$ that span the plane. (Think of $\mathbf{v}_1$ as the "x-axis" on a computer screen, $\mathbf{v}_2$ as the "y-axis," and $\mathbf{p}_0$ as the origin.) Thus the equation for the projection plane is

$$\mathbf{n} \cdot (\mathbf{x} - \mathbf{p}_0) = \mathbf{0}, \tag{1.7.6}$$

where $\mathbf{n} = \mathbf{v}_1 \times \mathbf{v}_2$. Now let $\mathbf{p}_1$ be the position vector of the projection point. Given a point on the object with position vector $\mathbf{a}$, we determine where the line from $\mathbf{a}$ to $\mathbf{p}_1$ intersects the plane. The line is parametrized by

$$\mathbf{x} = \mathbf{a} + t(\mathbf{p}_1 - \mathbf{a}).$$

We obtain the value of t that gives the intersection point by substituting into (1.7.6):

$$\begin{aligned}
\mathbf{n} \cdot (\mathbf{a} + t(\mathbf{p}_1 - \mathbf{a}) - \mathbf{p}_0) &= 0 \\
\mathbf{n} \cdot (\mathbf{p}_1 - \mathbf{a})t &= \mathbf{n} \cdot (\mathbf{p}_0 - \mathbf{a}) \\
t &= \frac{\mathbf{n} \cdot (\mathbf{p}_0 - \mathbf{a})}{\mathbf{n} \cdot (\mathbf{p}_1 - \mathbf{a})}.
\end{aligned}$$

Thus the point of intersection of the line with the plane has position vector

$$\mathbf{y} = \mathbf{a} + \frac{\mathbf{n} \cdot (\mathbf{p}_0 - \mathbf{a})}{\mathbf{n} \cdot (\mathbf{p}_1 - \mathbf{a})}(\mathbf{p}_1 - \mathbf{a}).$$

In order to obtain the $\mathbf{v}_1$- and $\mathbf{v}_2$-coordinate of the projected point (computer screen coordinates) we must find the components of $\mathbf{y} - \mathbf{p}_0$ in the directions of $\mathbf{v}_1$ and $\mathbf{v}_2$. By Formula (1.5.4), these are

$$c_1 = \operatorname{comp}_{\mathbf{v}_1}(\mathbf{y} - \mathbf{p}_0) = \frac{\mathbf{v}_1 \cdot (\mathbf{y} - \mathbf{p}_0)}{||\mathbf{v}_1||} = \mathbf{v}_1 \cdot (\mathbf{y} - \mathbf{p}_0) \tag{1.7.7}$$

and

$$c_2 = \operatorname{comp}_{\mathbf{v}_2}(\mathbf{y} - \mathbf{p}_0) = \mathbf{v}_2 \cdot (\mathbf{y} - \mathbf{p}_0) \tag{1.7.8}$$

since $\mathbf{v}_1$ and $\mathbf{v}_2$ are unit vectors. In the exercises, you are asked to use these formulas to determine the projected image of a simple three-dimensional figure. ♦

The analytical description of lines that we have given here is in a spirit different from that for planes. The line is described dynamically, in that if we think of the parameter t as time, the line gets traced by a "pencil" whose tip at time t is given by the parametric equations. On the other hand, the plane is described "statically." In a sense, the single equation for a plane does not reveal "how the plane got there" (i.e., how it might have been traced by some pencil or paintbrush), but rather it says, "This is where the plane sits."

A plane can also be represented parametrically by specifying the coordinates of a point on the plane as functions of two parameters. This is directly analogous to the way in which we represent lines. To see how to do this, consider first a given plane. Let $\mathbf{x}_0$ be a point in it, and let $\mathbf{a}$ and $\mathbf{b}$ be two

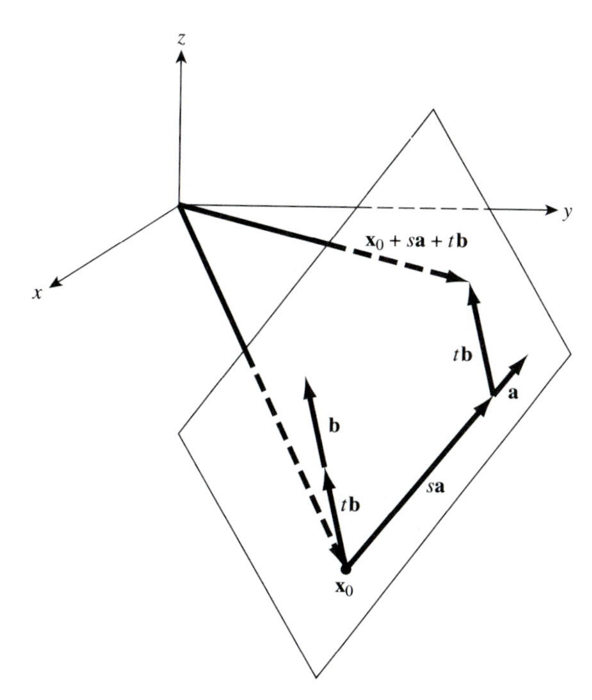

Figure 1.7.6 Parametric representation of a plane through $\mathbf{x}_0$ parallel to $\mathbf{a}$ and $\mathbf{b}$

linearly independent vectors that are parallel to the plane. We draw $\mathbf{a}$ and $\mathbf{b}$ so that their tails coincide at $\mathbf{x}_0$. See Figure 1.7.6. Let $\mathbf{x}$ be any other point in the plane. Because $\mathbf{x}$ and $\mathbf{x}_0$ form opposite corners of a parallelogram with edges parallel to $\mathbf{a}$ and $\mathbf{b}$, it follows that there are scalars s and t such that $\mathbf{x} = \mathbf{x}_0 + s\mathbf{a} + t\mathbf{b}$. Conversely, if we are given a point $\mathbf{x}_0$ and two vectors $\mathbf{a}$ and $\mathbf{b}$ that are linearly independent, every point $\mathbf{x}$ given by

$$\mathbf{x} = \mathbf{x}_0 + s\mathbf{a} + t\mathbf{b} \tag{1.7.9}$$

as s and t range over all real numbers lies on the plane through $\mathbf{x}_0$ perpendicular to $\mathbf{a} \times \mathbf{b}$ (verify this!). Consequently, we say that (1.7.9) is a **parametrization** of the plane through $\mathbf{x}_0$ parallel to $\mathbf{a}$ and $\mathbf{b}$.

Example 1.7.8 Find a parametrization for the plane through the three points (0, 1, 0), (8, 6, 4), and (4, 0, 7).

Solution We choose $\mathbf{x}_0$ to be (0, 1, 0). Two vectors that are parallel to the plane are $\mathbf{a} = (8 - 0,\ 6 - 1,\ -4 - 0) = (8,\ 7,\ -4)$ and $\mathbf{b} = (4,\ -1,\ 7)$. Thus a parametrization for the plane is

$$\begin{aligned} \mathbf{x} &= (0,\ 1,\ 0,\ ) + s(8,\ 7,\ -4) + t(4,\ -1,\ 7) \\ \mathbf{x} &= (8s + 4t,\ 1 + 7s - t,\ -4s + 7t). \quad \blacklozenge \end{aligned}$$

Example 1.7.9 Find the point in which the plane $\mathbf{x} = (2 + s - t,\ 6/7 + 2s,\ 1 + s/3 + t)$ intersects the z-axis.

Solution The point of intersection occurs when the x- and y-coordinates are zero. So we must solve the equations

$$\begin{aligned} 2 + s - t &= 0 \\ \frac{6}{7} + 2s &= 0. \end{aligned}$$

By the second, $s = -3/7$. Substituting into the first, we obtain $t = 2 - 3/7 = 11/7$. Thus we find that the z-intercept is $(0,\ 0,\ 1 + (1/3)(-3/7) + 11/7) = (0,\ 0, 0, 10/7)$. ♦

Exercises 1.7

1. Find an equation for the plane:

(a) through the point $(0,\ 2,\ 5)$ perpendicular to $6\,\mathbf{i} + \mathbf{j} - \mathbf{k}$

(b) through the origin perpendicular to $\left(-\frac{1}{2},\ \frac{1}{3},\ -\frac{1}{4}\right)$

(c) through the origin perpendicular to the segment from $(3,\ 8,\ 7)$ to $(4,\ 1,\ -9)$

(d) through the point $(3,\ 2,\ 3)$ parallel to the plane with equation $7x + 3y + z = 8$

(e) through the point $(2,\ -7,\ 7)$ perpendicular to the line with parametric equations $x = 7t + 1$, $y = -6t + 7$, and $z = 3$.

2. Find a parametrization of the line:

(a) through the point $(0,\ 5,\ -1)$ parallel to $-\mathbf{i} + 4\mathbf{j} + \mathbf{k}$

(b) through the points $(-3,\ -3,\ -2)$ and $(1,\ 1,\ 1)$

(c) through the point $(-8,\ 1,\ 4)$ perpendicular to the plane with equation $6y + 6z = 11$

(d) through the origin parallel to the line parametrized by $\mathbf{x} = (3t + 2,\ 13t + 17,\ 29t)$

3. Find the point in which the line $\mathbf{x} = (2+3t,\ -t,\ -5+4t)$ intersects the plane $4x + 3y = 0$.

4. Find the point in which the line with parametric equations

$$\begin{aligned} x &= 8t - 2 \\ y &= 3t + 8 \\ z &= 3t - 5 \end{aligned}$$

intersects the xz-plane.

5. Determine the points of intersection of the line $\mathbf{x} = (1+3t,\ -1+t,\ 5t)$ and the ellipsoid $x^2/9 + y^2/4 + z^2/16 = 1$.

6. Give an example of:

(a) three points on the plane $x + y - 5z = 1$

(b) two points on the line $\mathbf{x} = (1-t,\ 1+t,\ 6)$

7. Find the equation for the plane through the origin that is perpendicular to the planes $5x - y + z = 1$ and $2x + 2y - 13z = 2$.

8. Find an equation for the plane that passes through the points $(5,\ 5,\ 5)$, $(1,\ 2,\ 1)$, and $(4,\ 7,\ 4)$.

9. Find an equation for the plane containing the line with parametric equations $x = 4t - 1$, $y = 3t + 7$, and $z = -2t + 1$ and the point $(-10,\ 0,\ 11)$.

10. Verify that the lines parametrized by $\mathbf{x} = (t+1,\ 2t+2,\ 3t+3)$ and $\mathbf{x} = (8t+1,\ 2,\ -t+3)$ intersect. Then find an equation for the plane that contains both of them.

11. Show that the two lines $\mathbf{x} = \left(-\frac{1}{2}t,\ \frac{2}{3}t+1,\ t+\frac{3}{2}\right)$ and $\mathbf{x} = \left(t-\frac{3}{2},\ 2t+3,\ 3t+\frac{9}{2}\right)$ intersect. Find an equation for the plane that they determine.

12. Find parametric equations for the line in which the planes $x+y+z-2 = 0$ and $x + 2y + 3z + 4 = 0$ intersect.

13. Find a parametrization for the line through the origin parallel to the line in which the planes $x + z = 4$ and $y - z = 8$ intersect.

14. Find the point in which the planes

$$\begin{aligned} -3x + y + z &= 5 \\ x + y + z &= 18 \\ -x + y - z &= -6 \end{aligned}$$

intersect.

15. Show that the three planes

$$\begin{aligned} 3x - 2y &= 1 \\ 4y - 3z &= 1 \\ 2x - z &= 1 \end{aligned}$$

intersect in the line with parametric equations $x = 2t + 3$, $y = 3t + 4$, and $z = 4t + 5$.

16. Prove that the distance from a point $\mathbf{x}_0$ to the plane with equation $\mathbf{n} \cdot (\mathbf{x} - \mathbf{p}_0) = 0$ is

$$d = \frac{|\mathbf{n} \cdot (\mathbf{x}_0 - \mathbf{p}_0)|}{||\mathbf{n}||}.$$

17. Use the formula in Exercise 16 to find the distance from $(1, 5, -5)$ to the plane $8x + 3y - z = 1$.

18. Show that the distance between two parallel planes $ax + by + cz = d_1$ and $ax + by + cz = d_2$ is

$$\frac{|d_1 - d_2|}{\sqrt{a^2 + b^2 + c^2}}.$$

19. Show that the distance between two parallel lines $\mathbf{x} = \mathbf{x}_0 + t\,\mathbf{m}$ and $\mathbf{x} = \mathbf{x}_1 + t\,\mathbf{m}$ is

$$\frac{||(\mathbf{x}_1 - \mathbf{x}_0) \times \mathbf{m}||}{||\mathbf{m}||}.$$

20. Two lines $\mathbf{x} = \mathbf{x}_0 + t\,\mathbf{m}_0$ and $\mathbf{x} = \mathbf{x}_1 + t\,\mathbf{m}_1$ are said to be **skew** if they do not intersect. By observing that the lines lie in the two parallel planes, show that the distance between the two skew lines is

$$\frac{|\mathbf{m}_0 \times \mathbf{m}_1 \cdot (\mathbf{x}_1 - \mathbf{x}_0)|}{||\mathbf{m}_0 \times \mathbf{m}_1||}.$$

21. Let $\mathbf{v}_1 = (-1, 1, 0)/\sqrt{2}$, $\mathbf{v}_2 = (-1, -1, 1)/\sqrt{3}$, $\mathbf{p}_0 = (5, 5, 5)$, and $\mathbf{p}_1 = (10, 10, 10)$. Using Formulas (1.7.7) and (1.7.8), find the projection of the wire-frame tetrahedron with vertices $(1, 0, 0)$, $(-1/2, \sqrt{3}/2, 0)$, $(-1/2, -\sqrt{3}/2, 0)$, and $(0, 0, 1)$ onto the $\mathbf{v}_1\,\mathbf{v}_2$-plane. (*Suggestion*: You may want to use a programmable calculator or computer to do all of the arithmetic required.)

22. Find a parametrization for the plane through the point $(-6, 4, 7)$ parallel to the vectors $3\mathbf{i} + \mathbf{j} - \mathbf{k}$ and $\mathbf{i} - 5\mathbf{j} + 2\mathbf{k}$.

23. Find a parametrization for the plane that passes through the three points $(-4, -7, -11)$, $(-1, 2, 1)$, and $(0, 8, 0)$.

24. Find a parametrization for the plane containing the intersecting lines in Exercise 10.

25. Find a parametrization for the plane with the equation $12x - y + 17z = 13$.

26. Find a parametrization for the plane with equation $3(x-2)+9(y+1)-2(z-4) = 0$.

27. Find the equation for the plane with parametrization $\mathbf{x} = (1 + s + 3t,\ 5s,\ 8t + 1)$.

28. Find the equation for the plane with parametrization $\mathbf{x} = (1+s+t,\ 11+2s-t,\ s-t)$.

1.8 Vector-Valued Functions

The objective in Newtonian mechanics is to describe and predict the motion of objects that are influenced by forces. Part of what facilitates this description is some mathematical machinery that expresses precisely our intuitive notion of, for instance, how the position of a baseball thrown across a field changes from moment to moment, or how a planet follows a path through space. We introduce some of that machinery in this section. Although this is not a text on physics, it is nevertheless helpful to have physical motivations or interpretations for the concepts we define. Consequently, we will take time to introduce some basic and intuitive physical concepts as they are needed.

Definition 1.8.1 A **vector-valued function** of one variable is a rule, denoted by a **boldface** letter such as **f**, that associates with each number in an interval called the **domain** of **f** a vector in $\mathbb{R}^3$ or $\mathbb{R}^2$. If t is a number in the domain, then $\mathbf{f}(t)$ denotes the **value** of **f** at t. The components of **f** are real-valued functions, f_1, f_2, and f_3, say. Thus

$$\mathbf{f}(t) = f_1(t)\,\mathbf{i} + f_2(t)\,\mathbf{j} + f_3(t)\,\mathbf{k} \tag{1.8.1}$$

for each t in the domain of **f**.

If D denotes the domain of **f** then we write $\mathbf{f}: D \subset \mathbb{R} \to \mathbb{R}^3$ as shorthand for "**f** is a vector-valued function from D into $\mathbb{R}^3$." We may also abbreviate (1.8.1) by writing $\mathbf{f} = f_1\,\mathbf{i} + f_2\,\mathbf{j} + f_3\,\mathbf{k}$. We may refer to **f** as a **position function**.

Example 1.8.1 The function $\mathbf{f}\colon (0, \infty) \subset \mathbb{R} \to \mathbb{R}^3$ defined by

$$\mathbf{f}(t) = (3t + 1)\,\mathbf{i} - (5t - 2)\,\mathbf{j} + t\,\mathbf{k}$$

has as its components the functions $f_1(t) = (3t + 1)$, $f_2(t) = -(5t - 2)$, and $f_3(t) = t$; its domain is $D = (0, \infty)$.

We can visualize this vector-valued function as describing the path of a particle that moves along the half-line parametrized by $x = 3t + 1$, $y = -5t + 2$, $z = t$, where $t > 0$. Indeed, consistent with our earlier terminology we say that $\mathbf{x} = \mathbf{f}(t)$, $t > 0$ is a parametrization of the line, where $\mathbf{x}$ represents $x\,\mathbf{i} + y\,\mathbf{j} + z\,\mathbf{k}$. ♦

Example 1.8.2 Describe the path parametrized by $\mathbf{x} = (3\cos t)\,\mathbf{i} + (2\sin t)\,\mathbf{j}$.

Solution The most naive approach is simply to plot the points given by this equation for several choices of the parameter t. Since the cosine and sine functions are periodic with period 2π, the path (which lies entirely in the xy-plane) begins retracing itself when t reaches the value 2π. Thus it suffices to plot points for t in the interval $[0, 2\pi]$. Shown in Figure 1.8.1 are the position vectors corresponding to $t = 0$, $\pi/4$, $\pi/2$, $3\pi/4$, π, $5\pi/4$, $3\pi/2$, $7\pi/4$, and 2π.

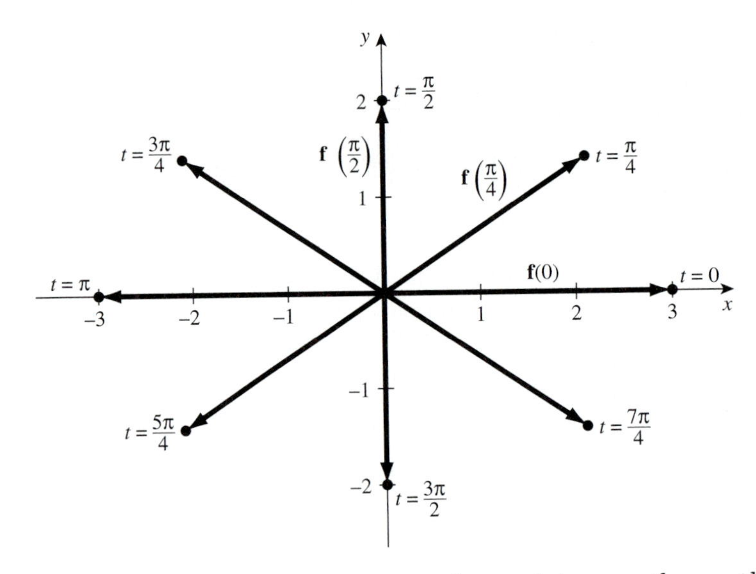

Figure 1.8.1 Position vectors for points on the path parametrized by $\mathbf{x} = 3\cos t\,\mathbf{i} + 2\sin t\,\mathbf{j}$

These suggest an elliptically shaped path. In this instance we can show that the path does follow an ellipse exactly. The parametrization is equivalent to the parametric equations

$$\begin{aligned} x &= 3\cos t \\ y &= 2\sin t, \end{aligned}$$

which can be written as

$$\begin{aligned} \frac{x}{3} &= \cos t \\ \frac{y}{2} &= \sin t. \end{aligned}$$

Squaring these, adding, and using the identity $\cos^2 t + \sin^2 t = 1$, we obtain

$$\frac{x^2}{9} + \frac{y^2}{4} = 1,$$

the equation for the ellipse suggested by Figure 1.8.1. ♦

Most often, a given parametrization will not be so easily analyzed as in Example 1.8.2. The tedious job of plotting points, however, can be relegated to a computer if it is important to have an accurate picture of a path. Vector-valued functions with one component constant are particularly easy to plot because the path lies in a plane, and most computer programs that do any graphics at all will produce such two-dimensional graphics.

Example 1.8.3 The path parametrized by $\mathbf{x} = (5\cos(t/5) + \cos(2t))\,\mathbf{i} + (5\sin(t/5) + \sin(3t))\,\mathbf{j}$ for $-20\pi \le t \le 20\pi$ appears in Figure 1.8.2. ♦

Many planar curves of historical interest have definitions that make them more easily represented by a parametrization than by a single equation in x and y. One such curve is the **cycloid**, defined as the path traced by a point on the circumference of a circle that rolls along a straight line. (In the Exercises, you are asked to investigate this curve.) Another is the **hypocycloid**, the path traced by a point on the circumference of a circle that rolls along the inside of a larger circle.

Example 1.8.4 Find a parametrization of the **hypocycloid** traced by a point on the circumference of a circle of radius r that rolls along inside another circle of larger radius R.

Solution We put the larger circle in the xy-plane with its center at the origin, and we put the smaller circle initially so that it and the larger one are tangent at the point $S = (R, 0)$. As the smaller circle rolls counterclockwise

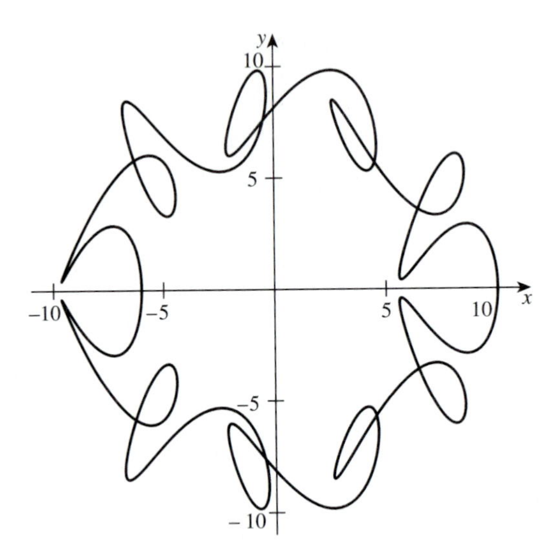

Figure 1.8.2 The path parametrized by $\mathbf{x} = (5\cos(t/5) + \cos(2t))\,\mathbf{i} + (5\sin(t/5) + \sin(3t))\,\mathbf{j}$ for $-20\pi \le t \le 20\pi$

(see Figure 1.8.3) the point P on it that was initially at $(R, 0)$ moves clockwise relative to the center C of the smaller circle. Let T be the (moving) point at which the circles are tangent, and let $t = \angle PCT$ be the parameter in terms of which we represent the coordinates of P.

The position vector of P is

$$\vec{OP} = \vec{OC} + \vec{CP},$$

so the problem reduces to finding the components of $\vec{OC}$ and $\vec{CP}$. With $\phi = \angle SOT$, we see, with the aid of Figure 1.8.3, that

$$\begin{aligned} \vec{OC} &= ||\vec{OC}||(\cos\phi\,\mathbf{i} + \sin\phi\,\mathbf{j}) \\ &= (R - r)(\cos\phi\,\mathbf{i} + \sin\phi\,\mathbf{j}). \end{aligned}$$

Also,

$$\begin{aligned} \vec{CP} &= ||\vec{CP}||(\cos(t-\phi)\,\mathbf{i} - \sin(t-\phi)\,\mathbf{j}) \\ &= r(\cos(t-\phi)\,\mathbf{i} - \sin(t-\phi)\,\mathbf{j}). \end{aligned}$$

The final ingredient we need is a relationship between t and ϕ, and this relies on the observation that the arcs $\overset{\frown}{TP}$ and $\overset{\frown}{TS}$ have the same length. From this it follows that $rt = R\phi$, so $\phi = (r/R)t$. Thus a parametrization is

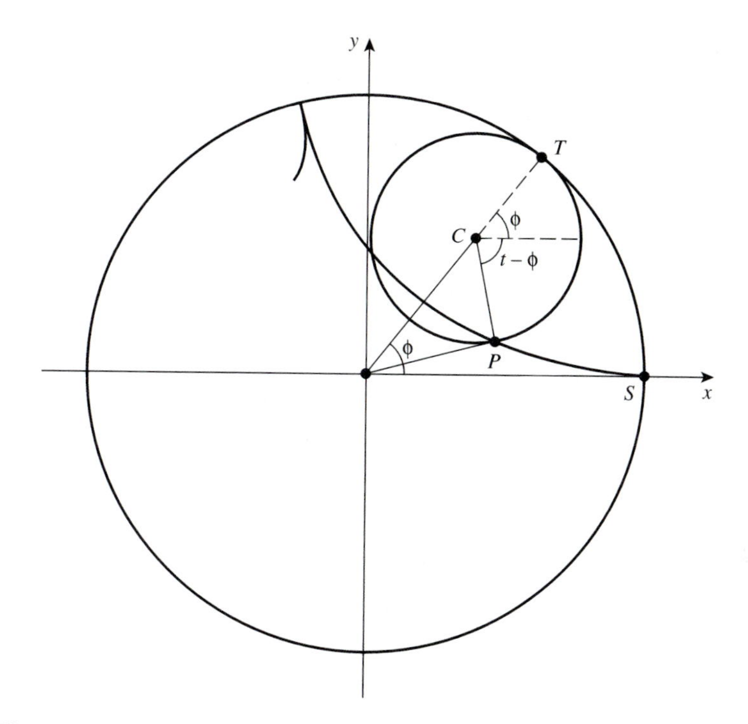

Figure 1.8.3 Obtaining parametric equations for the hypocycloid traced by P as the small circle rolls inside the large one

$\mathbf{x} = \mathbf{h}(t)$, where

$$\begin{aligned} \mathbf{h}(t) &= \vec{OP} = \vec{OC} + \vec{CP} \\ &= (R-r)(\cos\phi\,\mathbf{i} + \sin\phi\,\mathbf{j}) + r(\cos(t-\phi)\,\mathbf{i} - \sin(t-\phi)\,\mathbf{j}) \\ &= \left((R-r)\cos\left(\frac{r}{R}t\right) + r\cos\left(\left(1-\frac{r}{R}\right)t\right)\right)\mathbf{i} \\ &\qquad + \left((R-r)\sin\left(\frac{r}{R}t\right) - r\sin\left(\left(1-\frac{r}{R}\right)t\right)\right)\mathbf{j}. \end{aligned}$$

How could we check that this complicated-looking formula is correct? It is a simple matter to put these component functions into a graphics program with specific values for R and r. With $R = 5$ and $r = 1$, we obtain Figure 1.8.4. This is perfectly reasonable: The moving point on the inner circle will "bounce" off the outer circle four times before returning exactly to its starting point. The range of t-values that causes $\mathbf{h}$ to trace the path shown is $[0, 10\pi]$. Why must this interval be so long? How long an interval is needed if, for instance, $R = 5$ and $r = 2$? Try several different choices of R and

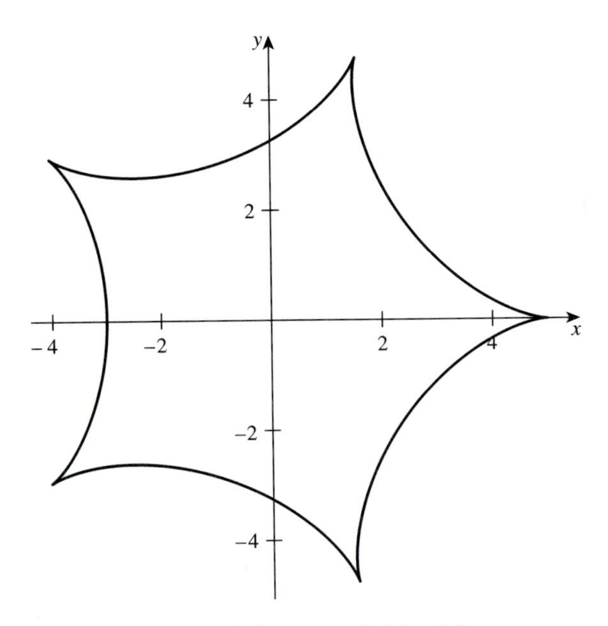

Figure 1.8.4 A hypocycloid of five cusps

r and see if you can make a general conjecture about how the relationship between R and r influences the shape of the path. ♦

We conclude this section by presenting two examples that deal with parametrized paths in three-dimensional space.

Example 1.8.5 Plot the path parametrized by

$$\mathbf{x} = \cos t\,\mathbf{i} + \sin t\,\mathbf{j} + \frac{t}{2\pi}\,\mathbf{k}.$$

Solution If the $\mathbf{k}$-component of $\mathbf{x}$ were identically zero, the path would follow the circle $x^2 + y^2 = 1$ counterclockwise as viewed from a point on the positive z-axis. This observation reveals that each point on the path must lie above or below this circle. As t increases from 0 to 2π, the $\mathbf{k}$-component increases to 1 while the projection of the point into the xy-plane goes once around the unit circle. As t increases past 2π, the $\mathbf{k}$-component continues to increase and the projection in the xy-plane once again goes around the unit circle. For t decreasing from 0, the $\mathbf{k}$-component decreases and the projection into the xy-plane goes clockwise around the circle. Thus the path is a **helix** shown in the computer-generated Figure 1.8.5. The range of t-values for this plot is the interval $[-2\pi, 4\pi]$. ♦

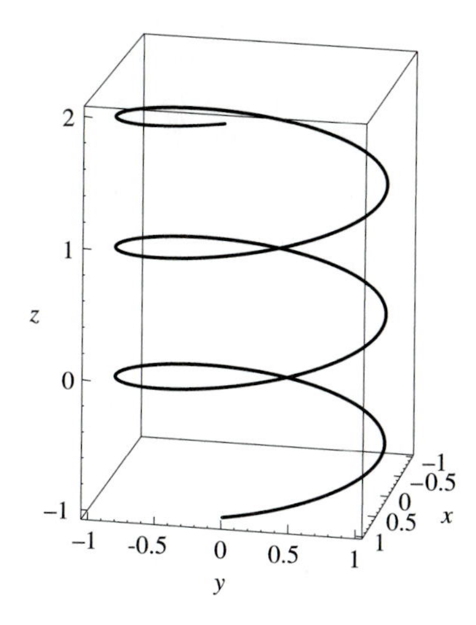

Figure 1.8.5 The path parametrized by $\mathbf{x} = \cos t\,\mathbf{i} + \sin t\,\mathbf{j} + t/(2\pi)\,\mathbf{k}$

Example 1.8.6 Find a parametrization for a path that has a helical shape as in Example 1.8.5 but which winds closer about the z-axis as t takes on larger positive values.

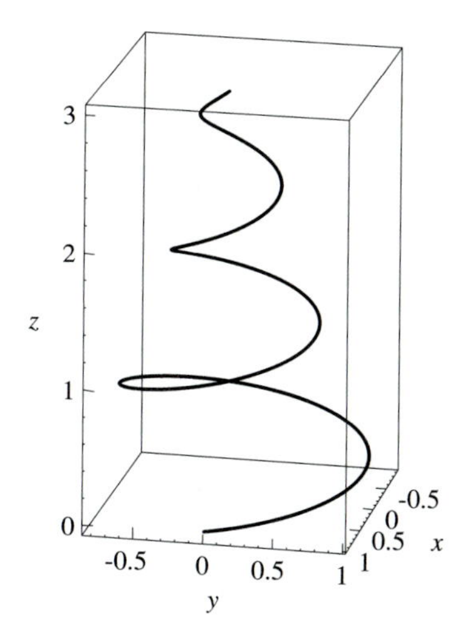

Figure 1.8.6 The path parametrized by (1.8.2)

Solution We can let the **k**-component increase in exactly the same way as in Example 1.8.5. The **i**- and **j**-components must decrease toward zero as t becomes more and more positive. So we want to scale the **i**- and **j**-components by a function f such that $f(t) \to 0$ as $t \to \infty$. Such examples as $1/t$, $1/(t^2+1)$, and e^{-t} come to mind. We select a slightly modified version of the second one: let $f(t) = 1/((t/10)^2+1)$ and examine the path parametrized by

$$\mathbf{x} = \frac{\cos t}{(t/10)^2+1}\,\mathbf{i} + \frac{\sin t}{(t/10)^2+1}\,\mathbf{j} + \frac{t}{2\pi}\,\mathbf{k}. \tag{1.8.2}$$

Figure 1.8.6 shows this path. You may want to use a computer program to plot a modification of this path by replacing $t/10$ by t or t^2. ♦

Exercises 1.8

In Exercises 1 through 4, determine the domain of the given function.

1. $\mathbf{f}(t) = \dfrac{1}{t}\,\mathbf{i} - \sqrt{t-1}\,\mathbf{j} + t^2\,\mathbf{k}$
2. $\mathbf{g}(t) = \tan t\,\mathbf{i} + \cot t\,\mathbf{j} + \sin t\,\mathbf{k}$
3. $\mathbf{h}(t) = \ln(t-2)\,\mathbf{i} + \ln(4-t)\,\mathbf{k}$
4. $\mathbf{u}(t) = \dfrac{1}{e^t-1}\,\mathbf{i} - \sqrt{1-t^2}\,\mathbf{j} + \dfrac{1}{2t-1}\,\mathbf{k}$

In Exercises 5 through 9, sketch by hand and describe the path parametrized.

5. $\mathbf{x} = (2t+1)\,\mathbf{i} + (-7t-4)\,\mathbf{j} + 8\,\mathbf{k}$
6. $\mathbf{x} = 3\sin t\,\mathbf{i} + 4\cos t\,\mathbf{j}$
7. $\mathbf{x} = 2t\,\mathbf{i} + \cos t\,\mathbf{j} + \sin t\,\mathbf{k}$
8. $\mathbf{x} = -\,\mathbf{i} + 2t\,\mathbf{j} - (t^2+2)\,\mathbf{k}$
9. $\mathbf{x} = e^{-t}\sin t\,\mathbf{i} + e^{-t}\cos t\,\mathbf{j}$
10. Match each of the following six functions with the path in Figure 1.8.7 that it parametrizes.

    (a) $\mathbf{f}(t) = \dfrac{t^3-t}{t^4+1}\,\mathbf{i} + \dfrac{t}{t^4+1}\,\mathbf{j}$

    (b) $\mathbf{c}(t) = (3\sin t + 1)\,\mathbf{i} + (\cos t - 1)\,\mathbf{j}$

    (c) $\mathbf{b}(t) = \left(6\cos\left(\frac{t}{7}\right) + \cos\left(\frac{6t}{7}\right)\right)\mathbf{i} + \left(6\sin\left(\frac{t}{7}\right) - \sin\left(\frac{6t}{7}\right)\right)\mathbf{j}$

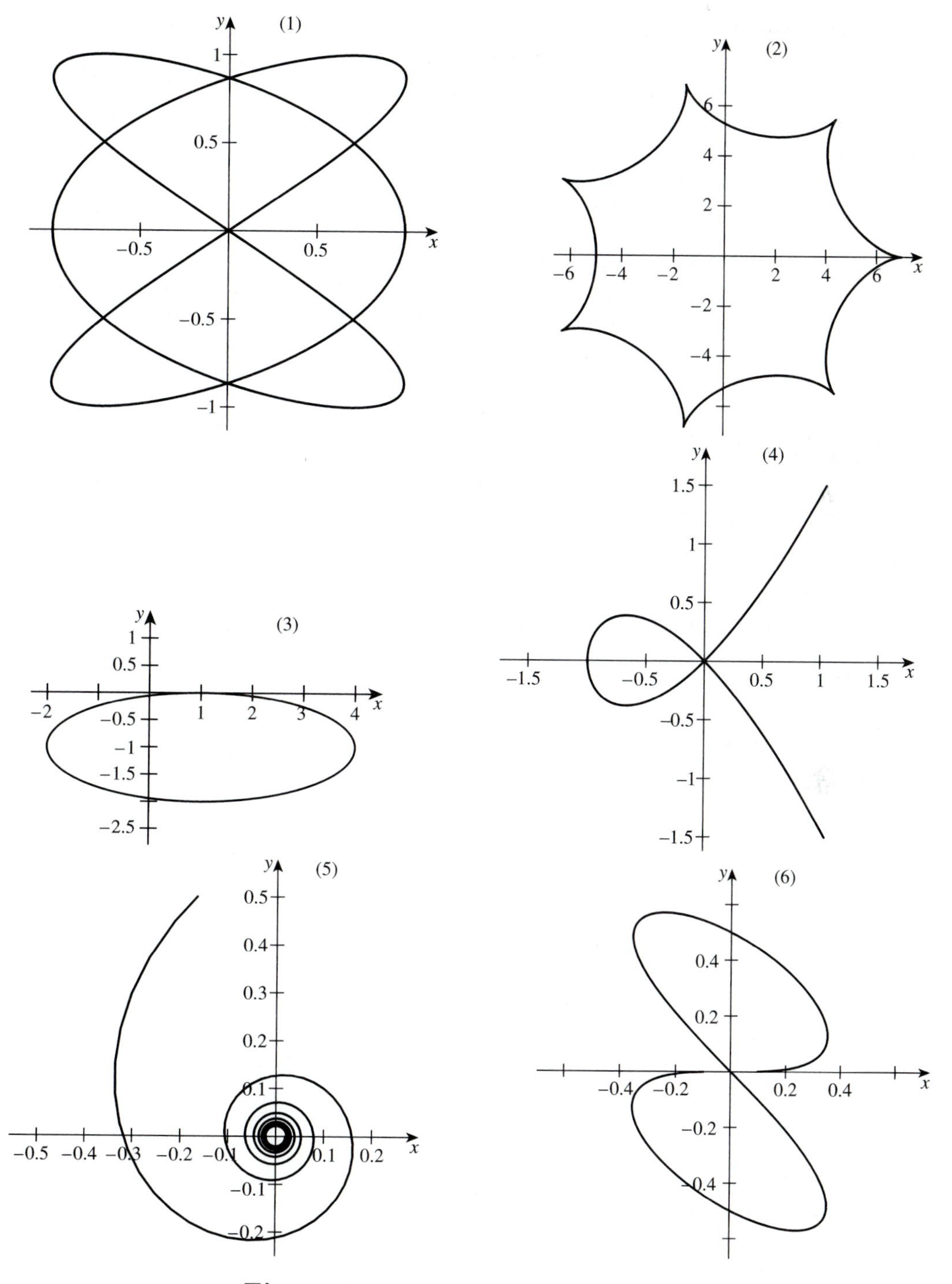

Figure 1.8.7 Graphs for Exercise 10

(d) $\mathbf{a}(t) = \cos 3t\,\mathbf{i} + \sin 2t\,\mathbf{j}$

(e) $\mathbf{d}(t) = t(t-1)\,\mathbf{i} + t(t-1)(t-2)\,\mathbf{j}$

(f) $\mathbf{e}(t) = \frac{\cos t}{t}\,\mathbf{i} + \frac{\sin t}{t}\,\mathbf{j}$

In Exercises 11 through 16, write a parametrization for the path and check yourself by plotting the path with a graphics program.

11. The line segment from $(0, 3, -2)$ to $(1, 5, 8)$

12. The semicircle $x^2 + y^2 = 100$, $y \geq 0$, in the xy-plane from $(10, 0)$ to $(-10, 0)$

13. A helix that winds counterclockwise about the y-axis twice as t varies from 0 to 4π

14. A circle of radius 3 lying in the plane $z = 4$ with center on the z-axis

15. An ellipse with semiaxes of lengths 4 and 5 lying in the plane $x = 1$ with center on the x-axis

16. A cycloid generated by a circle of radius 8 in the yz-plane that rolls along the y-axis.

17. In Figure 1.8.3 we can use the angle $\phi = \angle SOT$ as a parameter instead of the angle t. In this case show that a parametrization for the hypocycloid is

$$\begin{aligned} x &= (R-r)\cos\phi + r\cos\left(\frac{R-r}{r}\phi\right) \\ y &= (R-r)\sin\phi - r\sin\left(\frac{R-r}{r}\phi\right). \end{aligned}$$

18. If $R = 4r$ in Exercise 17, show that the hypocycloid is parametrized by

$$\mathbf{x} = (R\cos^3\phi)\,\mathbf{i} + (R\sin^3\phi)\,\mathbf{j}.$$

(*Hint*: Use the identities $\cos 3\theta = 4\cos^3\theta - 3\cos\theta$ and $\sin 3\theta = 3\sin\theta - 4\sin^3\theta$.) This is the **hypocycloid of four cusps**.

19. Show that the Cartesian equation for the curve in Exercise 18 is

$$x^{2/3} + y^{2/3} = R^{2/3}.$$

20. Find a parametrization for a **cycloid**, the curve traced by a point on the circumference of a circle as it rolls along a straight line. (*Hint*: See Figure 1.8.8. Let the parameter be t, the angle through which the circle has rotated.)

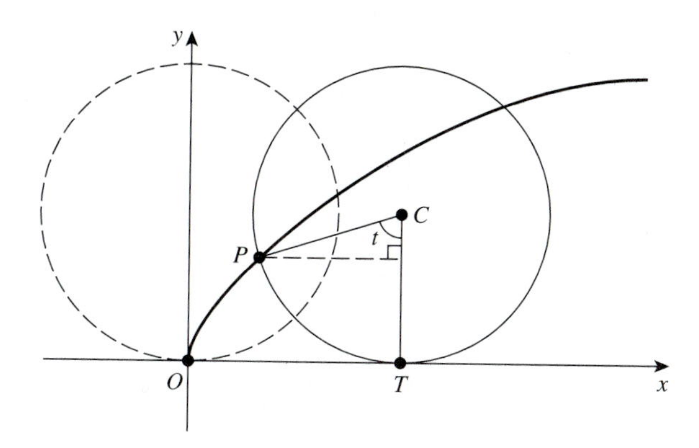

Figure 1.8.8 A cycloid is generated by a point P on a circle that rolls along a straight line

21. A **longbow curve** is defined as follows. Fix a circle in the xy-plane with radius a and center $(0, a)$, and fix the horizontal line $y = 2a$. Each non-horizontal line through the origin intersects the circle in a point A and the line in a point B. Let P be the midpoint of the segment from A to B. The set of points P obtained in this way is the longbow curve. (When you sketch it, you will see why it has this name.) By letting θ be the angle that a line through the origin forms with the positive x-axis, find a parametrization of the longbow curve in terms of the parameter θ.

22. An **epicycloid** is the path traced by a point on a circle of radius r that rolls on the outside of a fixed circle of radius R. By placing the moving circle initially so that it and the fixed circle are initially tangent at $(R, 0)$ and by having it roll counterclockwise in the xy-plane, show that the epicycloid is parametrized by

$$\mathbf{x} = \left[(R+r)\cos\left(\frac{rt}{R}\right) - r\cos\left(\left(\frac{r+R}{R}\right)t\right)\right]\mathbf{i} + \left[(R+r)\sin\left(\frac{rt}{R}\right) - r\sin\left(\left(\frac{r+R}{R}\right)t\right)\right]\mathbf{j}.$$

where t is the angle $\angle PCT$ shown in Figure 1.8.9.

23. Show that $\mathbf{x} = \tan t\,\mathbf{i} + \tan^2 t\,\mathbf{j}$, $-\pi/2 < t < \pi/2$ is a parametrization for the parabola $y = x^2$. Can you identify t geometrically?

24. The graph of the equation $x^3 + y^3 = 3axy$, where a is any nonzero constant, is called the **folium of Descartes**.

(a) Verify that $(0, 0)$ is a point on this curve.

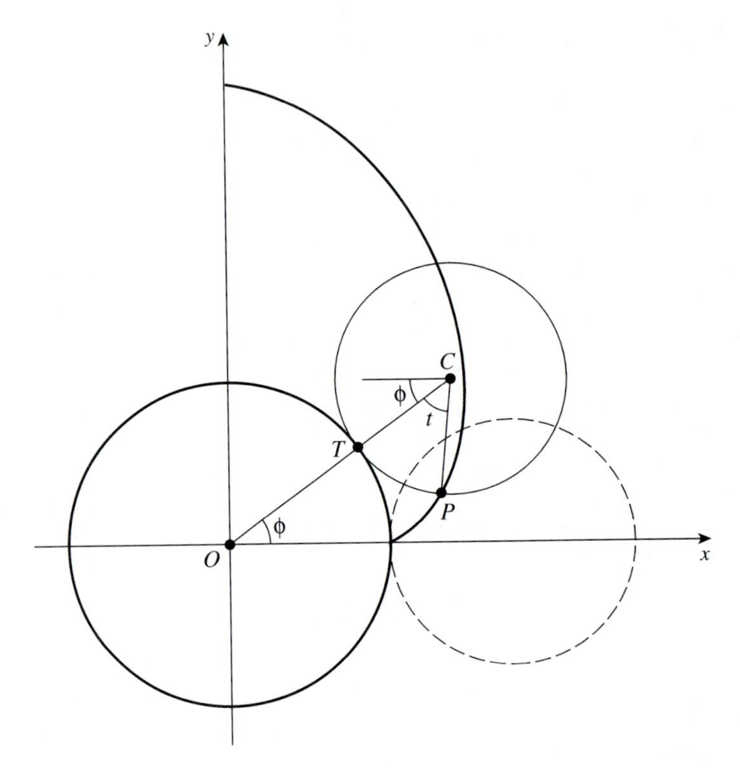

Figure 1.8.9 An epicycloid is generated by a point P on a circle that rolls on the outside of another

(b) Obtain a parametrization of the folium by considering the family $y = tx$ of lines through $(0, 0)$ with slopes t and finding, in terms of t, where each of these lines intersects the folium.

(c) Obtain a sketch, by hand or computer, using the parametrization in (b). Conjecture whether the folium has an asymptote.

25. By the same means as in Exercise 24, obtain

$$x = \frac{a(t^2 - 1)}{t^2 + 1}, \quad y = \frac{-2at}{t^2 + 1}$$

as a parametrization of the circle $x^2 + y^2 = a^2$. [*Hint*: Consider the family of lines through the point $(a, 0)$ on the circle.]

1.9 Derivatives and Motion

In this section we are concerned with "doing" calculus on vector-valued functions. For example, if $\mathbf{f}$ is the position of an object as a function of time, we may wish to measure the rate at which the magnitude and direction of $\mathbf{f}$ are changing at a given instant. We will call this rate the **derivative** or **velocity** of $\mathbf{f}$. The rate at which the velocity changes is important in describing and understanding how objects move as forces act on them. We will call the rate of change in the derivative the **second derivative** of $\mathbf{f}$ or **acceleration**. Before we give formal definitions for these quantities, we must first discuss the concept of the **limit** of a vector-valued function.

The idea of taking a limit of a vector-valued function carries over directly from scalar-valued functions. If $\mathbf{f}(t)$ gets arbitrarily close to a vector $\mathbf{L}$ as t gets arbitrarily close to a number a, then we will say, "The **limit** of $\mathbf{f}(t)$ as t approaches a is $\mathbf{L}$." This is expressed precisely in the following definition.

Definition 1.9.1 We write

$$\lim_{t \to a} \mathbf{f}(t) = \mathbf{L}$$

if for all choices of positive number ϵ there exists a positive number δ such that $||\mathbf{f}(t) - \mathbf{L}|| < \epsilon$ whenever t satisfies $0 < |t - a| < \delta$. If $\mathbf{L} = \mathbf{f}(a)$, then we say that $\mathbf{f}$ is **continuous** at a.

This leads us to a convenient way of calculating limits and understanding continuity of a vector-valued function.

Theorem 1.9.1 *If* $\mathbf{f}$ *has component functions* f_1, f_2, *and* f_3 *then*

$$\lim_{t \to a} \mathbf{f}(t) = \left[\lim_{t \to a} f_1(t)\right] \mathbf{i} + \left[\lim_{t \to a} f_2(t)\right] \mathbf{j} + \left[\lim_{t \to a} f_3(t)\right] \mathbf{k}$$

if and only if the three limits on the right-hand side exist.

Proof Suppose that the limit $\lim_{t \to a} \mathbf{f}(t)$ exists; call it $\mathbf{L} = L_1 \mathbf{i} + L_2 \mathbf{j} + L_3 \mathbf{k}$. Then by Definition 1.9.1, for all $\epsilon > 0$ there is a $\delta > 0$ such that

$$\sqrt{(f_1(t) - L_1)^2 + (f_2(t) - L_2)^2 + (f_3(t) - L_3)^2} < \epsilon$$

whenever $0 < |t - a| < \delta$. Then, evidently,

$$|f_1(t) - L_1| < \epsilon,$$

$$|f_2(t) - L_2| < \epsilon,$$

and

$$|f_3(t) - L_3| < \epsilon$$

whenever $0 < |t - a| < \delta$. That is, $\lim_{t \to a} f_1(t) = L_1$, $\lim_{t \to a} f_2(t) = L_2$, and $\lim_{t \to a} f_3(t) = L_3$.

The converse is an exercise. ♦

Example 1.9.1 Let $\mathbf{f}(t) = \dfrac{e^t - 1}{t}\,\mathbf{i} + \dfrac{\cos t}{t+1}\,\mathbf{j} + t\sin(1/t)\,\mathbf{k}$. Find $\lim_{t \to 0} \mathbf{f}(t)$.

Solution Theorem 1.9.1 says that the limit will be the vector with components the respective limits of the component functions of $\mathbf{f}$. By applying l'Hôpital's Rule to the $\mathbf{i}$-component, we get

$$\lim_{t \to 0} \frac{e^t - 1}{t} = \lim_{t \to 0} \frac{e^t}{1} = e^0 = 1.$$

The $\mathbf{j}$-component requires no sophisticated mathematical machinery:

$$\lim_{t \to 0} \frac{\cos t}{t+1} = \frac{\cos 0}{0+1} = \frac{1}{1} = 1.$$

For the $\mathbf{k}$-component we note first that $\sin(1/t)$ is bounded between -1 and 1 for all $t \neq 0$. Thus

$$-t \le t\sin(1/t) \le t$$

for $t \neq 0$. Since $\lim_{t \to 0}(-t) = 0$ and $\lim_{t \to 0} t = 0$, by the "Squeeze Theorem" of elementary calculus, $\lim_{t \to 0} t\sin(1/t) = 0$. Combining these results, we obtain

$$\lim_{t \to 0} \mathbf{f}(t) = 1\,\mathbf{i} + 1\,\mathbf{j} + 0\,\mathbf{k} = \mathbf{i} + \mathbf{j}. \quad ♦$$

If we regard a vector-valued function as the position of an object as a function of time, then it makes sense to ask such questions as, "Which way is it going?" and "How fast?" These are questions about direction and magnitude, quantities represented in vectors. Let $\mathbf{f}(t)$ be the position of an object at time t. For a small time increment h, during the time interval from t to $t+h$, the object is displaced by

$$\mathbf{f}(t+h) - \mathbf{f}(t).$$

If the motion were exactly along a straight line (which it is not in general) then

$$\frac{||\mathbf{f}(t+h) - \mathbf{f}(t)||}{h}$$

would be the average speed over the time interval and

$$\lim_{h \to 0} \frac{||\mathbf{f}(t+h) - \mathbf{f}(t)||}{h}$$

would be the instantaneous speed at time t, provided that the limit existed. This leads us to make the following definition.

Definition 1.9.2 The **derivative** of a vector-valued function $\mathbf{f}$ of one variable is the vector-valued function $\mathbf{f}'$ given by

$$\mathbf{f}'(t) = \lim_{h \to 0} \frac{\mathbf{f}(t+h) - \mathbf{f}(t)}{h}$$

defined whenever the limit exists. When this limit does exist, we say that $\mathbf{f}$ is **differentiable** at t. $\left(\text{We also write } \dfrac{d\mathbf{f}}{dt} \text{ or } \dfrac{d}{dt}\mathbf{f} \text{ for } \mathbf{f}'.\right)$

By using Theorem 1.9.1, we obtain a convenient formula for calculating derivatives.

Theorem 1.9.2 *If* $\mathbf{f}$ *has differentiable component functions* f_1, f_2, *and* f_3 *then*

$$\mathbf{f}'(t) = f_1'(t)\,\mathbf{i} + f_2'(t)\,\mathbf{j} + f_3'(t)\,\mathbf{k}$$

at each t *where* f_1, f_2, *and* f_3 *are differentiable.*

Proof Let t be a number at which $f_1'(t)$, $f_2'(t)$, and $f_3'(t)$ exist. Then by Definition 1.9.2 and our definitions of vector arithmetic,

$$\begin{aligned}
\mathbf{f}'(t) &= \lim_{h \to 0} \frac{\mathbf{f}(t+h) - \mathbf{f}(t)}{h} \\
&= \lim_{h \to 0} \frac{1}{h}\Big[f_1(t+h)\,\mathbf{i} + f_2(t+h)\,\mathbf{j} + f_3(t+h)\,\mathbf{k} \\
&\qquad\qquad - \left(f_1(t)\,\mathbf{i} + f_2(t)\,\mathbf{j} + f_3(t)\,\mathbf{k}\right)\Big] \\
&= \lim_{h \to 0} \left[\frac{f_1(t+h) - f_1(t)}{h}\,\mathbf{i} + \frac{f_2(t+h) - f_2(t)}{h}\,\mathbf{j} + \frac{f_3(t+h) - f_3(t)}{h}\,\mathbf{k}\right].
\end{aligned}$$

By Theorem 1.9.1 and our assumption that f_1, f_2, and f_3 are differentiable at t, this last limit is equal to

$$\begin{aligned}
&\left(\lim_{h \to 0} \frac{f_1(t+h) - f_1(t)}{h}\right)\mathbf{i} + \left(\lim_{h \to 0} \frac{f_2(t+h) - f_2(t)}{h}\right)\mathbf{j} \\
&\qquad\qquad + \left(\lim_{h \to 0} \frac{f_3(t+h) - f_3(t)}{h}\right)\mathbf{k} \\
&= f_1'(t)\,\mathbf{i} + f_2'(t)\,\mathbf{j} + f_3'(t)\,\mathbf{k}.
\end{aligned}$$

Thus $\mathbf{f}'(t) = f_1'(t)\,\mathbf{i} + f_2'(t)\,\mathbf{j} + f_3'(t)\,\mathbf{k}$. ♦

Example 1.9.2 Calculate the derivative of the function $\mathbf{f}$ given by

$$\mathbf{f}(t) = (t^2 + 2t + 5)\,\mathbf{i} + \left(\ln t + \frac{1}{t}\right)\mathbf{j} + (e^{3t} - e^{-t})\,\mathbf{k}.$$

Solution By Theorem 1.9.2, we just differentiate the components:

$$\mathbf{f}'(t) = (2t + 2)\mathbf{i} + \left(\frac{1}{t} - \frac{1}{t^2}\right)\mathbf{j} + (3e^{3t} + e^{-t})\,\mathbf{k}. \quad \blacklozenge$$

The geometry of the derivative vector is straightforward to understand. In Figure 1.9.1 we see that $\mathbf{f}(a)$ is the position vector of a point on the path

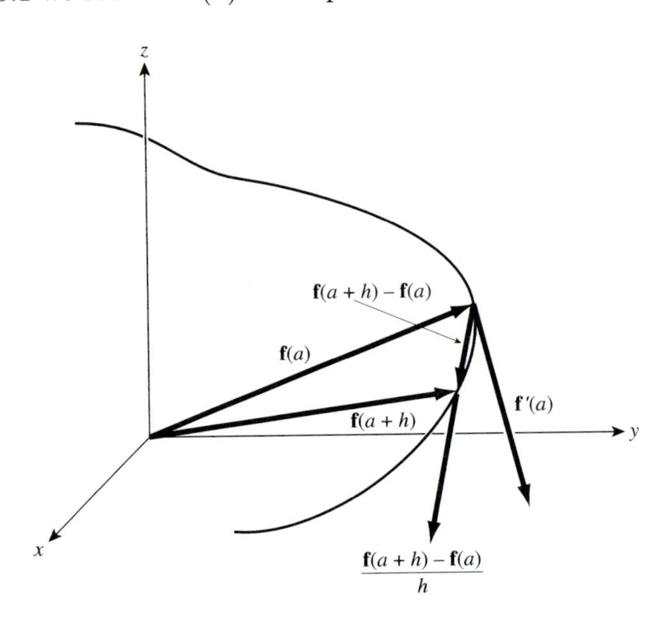

Figure 1.9.1 The derivative $\mathbf{f}'(a)$ is tangent to the path traced by $\mathbf{f}$ at the point $\mathbf{f}(a)$.

traced by $\mathbf{f}$ and that $\mathbf{f}(a + h)$ is the position of another such point. The difference $\mathbf{f}(a + h) - \mathbf{f}(a)$ is a vector joining two points on the object's path, and $\frac{1}{h}\left(\mathbf{f}(a + h) - \mathbf{f}(a)\right)$ is parallel to the difference. For smaller values of h, the endpoints of the vector are closer together, and in the limit as $h \to 0$, we see that $\mathbf{f}'(a)$ will "just graze" the path (i.e., $\mathbf{f}'(a)$ is tangent to the path). All this, of course, is under the assumption that $\mathbf{f}$ is differentiable at a and that $\mathbf{f}'(a) \neq \mathbf{0}$. In this case we call $\mathbf{f}'(a)$ a **tangent vector** to the path at $\mathbf{f}(a)$.

If we think of $\mathbf{f}(t)$ as the position of an object at time t, then we refer to $\mathbf{v}(t) = \mathbf{f}'(t)$ as the **velocity** at time t and $\mathbf{a}(t) = \mathbf{f}''(t) = \frac{d}{dt}\mathbf{f}'(t)$ as the **acceleration**. The magnitude $||\mathbf{v}(t)||$ of the velocity is the object's **speed**.

Example 1.9.3 Suppose that an object is moving through space along the path $\mathbf{x} = t\sin t\,\mathbf{i} + \cos t\,\mathbf{j} + \sin t\,\mathbf{k}$, and then at the moment when $t = \pi/4$ it suddenly begins traveling in the direction of its velocity vector with $||\mathbf{v}(\pi/4)||$ as its constant speed. What will be the object's position when $t = \pi/2$?

Solution Figure 1.9.2 shows schematically what happens. The particle follows the path traced by $\mathbf{f}(t) = t\sin t\,\mathbf{i} + \cos t\,\mathbf{j} + \sin t\,\mathbf{k}$ until $t = \pi/4$.

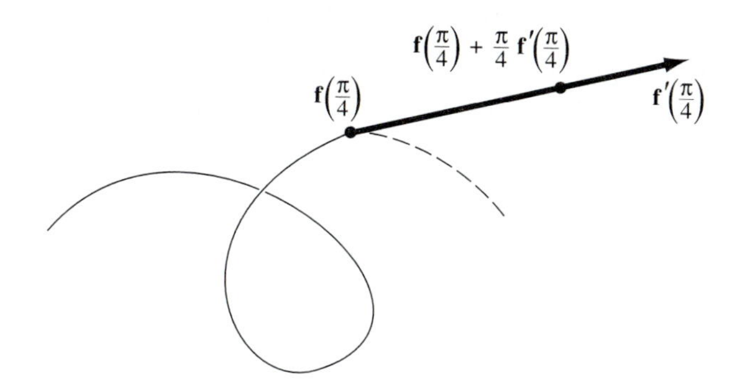

Figure 1.9.2 A particle leaves its appointed path to follow a tangent.

Then it follows the line through $\mathbf{f}(\pi/4)$ with direction vector $\mathbf{f}'(\pi/4)$. One parametrization of this line is

$$\mathbf{x} = \mathbf{f}(\pi/4) + t\mathbf{f}'(\pi/4).$$

However, when $t = \pi/4$, we see that $\mathbf{x} \neq \mathbf{f}(\pi/4)$. A parametrization of the same line with this property is

$$\mathbf{x} = \mathbf{f}\left(\frac{\pi}{4}\right) + \left(t - \frac{\pi}{4}\right)\mathbf{f}'\left(\frac{\pi}{4}\right). \tag{1.9.1}$$

Notice that the speed of the object with this position is

$$\left\|\frac{d}{dt}\left(\mathbf{f}\left(\frac{\pi}{4}\right) + \left(t - \frac{\pi}{4}\right)\mathbf{f}'\left(\frac{\pi}{4}\right)\right)\right\| = \left\|\mathbf{f}'\left(\frac{\pi}{4}\right)\right\| = \left\|\mathbf{v}\left(\frac{\pi}{4}\right)\right\|,$$

so (1.9.1) gives the position function required in the problem.

Now when $t = \pi/2$, the object's position is, by (1.9.1),

$$\begin{aligned}
&\mathbf{f}\left(\frac{\pi}{4}\right) + \left(\frac{\pi}{2} - \frac{\pi}{4}\right)\mathbf{f}'\left(\frac{\pi}{4}\right) \\
&= \left(\frac{\pi}{4}\cdot\frac{1}{\sqrt{2}}\mathbf{i} + \frac{1}{\sqrt{2}}\mathbf{j} + \frac{1}{\sqrt{2}}\mathbf{k}\right) + \frac{\pi}{4}\left((\sin t + t\cos t)\,\mathbf{i} - \sin t\,\mathbf{j} + \cos t\,\mathbf{k}\right)\Big|_{t=\pi/4} \\
&= \left(\frac{\pi}{4}\cdot\frac{1}{\sqrt{2}}\mathbf{i} + \frac{1}{\sqrt{2}}\mathbf{j} + \frac{1}{\sqrt{2}}\mathbf{k}\right) + \frac{\pi}{4}\left(\frac{1}{\sqrt{2}}\left(1 + \frac{\pi}{4}\right)\mathbf{i} - \frac{1}{\sqrt{2}}\mathbf{j} + \frac{1}{\sqrt{2}}\mathbf{k}\right) \\
&= \left(\frac{4\pi + 4\sqrt{2}\pi + \pi^2}{16\sqrt{2}}\right)\mathbf{i} + \left(\frac{16 - \pi^2}{16\sqrt{2}}\right)\mathbf{j} + \left(\frac{16 + \pi^2}{16\sqrt{2}}\right)\mathbf{k}. \quad \blacklozenge
\end{aligned}$$

The second derivative of a vector-valued function $\mathbf{f}$ is readily visualized. In Figure 1.9.3 we see that $\mathbf{v}(a)$ is tangent to the path at the point $\mathbf{f}(a)$, and

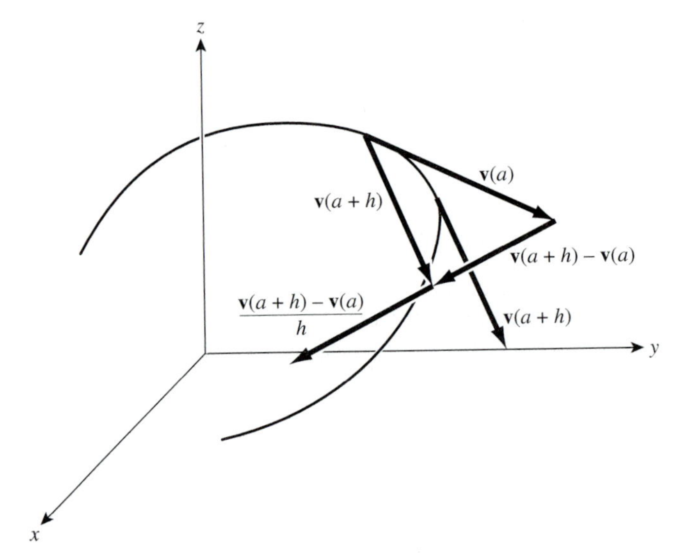

Figure 1.9.3 The acceleration vector $\mathbf{v}'(t) = \mathbf{f}''(t)$ points toward the concave side of the path traced by $\mathbf{f}$.

$\mathbf{v}(a+h)$ is tangent to the path at the point $\mathbf{f}(a+h)$; here we think of h as a small number. Now translate $\mathbf{v}(a+h)$ so that its tail lies at $\mathbf{f}(a)$. Because h is small, the difference $\mathbf{v}(a+h) - \mathbf{v}(a)$ will point in the general direction in which the curve is bending. So

$$\frac{\mathbf{v}(a+h) - \mathbf{v}(a)}{h}$$

points in the same general direction, and in the limit as $h \to 0$, this approaches $\mathbf{v}'(a) = \mathbf{f}''(a)$. Thus, as long as the direction of $\mathbf{v}$ is changing, $\mathbf{f}''$ points toward the concave side of the curve being traced.

Vector-valued functions obey differentiation rules that are almost exactly analogous to those for scalar-valued functions. Each rule listed in the following theorem is proved simply by writing the expressions on the left in terms of components and applying Theorem 1.9.2.

Theorem 1.9.3 *Let* $\mathbf{f}$ *and* $\mathbf{g}$ *be differentiable vector-valued functions of one variable, and let* h *be a differentiable scalar-valued function.*

1. $\dfrac{d}{dt}(\mathbf{f}+\mathbf{g}) = \mathbf{f}' + \mathbf{g}'$
2. $\dfrac{d}{dt}(\mathbf{f}-\mathbf{g}) = \mathbf{f}' - \mathbf{g}'$
3. $\dfrac{d}{dt}(c\mathbf{f}) = c\mathbf{f}'$ *for any constant* c
4. $\dfrac{d}{dt}(h\mathbf{f}) = h'\mathbf{f} + h\mathbf{f}'$
5. $\dfrac{d}{dt}(\mathbf{f}\cdot\mathbf{g}) = \mathbf{f}'\cdot\mathbf{g} + \mathbf{f}\cdot\mathbf{g}'$
6. $\dfrac{d}{dt}(\mathbf{f}\times\mathbf{g}) = \mathbf{f}'\times\mathbf{g} + \mathbf{f}\times\mathbf{g}'$
7. $\dfrac{d}{dt}\mathbf{f}(h(t)) = h'(t)\mathbf{f}'(h(t))$

Proof We prove part 4 as an illustration of the proof method for the rest.
Let $\mathbf{f} = f_1\,\mathbf{i} + f_2\,\mathbf{j} + f_3\,\mathbf{k}$. Then

$$\begin{aligned}
\frac{d}{dt}(h\mathbf{f}) &= \frac{d}{dt}[(hf_1)\,\mathbf{i} + (hf_2)\,\mathbf{j} + (hf_3)\,\mathbf{k}] \\
&= \frac{d}{dt}(hf_1)\,\mathbf{i} + \frac{d}{dt}(hf_2)\,\mathbf{j} + \frac{d}{dt}(hf_3)\,\mathbf{k} \\
&= (h'f_1 + hf_1')\,\mathbf{i} + (h'f_2 + hf_2')\,\mathbf{j} + (h'f_3 + hf_3')\,\mathbf{k} \\
&= (h'f_1\,\mathbf{i} + h'f_2\,\mathbf{j} + h'f_3\,\mathbf{k}) + (hf_1'\,\mathbf{i} + hf_2'\,\mathbf{j} + hf_3'\,\mathbf{k}) \\
&= h'(f_1\,\mathbf{i} + f_2\,\mathbf{j} + f_3\,\mathbf{k}) + h(f_1'\,\mathbf{i} + f_2'\,\mathbf{j} + f_3'\,\mathbf{k})
\end{aligned}$$

The steps are justified, respectively, by Theorem 1.9.2, the ordinary Product Rule, and definitions of vector arithmetic.

Proofs of the remaining parts are exercises. ♦

Since the derivative of a vector-valued function is just the vector with components the derivatives of the function's components, the most general **antiderivative** of a function $\mathbf{f} = f_1\,\mathbf{i} + f_2\,\mathbf{j} + f_3\,\mathbf{k}$ is the vector with components the general antiderivatives of f_1, f_2, and f_3.

Definition 1.9.3 For a continuous function $\mathbf{f} = f_1\,\mathbf{i} + f_2\,\mathbf{j} + f_3\,\mathbf{k}$,

$$\int \mathbf{f}(t)\,dt = \int f_1(t)\,dt\,\mathbf{i} + \int f_2(t)\,dt\,\mathbf{j} + \int f_3(t)\,dt\,\mathbf{k}.$$

Example 1.9.4 Find the position of a particle whose position is $\mathbf{i}+2\mathbf{j}-3\mathbf{k}$ when $t = 0$ and whose velocity is $\mathbf{v}(t) = e^t\,\mathbf{i} + (2+e^t)\,\mathbf{j} + e^{2t}\,\mathbf{k}$.

Solution By Definition 1.9.3, the position is

$$\begin{aligned}
\mathbf{f}(t) &= \int \mathbf{v}(t)\,dt \\
&= \int \left(e^t\,\mathbf{i} + (2+e^t)\,\mathbf{j} + e^{2t}\,\mathbf{k}\right)\,dt \\
&= \int e^t\,dt\,\mathbf{i} + \int (2+e^t)\,dt\,\mathbf{j} + \int e^{2t}\,dt\,\mathbf{k} \\
&= (e^t + c_1)\,\mathbf{i} + (2t + e^t + c_2)\,\mathbf{j} + \left(\frac{1}{2}e^{2t} + c_3\right)\,\mathbf{k},
\end{aligned}$$

where c_1, c_2, and c_3 are arbitrary constants. From the initial condition,

$$\mathbf{i} + 2\mathbf{j} - 3\,\mathbf{k} = (1+c_1)\mathbf{i} + (1+c_2)\mathbf{j} + \left(\frac{1}{2} + c_3\right)\,\mathbf{k}.$$

Thus $c_1 = 0$, $c_2 = 1$ and $c_3 = -7/2$, so

$$\mathbf{f}(t) = e^t\,\mathbf{i} + (2t + e^t + 1)\,\mathbf{j} + \left(\frac{1}{2}e^{2t} - \frac{7}{2}\right)\,\mathbf{k}. \quad \blacklozenge$$

Example 1.9.5 A particle moves so that its position vector has constant magnitude. Show that its position vector is always perpendicular to its velocity.

Solution Let $\mathbf{f}(t)$ be the particle's position at time t. Then

$$||\mathbf{f}(t)|| = c \quad \text{for all} \quad t, \tag{1.9.2}$$

where c is some positive constant. We want to show that $\mathbf{f}'(t)\cdot\mathbf{f}(t) = 0$ for all t. By squaring (1.9.2) and using Theorem 1.5.3, we can write

$$c^2 = ||\mathbf{f}(t)||^2 = \mathbf{f}(t)\cdot\mathbf{f}(t).$$

Differentiating and using Theorem 1.9.3, we obtain

$$\begin{aligned} 0 = \frac{d}{dt}c^2 &= \frac{d}{dt}(\mathbf{f}(t)\cdot\mathbf{f}(t)) \\ &= \mathbf{f}'(t)\cdot\mathbf{f}(t) + \mathbf{f}(t)\cdot\mathbf{f}'(t) \\ &= 2\mathbf{f}'(t)\cdot\mathbf{f}(t). \end{aligned}$$

Therefore, $\mathbf{f}'(t)\cdot\mathbf{f}(t) = 0$ for all t; position is perpendicular to velocity. ♦

Exercises 1.9

In Exercises 1 through 4, evaluate the limit.

1. $\lim\limits_{t\to 0}\left(e^{-t}\,\mathbf{i} + (2te^{-t}+1)\,\mathbf{j} + e^{-2t}\,\mathbf{k}\right)$

2. $\lim\limits_{t\to 5}\left(\dfrac{t^2-25}{t-5}\,\mathbf{i} + \sin t\,\mathbf{j} + \mathbf{k}\right)$

3. $\lim\limits_{t\to 0}\left(\dfrac{\sin t}{t}\,\mathbf{i} + \dfrac{1-\cos t}{t}\,\mathbf{j} + \dfrac{t}{\ln t}\,\mathbf{k}\right)$

4. $\lim\limits_{h\to 0}\left(\dfrac{(t+h)^3-t^3}{h}\,\mathbf{i} + \dfrac{(t+h)^2-t^2}{h}\,\mathbf{j}\right)$

In Exercises 5 through 8, calculate the derivative of the given function.

5. $\mathbf{A}(t) = (t^3-t)\,\mathbf{i} + (t^2+1)\,\mathbf{j} + (t-5)\,\mathbf{k}$

6. $\mathbf{B}(t) = (e^t - e^{-t})\,\mathbf{i} + 2e^t\,\mathbf{j} + e^{2t}\,\mathbf{k}$

7. $\mathbf{C}(t) = \ln(t^2+1)\,\mathbf{i} - 2\,\mathbf{j} + (4t-3)\,\mathbf{k}$

8. $\mathbf{D}(t) = \tan\left(\dfrac{t^2+1}{t^2-1}\right)\mathbf{i} + \dfrac{1-\sin t}{1+\cos t}\,\mathbf{j} + \csc t\cot t\,\mathbf{k}$

In the following two problems, find a parametrization for the line tangent to the path at the indicated parameter value.

9. $\mathbf{x} = e^t\cos t\,\mathbf{i} + e^t\sin t\,\mathbf{j} + e^{-t}\,\mathbf{k}$, $t = 0$

10. $\mathbf{x} = \dfrac{1}{t}\,\mathbf{i} + \dfrac{1}{t^3}\,\mathbf{j} + \ln t\,\mathbf{k},\ t = 2$

11. An object traveling along the path $\mathbf{x} = \sin t \cos t\,\mathbf{i} + \cos 3t\,\mathbf{j} - t\,\mathbf{k}$ suddenly at time $t = \pi/2$ begins traveling in the direction $\mathbf{v}(\pi/2)$ with constant speed $||\mathbf{v}(\pi/2)||$. What is its position when $t = \pi$? What is its acceleration when $t = \pi$?

12. For $\mathbf{f}(t) = \cos t\,\mathbf{i} + 2\sin t\,\mathbf{j}$, calculate the first and second derivatives when $t = \pi/4$. Make a careful sketch of $\mathbf{f}(\pi/4)$, $\mathbf{f}'(\pi/4)$, and $\mathbf{f}''(\pi/4)$ along with the curve traced by $\mathbf{f}(t)$.

13. For the position function $\mathbf{f}(t) = (t^3 + 3t)\,\mathbf{i} + (4t^2 - 1)\,\mathbf{j} + 6t\,\mathbf{k}$, calculate the velocity and acceleration. Then calculate $\mathbf{f}(0)$, $\mathbf{v}(0)$, and $\mathbf{a}(0)$ and use these to sketch a short section of the path traced by $\mathbf{f}(t)$ through $\mathbf{f}(0)$.

14. Prove that if $\lim_{t\to a} f_i(t) = L_i$, where $i = 1, 2, 3$, then $\lim_{t\to a} \mathbf{f}(t) = \mathbf{L}$, where $\mathbf{f} = f_1\,\mathbf{i} + f_2\,\mathbf{j} + f_3\,\mathbf{k}$ and $\mathbf{L} = L_1\,\mathbf{i} + L_2\,\mathbf{j} + L_3\,\mathbf{k}$.

15. Prove part 3 of Theorem 1.9.3.

16. Prove part 5 of Theorem 1.9.3.

17. Prove part 6 of Theorem 1.9.3.

18. Prove part 7 of Theorem 1.9.3.

19. By using Theorem 1.9.3, prove that

$$\frac{d}{dt} f(\mathbf{g}\cdot\mathbf{h}) = f'(\mathbf{g}\cdot\mathbf{h})(\mathbf{g}'\cdot\mathbf{h} + \mathbf{g}\cdot\mathbf{h}').$$

20. By using Theorem 1.9.3, prove that

$$\frac{d}{dt}||\mathbf{g}|| = \frac{\mathbf{g}\cdot\mathbf{g}'}{||\mathbf{g}||}.$$

21. By using Theorem 1.9.3, prove that

$$\frac{d}{dt} f(||\mathbf{g}||) = \frac{\mathbf{g}\cdot\mathbf{g}' f'(||\mathbf{g}||)}{||\mathbf{g}||}.$$

22. Prove that

$$\frac{d}{dt}\frac{1}{||\mathbf{g}||^2} = \frac{-2\mathbf{g}\cdot\mathbf{g}'}{||\mathbf{g}||^4}.$$

Calculate the following indefinite integrals.

23. $\displaystyle\int (4\,\mathbf{i} - 3\,\mathbf{j} + 2\,\mathbf{k})\,dt$

24. $\int \left((5t^3 + t + 1)\,\mathbf{i} + (t^2 + 3)\,\mathbf{j} + t\,\mathbf{k} \right)\,dt$

25. $\int \left(\frac{t}{t^2+1}\,\mathbf{i} + \frac{1}{t^2+1}\,\mathbf{j} + \frac{t^2+1}{t}\,\mathbf{k} \right)\,dt$

26. $\int \left(te^t\,\mathbf{i} + (e^{-5t} + 1)\,\mathbf{j} - \frac{e^{\sqrt{t}}}{\sqrt{t}}\,\mathbf{k} \right)\,dt$

27. $\int \left(\sin t \cos t\,\mathbf{i} + \cos^3 t\,\mathbf{j} - \sin^2 t\,\mathbf{k} \right)\,dt$

For the following two velocity functions, find the position function $\mathbf{r}$ that satisfies the given initial condition.

28. $\mathbf{v}(t) = (3t^2 + 1)\,\mathbf{i} + t\,\mathbf{j} + t^3\,\mathbf{k}$; $\mathbf{r}(0) = \mathbf{i} + \mathbf{j} + \mathbf{k}$

29. $\mathbf{v}(t) = \sin t\,\mathbf{i} + \cos 2t\,\mathbf{j} + \sin 3t\,\mathbf{k}$; $\mathbf{r}(0) = \mathbf{i} + \mathbf{j} + \mathbf{k}$

For each of the following three acceleration functions, find the position function $\mathbf{r}$ that satisfies the given conditions.

30. $\mathbf{a}(t) = (t + 1)\,\mathbf{i} - t^2\,\mathbf{j} + (2t + 3)\,\mathbf{k}$; $\mathbf{r}(0) = \mathbf{i} + \mathbf{k}$; $\mathbf{v}(0) = \mathbf{0}$

31. $\mathbf{a}(t) = \frac{1}{t+1}\,\mathbf{i} + (t^3 + t)\,\mathbf{j} + (t^2 + 1)\,\mathbf{k}$; $\mathbf{r}(1) = \mathbf{0}$, $\mathbf{v}(1) = \mathbf{0}$

32. $\mathbf{a}(t) = \mathbf{i} - 3\mathbf{j} + 2\,\mathbf{k}$; $\mathbf{r}(0) = \mathbf{i} - \mathbf{k}$, $\mathbf{r}(1) = 2\,\mathbf{i} + \mathbf{j} - \mathbf{k}$

33. Show that if a particle in three-dimensional space moves so that its velocity is always perpendicular to its position vector, the particle moves on a sphere centered at the origin.

34. Show that if a particle in three-dimensional space moves so that its velocity is always parallel to its position vector, the particle moves along a straight line through the origin.

35. Show that if a particle with position $\mathbf{f}(t)$ satisfies the conditions $\mathbf{f}'' = k\mathbf{f}$ for some scalar k and $(\mathbf{f} \times \mathbf{f}')' = \mathbf{0}$, the particle's motion is restricted to a plane.

36. Consider a parametrized curve $\mathbf{x} = \mathbf{f}(u)$ for values of u between a and t. The **pathlength function based at** a is given by

$$s(t) = \int_a^t ||\mathbf{f}'(u)||du. \tag{1.9.3}$$

This function measures the distance an object travels along its path as the parameter u varies from a to t. Here we regard t as varying and a as fixed. (See section4.1 for a more thorough discussion.)

(a) Find the pathlength function for $\mathbf{x} = \frac{1}{3}u^3\,\mathbf{i} + u^2\,\mathbf{j}$ based at 0.

(b) Find the pathlength function for the helix $\mathbf{x} = \cos u\ \mathbf{i}{+}\sin u\ \mathbf{j}{+}u\ \mathbf{k}$.

(c) Find the pathlength function for the line $\mathbf{x} = (au + x_0)\ \mathbf{i} + (bu + y_0)\ \mathbf{j} + (cu + z_0)\ \mathbf{k}$ based at 0.

37. Pathlength is often used as a parameter. For example, a circle given by $\mathbf{x} = a\cos u\,\mathbf{i} + a\sin u\,\mathbf{j}$ can be reparametrized so that pathlength s is the parameter instead of t:

$$s = \int_0^t ||d\mathbf{x}/du|| = \int_0^t || - a\sin u\,\mathbf{i} + a\cos u\ \mathbf{j}||\,du = at.$$

Substituting $t = s/a$ for u, we obtain $\mathbf{x}$ as a function of s:

$$\mathbf{x} = a\cos\left(\frac{s}{a}\right)\ \mathbf{i} + a\sin\left(\frac{s}{a}\right)\ \mathbf{j}.$$

Reparametrize the paths in Exercise 36 so that pathlength is the parameter.

1.10 Projects and Problems

In this section we pose questions that may, in part, be answered by using a graphing calculator, a graphics program, or a computer algebra system. Each of the projects or problems here is related to concepts introduced in Chapter 1.

1. Obtain graphs of $f(x, y) = \sin(\sqrt{x^2+y^2})$, $g(x, y) = \cos(\sqrt{x^2+y^2})$, $h(x, y) = 1/(\sqrt{x^2+y^2})$, and $k(x, y) = (\sqrt{x^2+y^2})/(1 + x^2 + y^2)$. From these, state in general how the graph of a function F defined by $F(x, y) = G(\sqrt{x^2+y^2})$, where G is a given function of one variable, is obtained from the graph of G.

2. Graphs of equations such as $x^2 + y^2 + z^2 = r^2$ cannot be generated as the graph of a function of two variables. For instance, if we choose $x = 0$ and $y = 0$, then $z^2 = r^2$ so $z = \pm r$, which are two distinct values. But we can sometimes use spherical coordinates to help *parametrize* the surface. In the same way as we can parametrize a plane, we can regard the points on the sphere of radius r as those points with rectangular coordinates (x, y, z) such that $x = r\sin\phi\cos\theta$, $y = r\sin\phi\sin\theta$, and $z = r\cos\phi$. Here the parameters θ and ϕ range, respectively, over the intervals $[0, 2\pi]$ and $[0, \pi]$.

(a) Use a parametric plotting function to graph the sphere $x^2 + y^2 + z^2 = r^2$ for $r = 1$.

(b) Verify that for any given positive numbers a, b, and c, the points (x, y, z) satisfying $x = a \sin\phi \cos\theta$, $y = b \sin\phi \sin\theta$, and $z = c \cos\phi$ lie on the ellipsoid $x^2/a^2 + y^2/b^2 + z^2/c^2 = 1$. Use this to obtain a parametrization of the ellipsoid and obtain a graph of the surface for a specific choice of a, b, and c.

(c) Produce a "movie" of the ellipsoids $x^2+y^2+z^2/c^2 = 1$ for values of c ranging from -4 to 4. Describe in words the geometric changes that take place as the parameter c changes.

3. By changing to cylindrical coordinates, show that the hyperboloid of one sheet $x^2/a^2 + y^2/b^2 - z^2/c^2 = 1$ can be parametrized by $x = a\sqrt{1 + w^2/c^2} \cos\theta$, $y = b\sqrt{1 + w^2/c^2} \sin\theta$, $z = w$. Make a "movie" of the family of hyperboloids $x^2/k^2 + y^2/k^2 - z^2/k^2 = 1$ for values of k decreasing from 1 toward 0. Describe the changes.

4. Obtain a graph of the spherical coordinate equation $\rho = \phi$.

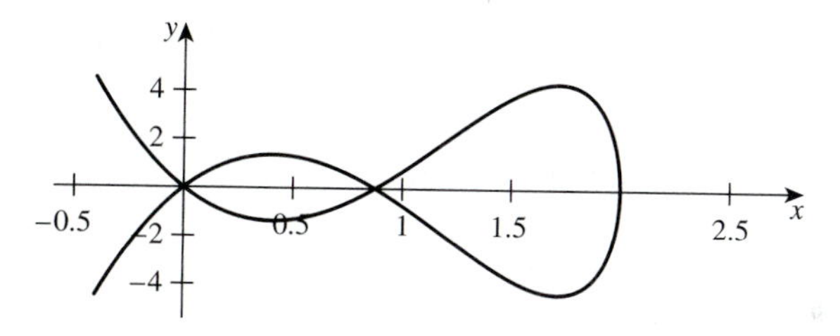

Figure 1.10.1 Figure for Exercise 5

5. (a) Obtain a graph of the path parametrized by

$$\mathbf{x} = (t(t-2),\, t(t-1)(t-2)), \quad -1 \le t \le 3$$

and find the points corresponding to the parameter values 0, 1, and 2.

(b) Find a parametrization similar to that in part (a) for a curve with two loops that is qualititatively similar to the one shown in Figure 1.10.1.

6. In the mechanical linkage shown in Figure 1.10.2, the bar OA of length a rotates counterclockwise about the fixed point O, and the bar AB of length b is linked to A at one end while B is constrained to slide along the x-axis. Assume that $b > a$, let M be the midpoint of AB, and let $t = \angle BOA$.

(a) Find a function $\mathbf{f}(t)$ that traces the path followed by M as t ranges from 0 to 2π.

(b) Obtain the graph of the path parametrized by $\mathbf{f}$ for various choices of a and b.

(c) Is the path parametrized by $\mathbf{f}$ an ellipse? If so, show it; if not, explain why not.

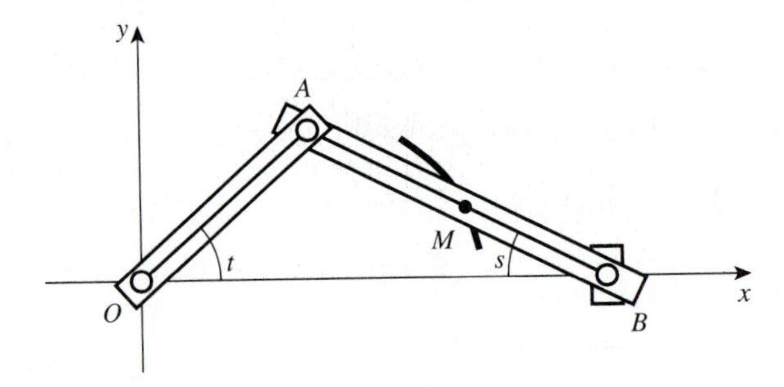

Figure 1.10.2 Mechanical linkage for Exercise 6

Chapter 2

Geometry and Linear Algebra in $\mathbb{R}^n$

Geometry in three dimensions is adequate for many applications. However, in physics, engineering, chemistry, and economics, and basically any discipline in which several independently varying numerical quantities must be considered simultaneously, higher-dimensional analogs of $\mathbb{R}^3$ become useful. For example, to formulate the two-body problem in physics—in which the goal is to determine the position in (three-dimensional) space of two bodies as a function of time from Newton's Second Law of Motion and the Universal Law of Gravitation—requires keeping track of six unknown time-dependent quantities: the x-, y-, and z-coordinates of both bodies. If we let (x_1, y_1, z_1) be the position of the first body and (x_2, y_2, z_2) the position of the second, it would make sense to write $(x_1, y_1, z_1, x_2, y_2, z_2)$ to represent the "state" of the two-body system and to regard this state as a point that moves in $\mathbb{R}^6$ as a function of time.

As another example, consider the robot arm shown schematically in Figure 2.0.1. Here the machine's pincers, "wrist," "elbow," and "shoulder" are all run by separate servomotors, and in addition, the base to which the shoulder is attached rotates about its axis. Thus the position of the tip of one pincer is dependent upon the angle through which each joint is flexed, and since there are five such angles, each to be monitored and changed simultaneously, it makes sense to represent the state or position of the robot arm in terms of $(\theta_1, \theta_2, \theta_3, \theta_4, \theta_5)$, a point in $\mathbb{R}^5$. Physical constraints on the arm's motion (e.g., the elbow cannot rotate the hand through the shoulder so θ_3 cannot be less than, say, 10° or greater than 340°) can be translated into statements about where in $\mathbb{R}^5$ the point $(\theta_1, \theta_2, \theta_3, \theta_4, \theta_5)$ may be. Studying the geometry of the set of possible arm positions can reveal, for instance, places in the robot's environment that are inaccessible to its grasp or positions in which the mechanism may "hang up."

These two examples and others that appear in the sequel are partly for those who are skeptical about discussions of "the fifth dimension" and who may regard such stuff as in the domain of science fiction. In popular culture such phrases as "five-dimensional space" are laden with fantastic and speculative meaning, but in mathematics and applications they have very precise definitions and very practical uses.

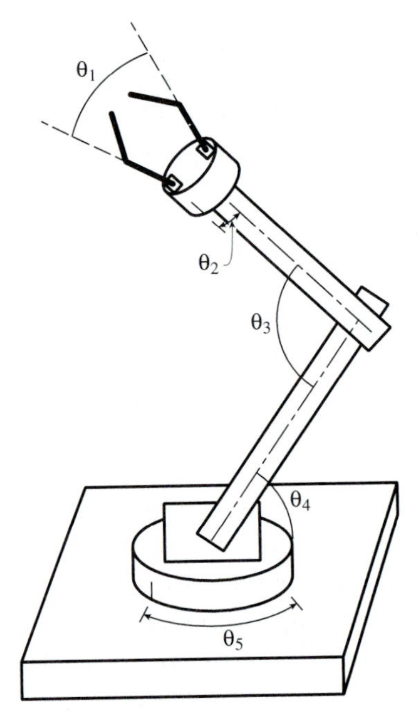

Figure 2.0.1 A robot arm

Quite aside from applications there is the purely mathematical temptation to try to generalize ideas in order to get a more complete perspective, or perhaps even just for the fun of it. Somehow, after having seen and talked about $\mathbb{R}^2$ and $\mathbb{R}^3$, it is almost irresistible to ask, what could "$\mathbb{R}^4$," "$\mathbb{R}^5$," and so on, really be? We will find that the natural way to define them leads to the conclusion that they are like $\mathbb{R}^2$ and $\mathbb{R}^3$ in many ways, so much so that we continue to use ordinary geometric terms such as "line," "plane," "sphere," and "angle," even when n is larger than 3. We do, however, give up drawing pictures.

In this chapter we also introduce and explore to a limited extent the concept of a **linear transformation** (or **linear function**). This idea is fundamental to developing calculus for functions of several variables because in very general terms, differential calculus provides a way of understanding non-linear functions by viewing them as approximately linear over small domains. Thus it is important to have a grasp on linear transformations themselves before embarking on a study in which they serve as a tool. In particular we will be interested in how they change the shape of geometric objects. The notational tool of representing linear transformations using **matrices** gives us a very convenient way of calculating their geometric properties.

2.1 Vectors and Coordinate Geometry in $\mathbb{R}^n$

We begin this section with a couple of omnibus definitions. In the first, we state exactly what we mean by a vector in $\mathbb{R}^n$, and in the second we describe basic vector arithmetic.

Definition 2.1.1 For any positive integer n, **n-dimensional Euclidean space** is

$$\mathbb{R}^n = \{(x_1, x_2, \ldots, x_n) | x_i \in \mathbb{R},\ i = 1, \ldots, n\}.$$

Elements of $\mathbb{R}^n$ are called **vectors** or **points** and are symbolized in print by boldface letters (e.g., **a**, **b**, **x**, **y**, etc.). It is conventional to use the same italic letter with subscripts for the components of the vector. For instance, if $\mathbf{x}$ is a vector in $\mathbb{R}^n$, then "$\mathbf{x}$" stands for $(x_1, x_2, x_3, \ldots, x_n)$. The **zero vector** is $(0, 0, \ldots, 0)$ and is denoted by $\mathbf{0}$.

Definition 2.1.2 Given two vectors $\mathbf{x}$ and $\mathbf{y}$ in $\mathbb{R}^n$ their **sum**, denoted $\mathbf{x}+\mathbf{y}$, is the vector in $\mathbb{R}^n$ whose components are the sums of the corresponding components in $\mathbf{x}$ and $\mathbf{y}$; if $\mathbf{x} = (x_1, \ldots, x_n)$ and $\mathbf{y} = (y_1, \ldots, y_n)$, then

$$\mathbf{x} + \mathbf{y} = (x_1 + y_1, x_2 + y_2, \ldots, x_n + y_n). \tag{2.1.1}$$

Multiplication of a vector by a scalar is defined as follows. For $\mathbf{x} \in \mathbb{R}^n$ and a real number k, $k\mathbf{x}$ is the vector whose components are k times the components of $\mathbf{x}$:

$$k\mathbf{x} = (kx_1, kx_2 \ldots, kx_n). \tag{2.1.2}$$

If $k \neq 0$, we define $\mathbf{x}/k$ to be $(1/k)\mathbf{x}$. Given $\mathbf{x}$ and $\mathbf{y}$ in $\mathbb{R}^n$, the **vector from x to y** (or the **difference** between $\mathbf{y}$ and $\mathbf{x}$) is $\mathbf{y} - \mathbf{x} = \mathbf{y} + (-1)\mathbf{x}$. The symbol $-\mathbf{x}$ denotes $(-1)\mathbf{x}$.

Example 2.1.1 Consider vectors in $\mathbb{R}^5$. If $\mathbf{x} = (1,\ -2,\ 3,\ 0,\ 7)$, $\mathbf{y} = (-3,\ 2,\ 10,\ 6,\ 0)$, and $k = 4$, then

$$\mathbf{x} + \mathbf{y} = (1 - 3,\ -2 + 2,\ 3 + 10,\ 0 + 6,\ 7 + 0) = (-2,\ 0,\ 13,\ 6,\ 7)$$

and

$$k\mathbf{y} = (4 \cdot (-3),\ 4 \cdot 2,\ 4 \cdot 10,\ 4 \cdot 6,\ 4 \cdot 0) = (-12,\ 8,\ 40,\ 24,\ 0).$$

Also,

$$\begin{aligned} \mathbf{y} - \mathbf{x} &= (-3,\ 2,\ 10,\ 6,\ 0) + (-1)(1,\ -2,\ 3,\ 0,\ 7) \\ &= (-3,\ 2,\ 10,\ 6,\ 0) + (-1,\ 2,\ -3,\ 0,\ -7) = (-4,\ 4,\ 7,\ 6,\ -7). \quad \blacklozenge \end{aligned}$$

All of the properties of vector arithmetic carry over from $\mathbb{R}^3$ to this context. They are summarized in the following theorem, and you are asked in the exercises to prove selected parts of it.

Theorem 2.1.1 *Let* $\mathbf{x}$, $\mathbf{y}$, *and* $\mathbf{z}$ *be arbitrary vectors in* $\mathbb{R}^n$, *and let* k *and* l *be arbitrary scalars. Then*

1. $\mathbf{x} + \mathbf{y} = \mathbf{y} + \mathbf{x}$
2. $\mathbf{x} + (\mathbf{y} + \mathbf{z}) = (\mathbf{x} + \mathbf{y}) + \mathbf{z}$
3. $\mathbf{0} + \mathbf{x} = \mathbf{x}$
4. $k(\mathbf{x} + \mathbf{y}) = k\mathbf{x} + k\mathbf{y}$
5. $(k + l)\mathbf{x} = k\mathbf{x} + l\mathbf{x}$
6. $(kl)\mathbf{x} = k(l\mathbf{x}) = l(k\mathbf{x})$
7. $\mathbf{x} + (-\mathbf{x}) = \mathbf{0}$
8. $1\mathbf{x} = \mathbf{x}$

Proof We give a proof for part 4 as an example of the proof method for all the other parts.

Let $\mathbf{x} = (x_1, x_2, \ldots, x_n)$ and $\mathbf{y} = (y_1, y_2, \ldots, y_n)$. Then

$$\begin{aligned} k(\mathbf{x} + \mathbf{y}) &= k(x_1 + y_1, x_2 + y_2, \ldots, x_n + y_n) \\ &= (k(x_1 + y_1), k(x_2 + y_2), \ldots, k(x_n + y_n)) \\ &= (kx_1 + ky_1, kx_2 + ky_2, \ldots, kx_n + ky_n) \\ &= (kx_1, kx_2, \ldots, kx_n) + (ky_1, ky_2, \ldots, ky_n) \\ &= k(x_1, x_2, \ldots, x_n) + k(y_1, y_2, \ldots, y_n) \\ &= k\mathbf{x} + k\mathbf{y}, \end{aligned}$$

where each step is justified either by the definition of scalar multiplication, the distributive property of scalar arithmetic, or the definition of vector addition. (Verify this!) ♦

The following definition extends a few other vector notions from $\mathbb{R}^3$ to $\mathbb{R}^n$.

Definition 2.1.3 The **magnitude** of a vector $\mathbf{x} = (x_1, x_2, \ldots, x_n)$ is

$$||\mathbf{x}|| = (x_1^2 + x_2^2 + \cdots + x_n^2)^{1/2}.$$

The **distance** between vectors $\mathbf{x}$ and $\mathbf{y}$ is $||\mathbf{x} - \mathbf{y}||$.

The **dot** (or **inner**) **product** of two vectors $\mathbf{x}$ and $\mathbf{y}$ in $\mathbb{R}^n$ is

$$\mathbf{x} \cdot \mathbf{y} = x_1 y_1 + x_2 y_2 + \ldots + x_n y_n. \tag{2.1.3}$$

Example 2.1.2 With $\mathbf{x}$ and $\mathbf{y}$ as in Example 2.1.1,

$$||\mathbf{x}|| = \sqrt{1^2 + (-2)^2 + 3^2 + 0^2 + 7^2} = \sqrt{63} = 3\sqrt{7},$$

$$||\mathbf{y}|| = \sqrt{(-3)^2 + 2^2 + 10^2 + 6^2 + 0^2} = \sqrt{149},$$

and the distance between $\mathbf{x}$ and $\mathbf{y}$ is

$$||\mathbf{x} - \mathbf{y}|| = \sqrt{4^2 + (-4)^2 + (-7)^2 + (-6)^2 + 7^2} = \sqrt{166}.$$

The dot product of $\mathbf{x}$ and $\mathbf{y}$ is

$$\mathbf{x} \cdot \mathbf{y} = 1 \cdot (-3) + (-2) \cdot 2 + 3 \cdot 10 + 0 \cdot 6 + 7 \cdot 0 = 23. \quad ♦$$

Note that $\mathbf{x} \cdot \mathbf{y}$ is always a *scalar*. The dot product has some important properties which are summarized in the following theorem.

Theorem 2.1.2 *Let* $\mathbf{x}, \mathbf{y}$, *and* $\mathbf{z}$ *be arbitrary vectors in* $\mathbb{R}^n$ *and* k *an arbitrary scalar. Then*

1. $||\mathbf{x}|| = \sqrt{\mathbf{x} \cdot \mathbf{x}}$

2. $\mathbf{x} \cdot \mathbf{y} = \mathbf{y} \cdot \mathbf{x}$

3. $\mathbf{x} \cdot (\mathbf{y} + \mathbf{z}) = \mathbf{x} \cdot \mathbf{y} + \mathbf{x} \cdot \mathbf{z}$

4. $(k\mathbf{x}) \cdot \mathbf{y} = \mathbf{x} \cdot (k\mathbf{y}) = k(\mathbf{x} \cdot \mathbf{y})$

5. $\mathbf{0} \cdot \mathbf{x} = 0$

6. $||k\mathbf{x}|| = |k| \, ||\mathbf{x}||$

The verification of these properties is an exercise.

Example 2.1.3 Prove that if $\mathbf{x} \cdot \mathbf{y} = 0$, then $||\mathbf{x} - \mathbf{y}||^2 = ||\mathbf{x}||^2 + ||\mathbf{y}||^2$. (Note that this is reminiscent of the Pythagorean Theorem.)

Solution By properties 1, 3, and 2 in Theorem 2.1.2, we can justify each step in the following chain of equalities:

$$\begin{aligned} ||\mathbf{x} - \mathbf{y}||^2 &= (\mathbf{x} - \mathbf{y}) \cdot (\mathbf{x} - \mathbf{y}) \\ &= (\mathbf{x} - \mathbf{y}) \cdot \mathbf{x} - (\mathbf{x} - \mathbf{y}) \cdot \mathbf{y} \\ &= \mathbf{x} \cdot \mathbf{x} - \mathbf{y} \cdot \mathbf{x} - (\mathbf{x} \cdot \mathbf{y} - \mathbf{y} \cdot \mathbf{y}) \\ &= ||\mathbf{x}||^2 - \mathbf{x} \cdot \mathbf{y} - \mathbf{x} \cdot \mathbf{y} + ||\mathbf{y}||^2 \\ &= ||\mathbf{x}||^2 - 2\mathbf{x} \cdot \mathbf{y} + ||\mathbf{y}||^2. \end{aligned}$$

Since $\mathbf{x} \cdot \mathbf{y} = 0$, this gives $||\mathbf{x} - \mathbf{y}||^2 = ||\mathbf{x}||^2 + ||\mathbf{y}||^2$. ♦

Note that our approach here in defining the dot product is *backward* from that used for vectors in $\mathbb{R}^3$. (See Definition 1.5.2.) In $\mathbb{R}^3$, viewing vectors as line segments made the angle between two vectors intuitive, and we defined the dot product in terms of the magnitudes of the vectors and the angle between them. By applying the law of cosines, we deduced that the dot product could also be computed in terms of the components. (See Theorem 1.5.1.) In $\mathbb{R}^n$, for $n > 3$, there is no ready picture and hence no immediate notion of angle between vectors, so we have instead defined the dot product in terms of components. There is still a chance that we can make sense of the notion of angle in $\mathbb{R}^n$ by first observing that, for nonzero vectors $\mathbf{x}, \mathbf{y} \in \mathbb{R}^3$, the angle between $\mathbf{x}$ and $\mathbf{y}$ is the number θ in the range 0 to π such that

$$\cos\theta = \frac{\mathbf{x} \cdot \mathbf{y}}{||\mathbf{x}|| \, ||\mathbf{y}||}. \tag{2.1.4}$$

If we can deduce that $-1 \leq \mathbf{x} \cdot \mathbf{y} / (||\mathbf{x}|| \, ||\mathbf{y}||) \leq 1$ for all nonzero vectors $\mathbf{x}$ and $\mathbf{y}$ *in* $\mathbb{R}^n$ then we can *define* the angle θ between $\mathbf{x}$ and $\mathbf{y}$ to be the number in the range 0 to π which satisfies Equation (2.1.4).

The content of the following theorem is essentially that the desired inequality does hold. The second part of the theorem is important in the example that follows.

Theorem 2.1.3 *(Cauchy-Schwarz Inequality) For two vectors* $\mathbf{x}$ *and* $\mathbf{y}$ *in* $\mathbb{R}^n$,

$$|\mathbf{x}\cdot\mathbf{y}| \le ||\mathbf{x}||\ ||\mathbf{y}||. \tag{2.1.5}$$

Moreover,

$$|\mathbf{x}\cdot\mathbf{y}| = ||\mathbf{x}||\,||\mathbf{y}|| \tag{2.1.6}$$

if and only if $\mathbf{y} = k\,\mathbf{x}$ *for some scalar* k.

Proof (optional) If we can show that $(\mathbf{x}\cdot\mathbf{y})^2 - ||\mathbf{x}||^2\ ||\mathbf{y}||^2 \le 0$ then we will have proved (2.1.5). Recall from elementary algebra that the **discriminant** of a quadratic polynomial $p(t) = at^2 + bt + c$ is $b^2 - 4ac$. The quantity that we want to show as negative has the form of a discriminant. Indeed, it *is* the discriminant of the quadratic polynomial $p(t) = \frac{1}{2}||\mathbf{x}||^2t^2 + (\mathbf{x}\cdot\mathbf{y})t + \frac{1}{2}||\mathbf{y}||^2$, and if we can show that $p(t)$ is always nonnegative then we will know that its discriminant is nonpositive.

By applying Theorem 2.1.2, we can "factor" $p(t)$ in the following way:

$$\begin{aligned}
p(t) &= \frac{1}{2}\Big((\mathbf{x}\cdot\mathbf{x})t^2 + 2(\mathbf{x}\cdot\mathbf{y})t + \mathbf{y}\cdot\mathbf{y}\Big) \\
&= \frac{1}{2}((t\,\mathbf{x})\cdot(t\,\mathbf{x}) + (t\,\mathbf{x})\cdot\mathbf{y} + (t\,\mathbf{x})\cdot\mathbf{y} + \mathbf{y}\cdot\mathbf{y}) \\
&= \frac{1}{2}((t\,\mathbf{x})\cdot(t\,\mathbf{x}+\mathbf{y}) + (t\,\mathbf{x}+\mathbf{y})\cdot\mathbf{y}) \\
&= \frac{1}{2}(t\,\mathbf{x}+\mathbf{y})\cdot(t\,\mathbf{x}+\mathbf{y}) \\
&= \frac{1}{2}||t\,\mathbf{x}+\mathbf{y}||^2.
\end{aligned}$$

This last expression is nonnegative for all choices of t, so $p(t) \ge 0$. Therefore its discriminant $(\mathbf{x}\cdot\mathbf{y})^2 - ||\mathbf{x}||^2||\mathbf{y}||^2$ is nonpositive; this is equivalent to Equation (2.1.5).

To prove that (2.1.6) holds if and only if $\mathbf{y} = k\,\mathbf{x}$, first consider the case in which $\mathbf{x} = \mathbf{0}$. In this case, $\mathbf{x}\cdot\mathbf{y} = 0$ and $||\mathbf{x}||\,||\mathbf{y}|| = 0$, so (2.1.6) holds. Now assume that $\mathbf{x} \ne \mathbf{0}$. We observe the equivalence of the following statements. Equation (2.1.6) holds if and only if the discriminant of p is zero. The discriminant of p is zero if and only if the equation $p(t) = 0$ has the (real) root

$$t = -\frac{\mathbf{x}\cdot\mathbf{y}}{||\mathbf{x}||^2}; \tag{2.1.7}$$

this is obtained simply by applying the quadratic formula. Since $p(t) = \frac{1}{2}||t\mathbf{x}+\mathbf{y}||^2$, the equation $p(t) = 0$ has the root given in (2.1.7) if and only if

$$\frac{1}{2}\left|\left|\left(-\frac{\mathbf{x}\cdot\mathbf{y}}{||\mathbf{x}||^2}\mathbf{x}+\mathbf{y}\right)\right|\right|^2 = 0. \tag{2.1.8}$$

Finally, (2.1.8) holds if and only if

$$\mathbf{y} = \frac{\mathbf{x}\cdot\mathbf{y}}{||\mathbf{x}||^2}\mathbf{x}.$$

This chain of equivalent statements shows that (2.1.6) holds if and only if $\mathbf{y} = k\mathbf{x}$, where $k = \mathbf{x}\cdot\mathbf{y}/||\mathbf{x}||^2$. ♦

The Cauchy-Schwarz inequality says that for two vectors $\mathbf{x}$ and $\mathbf{y}$,

$$-||\mathbf{x}||\ ||\mathbf{y}|| \le \mathbf{x}\cdot\mathbf{y} \le ||\mathbf{x}||\ ||\mathbf{y}||,$$

and if $\mathbf{x}$ and $\mathbf{y}$ are nonzero, dividing by $||\mathbf{x}||\ ||\mathbf{y}||$ gives

$$-1 \le \frac{\mathbf{x}\cdot\mathbf{y}}{||\mathbf{x}||\ ||\mathbf{y}||} \le 1.$$

Thus angles between vectors in $\mathbb{R}^n$ *do* make sense!

Definition 2.1.4 The **angle** between vectors $\mathbf{x}$ and $\mathbf{y}$ is the number θ in the range 0 to π that satisfies

$$\cos\theta = \frac{\mathbf{x}\cdot\mathbf{y}}{||\mathbf{x}||\ ||\mathbf{y}||}.$$

Example 2.1.4 The angle between the vectors $\mathbf{x} = (1, 2, 3, 4, 5)$ and $\mathbf{y} = (5, 4, 3, 2, 1)$ in $\mathbb{R}^5$ satisfies

$$\cos\theta = \frac{\mathbf{x}\cdot\mathbf{y}}{||\mathbf{x}||\ ||\mathbf{y}||} = \frac{5+8+9+8+5}{(\sqrt{1+4+9+16+25})^2} = \frac{35}{55} = \frac{7}{11}.$$

So $\theta \approx 0.881$ rad or $50.48°$. ♦

Definition 2.1.5 We say that two nonzero vectors $\mathbf{x}$ and $\mathbf{y}$ are **orthogonal** or **perpendicular** if their dot product is zero. Equivalently, they are orthogonal if the angle between them is $\pi/2$. We say that the vectors are

parallel if $\mathbf{x} = k\mathbf{y}$ for some scalar k. Equivalently, the vectors are parallel if the angle between them is 0 or π.

Example 2.1.5 Experimentalists frequently look for linear tendencies in data. For instance, a psychologist may be interested in the correlation between the number of hours students spend studying for a test and the average score on that test. By plotting the scores versus the amount of study time, she may observe that there is a "tendency" for the points to lie on a straight line. See Figure 2.1.1.

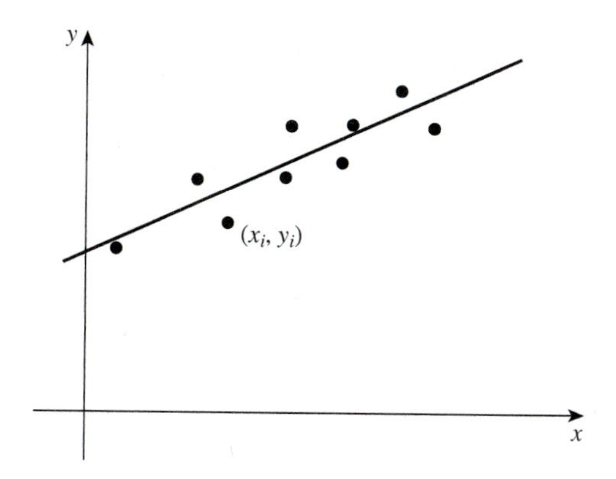

Figure 2.1.1 Test scores versus hours spent studying

How could such a tendency be measured? The **Pearson correlation coefficient**[1] is such a measure, and we shall see *why* it is a reasonable measure of linear tendency. It is defined as follows. Given n pairs of data values $(x_1, y_1), (x_2, y_2), \ldots, (x_n, y_n)$, calculate the average (or mean) and standard deviation of the x's and the y's, respectively. These are

$$\begin{aligned} \overline{x} &= \frac{1}{n}\sum_{i=1}^{n} x_i, \quad \overline{y} = \frac{1}{n}\sum_{i=1}^{n} y_i, \\ s_x &= \sqrt{\frac{1}{n}\sum_{i=1}^{n}(x_i - \overline{x})^2}, \text{ and } s_y = \sqrt{\frac{1}{n}\sum_{i=1}^{n}(y_i - \overline{y})^2}. \end{aligned}$$

The Pearson correlation coefficient is

[1]My colleague Ken Williams pointed out to me the relationship between correlation and angles between vectors in $\mathbb{R}^n$.

$$r = \frac{1}{ns_x s_y} \sum_{i=1}^{n} (x_i - \overline{x})(y_i - \overline{y}). \tag{2.1.9}$$

For example, if the data are (3, 8), (2, 8), (4, 11), and (1, 5), then

$$\begin{aligned}
\overline{x} &= \frac{3+2+4+1}{4} = \frac{5}{2}, \\
\overline{y} &= \frac{8+8+11+5}{4} = 8, \\
s_x &= \sqrt{\frac{1}{4}\left(\left(\frac{1}{2}\right)^2 + \left(-\frac{1}{2}\right)^2 + \left(\frac{3}{2}\right)^2 + \left(-\frac{3}{2}\right)^2\right)} = \frac{\sqrt{20}}{4} = \frac{\sqrt{5}}{2},
\end{aligned}$$

and

$$s_y = \sqrt{\frac{1}{4}\left(0^2 + 0^2 + 3^2 + (-3)^2\right)} = \frac{3\sqrt{2}}{2},$$

so by (2.1.9),

$$\begin{aligned}
r &= \frac{1}{4 \cdot (\sqrt{5}/2) \cdot 3\sqrt{2}/2}\left(\frac{1}{2} \cdot 0 + \left(-\frac{1}{2}\right) \cdot 0 + \frac{3}{2} \cdot 3 + \left(-\frac{3}{2}\right)(-3)\right) \\
&= \frac{3}{\sqrt{10}} \approx 0.949.
\end{aligned}$$

We shall see that r is always between -1 and 1 and that the data lie on a line if and only if $r = \pm 1$.

First, let

$$\mathbf{u} = (x_1 - \overline{x}, x_2 - \overline{x}, \ldots, x_n - \overline{x})$$

and

$$\mathbf{v} = (y_1 - \overline{y}, y_2 - \overline{y}, \ldots, y_n - \overline{y}).$$

Then the angle θ between these vectors in $\mathbb{R}^n$ satisfies

$$\begin{aligned}
\cos\theta &= \frac{\mathbf{u} \cdot \mathbf{v}}{||\mathbf{u}||||\mathbf{v}||} = \frac{\sum_{i=1}^{n} (x_i - \overline{x})(y_i - \overline{y})}{\sqrt{\sum_{i=1}^{n} (x_i - \overline{x})^2}\sqrt{\sum_{i=1}^{n} (y_i - \overline{y})^2}} \\
&= \frac{\sum_{i=1}^{n} (x_i - \overline{x})(y_i - \overline{y})}{n\sqrt{\frac{1}{n}\sum_{i=1}^{n} (x_i - \overline{x})^2}\sqrt{\frac{1}{n}\sum_{i=1}^{n} (y_i - \overline{y})^2}}
\end{aligned}$$

$$= \frac{1}{ns_x s_y} \sum_{i=1}^{n} (x_i - \overline{x})(y_i - \overline{y}) = r. \tag{2.1.10}$$

$$\tag{2.1.11}$$

That is, r is just the cosine of the angle between the vector of deviations from the mean in the x-values and the vector of deviations in the y-values. This shows that $-1 \leq r \leq 1$.

Now suppose that the points all lie exactly on the line $y = mx + b$. Then $y_i = mx_i + b$ for $i = 1, 2, \ldots n$, $\overline{y} = m\overline{x} + b$, and

$$\begin{aligned} \mathbf{v} &= (mx_1 + b - (m\overline{x} + b), mx_2 + b - (m\overline{x} + b), \ldots, mx_n + b - (m\overline{x} + b)) \\ &= (m(x_1 - \overline{x}), m(x_2 - \overline{x}), \ldots, m(x_n - \overline{x})) \\ &= m(x_1 - \overline{x}, \ldots, x_n - \overline{x}) \\ &= m\mathbf{u}. \end{aligned}$$

Thus $\mathbf{u}$ and $\mathbf{v}$ are parallel; indeed, by (2.1.6),

$$r = \frac{\mathbf{u} \cdot (m\mathbf{u})}{||\mathbf{u}||\ ||m\mathbf{u}||} = \frac{m}{|m|},$$

which is $+1$ if m is positive and -1 if m is negative.

On the other hand, if $r = \pm 1$, then, by (2.1.10), $\mathbf{u}$ and $\mathbf{v}$ satisfy

$$|\mathbf{u} \cdot \mathbf{v}| = |||\mathbf{u}||\ ||\mathbf{v}||\ r| = ||\mathbf{u}||\ ||\mathbf{v}||,$$

so by Theorem 2.1.3, $\mathbf{v} = k\mathbf{u}$ for some k. This means that $y_i - \overline{y} = k(x_i - \overline{x})$ for $i = 1, 2, \ldots, n$; in other words, the points (x_i, y_i) lie on the line $y - \overline{y} = k(x - \overline{x})$. ♦

Exercises 2.1

1. Calculate the sum, difference, and dot product of each of the following pairs of vectors:

(a) $(3, 0, 5, 0, 4, 2)$ and $\left(0, \frac{1}{2}, \frac{1}{10}, \frac{3}{2}, -\frac{1}{8}, 1\right)$

(b) $\left(\frac{1}{\sqrt{2}}, -\frac{1}{\sqrt{2}}, 0, \frac{1}{\sqrt{2}}\right)$ and $\left(-\frac{1}{\sqrt{2}}, 0, \frac{1}{\sqrt{2}}, -\frac{1}{\sqrt{2}}\right)$

(c) $\left(1, \frac{1}{2}, \frac{1}{3}, \ldots, \frac{1}{10}\right)$ and $\left(\frac{1}{2}, \frac{2}{3}, \frac{3}{4}, \ldots, \frac{10}{11}\right)$

(d) $(1, 2, 3, \ldots, n-1, n)$ and $(1, 1, 1, \ldots, 1, 1)$ [*Hint*: $\sum_{i=1}^{n} i = n(n+1)/2$ and $\sum_{i=1}^{n} i^2 = n(n+1)(2n+1)/6$.]

2. Find the magnitudes of the vectors in Exercise 1 and calculate the angle between the respective pairs of vectors.

3. Let $\mathbf{a} = (1,\ 0,\ 5,\ -3,\ 2)$, $\mathbf{b} = (4,\ 7,\ 0,\ 11,\ 10)$ and $\mathbf{c} = (6,\ 4,\ 0,\ 0,\ -2)$ be vectors in $\mathbb{R}^5$, and let $k = \frac{1}{2}$. Indicate which of the following expressions does not make sense, and for those that do, evaluate them.

(a) $\mathbf{a} + \mathbf{b} + \mathbf{c}$

(b) $k\mathbf{a} + \mathbf{c}$

(c) $\mathbf{ca}$

(d) $(\mathbf{a} \cdot \mathbf{c})\mathbf{b}$

(e) $\mathbf{b}/||k\mathbf{b}||$

(f) $\mathbf{c} \cdot (2\mathbf{b} + \mathbf{a})$

(g) $\mathbf{a}/\mathbf{b}$

(h) $\mathbf{c}/k$

4. Give an example of:

(a) A vector in $\mathbb{R}^{10}$ of magnitude $\sqrt{5}$

(b) Two nonzero vectors in $\mathbb{R}^6$ that are orthogonal

(c) Three nonzero vectors in $\mathbb{R}^4$ that sum to zero

5. Prove part 2 of Theorem 2.1.1.

6. Prove part 5 of Theorem 2.1.1.

7. Prove part 6 of Theorem 2.1.1.

8. Prove part 1 of Theorem 2.1.2.

9. Prove part 3 of Theorem 2.1.2.

10. Given $\mathbf{x}$ and $\mathbf{y}$ in $\mathbb{R}^n$, let $\mathbf{z} = \frac{1}{2}(\mathbf{x} + \mathbf{y})$. Show that $||\mathbf{z} - \mathbf{x}|| = ||\mathbf{z} - \mathbf{y}||$.

11. Prove that $\mathbf{x} \cdot \mathbf{y} = \frac{1}{4}\left(||\mathbf{x} + \mathbf{y}||^2 - ||\mathbf{x} - \mathbf{y}||^2\right)$ for all $\mathbf{x}, \mathbf{y} \in \mathbb{R}^n$. In the case $n = 2$ or $n = 3$, draw a picture and interpret this geometrically.

12. Prove that $||\mathbf{x}||^2 + ||\mathbf{y}||^2 = \frac{1}{2}(||\mathbf{x} + \mathbf{y}||^2 + ||\mathbf{x} - \mathbf{y}||^2)$ for all $\mathbf{x}$, $\mathbf{y}$ in $\mathbb{R}^n$. For vectors in $\mathbb{R}^2$, interpret this geometrically.

13. (a) Prove the **triangle inequality**:

$$||\mathbf{x} + \mathbf{y}|| \leq ||\mathbf{x}|| + ||\mathbf{y}|| \quad \text{for all } \mathbf{x}, \mathbf{y} \in \mathbb{R}^n.$$

(*Hint*: Use the Cauchy-Schwarz inequality and part 1 of Theorem 2.1.2.)

(b) Show that $||\mathbf{x}-\mathbf{y}|| \le ||\mathbf{x}-\mathbf{z}|| + ||\mathbf{z}-\mathbf{y}||$ for all $\mathbf{x}, \mathbf{y}, \mathbf{z} \in \mathbb{R}^n$.

14. Given a vector $\mathbf{u}$ in $\mathbb{R}^n$, the **(vector) projection** of a vector $\mathbf{b}$ in $\mathbb{R}^n$ onto $\mathbf{a}$ is the vector defined by

$$\mathrm{proj}_{\mathbf{a}}(\mathbf{b}) = \left(\frac{\mathbf{a}\cdot\mathbf{b}}{||\mathbf{a}||^2}\right)\mathbf{a}.$$

The **component of b in the direction** of **a** (or **scalar projection of b onto a**) is the scalar

$$\mathrm{comp}_{\mathbf{a}}(\mathbf{b}) = \frac{\mathbf{a}\cdot\mathbf{b}}{||\mathbf{a}||}.$$

Calculate the projection of $\mathbf{b}$ onto $\mathbf{a}$ and the component of $\mathbf{b}$ in the direction of $\mathbf{a}$ in each of the following:

(a) $\mathbf{a} = (1/\sqrt{3},\ 0,\ -1/\sqrt{3},\ 0,\ 1/\sqrt{3})$, $\mathbf{b} = (1,\ 1,\ 5,\ 4,\ 3)$

(b) $\mathbf{a} = (8,\ -1,\ 1,\ 2)$, $\mathbf{b} = (0,\ 9,\ 6,\ 6)$

(c) $\mathbf{a} = (1,\ -1,\ 1,\ -1, \ldots 1,\ -1) \in \mathbb{R}^{100}$,
$\mathbf{b} = (0,\ 1,\ 0,\ 2,\ 0,\ 3, \ldots 0,\ 50) \in \mathbb{R}^{100}$

Given a collection $\mathbf{x}_1, \mathbf{x}_2, \ldots, \mathbf{x}_k$ of vectors in $\mathbb{R}^n$, we say that the vectors form a **linearly dependent** set if one of them can be written as a linear combination of the others. That is, the set is linearly dependent if there is an index j in the range 1 to k such that

$$\mathbf{x}_j = \sum_{\substack{i=1\\ i\neq j}}^{k} c_i\mathbf{x}_i,$$

where the c_i's are given numbers.

15. Show that any set of vectors containing $\mathbf{0}$ is linearly dependent.

16. Show that $\mathbf{x}_1 = (1,\ 2,\ 3)$, $\mathbf{x}_2 = (-1,\ 0,\ 1)$ and $\mathbf{x}_3 = (5,\ 6,\ 7)$ form a linearly dependent set.

17. Show that any collection of three vectors in $\mathbb{R}^2$ is linearly dependent.

18. Prove that a set of vectors $\mathbf{x}_1, \mathbf{x}_2, \ldots, \mathbf{x}_k$ is linearly dependent if there exist constants $c_1, c_2, \ldots, c_k$ not all zero such that

$$c_1\mathbf{x}_1 + c_2\mathbf{x}_2 + \cdots + c_k\mathbf{x}_k = \mathbf{0}.$$

19. Show that the constant k in the last paragraph of Example 2.1.5 is positive if $r = 1$ and negative if $r = -1$.

20. Find the Pearson correlation coefficient for the following data.

i	1	2	3	4	5	6	7	8	9	10
x_i	2	4	3	4	1	5	6	0	8	7
y_i	4	7	6	6	4	6	7	2	9	9

21. A set A in $\mathbb{R}^n$ is **convex** if for any two points $\mathbf{x}$ and $\mathbf{y}$ in A, the points $\mathbf{x} + t(\mathbf{y} - \mathbf{x})$, $t \in [0, 1]$, all lie within A.

(a) Prove that the open ball $A = \{\mathbf{x} \in \mathbb{R}^n \mid ||\mathbf{x}|| < 1\}$ is convex.

(b) Prove that the sphere $S = \{||\mathbf{x}|| \in \mathbb{R}^n \mid ||\mathbf{x}|| = 1\}$ is not convex.

The Russian-German mathematician Hermann Minkowski (1864-1909) developed a four-dimensional geometry in which time is treated essentially as a fourth spatial coordinate. (Minkowski was a teacher of Albert Einstein, whose theories of special and general relativity make use of this geometry.) Writing points in $\mathbb{R}^4$ as (t, x, y, z), define the "distance" (i.e., time experienced in traveling at a uniform speed) between "events" $\mathbf{x}_1 = (t_1, x_1, y_1, z_1)$ and $\mathbf{x}_2 = (t_2, x_2, y_2, z_2)$ to be

$$d_M(\mathbf{x}_1, \mathbf{x}_2) = \sqrt{(t_2 - t_1)^2 - \left(\frac{x_2 - x_1}{c}\right)^2 - \left(\frac{y_2 - y_1}{c}\right)^2 - \left(\frac{z_2 - z_1}{c}\right)^2}, \tag{2.1.12}$$

where c is the speed of light measured in appropriate units (e.g., $c = 3.0 \times 10^8$ m/s).

22. Find the Minkowski distance between the points $\mathbf{x}_1 = (0, 0, 0, 0)$ and $\mathbf{x}_2 = (2, 0, 0, 0)$. (The segment between these two points represents a particle that is stationary for two time units.) Now, let $\mathbf{x}_3 = (1, 10^{-3}c, 10^{-3}c, 10^{-3}c)$ and calculate $d_M(\mathbf{x}_1, \mathbf{x}_3)$ and $d_M(\mathbf{x}_3, \mathbf{x}_2)$. How do these results compare with what happens when the Euclidean distance is used? This is a simple instance of what is called **time dilation**.

23. Photons are particles that travel in straight lines at the speed of light; if a photon passes through the origin, then its position in *three-dimensional* space is $(x, y, z) = (a_1t, a_2t, a_3t)$, where $\sqrt{a_1^2 + a_2^2 + a_3^2} = c$. Show that a photon lies on a "cone" in $\mathbb{R}^4$ (i.e., that $x^2 + y^2 + z^2 = kt^2$ for some constant k). Show that any particle traveling at a constant speed v less than c through the origin lies "inside" the light cone in the sense that $t > \sqrt{x^2 + y^2 + z^2}/c$. Why would the expression in (2.1.12) prohibit particles moving with a speed greater than that of light?

2.2 Matrices

It is helpful in any discussion of functions of several variables to have on hand some rudimentary tools from linear algebra. Thus the purpose of this section is to introduce some of these ideas. The first order of business is to discuss the notion of a matrix[2]. We can in many ways think of matrices as glorified vectors.

Definition 2.2.1 A **matrix** of size m by n ($m \times n$) is an arrangement of m rows of n numbers. The entries of the matrix are denoted a_{ij}, where the index i ranges from 1 to m and j ranges from 1 to n. We write the matrix as

$$\begin{bmatrix} a_{11} & a_{12} & \cdots & a_{1n} \\ a_{21} & a_{22} & \cdots & a_{2n} \\ \vdots & \vdots & & \vdots \\ a_{m1} & a_{m2} & \cdots & a_{mn} \end{bmatrix},$$

and abbreviate with $[a_{ij}]_{\substack{1 \le i \le m \\ 1 \le j \le n}}$, or $[a_{ij}]$, or A.

The **sum** of two $m \times n$ matrices A and B is the $m \times n$ matrix $A+B$ whose entries are the sums of the respective entries in A and B; in our shorthand notation,

$$A + B = [a_{ij}] + [b_{ij}] = [a_{ij} + b_{ij}].$$

If $m = n$ we say that A is **square**. The **product** of an $m \times n$ matrix A with scalar k is the $m \times n$ matrix, denoted kA or Ak, whose entries are the respective entries of A each multipled by k:

$$kA = [ka_{ij}].$$

The **difference** between two matrices A and B is the matrix $A - B$ given by

$$A - B = A + (-1)B.$$

Example 2.2.1 Let $A = \begin{bmatrix} 3 & 0 & 1 \\ -2 & 7 & 4 \end{bmatrix}$ and $B = \begin{bmatrix} 6 & -2 & 9 \\ 1 & 4 & 3 \end{bmatrix}$. Then for instance, $a_{11} = 3$, $a_{23} = 4$, and $b_{13} = 9$. Both matrices are of size 2×3, so

[2]The British mathematician J.J. Sylvester (1814-1897) applied this term to these mathematical objects. The Latin word *matrix* is derived from the Latin *mater*, which means *mother*; Sylvester used the term to suggest that these arrays of numbers were the "mothers" of *determinants*.

we can form their sum:

$$\begin{aligned} A+B &= \begin{bmatrix} 3+6 & 0-2 & 1+9 \\ -2+1 & 7+4 & 4+3 \end{bmatrix} \\ &= \begin{bmatrix} 9 & -2 & 10 \\ -1 & 11 & 7 \end{bmatrix}. \end{aligned}$$

Let $C = \begin{bmatrix} 12 & -1 \\ 1 & 3 \end{bmatrix}$. Then C has size 2×2, so $A + C$ is not defined. But we can, for instance, use the definition to calculate

$$\begin{aligned} \frac{2}{3}C &= \frac{2}{3}\begin{bmatrix} 12 & -1 \\ 1 & 3 \end{bmatrix} = \begin{bmatrix} 2/3 \cdot 12 & 2/3(-1) \\ 2/3 \cdot 1 & 2/3 \cdot 3 \end{bmatrix} \\ &= \begin{bmatrix} 8 & -2/3 \\ 2/3 & 2 \end{bmatrix}. \quad \blacklozenge \end{aligned}$$

Definition 2.2.2 The **product** of an $m \times n$ matrix $A = [a_{ij}]_{\substack{1 \le i \le m \\ 1 \le j \le n}}$ and an $n \times q$ matrix $B = [b_{ij}]_{\substack{1 \le i \le n \\ 1 \le j \le q}}$ is an $m \times q$ matrix, denoted AB, given by

$$AB = \left[\sum_{k=1}^{n} a_{ik} b_{kj} \right]_{\substack{1 \le i \le m \\ 1 \le j \le q}}.$$

A helpful way to think of the product of matrices A and B is first to regard the rows of A as vectors in $\mathbb{R}^n$ and the columns of B as vectors in $\mathbb{R}^n$. Then the ijth entry in the product AB is the dot product of row i in A with column j in B:

$$AB = \begin{bmatrix} & \vdots & & \\ a_{i1} & a_{i2} & \cdots & a_{in} \\ & \vdots & & \\ & \vdots & & \end{bmatrix} \begin{bmatrix} & b_{1j} & \\ \cdots & b_{2j} & \cdots \\ & \vdots & \\ & b_{nj} & \end{bmatrix}$$

Example 2.2.2 Let $A = \begin{bmatrix} 1 & -1 & 0 \\ 3 & 2 & 5 \end{bmatrix}$ and $B = \begin{bmatrix} 1 & -7 & 1 \\ 3 & 0 & 1 \\ 5 & 6 & 1 \end{bmatrix}$. Then

AB

$$= \begin{bmatrix} 1\cdot 1+(-1)\cdot 3+0 & 1\cdot(-7)+0+0 & 1\cdot 1+(-1)\cdot 1+0 \\ 3\cdot 1+2\cdot 3+5\cdot 5 & 3\cdot(-7)+0+5\cdot 6 & 3\cdot 1+2\cdot 1+5\cdot 1 \end{bmatrix}$$

$$= \begin{bmatrix} -2 & -7 & 0 \\ 34 & 9 & 10 \end{bmatrix}.$$

In this case, the product BA is not even defined because the number of columns in B does not match the number of rows in A. ♦

An **identity matrix** is a square matrix with 1 in each diagonal entry and zeros everywhere else:

$$I = \begin{bmatrix} 1 & 0 & \cdots & \cdots & 0 \\ 0 & 1 & 0 & \cdots & 0 \\ \vdots & & \ddots & & \vdots \\ 0 & \cdots & 0 & 1 & 0 \\ 0 & \cdots & 0 & 0 & 1 \end{bmatrix}. \tag{2.2.1}$$

The size of an identity matrix is usually understood from the context. An $n \times n$ identity matrix has the property that if A is an $n \times q$ matrix, then $IA = A$, and if A is a $q \times n$ matrix, then $AI = A$. In particular, if A is square, then $IA = AI = A$. Verify this!

A **zero matrix** is any matrix with all zero entries. (It is denoted by 0, and its size is usually understood from the context.)

Example 2.2.3 If we write $I\begin{bmatrix} 1 & 2 & 3 \\ 4 & 5 & 6 \end{bmatrix}$, then I stands for $\begin{bmatrix} 1 & 0 \\ 0 & 1 \end{bmatrix}$. If we write $\begin{bmatrix} 4 \\ 3 \\ 7 \end{bmatrix} + 0$, then 0 stands for $\begin{bmatrix} 0 \\ 0 \\ 0 \end{bmatrix}$. ♦

The $n \times n$ identity matrix is a special case of a **diagonal matrix**, that is, a square matrix A with zero entries off the diagonal of the matrix: $a_{ij} = 0$ for $i \neq j$. The general form of a diagonal matrix is

$$A = \begin{bmatrix} a_{11} & 0 & \cdots & \cdots & 0 \\ 0 & a_{22} & 0 & \cdots & 0 \\ \vdots & & \ddots & & \vdots \\ 0 & \cdots & 0 & a_{n-1,n-1} & 0 \\ 0 & \cdots & \cdots & 0 & a_{nn} \end{bmatrix}. \tag{2.2.2}$$

Example 2.2.4 With $A = \begin{bmatrix} 1/2 & 0 & 0 \\ 0 & 1/4 & 0 \\ 0 & 0 & 1/8 \end{bmatrix}$ and $B = \begin{bmatrix} 0 & 2 & 0 \\ -8 & 12 & 4 \\ -16 & 24 & -8 \end{bmatrix}$, we see that

$$AB = \begin{bmatrix} 0 & 1 & 0 \\ -2 & 3 & 1 \\ -2 & 3 & -1 \end{bmatrix}$$

and

$$BA = \begin{bmatrix} 0 & 1/2 & 0 \\ -4 & 3 & 1/2 \\ -8 & 6 & -1 \end{bmatrix}.$$

Notice that multiplying the diagonal matrix A by B on the left (i.e., forming the product AB) has rescaled the rows of B, and multiplying A by B on the right (i.e., forming the product BA) has rescaled the columns of B. Is it true in general that if A is a $n \times n$ diagonal matrix, then AB consists of the rows of B each scaled by the respective factors a_{11}, a_{22}, ..., a_{nn}? ♦

Having introduced matrices, we make the following identification between vectors in $\mathbb{R}^n$ and $n \times 1$ column matrices: For $\mathbf{x} \in \mathbb{R}^n$,

$$\mathbf{x} = (x_1, x_2, \ldots, x_n) = \begin{bmatrix} x_1 \\ x_2 \\ \vdots \\ x_n \end{bmatrix}. \tag{2.2.3}$$

In other words, we will regard an $n \times 1$ column matrix and a vector in $\mathbb{R}^n$ as the same. For instance $(1, 0, -1, 5) = \begin{bmatrix} 1 \\ 0 \\ -1 \\ 5 \end{bmatrix}$.

Also, we identify real numbers with 1×1 matrices: For $k \in \mathbb{R}$, $k = [k]$. This is so that, for example,

$$\begin{bmatrix} 2 & 4 & 1 & 2 \end{bmatrix} \begin{bmatrix} 1 \\ 2 \\ -2 \\ 1 \end{bmatrix} = [10] = 10$$

gives the same result as $(2, 4, 1, 2) \cdot (1, 2, -2, 1)$.

Matrix addition and multiplication possess properties that are formally the same as those of scalar addition and multiplication. We list these in the following theorem and indicate proofs for some. (Compare with Theorem 2.1.1.)

Theorem 2.2.1 *Let A, B, and C be matrices for which the indicated operations are defined, and let k and l be arbitrary scalars. Then*

1. $A+B=B+A$
2. $A+(B+C)=(A+B)+C$
3. $0+A=A$
4. $k(A+B)=kA+kB$
5. $(k+l)A=kA+lA$
6. $(kl)A=k(lA)=l(kA)$
7. $A+(-A)=0$
8. $IA=A$
9. $A(B+C)=AB+AC$
10. $(B+C)A=BA+CA$
11. $A(BC)=(AB)C$
12. $k(AB)=(kA)B=A(kB)$

Proof

4. By definition, $A+B$ is the matrix whose entries are $a_{ij}+b_{ij}$, where i ranges over the row indices and j over the column indices. Thus, by definition, $k(A+B)$ is the matrix with entries $k(a_{ij}+b_{ij})$. By the distributive property for scalars, these entries can be written as $ka_{ij}+kb_{ij}$. These, in turn, are the entries of the sum of the matrices with entries ka_{ij} and kb_{ij}, and these are the entries of the matrices kA and kB. Thus
$$k(A+B)=kA+kB.$$

Alternatively, using our shorthand notation $A = [a_{ij}]$, we could write the proof of (4):

$$\begin{aligned} k(A+B) &= k[a_{ij} + b_{ij}] \\ &= [k(a_{ij} + b_{ij})] \\ &= [ka_{ij} + kb_{ij}] \\ &= [ka_{ij}] + [kb_{ij}] \\ &= k[a_{ij}] + k[b_{ij}] \\ &= kA + kB. \end{aligned}$$

Indeed, we adopt the latter form for the other parts of the theorem proved here.

9. Let A have size $m \times n$, and B and C size $n \times q$. Then

$$\begin{aligned} A(B+C) &= [a_{ij}][b_{ij} + c_{ij}] \\ &= \left[\sum_{k=1}^{n} a_{ik}(b_{kj} + c_{kj})\right] \\ &= \left[\sum_{k=1}^{n} (a_{ik}b_{kj} + a_{ik}c_{kj})\right] \\ &= \left[\sum_{k=1}^{n} a_{ik}b_{kj} + \sum_{k=1}^{n} a_{ik}c_{kj}\right] \\ &= \left[\sum_{k=1}^{n} a_{ik}b_{kj}\right] + \left[\sum_{k=1}^{n} a_{ik}c_{kj}\right] \\ &= AB + AC. \end{aligned}$$

11. This identity is a bit harder to prove than others. Its proof is in an appendix to this section. ♦

Notice that properties 1 through 8 are exactly the same as those in Theorem 2.1.1. Matrices are just the same as vectors in these ways. Properties 9 through 12 do not correspond to any properties of vectors because vector multiplication is not defined. These last four are of great importance throughout the rest of this book.

Example 2.2.5 To calculate the product $\left(\begin{bmatrix} 1 \\ 3 \\ 5 \end{bmatrix} \begin{bmatrix} 1 & 1/3 & 1/5 \end{bmatrix}\right) \begin{bmatrix} 0 \\ 2 \\ -2 \end{bmatrix}$

we can, by property 11, first find

$$\begin{bmatrix} 1 & 1/3 & 1/5 \end{bmatrix} \begin{bmatrix} 0 \\ 2 \\ -2 \end{bmatrix} = \begin{bmatrix} 2/3 & -2/5 \end{bmatrix} = \begin{bmatrix} 4/15 \end{bmatrix} = \frac{4}{15}$$

so that the product is

$$\begin{bmatrix} 1 \\ 3 \\ 5 \end{bmatrix} \frac{4}{15} = \begin{bmatrix} 4/15 \\ 4/5 \\ 4/3 \end{bmatrix}.$$

Try calculating the products in the *given* order and compare the results. ♦

There are two operations we perform on matrices that will be of great importance a bit later: *transposing* and *inverting*.

Definition 2.2.3 The **transpose** of a matrix is $A = [a_{ij}]_{\substack{1 \leq i \leq m \\ 1 \leq j \leq n}}$ the matrix A^T whose ijth entry is a_{ji}, $1 \leq i \leq m$, $1 \leq j \leq n$. (The "T" is a superscript, not an exponent.)

Example 2.2.6 If $A = \begin{bmatrix} 1/2 & 2 & -3 \\ 0 & 1 & 3 \end{bmatrix}$ then $A^T = \begin{bmatrix} 1/2 & 0 \\ 2 & 1 \\ -3 & 3 \end{bmatrix}$. ♦

Definition 2.2.4 A square matrix A is **symmetric** if $A = A^T$.

Example 2.2.7 $\begin{bmatrix} 0 & \pi & 2\pi \\ \pi & 0 & 3\pi \\ 2\pi & 3\pi & 0 \end{bmatrix}$ is symmetric, whereas $B = \begin{bmatrix} 1 & 1 & 1 \\ -1 & 2 & 1 \\ 1 & 1 & 3 \end{bmatrix}$ is not since $b_{12} \neq b_{21}$. ♦

Example 2.2.8 Show that for matrices A and B of the same size,

$$(A + B)^T = A^T + B^T.$$

Solution For each choice of i and j, the ijth entry of $(A + B)^T$ is the jith entry of $A + B$, which is $a_{ji} + b_{ji}$. On the other hand, the ijth entry of $A^T + B^T$ is the sum of the ijth entry of A^T and that of B^T. These entries are a_{ji} and b_{ji} respectively, so the ijth entry of $A^T + B^T$ is $a_{ji} + b_{ji}$. Thus the entries of the matrices on each side of the identity agree, and the identity is proved. ◆

Definition 2.2.5 A square matrix A is **invertible** if there is a square matrix B such that $BA = I$ and $AB = I$. We call B the **inverse** of A and write A^{-1} for B.

Example 2.2.9 With $A = \begin{bmatrix} 5 & 7 \\ 2 & 3 \end{bmatrix}$, the matrix $B = \begin{bmatrix} 3 & -7 \\ -2 & 5 \end{bmatrix}$ is an inverse of A since

$$BA = \begin{bmatrix} 3 \cdot 5 - 7 \cdot 2 & 3 \cdot 7 - 7 \cdot 3 \\ -2 \cdot 5 + 5 \cdot 2 & -2 \cdot 7 + 5 \cdot 3 \end{bmatrix} = \begin{bmatrix} 1 & 0 \\ 0 & 1 \end{bmatrix}$$

and also, $AB = I$. ◆

Example 2.2.10 The matrix $A = \begin{bmatrix} 1 & 1 \\ 1 & 1 \end{bmatrix}$ does not have an inverse. To see this, suppose it in fact did have the inverse $B = \begin{bmatrix} a & b \\ c & d \end{bmatrix}$. Then $BA = I$; that is,

$$\begin{bmatrix} a+b & a+b \\ c+d & c+d \end{bmatrix} = \begin{bmatrix} 1 & 0 \\ 0 & 1 \end{bmatrix}.$$

From this we are forced to conclude that $a + b = 1$ and $a + b = 0$ (and also that $c + d = 0$ and $c + d = 1$). This is not possible for any choice of a, b, c, and d, so there is no such B: A is not invertible. ◆

If we are given a square matrix, how do we find its inverse? Let us start by examining a 2×2 matrix $A = \begin{bmatrix} a & b \\ c & d \end{bmatrix}$. If there is an inverse B, then it must satisfy $AB = \begin{bmatrix} 1 & 0 \\ 0 & 1 \end{bmatrix}$; in particular, row 1 of A dotted with column 2 of B must be 0, so column 2 of B must be a multiple of $\begin{bmatrix} -b \\ a \end{bmatrix}$. Similarly, row 2 of A dotted with column 1 of B must be 0, so column 1 of B must be a multiple of $\begin{bmatrix} d \\ -c \end{bmatrix}$. Thus B looks like $B = \begin{bmatrix} k_1 d & -k_2 b \\ -k_1 c & k_2 a \end{bmatrix}$ for some choice of k_1 and k_2. Now

$$AB = \begin{bmatrix} k_1(ad - bc) & 0 \\ 0 & k_2(ad - bc) \end{bmatrix} = \begin{bmatrix} k_1 \det(A) & 0 \\ 0 & k_2 \det(A) \end{bmatrix}. \quad (2.2.4)$$

From (2.2.4), AB is the identity if and only if $k_1 = k_2 = 1/\det(A)$, and this holds if and only if $\det(A) \neq 0$. In this case,

$$B = \begin{bmatrix} d/\det(A) & -b/\det(A) \\ -c/det(A) & a/\det(A) \end{bmatrix} = \frac{1}{\det(A)} \begin{bmatrix} d & -b \\ -c & a \end{bmatrix}. \quad (2.2.5)$$

Now B must also satisfy $BA = I$. You can easily check that B given by (2.2.5) does satisfy this identity, so B is an inverse of A. Is it the only one? Let C be any other inverse of A. Then $AC = CA = I$, so $B = BI = B(AC) = (BA)C = IC = C$. That is, C must be the same as B: the inverse is unique. This justifies our writing A^{-1} for *the* inverse of a 2×2 matrix A.

When $\det(A) = 0$, there is no choice of k_1 or k_2 that makes AB the identity: In this case, A is not invertible.

This result for 2×2 matrices hints at a general result for inverses of $n \times n$ matrices. For a proof of the following theorem, consult a text on linear algebra.

Theorem 2.2.2 *An $n \times n$ matrix $A = [a_{ij}]$ is invertible if and only if $\det(A) \neq 0$. When $\det(A) = |A| \neq 0$, the inverse is given by*

$$A^{-1} = \frac{1}{|A|} \begin{bmatrix} +|A_{11}| & -|A_{12}| & +|A_{13}| & \cdots & (-1)^{1+n}|A_{1n}| \\ -|A_{21}| & +|A_{22}| & -|A_{23}| & \cdots & (-1)^{2+n}|A_{2n}| \\ +|A_{31}| & -|A_{32}| & +|A_{33}| & \cdots & (-1)^{3+n}|A_{3n}| \\ \vdots & \vdots & \vdots & & \vdots \\ (-1)^{n+1}|A_{n1}| & (-1)^{n+2}|A_{n2}| & \cdots & \cdots & (-1)^{2n}|A_{nn}| \end{bmatrix}^T, \quad (2.2.6)$$

where A_{ij} is the $(n-1) \times (n-1)$ matrix obtained by deleting row i and column j from A.

The matrix on the right-hand side of (2.2.6) is called the **adjoint** of A and is sometimes denoted adj (A).

Although this is not the only means of calculating inverses, and in fact it is terribly inefficient for even moderately large n, the formula is easy to use for values of n such as 2 or 3. Moreover, the formula displays the close connection between invertibility and determinants.

Example 2.2.11 Calculate the inverse of $A = \begin{bmatrix} 0 & 1 & 0 \\ 0 & 0 & 1 \\ 1 & 2 & 3 \end{bmatrix}$. Here

$$A_{11} = \begin{bmatrix} 0 & 1 \\ 2 & 3 \end{bmatrix}, A_{12} = \begin{bmatrix} 0 & 1 \\ 1 & 3 \end{bmatrix}, \ldots, A_{33} = \begin{bmatrix} 0 & 1 \\ 0 & 0 \end{bmatrix},$$

so

$$|A_{11}| = -2, |A_{12}| = -1, \ldots, |A_{33}| = 0.$$

Expanding along the top row of A, we see that $|A| = -\begin{vmatrix} 0 & 1 \\ 1 & 3 \end{vmatrix} = 1$, so by Theorem 2.2.2,

$$\begin{aligned} A^{-1} &= \frac{1}{1}\begin{bmatrix} (-2) & -(-1) & +0 \\ -3 & +0 & -(-1) \\ +1 & -(0) & +0 \end{bmatrix}^T = \begin{bmatrix} -2 & 1 & 0 \\ -3 & 0 & 1 \\ 1 & 0 & 0 \end{bmatrix}^T \\ &= \begin{bmatrix} -2 & -3 & 1 \\ 1 & 0 & 0 \\ 0 & 1 & 0 \end{bmatrix}. \quad \blacklozenge \end{aligned}$$

Example 2.2.12 Show that $(AB)^{-1} = B^{-1}A^{-1}$, where A and B are any two invertible matrices.

Solution The objective is to show that when $B^{-1}A^{-1}$ is multiplied by AB on the left and on the right, we obtain the identity matrix I. It is certainly possible to show this by writing a proof using the entries of the matrices, but in fact it is far more desirable to have a proof that only makes use of the properties of matrix multiplication. We do the latter:

$$\begin{aligned} (B^{-1}A^{-1})(AB) &= ((B^{-1}A^{-1})A)B \\ &= (B^{-1}(A^{-1}A))B \\ &= (B^{-1}I)B \\ &= B^{-1}B \\ &= I, \end{aligned}$$

and by the same reasoning, $(AB)(B^{-1}A^{-1}) = I$. Thus $B^{-1}A^{-1}$ is the inverse of AB. ♦

Appendix: proof of part 11 of Theorem 2.2.1

The basic problem is, of course, to show that the ijth entry of the matrix $A(BC)$ is the same as the ijth entry of $(AB)C$. Let A, B, and C have sizes $m \times n$, $n \times p$, and $p \times q$ respectively. Now the ijth entry of the $n \times q$ matrix BC is

$$e_{ij} = \sum_{k=1}^{p} b_{ik}c_{kj}, \quad 1 \le i \le n, \quad 1 \le j \le q.$$

So the ijth entry of $A(BC)$ is

$$\begin{aligned} \sum_{l=1}^{n} a_{il}e_{lj} &= \sum_{l=1}^{n} a_{il} \sum_{k=1}^{p} b_{lk}c_{kj} \\ &= \sum_{l=1}^{n}\sum_{k=1}^{p} a_{il}b_{lk}c_{kj}. \end{aligned} \tag{2.2.7}$$

On the other hand, the ijth entry of AB is

$$f_{ij} = \sum_{l=1}^{n} a_{il}b_{lj}, \quad 1 \le i \le m, \quad 1 \le j \le p$$

so the ijth entry of $(AB)C$ is

$$\begin{aligned} \sum_{k=1}^{p} f_{ik}c_{kj} &= \sum_{k=1}^{p}\sum_{l=1}^{n} (a_{il}b_{lk})c_{kj} \\ &= \sum_{k=1}^{p}\sum_{l=1}^{n} a_{il}b_{lk}c_{kj}. \end{aligned} \tag{2.2.8}$$

By reversing the order of the sums in (2.2.8), we see that the ijth entry of $(AB)C$ is

$$\sum_{l=1}^{n}\sum_{k=1}^{p} a_{il}b_{lk}c_{kj},$$

which is, by (2.2.7), the ijth entry of $A(BC)$. ♦

Exercises 2.2

With $A = \begin{bmatrix} 2 & 3 \\ -1 & 0 \end{bmatrix}$, $B = \begin{bmatrix} 6 & 7 & 1 \\ 0 & 4 & 2 \end{bmatrix}$, $C = \begin{bmatrix} 10 \\ 1 \end{bmatrix}$, and $D = \begin{bmatrix} 2 & 5 \end{bmatrix}$, evaluate the following, if the expression is defined:

1. $AC + D^T$

2. $DA + C^T$

3. AB

4. BA

5. $B^T C$

6. $CD + 3I$

7. $-2A + CDA$

8. $(A - \frac{4}{5}I)C$

9. Suppose that A is a 3×3 matrix such that $A \begin{bmatrix} 2 \\ 0 \\ 0 \end{bmatrix} = \begin{bmatrix} 6 \\ -4 \\ 0 \end{bmatrix}$. What is the first column of A? If $A \begin{bmatrix} 0 \\ 0 \\ 3 \end{bmatrix} = \begin{bmatrix} -3 \\ 9 \\ 7 \end{bmatrix}$, what is the third column of A? Suppose that the second column of A is $\begin{bmatrix} 1 \\ 2 \\ -3 \end{bmatrix}$. What is $A \begin{bmatrix} 0 \\ 1 \\ 0 \end{bmatrix}$?

10. Verify for 3×3 matrices A and B, where the columns of B are the vectors $\mathbf{b}_1$, $\mathbf{b}_2$, and $\mathbf{b}_3$, that the columns of AB are $A\mathbf{b}_1$, $A\mathbf{b}_2$, and $A\mathbf{b}_3$. In other words, show that $A \begin{bmatrix} \mathbf{b}_1 & \mathbf{b}_2 & \mathbf{b}_3 \end{bmatrix} = \begin{bmatrix} A\mathbf{b}_1 & A\mathbf{b}_2 & A\mathbf{b}_3 \end{bmatrix}$.

Find the inverse of the following, if possible:

11. $\begin{bmatrix} 1 & 3 \\ 1 & 2 \end{bmatrix}$

12. $\begin{bmatrix} 4 & 1 \\ -8 & 2 \end{bmatrix}$

13. $\begin{bmatrix} 1/3 & 1/3 & 0 \\ 1/3 & 1/3 & -2/3 \\ 0 & -2/3 & 1 \end{bmatrix}$

14. $\begin{bmatrix} 1 & 2 & 3 \\ 0 & 1 & 2 \\ 1 & 0 & -1 \end{bmatrix}$

15. $\begin{bmatrix} 1/2 & 0 & 0 & 0 \\ 0 & 1/3 & 0 & 0 \\ 0 & 0 & 1/4 & 0 \\ 0 & 0 & 0 & 1/5 \end{bmatrix}$

16. $\begin{bmatrix} 0 & 1 & 0 & 0 \\ 0 & 0 & 0 & 1 \\ 1 & 0 & 0 & 0 \\ 0 & 0 & 1 & 0 \end{bmatrix}$

17. $\begin{bmatrix} -1 & 0 & -1 \\ 0 & -1 & 0 \\ -1 & 0 & -1 \end{bmatrix}$

Given an $n \times n$ invertible matrix A and and $n \times 1$ column matrix $\mathbf{b}$, we can solve the equation

$$A\mathbf{x} = \mathbf{b}$$

for the unknown vector $\mathbf{x}$ simply by multiplying both sides by A^{-1} and using various properties of matrix arithmetic:

$$\begin{aligned} A^{-1}(A\mathbf{x}) &= A^{-1}\mathbf{b} \\ (A^{-1}A)\mathbf{x} &= A^{-1}\mathbf{b} \\ I\mathbf{x} &= A^{-1}\mathbf{b} \\ \mathbf{x} &= A^{-1}\mathbf{b}. \end{aligned} \tag{2.2.9}$$

Use (2.2.9) to solve the following systems of equations.

18. $\begin{aligned} x + y &= 6 \\ x - 2y &= 4 \end{aligned}$

19. $\begin{aligned} -x + 2y \phantom{{}+2z} &= 1 \\ 2x - y + 2z &= 1 \\ 2y - z &= 1 \end{aligned}$

20.
$$\begin{array}{rcrcrcrcl} x_1 & + & x_2 & & & & & = & 3 \\ & & x_2 & + & x_3 & & & = & 4 \\ & & & & x_3 & + & x_4 & = & 5 \\ & & & & & & x_4 & = & 6 \end{array}$$

In Exercises 21 through 24, prove by direct computations with matrix entries:

21. For 2×2 matrices A and B, $\det(AB) = \det(A)\det(B)$.

22. For an invertible 2×2 matrix A, $\det(A^{-1}) = \dfrac{1}{\det(A)}$.

23. For a 2×2 matrix A, $\det(A^T) = \det(A)$.

24. For 2×2 matrices A and B, $(AB)^{-1} = B^{-1}A^{-1}$.

25. Prove 5 in Theorem 2.2.1.

26. Prove 10 in Theorem 2.2.1.

27. Prove 12 in Theorem 2.2.1.

28. Verify the formula in Theorem 2.2.2 for 3×3 matrices.

29. Prove that AA^T is symmetric for any square matrix A.

30. Prove that for any square matrix A, $\frac{1}{2}(A + A^T)$ is symmetric.

31. A square matrix A is **skew-symmetric** if $A^T = -A$.

(a) What are the diagonal entries of a skew-symmetric matrix?

(b) Show that $\frac{1}{2}(A - A^T)$ is skew-symmetric for any square matrix A.

32. Prove that any square matrix can be written as a sum of a symmetric and a skew-symmetric matrix. (This fact will be important later when we discuss the first derivative of vector fields and the second derivatives of scalar-valued functions. It is also important in connection with divergence and curl.)

33. Let A be a skew-symmetric 3×3 matrix. Find the vector $\mathbf{a}$ such that $\mathbf{a} \times \mathbf{x} = A\mathbf{x}$ for all $\mathbf{x}$ in $\mathbb{R}^3$.

34. Let A represent a square matrix. Prove that $A\mathbf{x} = \mathbf{x}$ for all $\mathbf{x} \in \mathbb{R}^n$ if and only if $A = I$.

35. Show that no 3×3 skew-symmetric matrix is invertible.

36. Give an example of the following:

(a) A nonzero 2×2 matrix A such that $A^2 = 0$

(b) a nonzero, nonidentity 2×2 matrix A such that $A^2 = A$

37. Prove that if Q is invertible and A is square, then with $B = Q^{-1}AQ$:

(a) $B^{-1} = Q^{-1}A^{-1}Q$

(b) $B^2 = Q^{-1}A^2Q$

38. The **norm** of a square matrix $A = [a_{ij}]$ is

$$||A|| = \sqrt{\sum_{i=1}^{n}\sum_{j=1}^{n} a_{ij}^2}.$$

Calculate the norm of the following matrices.

(a) $\begin{bmatrix} 3 & 1 \\ 7 & -2 \end{bmatrix}$ (b) $\begin{bmatrix} 8 & 0 & 4 \\ 2 & 1 & 0 \\ 5 & 3 & 3 \end{bmatrix}$

(c) $\begin{bmatrix} 1 & 0 & 0 & 0 \\ 0 & 2 & 0 & 0 \\ 0 & 0 & 3 & 0 \\ 0 & 0 & 0 & 4 \end{bmatrix}$ (d) $\begin{bmatrix} 6 & 1 & -3 \\ 0 & 6 & 1 \\ 0 & 0 & 6 \end{bmatrix}$

39. Show that $A = B$ if and only if $||A - B|| = 0$.

40. The **trace** $\operatorname{tr}(A)$ of a square matrix $A = [a_{ii}]$ is the sum of the diagonal entries:

$$\operatorname{tr}(A) = \sum_{i=1}^{n} a_{ii}.$$

Calculate the trace of each of the matrices in Exercise 38.

41. (a) Prove that $\operatorname{tr}(A + B) = \operatorname{tr}(A) + \operatorname{tr}(B)$.

(b) Prove that $\operatorname{tr}(A^T) = \operatorname{tr}(A)$.

42. For an $n \times n$ square matrix A, show that the value of r that minimizes $||A - rI||^2$ is $r = \operatorname{tr}(A)/n$.

2.3 Linear Transformations

It is not much of an exaggeration to say that the notion of linearity is one of the most fundamental in scientific and everyday thinking. In simple forms, linear relations are simply proportionalities. For instance, sales tax in a given locality is a constant proportion of the amount of a purchase. So if P represents the (variable) amount of a purchase and T represents the sales tax on that purchase, then $T = rP$, where r is the tax rate (e.g., $r = 0.07$).

The graph of this equation is a straight line. A change in P gives rise to a change in T that is proportional to the change in P.

Linear relations may be somewhat more complex. For instance, suppose that a person has two jobs, one in a state where the income tax rate is 2% and one in a state where the tax rate is 3%. If she earns x dollars per year in the first state and y dollars per year in the second, then the total amount of state income tax she will pay is

$$T = 0.02x + 0.03y.$$

Here too we will say that T is a linear function of x and y. A change in either x or y is met with a proportional change in T. By rewriting the formula as

$$T = (0.02, 0.03) \cdot (x, y) = \begin{bmatrix} 0.02 & 0.03 \end{bmatrix} \begin{bmatrix} x \\ y \end{bmatrix}$$

we see that the function is a "proportionality" where the multiplier is a matrix.

Definition 2.3.1 A **linear transformation** from $\mathbb{R}^n$ to $\mathbb{R}^m$ is a function of the form

$$\begin{aligned} \mathbf{T}(x_1, x_2, \ldots, x_n) \quad = \quad & (a_{11}x_1 + a_{12}x_2 + \ldots + a_{1n}x_n, \\ & a_{21}x_1 + a_{22}x_2 + \ldots + a_{2n}x_n, \\ & \vdots \\ & a_{m1}x_1 + a_{m2}x_2 + \ldots + a_{mn}x_n) \end{aligned}$$

where a_{ij}, $1 \le i \le m$, $1 \le j \le n$ are given numbers and x_1, x_2, ..., x_n are variables. We abbreviate sometimes by writing $\mathbf{T} : \mathbb{R}^n \to \mathbb{R}^m$.

Example 2.3.1 $\mathbf{T}(x_1, x_2) = (x_1 - x_2, 3x_1 + 4x_2, x_2 - x_1)$ is a linear transformation from $\mathbb{R}^2$ to $\mathbb{R}^3$. ♦

Matrices provide a convenient way of representing linear transformations. The list of variables $x_1, \ldots, x_n$ can be thought of as a vector and hence as a $n \times 1$ column matrix

$$\mathbf{x} = \begin{bmatrix} x_1 \\ x_2 \\ \vdots \\ x_n \end{bmatrix}.$$

The vector

$$\begin{array}{ccccc}(a_{11}x_1 & + & \cdots & + & a_{1n}x_n, \\ & & \vdots & & \\ a_{m1}x_1 & + & \cdots & + & a_{mn}x_n)\end{array}$$

can be written as a $m \times 1$ column matrix

$$\begin{bmatrix} a_{11}x_1 & + & \cdots & + & a_{1n}x_n \\ & & \vdots & & \\ a_{m1}x_1 & + & \cdots & + & a_{mn}x_n \end{bmatrix},$$

which is recognizable as the matrix product $A\mathbf{x}$, where

$$A = \begin{bmatrix} a_{11} & \cdots & a_{1n} \\ \vdots & & \vdots \\ a_{m1} & \cdots & a_{mn} \end{bmatrix}.$$

Thus we can represent the transformation as $\mathbf{T}(\mathbf{x}) = A\mathbf{x}$, an expression that is considerably less trouble to write than the representation given in the definition!

Linear transformations possess two properties of profound importance. These are given in the following theorem.

Theorem 2.3.1 *For any linear transformation* $\mathbf{T} : \mathbb{R}^n \to \mathbb{R}^m$*:*

1. $\mathbf{T}(\mathbf{x} + \mathbf{y}) = \mathbf{T}(\mathbf{x}) + \mathbf{T}(\mathbf{y})$ *for all* $\mathbf{x}$, $\mathbf{y} \in \mathbb{R}^n$.

2. $\mathbf{T}(k\mathbf{x}) = k\mathbf{T}(\mathbf{x})$ *for all* $\mathbf{x} \in \mathbb{R}^n$ *and* $k \in \mathbb{R}$.

Proof All of the hard work has been done in Section 2.2. These properties come directly from Theorem 2.2.1, parts 9 and 12 respectively.

1. Let A be the matrix that represents $\mathbf{T}$. Then

$$\begin{aligned} \mathbf{T}(\mathbf{x} + \mathbf{y}) &= A(\mathbf{x} + \mathbf{y}) \\ &= A\mathbf{x} + A\mathbf{y} \\ &= \mathbf{T}(\mathbf{x}) + \mathbf{T}(\mathbf{y}). \end{aligned}$$

2. To prove part 2, simply note that

$$\mathbf{T}(k\mathbf{x}) = A(k\mathbf{x}) = kA\mathbf{x} = k\,\mathbf{T}(\mathbf{x}). \quad \blacklozenge$$

Example 2.3.2 The transformation $\mathbf{T}(x_1, x_2) = (x_1^2 + x_2, \sqrt{x_1 x_2})$ is not linear. If it were then, for instance, $\mathbf{T}(2, 4)$ should equal $2\mathbf{T}(1, 2)$, by Theorem 2.3.1, property 2. But $\mathbf{T}(2, 4) = (8, \sqrt{8})$, whereas $2\mathbf{T}(1, 2) = 2(3, \sqrt{2}) = (6, 2\sqrt{2})$. ♦

In most of this book we will be concerned with linear transformations $\mathbf{T}$ that take $\mathbb{R}^n$ to $\mathbb{R}^n$ or $\mathbb{R}^n$ to $\mathbb{R}$. In the former case, the matrix representing $\mathbf{T}$ is square. What size does the representative matrix have in the latter case?

Definition 2.3.2 A linear transformation $\mathbf{T} : \mathbb{R}^n \to \mathbb{R}^n$ is **invertible** if there is a linear transformation $\mathbf{U} : \mathbb{R}^n \to \mathbb{R}^n$ such that

$$\mathbf{T}(\mathbf{U}(\mathbf{x})) = \mathbf{U}(\mathbf{T}(\mathbf{x})) = \mathbf{x}$$

for all $\mathbf{x} \in \mathbb{R}^n$. The transformation $\mathbf{U}$ is called the **inverse** of $\mathbf{T}$.

The definition simply requires a transformation $\mathbf{U}$ that exactly undoes $\mathbf{T}$ and such that $\mathbf{T}$ undoes $\mathbf{U}$. How do we find $\mathbf{U}$? Given $\mathbf{T}$, let A be the matrix that represents it. Suppose that $\mathbf{T}$ is invertible with inverse $\mathbf{U}$ given by $\mathbf{U}(\mathbf{x}) = B\mathbf{x}$. Then for all $\mathbf{x} \in \mathbb{R}^n$,

$$\mathbf{x} = \mathbf{T}(\mathbf{U}(\mathbf{x})) = A(B\mathbf{x}) = (AB)\mathbf{x}$$

and

$$\mathbf{x} = \mathbf{U}(\mathbf{T}(\mathbf{x})) = B(A\mathbf{x}) = (BA)\mathbf{x}.$$

By Exercise 34 in Section 2.2, $AB = BA = I$, so A is invertible with inverse B. On the other hand, suppose that A is invertible. Then $\mathbf{U}$ given by $\mathbf{U}(\mathbf{x}) = A^{-1}\mathbf{x}$ satisfies

$$\mathbf{U}(\mathbf{T}(\mathbf{x})) = A^{-1}(A\mathbf{x}) = (A^{-1}A)\mathbf{x} = I\mathbf{x} = \mathbf{x}$$

and

$$\mathbf{T}(\mathbf{U}(\mathbf{x})) = A(A^{-1}\mathbf{x}) = \mathbf{x}$$

for all $\mathbf{x} \in \mathbb{R}^n$. Thus $\mathbf{T}$ is invertible. This proves the following theorem.

Theorem 2.3.2 *A linear transformation $\mathbf{T} : \mathbb{R}^n \to \mathbb{R}^n$ is invertible if and only if the matrix A representing it is invertible. In this case the inverse transformation is $\mathbf{T}^{-1}(\mathbf{x}) = A^{-1}\mathbf{x}$.*

Example 2.3.3 Find the inverse of

$$\mathbf{T}(x_1, x_2, x_3) = (x_1 - x_3,\ 2x_1 + x_2,\ x_2 + x_3).$$

Solution The matrix that represents $\mathbf{T}$ is $A = \begin{bmatrix} 1 & 0 & -1 \\ 2 & 1 & 0 \\ 0 & 1 & 1 \end{bmatrix}$. By Theorem 2.3.2, $\mathbf{T}$ has an inverse if A is invertible. You can check, via hand calculations, calculator, or computer that A is invertible with

$$A^{-1} = \begin{bmatrix} -1 & 1 & -1 \\ 2 & -1 & 2 \\ -2 & 1 & -1 \end{bmatrix}.$$

Thus $\mathbf{T}^{-1}(\mathbf{x}) = (-x_1 + x_2 - x_3,\ 2x_1 - x_2 + 2x_3,\ -2x_1 + x_2 - x_3)$. ♦

In some cases it is easier to invert a transformation by solving a few equations rather than working with inverses of matrices.

Example 2.3.4 Let $\mathbf{T}(x_1, x_2, x_3, x_4) = (x_3, x_1, x_4, x_2)$. Find the inverse of $\mathbf{T}$.

Solution For any given vector (x_1, x_2, x_3, x_4), we want to find formulas for the components of the vector $(y_1, y_2, y_3, y_4) = \mathbf{T}^{-1}(x_1, x_2, x_3, x_4)$. Applying $\mathbf{T}$ to both sides, we obtain

$$\mathbf{T}(y_1, y_2, y_3, y_4) = (x_1, x_2, x_3, x_4).$$

Since $\mathbf{T}(y_1, y_2, y_3, y_4) = (y_3, y_1, y_4, y_2)$, this gives

$$(y_3, y_1, y_4, y_2) = (x_1, x_2, x_3, x_4).$$

Equating components, we obtain $y_1 = x_2$, $y_2 = x_4$, $y_3 = x_1$, and $y_4 = x_3$, so

$$\mathbf{T}^{-1}(x_1, x_2, x_3, x_4) = (x_2, x_4, x_1, x_3). \quad ♦$$

Exercises 2.3

Find the matrices for the following linear transformations.

1. $\mathbf{T}(x_1, x_2) = (3x_1 + 5x_2,\ x_2,\ -x_1 + 4x_2)$

2. $\mathbf{T}(x_1, x_2, x_3, x_4) = (x_1 - x_2,\ x_3 + x_4)$

3. $\mathbf{T}(x_1, x_2, x_3) = (4x_1 + x_2/2 + x_3/3,\ x_1/2 + 3x_2 - x_3/4,\ x_1/3 + x_2/4 + 2x_3)$

4. $\mathbf{T}(x_1, x_2, x_3 \ldots x_{20}) = x_1 + x_2 + x_3 \cdots + x_{20}$

5. Explain why $\mathbf{T}(x_1,\, x_2,\, x_3) = (x_1 - x_2x_3,\, x_3 - x_1^2 + x_2)$ is not linear.

6. Explain why $\mathbf{T}(x_1,\, x_2) = \dfrac{x_1^2 x_2}{x_1 + x_2}$ is not linear.

For each of the linear transformations in Exercises 7 through 12, determine whether it is invertible. If so, find the inverse.

7. $\mathbf{T}(x_1,\, x_2) = (2x_1 + x_2,\, x_1 + 5x_2)$

8. $\mathbf{T}(x_1,\, x_2) = (2x_1 - 3x_2,\, -4x_1 + 6x_2)$

9. $\mathbf{T}(x_1,\, x_2,\, x_3) = \left(\frac{1}{3}(x_1 + x_2),\, \frac{1}{3}(x_1 + x_2 - 2x_3),\, -\frac{1}{3}(2x_2 + 3x_3)\right)$

10. $\mathbf{T}(x_1,\, x_2,\, \ldots,\, x_{10}) = (10x_1,\, 9x_2,\, 8x_3,\, \ldots,\, 2x_9,\, x_{10})$

11. $\mathbf{T}(x_1,\, x_2,\, x_3,\, x_4,\, x_5) = (x_5,\, x_4,\, x_3,\, x_2,\, x_1)$

12. $\mathbf{T}(x_1,\, x_2,\, x_3,\, x_4) = (x_2,\, x_3,\, x_4,\, 0)$

13. Show that the composition of two linear transformations is a linear transformation. That is, if $\mathbf{S} : \mathbb{R}^n \to \mathbb{R}^m$ and $\mathbf{T} : \mathbb{R}^m \to \mathbb{R}^k$, then the transformation

$$\mathbf{U}(\mathbf{x}) = \mathbf{T}(\mathbf{S}(\mathbf{x}))$$

is linear. (We also write $\mathbf{T} \circ \mathbf{S}$ for the composition $\mathbf{S}$ followed by $\mathbf{T}$.)

14. Find the composition $\mathbf{T} \circ \mathbf{S}$ in each of the following.

(a) $\mathbf{S}(x_1,\, x_2) = (2x_1 + x_2,\, -2x_1 + 5x_2)$;
$\mathbf{T}(x_1,\, x_2) = (\frac{1}{2}x_1 + x_2,\, -\frac{1}{2}x_1 + \frac{1}{5}x_2)$

(b) $\mathbf{S}(x_1,\, x_2,\, x_3) = (x_1 + x_3,\, x_2,\, -x_3 + x_2)$;
$\mathbf{T}(x_1,\, x_2,\, x_3) = (x_3,\, x_1,\, x_2)$

(c) $\mathbf{S}(x_1,\, x_2,\, x_3,\, x_4) = (x_1 + x_3 + x_4,\, x_1 - x_2 - x_3 + x_4)$;
$\mathbf{T}(x_1,\, x_2) = (6x_1 - 7x_2,\, 7x_1 + 7x_2)$

(d) $\mathbf{S}(x) = (10x,\, -5x,\, x/5,\, -x/10)$;
$\mathbf{T}(x_1,\, x_2,\, x_3,\, x_4) = \frac{1}{2}x_1 + \frac{1}{5}x_2 - \frac{1}{5}x_3 + \frac{1}{5}x_4$

15. For which pairs of linear transformations in Exercise 14 is the composition $\mathbf{S} \circ \mathbf{T}$ defined? Calculate $\mathbf{S} \circ \mathbf{T}$ in these cases.

16. Given a linear transformation $\mathbf{T} : \mathbb{R}^n \to \mathbb{R}^n$, the **iterates** of $\mathbf{T}$ are the transformations given by

$$\begin{aligned}
\mathbf{T}^1(\mathbf{x}) &= \mathbf{T}(\mathbf{x}) \\
\mathbf{T}^2(\mathbf{x}) &= \mathbf{T}(\mathbf{T}(\mathbf{x})) \\
\mathbf{T}^3(\mathbf{x}) &= \mathbf{T}(\mathbf{T}^2\,\mathbf{x})) \\
&\vdots
\end{aligned}$$

In general, the ith iterate of $\mathbf{T}$ is defined by $\mathbf{T}^i(\mathbf{x}) = \mathbf{T}(\mathbf{T}^{i-1}(\mathbf{x}))$, $i = 1$, 2, ..., where $\mathbf{T}^0(\mathbf{x}) = \mathbf{x}$.

(a) Find a formula for the ith iterate of $\mathbf{T}(x_1, x_2) = (x_2, -x_1)$.

(b) Find a formula for the ith iterate of $\mathbf{T}(x_1, x_2, x_3) = (x_2, x_3, 0)$.

(c) Find a formula for $\mathbf{T}^i(\mathbf{x})$ where $\mathbf{T}(x_1, x_2, x_3) = \left(\frac{7}{8}x_1, -\frac{2}{3}x_2, \frac{1}{4}x_3\right)$.

17. Let Q be an invertible matrix and D a diagonal matrix. Show that if

$$\mathbf{T}(\mathbf{x}) = Q^{-1}DQ\,\mathbf{x},$$

then $\mathbf{T}^n(\mathbf{x}) = Q^{-1}D^nQ\,\mathbf{x}$.

18. Given $\mathbf{a} \in \mathbb{R}^3$, show that $\mathbf{T}(\mathbf{x}) = \mathbf{a} \times \mathbf{x}$ is a linear transformation. What is its matrix?

19. Given $\mathbf{a} \in \mathbb{R}^3$ show that $\mathbf{T}(\mathbf{x}) = (\mathbf{a} \cdot \mathbf{x})\mathbf{a}$ is a linear transformation. (*Hint*: Write the right-hand side using matrix notation and use the fact that $k\,\mathbf{a} = \mathbf{a}k$ for a scalar k and vector $\mathbf{a}$.)

2.4 Geometry of Linear Transformations

The purpose of this section is to explore the questions, "What does a linear transformation 'do' to sets of points?" and in the cases of $\mathbb{R}^2$ and $\mathbb{R}^3$, "What effect does a linear transformation have on area or volume?" The importance of satisfactory answers to these questions will become more apparent when we study the derivatives of vector fields and changes of variable in multiple integrals. Broadly speaking, however, these geometric ideas are important to calculus of several variables because we can understand the geometry of a nonlinear vector field by regarding it as locally linear. In other words, we think of a nonlinear vector field as well approximated by a linear vector field (transformation) over small ranges.

Given a linear transformation $\mathbf{T} : \mathbb{R}^n \to \mathbb{R}^n$ and a point $\mathbf{x} \in \mathbb{R}^n$, $\mathbf{T}(\mathbf{x})$ is just another point in $\mathbb{R}^n$. If we have a collection P of points in $\mathbb{R}^n$ and apply $\mathbf{T}$ to each point in P, then we obtain a new collection of points given by

$$\mathbf{T}(P) = \{\mathbf{T}(\mathbf{x})\,|\,\mathbf{x} \in P\}, \tag{2.4.1}$$

called the **image of P under $\mathbf{T}$**. See Figure 2.4.1.

Let's examine a few of the properties of linear transformations that allow us to visualize $\mathbf{T}(P)$ whenever P is a parallelogram or parallelepiped. We will discover that when P is a parallelogram in $\mathbb{R}^2$, then, for an invertible $\mathbf{T}$,

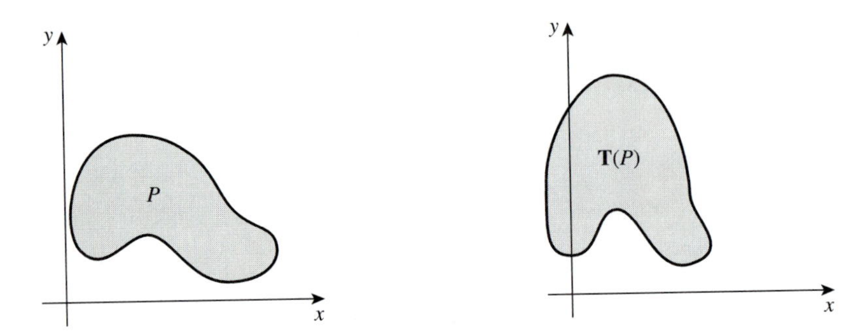

Figure 2.4.1 A set P in $\mathbb{R}^2$ and its image $\mathbf{T}(P)$

the image of P under $\mathbf{T}$ is also a parallelogram whose area is directly related to the determinant of the matrix that represents $\mathbf{T}$. A **parallelogram** in $\mathbb{R}^n$ is a set of the form

$$P = \{\mathbf{x} | \mathbf{x} = s\,\mathbf{a} + t\,\mathbf{b} + \mathbf{x}_0, \quad 0 \leq s \leq 1, \quad 0 \leq t \leq 1\}, \tag{2.4.2}$$

where $\mathbf{a}$, $\mathbf{b}$, and $\mathbf{x}_0$ are given vectors. See Figure 2.4.2.

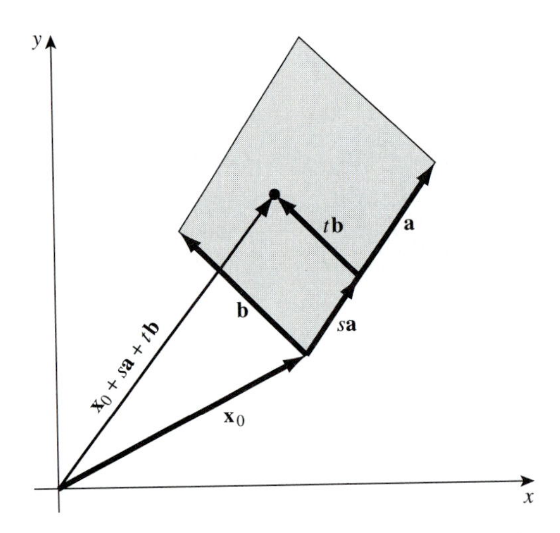

Figure 2.4.2 A parallelogram in $\mathbb{R}^2$

We say that $\mathbf{a}$ and $\mathbf{b}$ **span** the parallelogram and that $\mathbf{x}_0$ is a **corner**. In particular we focus on parallelograms in $\mathbb{R}^2$.

Theorem 2.4.1 *Let* $\mathbf{T}(\mathbf{x}) = A\mathbf{x}$ *be an invertible linear transformation on* $\mathbb{R}^2$.

1. *The image of a parallelogram P under $\mathbf{T}$ is a parallelogram.*

2. $\text{Area}\,(\mathbf{T}(P)) = |\det(A)|\text{Area}\,(P)$.

Proof 1. Let $P = \{\mathbf{x}|\mathbf{x} = s\,\mathbf{a} + t\,\mathbf{b} + \mathbf{x}_0, \quad 0 \le s \le 1, \quad 0 \le t \le 1\}$. For any point $\mathbf{x} = s\,\mathbf{a} + t\,\mathbf{b} + \mathbf{x}_0 \in P$,

$$\begin{aligned} \mathbf{T}(\mathbf{x}) &= \mathbf{T}(s\mathbf{a} + t\,\mathbf{b} + \mathbf{x}_0) \\ &= s\,\mathbf{T}(\mathbf{a}) + t\,\mathbf{T}(\mathbf{b}) + \mathbf{T}(\mathbf{x}_0), \end{aligned} \tag{2.4.3}$$

by the linearity of $\mathbf{T}$. Since $\mathbf{T}$ is invertible, $\mathbf{T}(\mathbf{a})$ and $\mathbf{T}(\mathbf{b})$ are nonzero. The result in (2.4.3) says that $\mathbf{T}(\mathbf{x})$ is in the parallelogram with corner $\mathbf{T}(\mathbf{x}_0)$ spanned by $\mathbf{T}(\mathbf{a})$ and $\mathbf{T}(\mathbf{b})$. Conversely, if $\mathbf{y}$ is a point in the parallelogram with corner $\mathbf{T}(\mathbf{x}_0)$ spanned by $\mathbf{T}(\mathbf{a})$ and $\mathbf{T}(\mathbf{b})$, then

$$\begin{aligned} \mathbf{y} &= s\,\mathbf{T}(\mathbf{a}) + t\,\mathbf{T}(\mathbf{b}) + \mathbf{T}(\mathbf{x}_0) \\ &= \mathbf{T}(s\,\mathbf{a} + t\,\mathbf{b} + \mathbf{x}_0) \end{aligned}$$

for some choice of $0 \le s \le 1$ and $0 \le t \le 1$. Thus $\mathbf{y} = \mathbf{T}(\mathbf{x})$ for some $\mathbf{x} \in P$. We have proved, in fact, that

$$\mathbf{T}(P) = \{\mathbf{x}\,|\,\mathbf{x} = s\,\mathbf{T}(\mathbf{a}) + t\,\mathbf{T}(\mathbf{b}) + \mathbf{T}\mathbf{x}_0, \quad 0 \le s \le 1, \quad 0 \le t \le 1\}.$$

2. The area of a parallelogram in $\mathbb{R}^2$ spanned by two vectors is the absolute value of the determinant of the 2×2 matrix whose rows (or columns) are the vectors. Thus

$$\begin{aligned} \text{Area}\,(\mathbf{T}(P)) &= \left|\det\left(\begin{bmatrix} A\mathbf{a} & A\mathbf{b} \end{bmatrix}\right)\right| \\ &= \left|\det\left(A\begin{bmatrix} \mathbf{a} & \mathbf{b} \end{bmatrix}\right)\right| \\ &= \left|\det(A)\det\left(\begin{bmatrix} \mathbf{a} & \mathbf{b} \end{bmatrix}\right)\right| \\ &= |\det(A)| \cdot \left|\det\left(\begin{bmatrix} \mathbf{a} & \mathbf{b} \end{bmatrix}\right)\right| \\ &= |\det(A)| \cdot \text{Area}\,(P). \end{aligned}$$

Here we have used Exercise 21 from Section 2.2. ♦

Example 2.4.1 Find the image of the parallelogram in $\mathbb{R}^2$ with vertices $(2, 3)$, $\left(6, \frac{11}{2}\right)$, $\left(7, \frac{17}{2}\right)$, and $(3, 6)$ under the linear transformation

$$\mathbf{T}(\mathbf{x}) = \begin{bmatrix} -1 & 2 \\ 3 & -2 \end{bmatrix} \mathbf{x}.$$

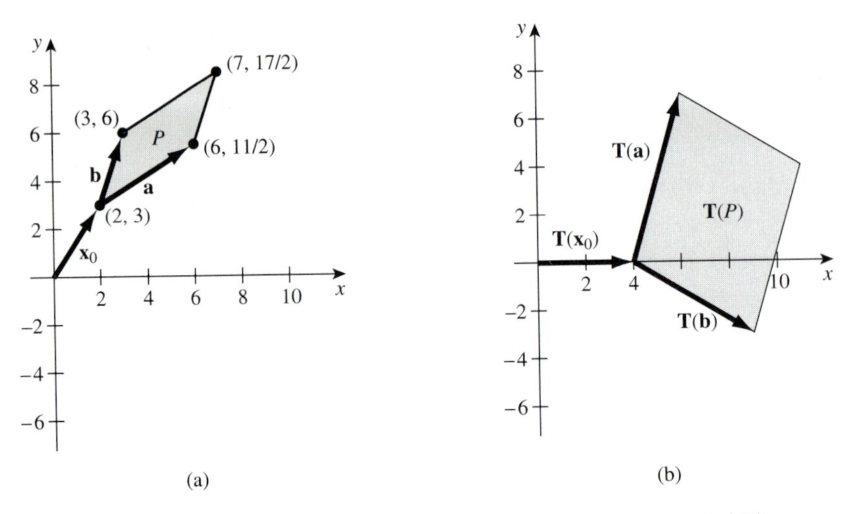

Figure 2.4.3 A parallelogram P and its image $\mathbf{T}(P)$

Solution The parallelogram P has a corner at $\mathbf{x}_0 = \begin{bmatrix} 2 \\ 3 \end{bmatrix}$. By consulting Figure 2.4.3(a), we see that it is spanned by

$$\mathbf{a} = \begin{bmatrix} 6-2 \\ 11/2-3 \end{bmatrix} = \begin{bmatrix} 4 \\ 5/2 \end{bmatrix} \quad \text{and} \quad \mathbf{b} = \begin{bmatrix} 3-2 \\ 6-3 \end{bmatrix} = \begin{bmatrix} 1 \\ 3 \end{bmatrix}.$$

By Theorem 2.4.1, the image of P is the parallelogram with corner

$$\mathbf{T}(\mathbf{x}_0) = \begin{bmatrix} -1 & 2 \\ 3 & -2 \end{bmatrix} \begin{bmatrix} 2 \\ 3 \end{bmatrix} = \begin{bmatrix} 4 \\ 0 \end{bmatrix}$$

spanned by

$$\mathbf{T}(\mathbf{a}) = \begin{bmatrix} -1 & 2 \\ 3 & -2 \end{bmatrix} \begin{bmatrix} 4 \\ 5/2 \end{bmatrix} = \begin{bmatrix} 1 \\ 7 \end{bmatrix}$$

and

$$\mathbf{T}(\mathbf{b}) = \begin{bmatrix} -1 & 2 \\ 3 & -2 \end{bmatrix} \begin{bmatrix} 1 \\ 3 \end{bmatrix} = \begin{bmatrix} 5 \\ -3 \end{bmatrix}.$$

Thus

$$\mathbf{T}(P) = \left\{ \mathbf{x} \,\middle|\, \mathbf{x} = s \begin{bmatrix} 1 \\ 7 \end{bmatrix} + t \begin{bmatrix} 5 \\ -3 \end{bmatrix} + \begin{bmatrix} 4 \\ 0 \end{bmatrix},\ 0 \le s \le 1,\ 0 \le t \le 1 \right\}.$$

See Figure 2.4.3(b). ♦

A **parallelepiped** in $\mathbb{R}^n$ $(n \geq 3)$ is a set of the form

$$P = \{\mathbf{x} \in \mathbb{R}^n | \mathbf{x} = s\,\mathbf{a} + t\,\mathbf{b} + u\,\mathbf{c} + \mathbf{x}_0, \ \ 0 \leq s \leq 1, \ \ 0 \leq t \leq 1, \ \ 0 \leq u \leq 1\}$$

where $\mathbf{a}$, $\mathbf{b}$, and $\mathbf{c}$ are given vectors that **span** the parallelepiped and $\mathbf{x}_0$ is a **corner**.

The next theorem is a three-dimensional analog to Theorem 2.4.1.

Theorem 2.4.2 *Let* $\mathbf{T}(\mathbf{x}) = A\,\mathbf{x}$ *be an invertible linear transformation on* $\mathbb{R}^3$.

1. *The image of a parallelepiped* P *under* $\mathbf{T}$ *is a parallelepiped.*

2. Volume $(\mathbf{T}(P)) = |\det(A)|$ Volume (P).

Proof 1. Let $P = \{\mathbf{x} \in \mathbb{R}^3 \,|\, \mathbf{x} = s\,\mathbf{a} + t\,\mathbf{b} + u\,\mathbf{c} + \mathbf{x}_0, 0 \leq s \leq 1, 0 \leq t \leq 1, 0 \leq u \leq 1\}$. Then $\mathbf{x} \in P$ if and only if

$$\begin{aligned} \mathbf{T}(\mathbf{x}) &= \mathbf{T}(s\,\mathbf{a} + t\,\mathbf{b} + u\,\mathbf{c} + \mathbf{x}_0) \\ &= s\,\mathbf{T}(\mathbf{a}) + t\,\mathbf{T}(\mathbf{b}) + u\,\mathbf{T}(\mathbf{c}) + \mathbf{T}(\mathbf{x}_0) \end{aligned}$$

for some $0 \leq s \leq 1, 0 \leq t \leq 1, 0 \leq u \leq 1$. That is, $\mathbf{x} \in P$ if and only if $\mathbf{T}(\mathbf{x})$ is a point in the parallelepiped with corner $\mathbf{T}(\mathbf{x}_0)$ spanned by $\mathbf{T}(\mathbf{a})$, $\mathbf{T}(\mathbf{b})$, and $\mathbf{T}(\mathbf{c})$. This is true for each $\mathbf{x} \in P$, so $\mathbf{T}(P)$ is a parallelepiped.

2. By Theorem 1.6.4, the volume of a parallelepiped is the absolute value of the determinant of the matrix with columns the vectors that span the parallelepiped. So

$$\begin{aligned} \text{Volume}\,(\mathbf{T}(P)) &= \left|\det\left(\begin{bmatrix} A\,\mathbf{a} & A\,\mathbf{b} & A\,\mathbf{c} \end{bmatrix}\right)\right| \\ &= \left|\det\left(A\begin{bmatrix} \mathbf{a} & \mathbf{b} & \mathbf{c} \end{bmatrix}\right)\right| \\ &= \left|\det(A)\det\left(\begin{bmatrix} \mathbf{a} & \mathbf{b} & \mathbf{c} \end{bmatrix}\right)\right| \\ &= |\det(A)|\,\text{Volume}\,(P). \quad \blacklozenge \end{aligned}$$

Example 2.4.2 Suppose that a cloud of dust in the first octant is moving so that each particle travels in a straight line parallel to the y-axis with a velocity that is proportional to $x + z$. Physically, this means that the material near the "wall" $x = 0$ and the "floor" $z = 0$ is moving slowly, whereas the material farther "up" and "out" is moving faster. Further, suppose that the cloud is initially cubical in shape: All of the dust particles lie in the cube C with vertices (0, 0, 0), (1, 0, 0), (1, 1, 0), (0, 1, 0), (0, 0, 1),

(1, 0, 1), (1, 1, 1), (0, 1, 1). What shape will the cloud have after 15 time units? What will its volume be at that time?

Solution After 15 time units, a dust particle with initial position (x, y, z) will have moved $15 \cdot k(x+z)$ units in the positive y-direction $(k > 0)$. Thus its new position will be

$$(x,\ y + 15k(x+z),\ z).$$

The new position as a function of the old then is given by the *linear* transformation

$$\mathbf{T}(x, y, z) = (x,\ 15kx + y + 15kz,\ z) = \begin{bmatrix} 1 & 0 & 0 \\ 15k & 1 & 15k \\ 0 & 0 & 1 \end{bmatrix} \begin{bmatrix} x \\ y \\ z \end{bmatrix},$$

so the shape of cubical dust cloud C after 15 time units is just $\mathbf{T}(C)$. Since C has corner $\mathbf{x}_0 = (0, 0, 0)$ and is spanned by

$$\mathbf{a} = \begin{bmatrix} 1 \\ 0 \\ 0 \end{bmatrix}, \quad \mathbf{b} = \begin{bmatrix} 0 \\ 1 \\ 0 \end{bmatrix}, \quad \text{and} \quad \mathbf{c} = \begin{bmatrix} 0 \\ 0 \\ 1 \end{bmatrix},$$

the shape after 15 time units will be, by Theorem 2.4.2, the parallelepiped with corner $\mathbf{T}(\mathbf{0}) = \mathbf{0}$ spanned by $\mathbf{T}(\mathbf{a}) = \begin{bmatrix} 1 \\ 15k \\ 0 \end{bmatrix}$, $\mathbf{T}(\mathbf{b}) = \begin{bmatrix} 0 \\ 1 \\ 0 \end{bmatrix}$, and $\mathbf{T}(\mathbf{c}) = \begin{bmatrix} 0 \\ 15k \\ 1 \end{bmatrix}$. Figure 2.4.4 illustrates the change in the shape of this dust cloud over 15 time units when $k = \frac{1}{10}$.

The volume of the cloud after 15 time units is, by Theorem 2.4.2,

$$\left| \det\left(\begin{bmatrix} 1 & 0 & 0 \\ 15k & 1 & 15k \\ 0 & 0 & 1 \end{bmatrix} \right) \right| \cdot \text{Volume}(C) = 1 \cdot 1 = 1;$$

the volume doesn't change! ♦

Example 2.4.3 Consider a dust cloud in which each particle is moving in uniform circular motion counterclockwise about the z-axis with constant angular velocity ω. For a particle that has initial position (x, y, z) and that

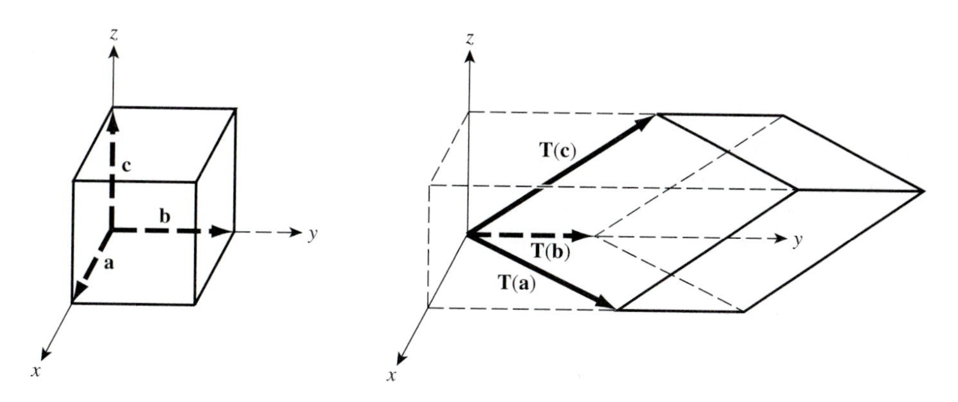

Figure 2.4.4 A shear transformation at work

moves in this way for τ time units, find the new position. Show that this new position is a linear function of $(x,\ y,\ z)$.

Solution For a particle with position $(x,\ y,\ z)$, let $(r,\ \theta,\ z)$ be its cylindrical coordinates. Then after time τ, its new position in cylindrical coordinates is $(r,\ \theta+\omega\tau,\ z)$. The rectangular coordinates of this new point are

$$
\begin{aligned}
&(r\cos(\theta+\omega\tau),\ r\sin(\theta+\omega\tau),\ z)\\
&\quad= (r(\cos\theta\cos\omega\tau-\sin\theta\sin\omega\tau),\ r(\sin\theta\cos\omega\tau+\cos\theta\sin\omega\tau),\ z)\\
&\quad= (\cos\omega\tau(r\cos\theta)-\sin\omega\tau(r\sin\theta),\ \cos\omega\tau(r\sin\theta)+\sin\omega\tau(r\cos\theta),\ z)\\
&\quad= ((\cos\omega\tau)x-(\sin\omega\tau)y,\ (\cos\omega\tau)y+(\sin\omega\tau)x,\ z)\\
&\quad= \begin{bmatrix} \cos\omega\tau & -\sin\omega\tau & 0\\ \sin\omega\tau & \cos\omega\tau & 0\\ 0 & 0 & 1\end{bmatrix}\begin{bmatrix} x\\ y\\ z\end{bmatrix}.
\end{aligned}
$$

This new position is a linear function of $\begin{bmatrix} x\\ y\\ z\end{bmatrix}$ since it has the form $A\begin{bmatrix} x\\ y\\ z\end{bmatrix}$, where A is a 3×3 matrix. ♦

As Example 2.4.3 shows, a matrix of the form

$$
A=\begin{bmatrix} \cos\theta & -\sin\theta & 0\\ \sin\theta & \cos\theta & 0\\ 0 & 0 & 1\end{bmatrix} \tag{2.4.4}
$$

represents a linear transformation whose action can be thought of as a rotation counterclockwise by the angle θ about the z-axis. Consequently, a matrix of the form (2.4.4) is called a **rotation matrix**.

By changing the roles of the variables x, y, and z, we can write down matrices for rotations about the other coordinate axes. For instance, a counterclockwise rotation about the y-axis (as viewed from the positive y-axis) by an angle θ is represented by a matrix of the form

$$\begin{bmatrix} \cos\theta & 0 & \sin\theta \\ 0 & 1 & 0 \\ -\sin\theta & 0 & \cos\theta \end{bmatrix}.$$

Example 2.4.4 Find the (perpendicular) projection of a point (x, y, z) onto a plane through the origin. Show that this is a linear transformation.

Solution Let $\mathbf{n} = (a, b, c)$ be a normal to the plane. Since the plane passes through the origin, its equation is

$$\mathbf{x} \cdot \mathbf{n} = 0. \tag{2.4.5}$$

See Figure 2.4.5. For a given point (x, y, z), $\mathbf{T}(x, y, z)$ is the point on the

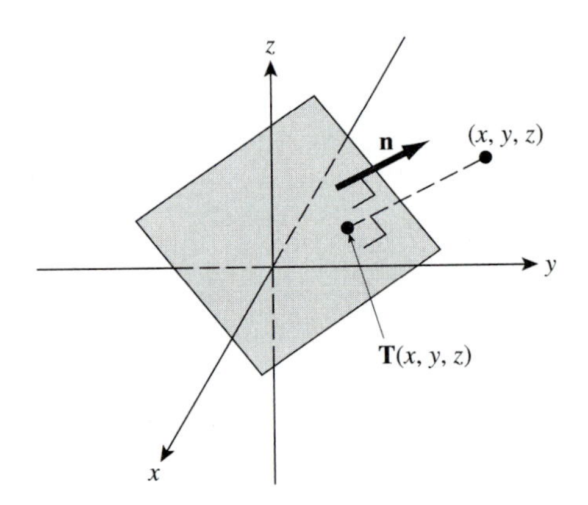

Figure 2.4.5 The projection of a point onto a plane through the origin

plane in which the line through (x, y, z) parallel to $\mathbf{n}$ intersects the plane. That line is

$$\mathbf{x} = (x, y, z) + t\,\mathbf{n}, \tag{2.4.6}$$

so by substituting (2.4.6) into (2.4.5) we can solve for t:

$$((x, y, z) + t\,\mathbf{n}) \cdot \mathbf{n} \quad = \quad 0$$

$$\begin{aligned} t||\mathbf{n}||^2 &= -(x,\, y,\, z)\cdot\mathbf{n} \\ t &= \frac{-(x,\, y,\, z)\cdot\mathbf{n}}{||\mathbf{n}||^2}. \end{aligned} \tag{2.4.7}$$

Thus $\mathbf{T}(x, y, z)$ is the point on the line obtained by substituting the value of t from (2.4.7) into (2.4.6):

$$(x, y, z) - \frac{(x,\, y,\, z)\cdot\mathbf{n}}{||\mathbf{n}||^2}\mathbf{n} = \frac{1}{||\mathbf{n}||^2}\left(||\mathbf{n}||^2(x,\, y,\, z) - (\mathbf{n}\cdot(x,\, y,\, z))\mathbf{n}\right).$$

Using the components of $\mathbf{n}$, we have

$$\begin{aligned} \mathbf{T}(x,\, y,\, z) &= \frac{1}{||\mathbf{n}||^2}\left(||\mathbf{n}||^2(x,\, y,\, z) - (ax+by+cz)(a,\, b,\, c)\right) \\ &= \frac{1}{||\mathbf{n}||^2}(||\mathbf{n}||^2x - a(ax+by+cz), ||\mathbf{n}||^2y - b(ax+by+cz), \\ &\quad ||\mathbf{n}||^2z - c(ax+by+cz)) \\ &= \frac{1}{||\mathbf{n}||^2}((b^2+c^2)x - aby - acz, -abx + (a^2+c^2)y - bcz, \\ &\quad -acx - bcy + (a^2+b^2)z) \\ &= \frac{1}{a^2+b^2+c^2}\begin{bmatrix} b^2+c^2 & -ab & -ac \\ -ab & a^2+c^2 & -bc \\ -ac & -bc & a^2+b^2 \end{bmatrix}\begin{bmatrix} x \\ y \\ z \end{bmatrix}. \end{aligned}$$

Thus the projection operation is linear. ♦

Example 2.4.5 Consider a dust cloud once more. This time, think of the particles as moving along straight lines emanating from the origin. As they move, their speed increases in proportion to the distance from the origin. Find an expression for the position of these dust particles after a time τ.

Solution The speed changes from point to point, so to find the new position, we must integrate. Let (x, y, z) be the position of a particle initially, let $\mathbf{x}(t) = \begin{bmatrix} x_1(t) \\ x_2(t) \\ x_3(t) \end{bmatrix}$ represent the position at subsequent times t, and let $\mathbf{T}(x, y, z)$ be the position at time τ. To say that the particles move outward from $\mathbf{0}$ with velocity proportional to distance from the origin means that the velocity is $k||\mathbf{x}(t)|| \cdot \mathbf{u}$, where k is some constant and $\mathbf{u}$ is a unit vector parallel to $\mathbf{x}(t)$. Since the velocity at time t is $\mathbf{x}'(t)$, this gives

$$\begin{aligned} \mathbf{x}'(t) &= k||\mathbf{x}(t)||\frac{\mathbf{x}(t)}{||\mathbf{x}(t)||} \\ \mathbf{x}'(t) &= k\,\mathbf{x}(t). \end{aligned} \tag{2.4.8}$$

See Figure 2.4.6. In terms of components, (2.4.8) says that

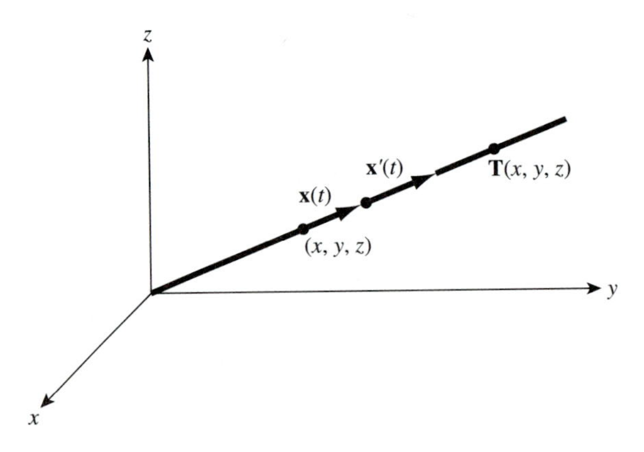

Figure 2.4.6 Dust particles moving outward from the origin with speed proportional to displacement

$$\begin{aligned} x_1'(t) &= kx_1(t) \\ x_2'(t) &= kx_2(t) \\ x_3'(t) &= kx_3(t). \end{aligned}$$

The general solutions to these three differential equations are multiples of exponential functions:

$$\begin{aligned} x_1(t) &= C_1e^{kt} \\ x_2(t) &= C_2e^{kt} \\ x_3(t) &= C_3e^{kt}, \end{aligned}$$

where C_1, C_2, and C_3 are constants. Since we are taking $x_1(0) = x$, $x_2(0) = y$, and $x_3(0) = z$, this gives $C_1 = x$, $C_2 = y$, and $C_3 = z$. Thus

$$\begin{aligned} x_1(t) &= xe^{kt} \\ x_2(t) &= ye^{kt} \\ x_3(t) &= ze^{kt}. \end{aligned}$$

After a time τ, the position of a dust particle initially with position $(x,\, y,\, z)$ will have position

$$\begin{aligned}\mathbf{T}(x,\, y,\, z) &= (xe^{k\tau},\, ye^{k\tau},\, ze^{k\tau}) \\ &= e^{k\tau}(x,\, y,\, z) \\ &= \begin{bmatrix} e^{k\tau} & 0 & 0 \\ 0 & e^{k\tau} & 0 \\ 0 & 0 & e^{k\tau} \end{bmatrix}\begin{bmatrix} x \\ y \\ z \end{bmatrix}.\end{aligned}$$

This is a linear transformation. ♦

Generally, a linear transformation $\mathbf{T}$ on $\mathbb{R}^n$ is called a **dilation** if $\mathbf{T}(\mathbf{x}) = r\,\mathbf{x}$ for some real number r. The preceding example captures some of the motivation for naming such transformations in this way. The transformation dilates the dust cloud by a factor of $e^{k\tau}$ in time τ.

A dilation is a special case of a **coordinate rescaling**, that is, a linear transformation of the form $\mathbf{T}(\mathbf{x}) = D\,\mathbf{x}$, where D is diagonal. We can think of a coordinate rescaling as a "directional dilation." For instance, if $\mathbf{T}(\mathbf{x}) = \begin{bmatrix} a & 0 & 0 \\ 0 & b & 0 \\ 0 & 0 & c \end{bmatrix}\mathbf{x}$ is applied to the unit ball $B = \{(x,\, y,\, z) \,|\, x^2 + y^2 + z^2 \leq 1\}$, then

$$\mathbf{T}(B) = \left\{(x,\, y,\, z) \,\middle|\, \frac{x^2}{a^2} + \frac{y^2}{b^2} + \frac{z^2}{c^2} \leq 1\right\}.$$

That is, B is stretched by a in the x-direction, by b in the y-direction, and c in the z-direction into an ellipsoid.

Exercises 2.4

For each of the following, (a) find the image of the indicated set under the given linear transformation, and (b) find the area or volume of the image.

1. P is the parallelogram in $\mathbb{R}^2$ with corner $(0,\, 0)$ spanned by $(1,\, -1)$ and $(4,\, 7)$; $\mathbf{T}(\mathbf{x}) = (3x_1 + 4x_2,\, 4x_1 + 5x_2)$.

2. P is the parallelogram in $\mathbb{R}^2$ with vertices $(-1,\, -1)$, $(-2,\, 2)$, $(-2,\, 4)$, and $(-1,\, 1)$; $\mathbf{T}(\mathbf{x}) = \left(-\frac{1}{4}x_1 + \frac{2}{3}x_2,\, \frac{2}{5}x_1 - \frac{1}{7}x_2\right)$.

3. P is the parallelepiped with corner $(1,\, 1,\, 1)$ spanned by $(1,\, 1,\, 2)$, $(1,\, 2,\, 1)$, and $(2,\, 1,\, 1)$; $\mathbf{T}(\mathbf{x}) = (3x_2,\, -4x_1,\, 5x_3)$.

4. P is the parallelepiped with vertices $(-5, 0, -1)$, $(-4, 1, 0)$, $(-5, 2, 0)$, $(-6, 1, -1)$, $(-5, 3, 1)$, $(-4, 4, 2)$, $(-5, 5, 2)$, and $(-6, 4, 1)$;

$$\mathbf{T}(\mathbf{x}) = \begin{bmatrix} 1/\sqrt{2} & -1/\sqrt{2} & 0 \\ 1/\sqrt{2} & 1/\sqrt{2} & 0 \\ 0 & 0 & 1 \end{bmatrix} \mathbf{x}.$$

5. The **triangle** in $\mathbb{R}^n$ with corner $\mathbf{x}_0$ spanned by $\mathbf{a} \neq \mathbf{0}$ and $\mathbf{b} \neq \mathbf{0}$ is the set

$$P = \{\mathbf{x} \in \mathbb{R}^n | \mathbf{x} = \mathbf{x}_0 + v(\mathbf{a} + u(\mathbf{b} - \mathbf{a})),\ 0 \leq u \leq 1,\ 0 \leq v \leq 1\}. \quad (2.4.9)$$

See Figure 2.4.7 for a three-dimensional motivation for this definition. By making the substitutions $s = v(1 - u)$ and $t = uv$, show that this definition is equivalent to

$$P = \{\mathbf{x} \in \mathbb{R}^n | \mathbf{x} = \mathbf{x}_0 + s\,\mathbf{a} + t\,\mathbf{b},\quad 0 \leq s,\quad 0 \leq t,\quad s + t \leq 1\}.$$

6. Show that the image of a triangle under an invertible linear transformation is another triangle.

7. Show that for a triangle P in $\mathbb{R}^2$ and a linear transformation $\mathbf{T}$ on $\mathbb{R}^2$,

$$\text{Area}(\mathbf{T}(P)) = |\det(A)|\,\text{Area}\,(P),$$

where A is the matrix of $\mathbf{T}$.

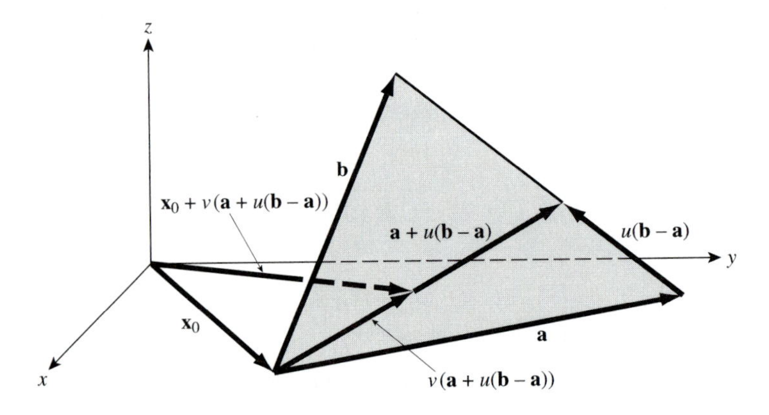

Figure 2.4.7 A triangle in $\mathbb{R}^3$

8. Find the area of the triangle with corner $(-1, -3)$ spanned by $(2, 5)$ and $(3, -2)$. Find the area of its image under $\mathbf{T}(\mathbf{x}) = \begin{bmatrix} 1 & -1 \\ 1 & 1 \end{bmatrix} \mathbf{x}$.

9. Find the area of the triangle with vertices $(7, 11)$, $(10, 4)$, and $(1, 2)$. Find the area of its image under $\mathbf{T}(\mathbf{x}) = \begin{bmatrix} 3 & 5 \\ -5 & 2 \end{bmatrix} \mathbf{x}$.

Find a formula for the linear transformation on $\mathbb{R}^3$ that represents the following geometric operations.

10. Dilation by a factor of 6.

11. Dilation by a factor of 2/3.

12. Counterclockwise rotation about the z-axis by $60°$.

13. Clockwise rotation about the z-axis by $45°$.

14. Clockwise rotation by $30°$ about the x-axis.

15. Projection onto the yz-plane.

16. Projection onto the plane $x + y + z = 0$.

17. Projection onto the plane $3x - 5z = 0$.

18. Counterclockwise rotation about the z-axis by $120°$ followed by projection onto the plane $2x + 3y + 4z = 0$.

19. Projection onto the xz-plane followed by counterclockwise rotation by $45°$ about the y-axis.

20. Particles of dust in the upper half-space $z > 0$ are moving parallel to the y-axis, each with constant speed proportional to its distance from the xy-plane. Find a formula for the position of a particle initially at (x, y, z) after τ time units have elapsed.

21. Describe the image of the unit square in $\mathbb{R}^2$ under the coordinate rescaling $\mathbf{T}(\mathbf{x}) = \begin{bmatrix} 3 & 0 \\ 0 & 2/3 \end{bmatrix} \mathbf{x}$.

22. Describe the image of the unit cube in $\mathbb{R}^3$ under

$$\mathbf{T}(\mathbf{x}) = \begin{bmatrix} 4 & 0 & 0 \\ 0 & -1/2 & 0 \\ 0 & 0 & -1/3 \end{bmatrix} \mathbf{x}.$$

23. Describe the image of the unit sphere under the transformation in Exercise 22.

24. Given mutually orthogonal unit vectors $\mathbf{v}_1, \mathbf{v}_2, \mathbf{v}_3, \ldots, \mathbf{v}_n$ in $\mathbb{R}^n$, let $\mathbf{T}(\mathbf{x})$ be the vector with entries the components of $\mathbf{x}$ in the directions of $\mathbf{v}_1, \mathbf{v}_2, \mathbf{v}_3, \ldots, \mathbf{v}_n$ respectively. See Exercise 14 in Section 2.1. Show that the matrix of $\mathbf{T}$ is

$$A = \begin{bmatrix} -- & \mathbf{v}_1 & -- \\ -- & \mathbf{v}_2 & -- \\ & \vdots & \\ -- & \mathbf{v}_n & -- \end{bmatrix}.$$

Such a transformation (and matrix) are called **orthogonal.**

25. Show that if A is orthogonal, then $A^{-1} = A^T$.

26. Show that if A is orthogonal and D is diagonal, then ADA^T is symmetric. (The converse of this is also true but much harder to prove: If B is symmetric, then there exists an orthogonal matrix A and a diagonal matrix D such that $B = ADA^T$. This says that if $\mathbf{T}(\mathbf{x}) = B\mathbf{x}$ where B is symmetric, then there is a coordinate system $\mathbf{v}_1, \mathbf{v}_2, \mathbf{v}_3, \ldots, \mathbf{v}_n$ such that $\mathbf{T}$ is a coordinate rescaling with respect to that system. Consult a linear algebra text for more details.)

27. Let

$$\mathbf{T}(\mathbf{x}) = \begin{bmatrix} 1/2 & \sqrt{3}/2 & 0 \\ -\sqrt{3}/2 & 1/2 & 0 \\ 0 & 0 & 1 \end{bmatrix} \begin{bmatrix} 2 & 0 & 0 \\ 0 & -4 & 0 \\ 0 & 0 & 6 \end{bmatrix} \begin{bmatrix} 1/2 & -\sqrt{3}/2 & 0 \\ \sqrt{3}/2 & 1/2 & 0 \\ 0 & 0 & 1 \end{bmatrix} \mathbf{x}$$

Describe geometrically the effect of applying $\mathbf{T}$.

28. Show that for any skew-symmetric 3×3 matrix A, there is a vector $\mathbf{a} \in \mathbb{R}^3$ such that the linear transformation $\mathbf{T}(\mathbf{x}) = A\mathbf{x}$ is the composition of

projection onto the plane $\mathbf{a} \cdot \mathbf{x} = 0$, followed by
rotation by 90° about $\mathbf{a}$, followed by
dilation by $\pm||\mathbf{a}||$.

(Hint: see Exercise 33 in Section 2.2.)

29. Show that for any nonzero skew-symmetric 2×2 matrix A, the transformation $\mathbf{T}(\mathbf{x}) = A\mathbf{x}$ is the composition of a counterclockwise rotation by 90° followed by a dilation (by a positive or negative factor).

2.5 Quadratic Forms

The purpose of this section is to examine a special category of real-valued functions that play an important role in understanding calculus of functions of several variables. **Quadratic forms** are functions that generalize the simplest single-variable quadratic functions $f(x) = kx^2$, where k is a constant. In rough terms, quadratic forms are to kx^2 what linear transformations are to kx.

Definition 2.5.1 A **quadratic form** is a function $p : \mathbb{R}^n \to \mathbb{R}$ of the form

$$\begin{array}{l} p(x_1, x_2, \ldots, x_n) = \\ \begin{array}{ccccccccc} c_{11}x_1^2 & + & c_{12}x_1x_2 & + & \ldots & + & c_{1,n-1}x_1x_{n-1} & + & c_{1n}x_1x_n \\ & & c_{22}x_2^2 & + & \ldots & + & c_{2,n-1}x_2x_{n-1} & + & c_{2n}x_2x_n \\ & & & & \ddots & & & & \vdots \\ & & & & & + & c_{n-1,n-1}x_{n-1}^2 & + & c_{n-1,n}x_{n-1}x_n \\ & & & & & & & + & c_{nn}x_n^2 \end{array} \end{array} \quad (2.5.1)$$

In other words, a quadratic form is a second degree polynomial in which all the terms have degree two.

Example 2.5.1 The functions $p(x, y) = x^2 - y^2$, $q(x, y, z) = x^2 + 3xy - 2xz + y^2 + yz + 6z^2$, and $r(x_1, x_2, x_3, x_4) = x_1^2 + x_2^2 + x_3^2 + x_4^2$ are all quadratic forms. The functions $s(x, y) = x - 3y^3$, and $t(x, y) = x\sqrt{y} + xy + y^2$ are not. ♦

Notice that the function in (2.5.1) can be written as

$$p(x_1, x_2, \ldots, x_n) = \begin{bmatrix} x_1 & x_2 & \cdots & x_n \end{bmatrix} U \begin{bmatrix} x_1 \\ x_2 \\ \vdots \\ x_n \end{bmatrix} \quad (2.5.2)$$

where U is the **upper triangular matrix** given by

$$U = \begin{bmatrix} c_{11} & c_{12} & \cdots & c_{1n} \\ 0 & c_{22} & \cdots & c_{2n} \\ \vdots & & \ddots & \vdots \\ 0 & 0 & \cdots & c_{nn} \end{bmatrix}.$$

If we let $\mathbf{x} = (x_1, x_2, \ldots, x_n)$ then (2.5.2) can be written even more succinctly as $p(\mathbf{x}) = \mathbf{x}^T U \mathbf{x}$.

If $A = [a_{ij}]$ is any $n \times n$ matrix, the function $q : \mathbb{R}^n \to \mathbb{R}$ defined by

$$q(\mathbf{x}) = \mathbf{x}^T A\mathbf{x} \tag{2.5.3}$$

is a quadratic form. To see this, we just write (2.5.3) in terms of components:

$$\begin{aligned}
\mathbf{x}^T A\mathbf{x} &= \begin{bmatrix} x_1 & x_2 & \cdots & x_n \end{bmatrix} A \begin{bmatrix} x_1 \\ x_2 \\ \vdots \\ x_n \end{bmatrix} \\
&= \begin{bmatrix} x_1 & x_2 & \cdots & x_n \end{bmatrix} \begin{bmatrix} \sum_{k=1}^n a_{1k}x_k \\ \sum_{k=1}^n a_{2k}x_k \\ \vdots \\ \sum_{k=1}^n a_{nk}x_k \end{bmatrix} \\
&= x_1 \sum_{k=1}^n a_{1k}x_k + x_2 \sum_{k=1}^n a_{2k}x_k + \ldots + x_n \sum_{k=1}^n a_{nk}x_k \\
&= \sum_{k=1}^n (a_{1k}x_1x_k + a_{2k}x_2x_k + \ldots + a_{nk}x_nx_k).
\end{aligned}$$

Each term in this last sum is of degree 2 in the variables $x_1, x_2, \ldots, x_n$.

There are many different matrices that can represent a quadratic form, but it turns out that there is only one *symmetric matrix* that represents a given quadratic form. The general quadratic form in (2.5.1) can be written as

$$p(\mathbf{x}) = \mathbf{x}^T S\mathbf{x}$$

where S is the symmetric matrix

$$S = \begin{bmatrix} c_{11} & c_{12}/2 & c_{13}/2 & \cdots & c_{1n}/2 \\ c_{12}/2 & c_{22} & c_{23}/2 & \cdots & c_{2n}/2 \\ c_{13}/2 & c_{23}/2 & c_{33} & \cdots & c_{3n}/2 \\ \vdots & & & \ddots & \vdots \\ c_{1n}/2 & c_{2n}/2 & c_{3n}/2 & & c_{nn} \end{bmatrix}.$$

If T is any other symmetric matrix such that $p(\mathbf{x}) = \mathbf{x}^T T\mathbf{x}$ then it can be shown that $S = T$.

Theorem 2.5.1 *For any quadratic form p there exists a unique symmetric matrix S such that*

$$p(\mathbf{x}) = \mathbf{x}^T S\mathbf{x}. \tag{2.5.4}$$

Example 2.5.2 Write $p(x, y, z) = 4x^2 - 2xy + 6xz + z^2$ in the form of Equation 2.5.4.

Solution The diagonal entries of S are the coefficients on the squared terms; the off-diagonal entries are half the coefficients in the mixed terms:

$$p(x, y, z) = \begin{bmatrix} x & y & z \end{bmatrix} \begin{bmatrix} 4 & -1 & 3 \\ -1 & 0 & 0 \\ 3 & 0 & 1 \end{bmatrix} \begin{bmatrix} x \\ y \\ z \end{bmatrix}. \quad \blacklozenge$$

Definition 2.5.2 A quadratic form p is said to be **positive definite** if $p(\mathbf{0}) = 0$ and $p(\mathbf{x}) > 0$ for all $\mathbf{x} \neq \mathbf{0}$; p is said to be **negative definite** if $-p$ is positive definite; p is said to be **indefinite** if there are points $\mathbf{x}_1 \neq \mathbf{0}$ and $\mathbf{x}_2 \neq \mathbf{0}$ such that $p(\mathbf{x}_1) > 0$ and $p(\mathbf{x}_2) < 0$.

In Chapter 3, it will be important for us to be able to determine whether a quadratic form is positive or negative definite. In a few moments we will see how to tell this by examining the symmetric matrix that represents the form. But first let's consider an example.

Example 2.5.3 The quadratic form

$$p(x, y) = Ax^2 + By^2 \tag{2.5.5}$$

is positive definite for $A > 0$ and $B > 0$: $p(0, 0) = 0$, and for $(x, y) \neq (0, 0)$, the values of p are positive. The graph of p in this case is an elliptic paraboloid that opens "upward" along the positive z-axis.

When $A < 0$ and $B < 0$ in (2.5.5), the form is negative definite because $-p(x, y) = -(Ax^2 + By^2) = (-A)x^2 + (-B)y^2$ is positive definite. The graph of p in this case is a down-turned elliptic paraboloid.

When A and B differ in sign, the form in (2.5.5) is indefinite. For instance, if $A < 0$ and $B > 0$ then $p(0, y) = By^2 > 0$ for all $y \neq 0$ and $p(x, 0) = Ax^2 < 0$ for all $x \neq 0$. The graph in this case is a hyperbolic paraboloid. $\blacklozenge$

Theorem 2.5.2 *(Sylvester's[3] Theorem) Let p be a quadratic form given by $p(\mathbf{x}) = \mathbf{x}^T A\mathbf{x}$, where $A = [a_{ij}]$ is an $n \times n$ symmetric matrix.*

---

[3]This is the same J. J. Sylvester who coined the term *matrix*.

1. *If* $a_{11} > 0$, $\begin{vmatrix} a_{11} & a_{12} \\ a_{21} & a_{22} \end{vmatrix} > 0$,

$$\begin{vmatrix} a_{11} & a_{12} & a_{13} \\ a_{21} & a_{22} & a_{23} \\ a_{31} & a_{32} & a_{33} \end{vmatrix} > 0, \cdots, \begin{vmatrix} a_{11} & a_{12} & \cdots & a_{1n} \\ a_{21} & a_{22} & \cdots & a_{2n} \\ \vdots & & & \vdots \\ a_{n1} & a_{n2} & \cdots & a_{nn} \end{vmatrix} > 0,$$

then p *is positive definite.*

2. *If* $a_{11} < 0$, $\begin{vmatrix} a_{11} & a_{12} \\ a_{21} & a_{22} \end{vmatrix} > 0$,

$$\begin{vmatrix} a_{11} & a_{12} & a_{13} \\ a_{21} & a_{22} & a_{23} \\ a_{31} & a_{32} & a_{33} \end{vmatrix} < 0, \cdots, (-1)^n \begin{vmatrix} a_{11} & a_{12} & \cdots & a_{1n} \\ a_{21} & a_{22} & \cdots & a_{2n} \\ \vdots & & & \vdots \\ a_{n1} & a_{n2} & \cdots & a_{nn} \end{vmatrix} > 0,$$

then p *is negative definite.*

3. *If all the determinants in (1) and (2) are nonzero, and they are neither all positive nor alternating in sign as in (2), then* p *is indefinite.*

A proof of Theorem 2.5.2 is beyond the scope of this book, but you can look one up in many linear algebra texts. It is possible to get some insight in the two-dimensional case without much trouble. See the Exercises.

If one or more of the determinants in Sylvester's Theorem is zero, it can be shown that the quadratic form is neither positive definite nor negative definite. In this situation, it *may* be indefinite, but there are other possibilities. It may be **positive semidefinite**—p is nonnegative but equal to zero at points other than the origin—or **negative semidefinite**—p is nonpositive but equal to zero at points other than the origin. For instance $p(x, y) = 5x^2$ is positive semidefinite.

Example 2.5.4 Determine the definiteness of the quadratic form $p(x, y, z) = x^2 + 5y^2 - 10z^2 + 6xy - 4xz + 2yz$.

Solution The symmetric matrix that represents p is $A = \begin{bmatrix} 1 & 3 & -2 \\ 3 & 5 & 1 \\ -2 & 1 & -10 \end{bmatrix}$.

Since $a_{11} = 1 > 0$, $\begin{vmatrix} a_{11} & a_{12} \\ a_{21} & a_{22} \end{vmatrix} = \begin{vmatrix} 1 & 3 \\ 3 & 5 \end{vmatrix} = -4 < 0$, and $\det(A) = 7 > 0$, by Theorem 2.5.2, the form is indefinite. ♦

Example 2.5.5 Determine the definiteness of the quadratic form $p(x,\ y,\ z) = -x^2 - 4y^2 - 3z^2 + 2xy - 2xz$.

Solution The symmetric matrix that represents the quadratic form is $A = \begin{bmatrix} -1 & 1 & -1 \\ 1 & -4 & 0 \\ -1 & 0 & -3 \end{bmatrix}$. Since $a_{11} = -1 < 0$, $\begin{vmatrix} a_{11} & a_{12} \\ a_{21} & a_{22} \end{vmatrix} = \begin{vmatrix} -1 & 1 \\ 1 & -4 \end{vmatrix} = 3 > 0$, and $\det(A) = -5 < 0$, by Theorem 2.5.2, the form is negative definite. ♦

Exercises 2.5

In Exercises 1 through 5, determine the symmetric matrix that represents the quadratic form.

1. $p(x, y) = -\frac{2}{3}x^2 + \frac{5}{3}y^2 + \frac{4}{3}xy$

2. $q(x, y) = x^2 + y^2 - z^2 + 5xy - 3yz$

3. $r(x_1,\ x_2,\ x_3,\ x_4) = x_3^2 - x_2x_3 + x_1x_4$

4. $s(\mathbf{x}) = ||\mathbf{x}||^2,\ \mathbf{x} \in \mathbb{R}^n$

5. $t(\mathbf{x}) = \mathbf{x}^T \begin{bmatrix} 6 & 1 & 8 & -2 \\ 0 & 5 & 1 & 9 \\ 0 & 0 & -2 & 0 \\ 0 & 0 & 0 & 0 \end{bmatrix} \mathbf{x}$

Use Theorem 2.5.2 to categorize the quadratic forms in Exercises 6 through 13 as positive definite, negative definite, indefinite, or none of these.

6. $p(x,\ y) = \begin{bmatrix} x & y \end{bmatrix} \begin{bmatrix} 3 & -4 \\ -4 & 6 \end{bmatrix} \begin{bmatrix} x \\ y \end{bmatrix}$

7. $p(x,\ y) = \begin{bmatrix} x & y \end{bmatrix} \begin{bmatrix} -6 & 2 \\ 2 & -3 \end{bmatrix} \begin{bmatrix} x \\ y \end{bmatrix}$

8. $p(x,\ y) = 7x^2 + 14xy + 7y^2$

9. $q(x,\ y,\ z) = x^2 + 4y^2 + 9z^2 + 2xy + 2xz$

10. $q(x,\ y,\ z) = \begin{bmatrix} x & y & z \end{bmatrix} \begin{bmatrix} 12 & 5 & 4 \\ 5 & 3 & 3 \\ 4 & 3 & 4 \end{bmatrix} \begin{bmatrix} x \\ y \\ z \end{bmatrix}$

11. $q(x, y, z) = \begin{bmatrix} x & y & z \end{bmatrix} \begin{bmatrix} 1 & 0 & 0 \\ 0 & -1 & 0 \\ 0 & 0 & -10 \end{bmatrix} \begin{bmatrix} x \\ y \\ z \end{bmatrix}$

12. $r(\mathbf{x}) = \mathbf{x}^T \begin{bmatrix} -1 & -1 & 0 & 0 \\ -1 & -2 & 0 & 0 \\ 0 & 0 & -4 & 0 \\ 0 & 0 & 0 & -8 \end{bmatrix} \mathbf{x}$

13. $r(\mathbf{x}) = \mathbf{x}^T \begin{bmatrix} 0 & 0 & 0 & 1 \\ 0 & 0 & 1 & 0 \\ 0 & 1 & 0 & 0 \\ 1 & 0 & 0 & 0 \end{bmatrix} \mathbf{x}$

14. Provided that $b^2 - 4ac \neq 0$, show that $p(x, y) = ax^2 + bxy + cy^2 + dx + ey + f$ can be written as $q(x - h, y - k) + l$, where q is a quadratic form.

15. Let $A = \begin{bmatrix} a & b \\ b & c \end{bmatrix}$ be a symmetric 2×2 matrix. Show that if $a > 0$ and $\det(A) > 0$, then the quadratic form $p(\mathbf{x}) = \mathbf{x}^T A\mathbf{x}$ is positive definite. (*Hint*: Write out p in terms of the variables x and y and complete the square.)

16. Show that if a, b, and c are all nonzero and $A = \begin{bmatrix} 0 & a & b \\ a & 0 & c \\ b & c & 0 \end{bmatrix}$, then $p(\mathbf{x}) = \mathbf{x}^T A\mathbf{x}$ is indefinite.

17. Show that if A is a $n \times n$ symmetric matrix such that $\operatorname{tr}(A) = 0$, then $p(\mathbf{x}) = \mathbf{x}^T A\mathbf{x}$ is indefinite.

18. Find conditions on a and b so that the quadratic form

$$p(\mathbf{x}) = \mathbf{x}^T \begin{bmatrix} -1 & 1 & b \\ 1 & -4 & 0 \\ b & 0 & a \end{bmatrix} \mathbf{x}$$

is negative definite.

Chapter 3

Differentiation

In this chapter we are concerned with measuring the way in which a function of several variables changes as those variables change. In single-variable calculus, we know that the derivative provides such a measure. Differentiation of functions in the multivariable context is analogous to that of single-variable functions.

Remember from introductory calculus that if f is a given real-valued function of one variable, then the derivative of f at a number a is the number we denote by $f'(a)$ or $Df(a)$ that is calculated by the formula

$$f'(a) = \lim_{x \to a} \frac{f(x) - f(a)}{x - a}$$

if the limit exists. This number, you learned, can be interpreted in various ways. For instance, it represents the instantaneous rate of change in f at a; when f is the position of an object that moves along a straight line and x is time then $f'(a)$ is the object's velocity at time a.

The derivative of f at a can also represent the slope of the line tangent to the graph of f at the point $(a, f(a))$. With this interpretation, the derivative of f at a gives rise to a new *function*, sometimes called the **linear approximation** of f at a, whose graph is the line through $(a, f(a))$ that most closely resembles the graph of f very near this point. The linear approximation is the **affine** function $L(x) = f(a) + f'(a)(x - a)$.

The latter view of the derivative is one that generalizes to functions that depend on more than one variable. Instead of a single number to represent the derivative of a function $\mathbf{f}$ at a point $\mathbf{a}$, we will have a *matrix*, called the **Jacobian matrix**,that represents the "total derivative" of $\mathbf{f}$ at $\mathbf{a}$. Each entry in the matrix will be a "partial derivative" of one component of the function of interest. Both partial and total derivatives will be defined in terms of a limiting process analogous to that for calculating ordinary derivatives.

We start this chapter by studying functions and ways of visualizing them (when that is possible): graphs, level sets, and vector fields on $\mathbb{R}^2$ and $\mathbb{R}^3$. Then we establish how to "take a limit" of a function of several variables. This provides us with language we can use to define the partial and total derivative of a function of more than one variable. In examples we will see that some questions in the sciences can be phrased in terms of partial derivatives and that they can sometimes be answered by finding solutions to partial differential equations. We will examine other concepts of derivative, such as gradient, directional derivative, divergence, and curl: how they arise naturally in physical situations and how they are related to the total derivative. Finally, we focus on real-valued functions of several variables and use a multivariable second derivative to determine when such functions attain local maximum or local minimum values.

3.1 Graphs, Level Sets, and Vector Fields: Geometry

In Section 1.2 we discussed functions of two variables. Such functions associate an ordered pair (or point) (x, y) in $\mathbb{R}^2$ with a number (or point) in $\mathbb{R}$. The function f can be visualized by its graph, the set of points $(x,\ y,\ z)$ such that $z = f(x,\ y)$. Similarly, in Section 1.8 we defined a vector-valued function of one variable. Such a function associates a number in $\mathbb{R}$ with an ordered triple in $\mathbb{R}^3$. We visualized one of these functions as the curve in $\mathbb{R}^3$ that it parametrized. In Sections 2.3 and 2.4 we encountered a specialized type of vector-valued function called a linear transformation. These are specific cases of a general idea: functions from $\mathbb{R}^n$ to $\mathbb{R}^m$ where n and m are any positive integers.

When n and m have values such as 2 and 2, there are pictures to be drawn and physical interpretations to give. For other values of n and m, pictures are not possible, but analogies to the lower-dimensional pictures are. In this section we talk about the general case but concentrate more on specific pairings of n and m: 2 and 1, 2 and 2, 2 and 3, 3 and 1, and 3 and 3.

Definition 3.1.1 **function** from a subset D of $\mathbb{R}^n$ to $\mathbb{R}^m$ is a rule $\mathbf{f}$ by which each vector in D is associated with exactly one vector in $\mathbb{R}^m$. The set D is its **domain**. We sometimes say that **f maps**[1] D into $\mathbb{R}^m$, and we refer to $\mathbf{f}(D) = \{\mathbf{f}(\mathbf{x}) | \mathbf{x} \in D\}$ as the **range** of **f**. For short, we write $\mathbf{f} : D \to \mathbb{R}^m$,

[1]The analogy with ordinary road maps is strong: A map of the United States represents a function that has domain the actual countryside and range the xy-plane.

or $\mathbf{f} : D \subset \mathbb{R}^n \to \mathbb{R}^m$. Whenever $m > 1$, we write a boldface $\mathbf{f}$ (f with an arrow overhead by hand, like this: $\vec{f}$), and when $m = 1$, we write an ordinary italic f.

Example 3.1.1 By writing $f(x, y) = x^2y^3 + x^3y^2$, we define a function f from $\mathbb{R}^2$ to $\mathbb{R}$. Its domain is all of $\mathbb{R}^2$ since there are no restrictions on the x- and y-values that can be substituted into the formula on the right. Strictly speaking, instead of writing "$f(x, y)$," we should write "$f((x, y))$" or "$f\left(\begin{bmatrix} x \\ y \end{bmatrix}\right)$," but it is clear that the extra parentheses or brackets would be redundant, so they are omitted. ♦

Example 3.1.2 By writing $\mathbf{f}(x_1,\, x_2,\, x_3) = (x_1/x_2,\, x_1/x_3,\, x_2/x_3)$ we define a function from $\mathbb{R}^3$ to $\mathbb{R}^3$. The formula for $\mathbf{f}$ could also be written as

$$\mathbf{f}\left(\begin{bmatrix} x_1 \\ x_2 \\ x_3 \end{bmatrix}\right) = \begin{bmatrix} x_1/x_2 \\ x_1/x_3 \\ x_2/x_3 \end{bmatrix}.$$

The domain of this function is the set of all points (x_1, x_2, x_3) such that $x_2 \neq 0$ and $x_3 \neq 0$. Since the set of points for which $x_2 = 0$ or $x_3 = 0$ consists of the x_1x_3-plane and the x_1x_2-plane, this domain can be described geometrically as all points that do not lie on either of these planes. ♦

Example 3.1.3 Find the domain of the function $\mathbf{f}$ given by

$$\mathbf{f}(x,\, y) = \left(\frac{1}{\sqrt{4 - x^2 - y^2}},\, \frac{1}{\sqrt{x^2 + y^2 - 1}},\, \frac{1}{x^2 + y^2}\right).$$

Solution This function maps $\mathbb{R}^2$ to $\mathbb{R}^3$, so its domain is a subset of $\mathbb{R}^2$. The variables must satisfy $4 - x^2 - y^2 > 0$, $x^2 + y^2 - 1 > 0$, and $x^2 + y^2 > 0$. The first two inequalities require $(x,\, y)$ to lie inside the circle of radius 2 centered at the origin and outside the circle of radius 1 centered at the origin. Any point that satisfies the second inequality automatically satisfies the third, so the domain is

$$D = \{(x,\, y) \,|\, x^2 + y^2 < 4,\ 1 < x^2 + y^2\}. \quad ♦$$

See Figure 3.1.1.

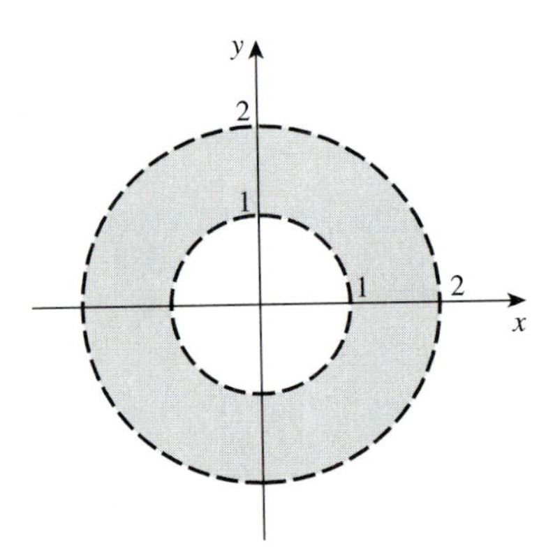

Figure 3.1.1 The domain of the function in Example 3.1.3

Definition 3.1.2 The **graph** of a function $f : D \subset \mathbb{R}^n \to \mathbb{R}$ is the set of points $(x_1, x_2, \ldots, x_n, x_{n+1})$ in $\mathbb{R}^{n+1}$ such that $x_{n+1} = f(x_1, \ldots, x_n)$. More generally, the graph of a function $\mathbf{f} : \mathbb{R}^n \to \mathbb{R}^m$ is the set of points

$$(x_1, x_2, \ldots, x_n, x_{n+1}, x_{n+2}, \ldots x_{n+m})$$

in $\mathbb{R}^{n+m}$ such that

$$(x_{n+1}, x_{n+2}, \ldots, x_{n+m}) = \mathbf{f}(x_1, x_2, \ldots, x_n).$$

We have already encountered graphs of functions from $\mathbb{R}^2$ to $\mathbb{R}$ in Chapter 1. They are "surfaces" in $\mathbb{R}^3$, usually two-dimensional subsets that pass the "vertical line test." The graph of a function from, for example,$\mathbb{R}^3$ to $\mathbb{R}$, would require our plotting points in $\mathbb{R}^4$, a task that is beyond the abilities of textbook writers, students, and other mortals. We can visualize the graph of a function $\mathbf{f}$ from $\mathbb{R}$ to $\mathbb{R}^2$: This is the set of points (x, y, z) in $\mathbb{R}^3$ such that $(y, z) = \mathbf{f}(x)$.

Example 3.1.4 The graph of $\mathbf{f} : \mathbb{R} \to \mathbb{R}^2$, where $\mathbf{f}$ is given by

$$\mathbf{f}(x) = (e^{-x} \cos 2\pi x, e^{-x} \sin 2\pi x),$$

consists of the points (x, y, z) in $\mathbb{R}^3$ such that

$$(y, z) = (e^{-x} \cos 2\pi x, e^{-x} \sin 2\pi x).$$

From what we saw in Section 1.8, the graph of this function is a helical curve that winds about the x-axis more and more tightly as x increases.

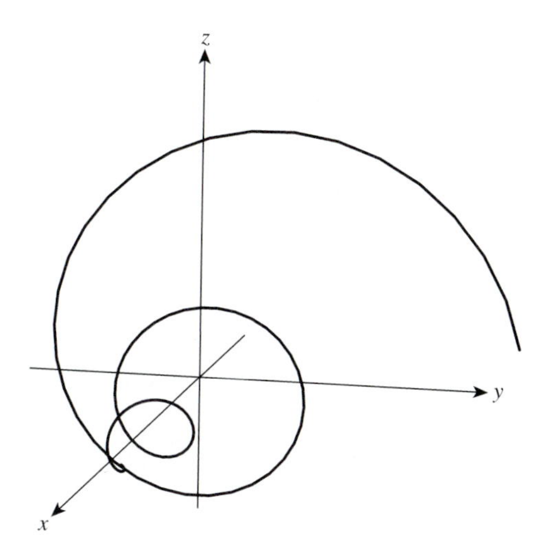

Figure 3.1.2 Graph of $\mathbf{f}(x) = (e^{-x}\cos 2\pi x,\ e^{-x}\sin 2\pi x)$

The computer-generated image in Figure 3.1.2 gives a reasonable idea of its shape for values of x between -1 and 2.

Note that the graph of this function is different from the curve in $\mathbb{R}^2$ parametrized by $\mathbf{f}$. What is the connection between the two? ♦

To visualize the graph of a function from $\mathbb{R}^2$ to $\mathbb{R}^2$ we would need $2+2=4$ spatial dimensions. Here again we run up against the limitations of drawing pictures. Indeed, obtaining graphs of functions from $\mathbb{R}$ into $\mathbb{R}^m$ with $m > 2$ or from $\mathbb{R}^n$ to $\mathbb{R}^m$ with $n > 2$ is a hopeless task. But there are ways of studying the geometry of functions besides graphing them, and in fact these other techniques often reveal information that would not be obvious from a graph.

Definition 3.1.3 For a function $f : \mathbb{R}^n \to \mathbb{R}$ and a constant $c \in \mathbb{R}$, the set of points $(x_1, x_2, \ldots, x_n) \in \mathbb{R}^n$ that satisfy the equation $f(x_1, x_2, \ldots, x_n) = c$ is called a **level set of** f.

In particular, the level sets of a function f from $\mathbb{R}^2$ to $\mathbb{R}$ consist of points (x, y) in $\mathbb{R}^2$ that satisfy $f(x, y) = c$. For many such functions, the level set is a curve in the xy-plane.

To obtain a level curve from the graph of f, think of intersecting the graph of f [the surface in $\mathbb{R}^3$ given by $z = f(x, y)$] with the horizontal plane $z = c$; this is often a curve lying in the plane $z = c$. Then project this curve into the xy-plane and forget about the z-axis. See Figure 3.1.3. Conversely,

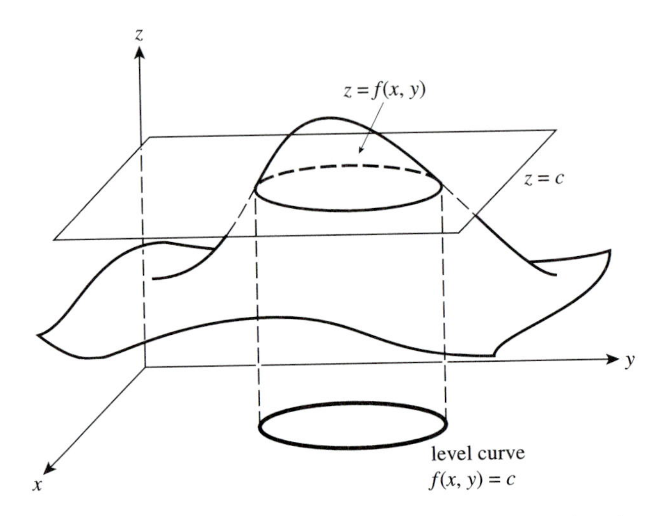

Figure 3.1.3 A level curve $f(x, y) = c$ is the projection into the xy-plane of the intersection of the surface $z = f(x, y)$ and the plane $z = c$.

given several level curves for a function, information about the graph of the function can be obtained by reversing these steps.

Example 3.1.5 Discuss and sketch some of the level sets of $f(x, y) = x^2 - y^2$.

Solution We consider the c-values 0, 1, -1, 2, and -2. This will be enough to reveal some of the general features of the level sets.

When $c = 0$, the level set consists of (x, y) such that $x^2 - y^2 = 0$. This is equivalent to $(x - y)(x + y) = 0$, so the level set is all points (x, y) such that either $x = y$ or $x = -y$ (i.e., the two lines of slope $+1$ and -1 that intersect at the origin). See Figure 3.1.4.

When $c = 1$, the level set is the graph of the equation $x^2 - y^2 = 1$, a hyperbola that has vertices (1, 0) and $(-1, 0)$ and asymptotes the lines $y = x$ and $y = -x$. Similarly, when $c = -1$, the level set is the graph of $y^2 - x^2 = 1$, a hyperbola with vertices (0, 1) and $(0, -1)$ and asymptotes the same two lines.

When $c = 2$, the level set is the graph of $x^2/2 - y^2/2 = 1$, a hyperbola with vertices $(\sqrt{2}, 0)$ and $(-\sqrt{2}, 0)$ and asymptotes $y = x$ and $y = -x$. Also, when $c = -2$ we obtain a hyperbola with the same asymptotes and with vertices $(0, \sqrt{2})$ and $(0, -\sqrt{2})$. Again, see Figure 3.1.4. For c-values going from negative values to zero, the level curves are hyperbolas with vertices on the y-axis that get closer and closer to the origin, and for c increasing from

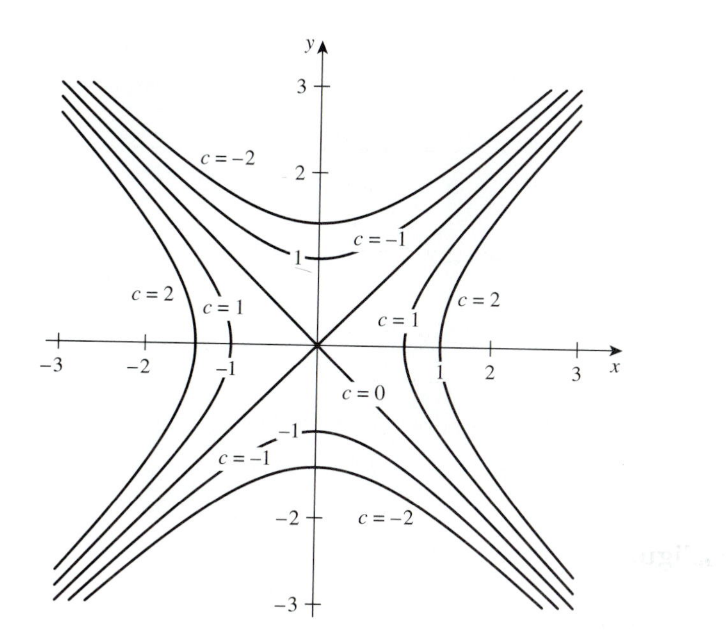

Figure 3.1.4 Level curves of $f(x, y) = x^2 - y^2$

zero to positive values, the level curves go from the degenerate hyperbola of two intersecting lines to hyperbolas with vertices on the x-axis. ♦

Example 3.1.6 Discuss the graph of the function whose level curves are shown in Figure 3.1.5. (The darker shading indicates smaller c-values, and lighter indicates larger c-values.)

Solution The closed curves encircling the point $(-2, 3)$ corresponding to smaller values of c indicate that for (x, y) on these curves, $f(x, y)$ is smaller. Thus the graph of f has a "pit" centered at $(-2, 3)$. Similarly, for (x, y) near $(0, 0)$ the c-values are larger, so there is a peak at $(0, 0)$. Near $(1, 2)$ the c-values are even larger than those near $(0, 0)$, so there is another peak at $(1, 2)$ that is higher than the one at $(0, 0)$.

Of course this example is rigged (as are most examples in mathematics texts), so that it is easy to show the graph of the function. See Figure 3.1.6 for the graph. ♦

A level set of $f : \mathbb{R}^3 \to \mathbb{R}$ is the graph of the equation $f(x, y, z) = c$, and from our experience in Section 1.2 we know that such a graph is often a **surface**.

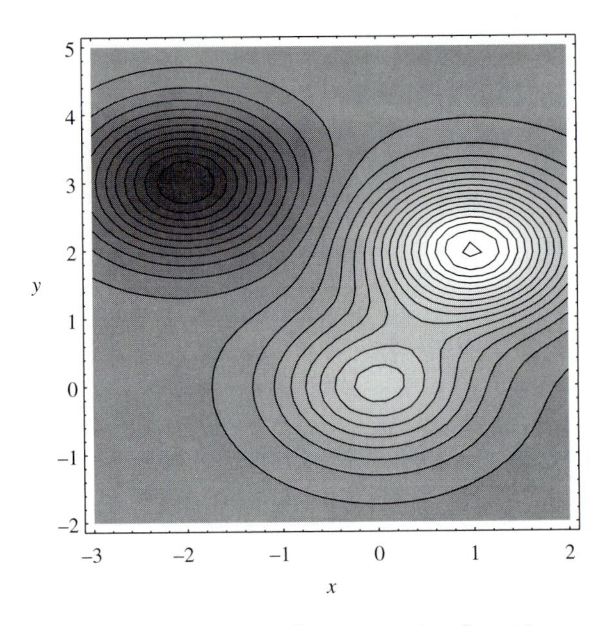

Figure 3.1.5 Level curves of a function

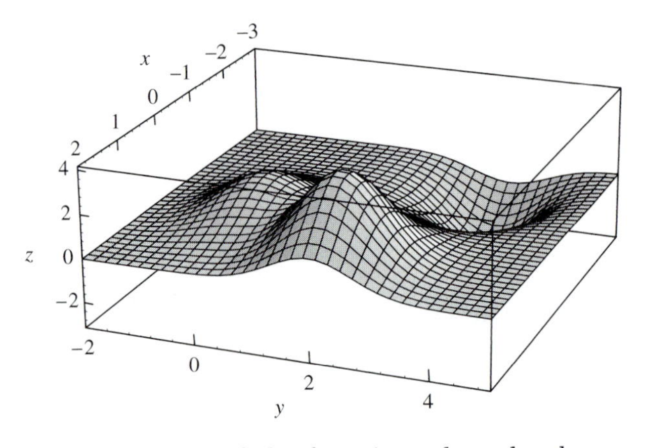

Figure 3.1.6 Graph of the function whose level curves are shown in Figure 3.1.5

Example 3.1.7 Suppose that the function $T(x, y, z) = 100/(1 + x^2 + y^2 + z^2)$ represents the temperature of air at the point (x, y, z). Describe the **isotherms** (surfaces on which the temperature is constant.)

Solution The isotherms are just the level sets of T, the graphs of the equations $100/(1 + x^2 + y^2 + z^2) = c$ for various values of c. We can rewrite this as

$$x^2 + y^2 + z^2 = \frac{100}{c} - 1$$

to see that the isotherms are the spheres with radii $\sqrt{100/c - 1}$ and center $(0, 0, 0)$. The quantity under the radical must be nonnegative, so c must be less than or equal to 100. Such a temperature distribution is qualitatively reasonable: It might model the situation in which a small object at the origin with temperature 100 heats the surrounding air. The temperature of nearby air is close to 100, and air near points (x, y, z) far from the origin will have temperature close to zero. ♦

Definition 3.1.4 A function **f** that maps a subset D of $\mathbb{R}^n$ *back into* $\mathbb{R}^n$ is called a **vector field on** $\mathbb{R}^n$.

The reason for this terminology is that physical phenomena such as gravitational, electrical, and magnetic fields can be regarded as functions that assign a vector (typically, a force or velocity) to points in space. Vector fields on $\mathbb{R}^2$ and $\mathbb{R}^3$ can be studied graphically. Select several points in the underlying space and for each such point, draw the vector which is the value of the field at that point so that its tail lies on the point.

Example 3.1.8 Discuss the vector field $\mathbf{f} : \mathbb{R}^2 \to \mathbb{R}^2$ given by $\mathbf{f}(x, y) = (-y, x)$.

y					
1	$(-1, 0)$	$(-1, 1/3)$	$(-1, 2/3)$	$(-1, 1)$	
2/3	$(-2/3, 0)$	$(-2/3, 1/3)$	$(-2/3, 2/3)$	$(-2/3, 1)$	
1/3	$(-1/3, 0)$	$(-1/3, 1/3)$	$(-1/3, 2/3)$	$(-1/3, 1)$	
0	$(0, 0)$	$(0, 1/3)$	$(0, 2/3)$	$(0, 1)$	
	0	1/3	2/3	1	x

Table 3.1.1 Sixteen values of the vector field $\mathbf{f}(x, y) = (-y, x)$

Solution It is a simple matter to calculate the values at a few points. In Table 3.1.1 the values of the vector field are shown corresponding to the point of evaluation, and in Figure 3.1.7 the vectors are sketched with the same scale as the coordinate axes. This figure is somewhat cluttered, so it would be helpful to draw the vectors to a smaller scale so that they do not overlap.

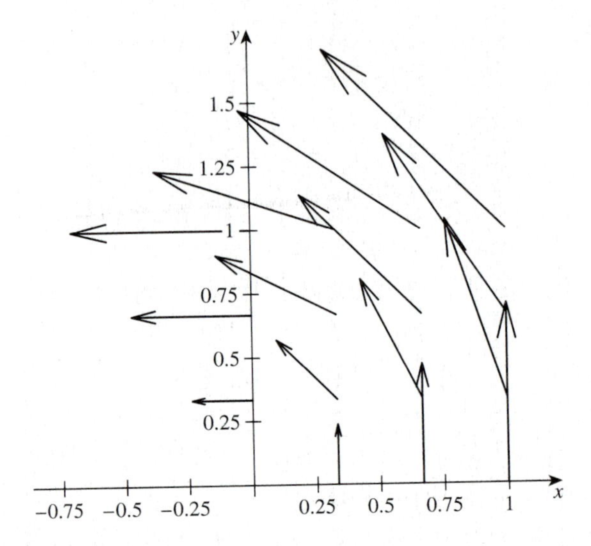

Figure 3.1.7 Plot of values in Table 3.1.1

In Figure 3.1.8 appears the same vector field plotted for (x, y) values lying in the square $-1 \le x \le 1$, $-1 \le y \le 1$ and with the vectors scaled by a factor of about 1/3. There is a strong suggestion of motion in this figure. Indeed, if $\mathbf{f}(x, y)$ represented the velocity of water at (x, y), then a small light object such as a leaf floating on the water would move so that

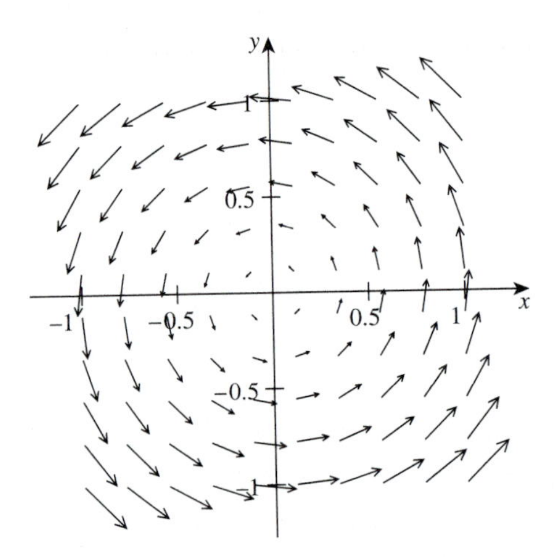

Figure 3.1.8 Computer plot of $\mathbf{f}(x, y) = (-y, x)$

the vector tangent to its path at $(x,\ y)$ is parallel to the $\mathbf{f}(x,\ y)$. The figure suggests that the path would be circular. ♦

Example 3.1.9 Consider the vector field $\mathbf{f}(x, y, z) = (y/z, -x/z, z/5)$. Certainly, calculating values of $\mathbf{f}$ at various points in $\mathbb{R}^3$ is no harder than it was in Example 3.1.7. Drawing the vectors is another matter. Many computer graphics programs provide automatic vector field sketching. Figure 3.1.9 is such a computer-generated picture. Once we have the drawing, we can get

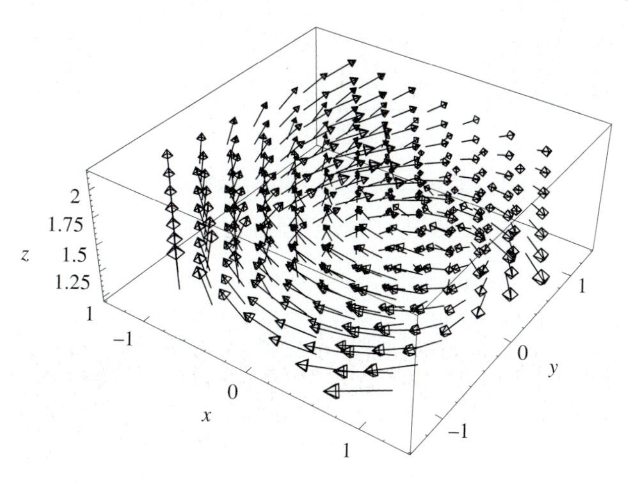

Figure 3.1.9 Computer plot of the vector field $\mathbf{f}(x, y, z) = (y/z, -x/z, z/5)$

some general information. If this is a velocity field for a fluid of some sort, then the picture reveals that a small particle introduced into the field will follow a path that spirals upward clockwise in a vortex centered about the z-axis. ♦

A function $\mathbf{f} : \mathbb{R}^n \to \mathbb{R}^n$ can be studied by considering how it deforms subsets of its domain. By this we mean, given a set of points A in the domain of $\mathbf{f}$, what does the set of points $\mathbf{f}(A) = \{\mathbf{f}(\mathbf{x})|\mathbf{x} \in A\}$ look like? The set $\mathbf{f}(A)$ is called the **image** of A **under f**. We have already done some of this in Section 2.4, when $\mathbf{f}$ was a linear transformation. Here again, it is feasible to try to do this for $n = 2$, or perhaps $n = 3$ if one is very optimistic.

Example 3.1.10 Determine the image of the unit square $\{(x, y)|0 \leq x \leq 1, 0 \leq y \leq 1\}$ under the function $\mathbf{f}(x, y) = (x^2 - y^2,\ 2xy)$.

Solution In the following discussion, refer to Figure 3.1.10. A point on the line segment from (0, 0) to (0, 1) has the form $(0,\ y)$, where $0 \leq y \leq 1$.

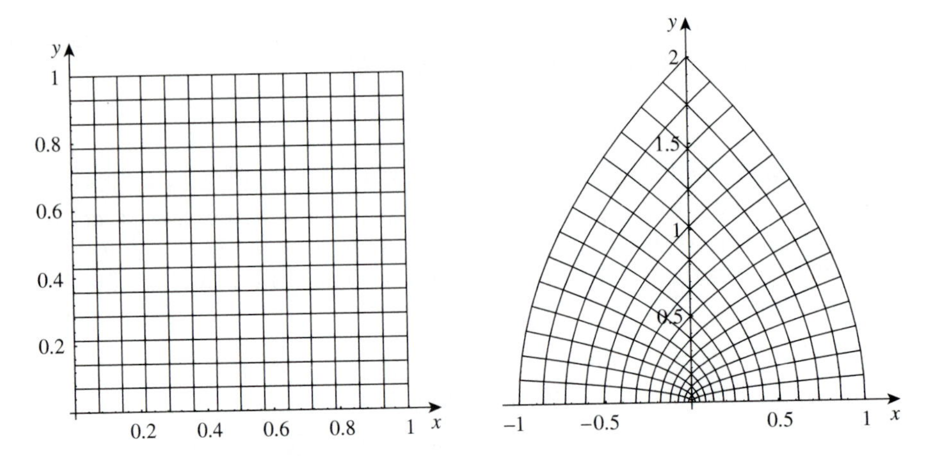

Figure 3.1.10 The unit square and its image under the vector field $\mathbf{f}(x, y) = (x^2 - y^2,\, 2xy)$

Thus each point in the image of this segment has the form $\mathbf{f}(0,\, y) = (-y^2,\, 0)$, where $0 \le y \le 1$. As y ranges from 0 to 1, the point $(-y^2,\, 0)$ moves along the x-axis from $(0,\, 0)$ to $(-1,\, 0)$, so this segment is the image of the left edge of the square.

In a similar fashion, we can see that the image of the bottom edge of the square (the set $\{(x,\, 0) | 0 \le x \le 1\}$) is $\{(x^2,\, 0) | 0 \le x \le 1\}$. As x ranges from 0 to 1, the point $(x^2,\, 0)$ moves along the line segment from $(0,\, 0)$ to $(1,\, 0)$, so the image of the bottom edge is the segment $\{(x,\, 0) | 0 \le x \le 1\}$.

For the remaining two edges, we proceed somewhat more formally. The image of the right edge of the square is

$$\begin{aligned} \mathbf{f}(\{(1, y) \,|\, 0 \le y \le 1\}) &= \left\{(1 - y^2, 2y) \,\middle|\, 0 \le y \le 1\right\} \\ &= \left\{\left(1 - \left(\frac{y}{2}\right)^2, y\right) \,\middle|\, 0 \le y \le 2\right\}, \end{aligned}$$

which is the portion of the parabola $x = 1 - y^2/4$ between $y = 0$ and $y = 2$. Similarly, the image of the top of the square is

$$\begin{aligned} \mathbf{f}(\{(x, 1) \,|\, 0 \le x \le 1\}) &= \left\{(x^2 - 1, 2x) \,\middle|\, 0 \le x \le 1\right\} \\ &= \left\{\left(\frac{y^2}{4} - 1, y\right) \,\middle|\, 0 \le y \le 2\right\}, \end{aligned}$$

which is the portion of the parabola $x = y^2/4 - 1$ between $y = 0$ and $y = 2$. Again, see Figure 3.1.10. Roughly speaking, the function holds the bottom of the square in place while it rotates counterclockwise and pulls other points in the square outward (away from the origin). ♦

Exercises 3.1

In Exercises 1 through 8, find the domain of the given function. If the domain is a subset of $\mathbb{R}^2$ or $\mathbb{R}^3$, make a sketch or write a brief description.

1. $f(x, y) = \ln(x + y)$

2. $h(x, y) = \tan(x - y)$

3. $\mathbf{g}(x, y) = (y/x, x/y)$

4. $\mathbf{h}(x, y) = \left(\sqrt{\dfrac{x^2 - 1}{y^2 - 1}}, xy, \dfrac{1}{x^2 + y^2} \right)$

5. $G(x, y, z) = \sqrt{25 - x^2 - y^2 - z^2}$

6. $F(x_1, x_2, x_3) = \dfrac{e^{x_3}}{2e^{x_1} - e^{x_2}}$

7. $p(x_1, x_2, x_3, x_4) = \dfrac{x_1 - x_4}{6x_3 + x_2}$

8. $\mathbf{H}(x_1, x_2, x_3, x_4) = \left(\sec(x_1 x_2 x_3), \dfrac{1}{x_4} \right)$

In Exercises 9 through 14, sketch the level curves for the indicated values of c.

9. $f(x, y) = \dfrac{x^2}{4} + \dfrac{y^2}{9}$; $c = 0, 1, 4, 9$

10. $g(x, y) = x^2 - 4y^2$; $c = -2, -1, 0, 1, 2$

11. $f(x, y) = \sqrt{4 - x^2 - y^2}$; $c = 0, 1, \sqrt{2}, \sqrt{3}$

12. $f(x, y) = 1 - \sqrt{x^2 + y^2}$; $c = 0, 1/4, 1/2, 1$

13. $f(x, y) = \cos y$; $c = 0, 1/2, \dfrac{1}{\sqrt{2}}, \dfrac{\sqrt{3}}{2}$

14. $f(x, y) = 8x + 4y - 8$; $c = 0, 1, 2, 3, 4$

In Exercises 15 through 18, describe and sketch the level surfaces $f(x, y, z) = c$ for the indicated values of c.

15. $f(x, y, z) = x + y + z$, $c = -3, -2, -1, 0$

16. $f(x, y, z) = \sqrt{x^2 + y^2 + z^2}$, $c = 0, 1, 2$

17. $f(x, y, z) = z^2 - x^2 - y^2$, $c = -2, -1, 0, 1, 2$

18. $f(x, y, z) = \dfrac{z}{\sqrt{x^2 + y^2}}$, $c = 0, 1, 2, 3$

19. What is the level set of $f(x, y) = \begin{cases} 0, & x^2 + y^2 < 1 \\ x^2 + y^2 - 1, & x^2 + y^2 \geq 1 \end{cases}$ corresponding to $c = 0$?

In Exercises 20 through 23, sketch several values of the given vector field, and assuming that it represents a velocity field, describe the motion of a particle that starts at the points indicated. A computer graphics program may be helpful for some of these.

20. $\mathbf{f}(x, y) = (1, 2)$; starting points $(0, 0)$, $(3, 1)$.

21. $\mathbf{f}(x, y) = \left(\frac{1}{2}x, y\right)$; starting points $(0, 0)$, $(-1, 2)$.

22. $\mathbf{f}(x, y) = (x, y)$; starting points $(1, 1)$, $(-1, -1)$.

23. $\mathbf{f}(x, y) = (x^2, -1)$; starting points $(-1, 1)$, $(1, -1)$.

24. Sketch several values of $\mathbf{f}(x, y) = (x^2 - y^2, 2xy)$ for $0 \leq x \leq 1$ and $0 \leq y \leq 1$. If this is a velocity field, sketch the paths followed by particles that start at $(1/4, 0)$ and $(1/4, 1/4)$.

25. Sketch the image of the unit square $0 \leq x \leq 1, 0 \leq y \leq 1$ under the function in Exercise 21.

26. Sketch the image of the unit square under $\mathbf{f}(x, y) = (x + y, x - y)$.

27. Sketch the image of the unit square under $\mathbf{f}(x, y) = (x - y^2, y)$.

28. Sketch the image of the rectangle $0 \leq x \leq 1, 0 \leq y \leq 2\pi$ under $\mathbf{f}(x, y) = (x \cos y, x \sin y)$.

29. Explain why two level curves of a function $f(x, y)$ cannot intersect.

3.2 Continuity

In this section our objective is to make precise what we mean by saying that a function from $\mathbb{R}^n$ to $\mathbb{R}^m$ is *continuous*. In elementary calculus, there are easy (and accurate) intuitive descriptions of continuity. One, of course, is that f

is continuous if you can draw its graph without lifting your pencil from the paper. Such an intuitive definition, however, is completely inadequate for, say, a function **f** from $\mathbb{R}^2$ to $\mathbb{R}^3$: We have essentially no way to draw its graph or visualize it geometrically. This impasse brings us to define continuity in analytical rather than geometric terms.

Example 3.2.1 Consider the function $f : \mathbb{R}^3 \to \mathbb{R}$ given by

$$f(x, y, z) = \begin{cases} x^2 + y^2 + z^2, & x < 0 \\ x^2 + y^2 + z^2 + 1, & x \geq 0 \end{cases}$$

We see that $f(0, 0, 0) = 1$. But if (x, y, z) is very close to $(0, 0, 0)$ and $x < 0$, then $f(x, y, z) = x^2 + y^2 + z^2$, which is nearly 0. Intuitively, we would say that f has a jump at the point $(0, 0, 0)$.

On the other hand, $f(1, 1, 1) = 4$, and if (x, y, z) is very close to $(1, 1, 1)$, then $f(x, y, z) = x^2 + y^2 + z^2 + 1$ is very close to 4. We would say f is continuous at $(1, 1, 1)$. ♦

The next couple of pages are a bit technical, but they contain the essential building blocks for our understanding of limits, continuity, and the total derivative of a function of several variables. Although you should read carefully all of the definitions, theorems, and examples, you may—on a first reading—skip the proofs.

Definition 3.2.1 Given a point **a** in $\mathbb{R}^n$ and a positive number r, the **deleted open ball of radius r centered at a** is

$$B_r^0(\mathbf{a}) = \{\mathbf{x} \in \mathbb{R}^n | 0 < \|\mathbf{x} - \mathbf{a}\| < r\}.$$

The **open ball of radius r centered at a** $\in \mathbb{R}^n$ is simply the deleted open ball with the center included:

$$B_r(\mathbf{a}) = \{\mathbf{x} \in \mathbb{R}^n \mid \|\mathbf{x} - \mathbf{a}\| < r\}.$$

The term *open ball* of course comes from the setting of $\mathbb{R}^3$, where such objects are readily visualized.

Definition 3.2.2 Let $f : D \subset \mathbb{R}^n \to \mathbb{R}$.

1. We say that f **has limit L as x approaches a** if f is defined in some deleted open ball centered at **a** and for each positive number ε there

corresponds a positive number δ such that $0 < ||\mathbf{x} - \mathbf{a}|| < \delta$ implies that $|f(\mathbf{x}) - L| < \varepsilon$. We write

$$\lim_{\mathbf{x} \to \mathbf{a}} f(\mathbf{x}) = L.$$

If there is no number L satisfying these conditions, we say $\lim_{\mathbf{x} \to \mathbf{a}} f(\mathbf{x})$ **does not exist**.

2. We say that $f : \mathbb{R}^n \to \mathbb{R}$ is **continuous at the point a** $\in D$ if

$$\lim_{\mathbf{x} \to \mathbf{a}} f(\mathbf{x}) = f(\mathbf{a}).$$

We say that $f : D \subset \mathbb{R}^n \to \mathbb{R}$ is **continuous on a set** S in D if it is continuous at each point in S. Finally, we simply say that f is **continuous** if it is continuous on its domain.

The following theorem will be important at various points throughout this book.

Theorem 3.2.1 *If $f : D \subset \mathbb{R}^n \to \mathbb{R}$ is continuous at $\mathbf{a}$ in D and $f(\mathbf{a}) > 0$ then there exists $r > 0$ such that $f(\mathbf{x}) > 0$ for all $\mathbf{x} \in B_r(\mathbf{a})$. (In other words, if a continuous function is positive at a point then it must be positive on an open ball about that point.)*

Proof For a contradiction, suppose that in fact there is no positive r such that $f(\mathbf{x}) > 0$ for all $\mathbf{x} \in B_r(\mathbf{a})$. Then for each $r > 0$ there exists a point $\mathbf{x}_r \in B_r(\mathbf{a})$ such that $f(\mathbf{x}_r) \leq 0$. But f is continuous at $\mathbf{a}$, so for $\varepsilon = f(\mathbf{a})$ there exists $\delta > 0$ such that $|f(\mathbf{a}) - f(\mathbf{x})| < \varepsilon$ whenever $||\mathbf{x} - \mathbf{a}|| < \delta$. In particular, when $\mathbf{x} = \mathbf{x}_r$ for $r < \delta$,

$$\begin{aligned} |f(\mathbf{a}) - f(\mathbf{x}_r)| &< f(\mathbf{a}) \\ |f(\mathbf{a}) - 0| &< f(\mathbf{a}) \\ f(\mathbf{a}) &< f(\mathbf{a}), \end{aligned}$$

which is impossible. Thus the original supposition was false; There must exist $r > 0$ such that $f(\mathbf{x}) > 0$ for all $\mathbf{x}$ satisfying $||\mathbf{x} - \mathbf{a}|| < r$. ♦

Of course, it follows immediately from Theorem 3.2.1 that if $f(\mathbf{a}) < 0$, then there exists $r > 0$ such that $f(\mathbf{x}) < 0$ for $\mathbf{x} \in B_r(\mathbf{a})$.

Theorem 3.2.2 *Suppose that $\lim_{\mathbf{x} \to \mathbf{a}} f(\mathbf{x}) = L$ and $\lim_{\mathbf{x} \to \mathbf{a}} g(\mathbf{x}) = M$. Then*

1. $\lim_{\mathbf{x}\to\mathbf{a}} (f(\mathbf{x}) \pm g(\mathbf{x})) = L \pm M$

2. $\lim_{\mathbf{x}\to\mathbf{a}} (f(\mathbf{x})g(\mathbf{x})) = LM$

3. $\lim_{\mathbf{x}\to\mathbf{a}} \dfrac{f(\mathbf{x})}{g(\mathbf{x})} = \dfrac{L}{M}$ *provided that* $M \neq 0$.

Proof We prove part 2 to illustrate the reasoning that goes into such proofs. Let $\varepsilon > 0$ be given. We require $\delta > 0$ so that $|f(\mathbf{x})g(\mathbf{x}) - LM| < \varepsilon$ whenever $|\mathbf{x} - \mathbf{a}| < \delta$. By the triangle inequality[2],

$$\begin{aligned} |f(\mathbf{x})g(\mathbf{x}) - LM| &= |f(\mathbf{x})g(\mathbf{x}) - f(\mathbf{x})M + f(\mathbf{x})M - LM| \\ &\leq |f(\mathbf{x})||g(\mathbf{x}) - M| + |f(\mathbf{x}) - L||M|. \end{aligned} \tag{3.2.1}$$

Since $\lim_{\mathbf{x}\to\mathbf{a}} f(x)$ exists and equals L, there is a positive number δ_f such that

$$|f(\mathbf{x}) - L| < \varepsilon \tag{3.2.2}$$

whenever $0 < \|\mathbf{x} - \mathbf{a}\| < \delta_f$. This gives an estimate for the second term on the right side of (3.2.1). Also, by (3.2.2),

$$L - \varepsilon < f(\mathbf{x}) < \varepsilon + L$$

for $0 < \|\mathbf{x} - \mathbf{a}\| < \delta_f$, so with $B = \max\{|L - \varepsilon|, |L + \varepsilon|\}$, it follows that

$$|f(\mathbf{x})| < B \tag{3.2.3}$$

for $0 < \|\mathbf{x} - \mathbf{a}\| < \delta_f$. (Verify this.) This gives an estimate on the first factor in the first term on the right side of (3.2.1) which is independent of $\mathbf{x}$. Finally, since $\lim_{\mathbf{x}\to\mathbf{a}} g(\mathbf{x}) = M$, there exists $\delta_g > 0$ such that

$$|g(\mathbf{x}) - M| < \varepsilon \tag{3.2.4}$$

whenever $0 < \|\mathbf{x} - \mathbf{a}\| < \delta_g$. Thus with $\delta = \min\{\delta_f, \delta_g\}$, we have (3.2.3), (3.2.2), and (3.2.4) all satisfied whenever $0 < \|\mathbf{x} - \mathbf{a}\| < \delta$. By (3.2.1), then,

$$|f(\mathbf{x})g(\mathbf{x}) - LM| \leq B \cdot \varepsilon + \varepsilon \cdot |M| = (B + |M|)\varepsilon$$

for $0 < \|\mathbf{x} - \mathbf{a}\| < \delta$. Thus

$$\lim_{\mathbf{x}\to\mathbf{a}} f(\mathbf{x})g(\mathbf{x}) = LM. \quad \blacklozenge$$

[2] $|a + b| \leq |a| + |b|$ for any two numbers a and b.

A few other observations make our groundwork complete.

Theorem 3.2.3

1. $\lim_{\mathbf{x}\to\mathbf{a}} c = c$ *for any constant* c.

2. *If* $\mathbf{x} = (x_1, x_2 \ldots, x_n)$ *and* $\mathbf{a} = (a_1, a_2 \ldots a_n)$ *then* $\lim_{\mathbf{x}\to\mathbf{a}} x_i = a_i$, $i = 1$, $2, \ldots, n$.

Theorem 3.2.3 can, along with Theorem 3.2.2, be used to show that a function such as $f(x_1, x_2, x_3, x_4) = x_1x_3 + 5x_2^2 + x_4 + 1$ is continuous at any point $\mathbf{a}$ in $\mathbb{R}^4$ and that one such as $g(x_1, x_2, x_3) = x_1x_2x_3/(x_2 - 3)$ is continuous at any point (x_1, x_2, x_3) where $x_2 \neq 3$. The general result is summarized in the following theorem.

Theorem 3.2.4

1. *If* $p(\mathbf{x})$ *is a* **polynomial function** *(i.e. one of the form*

$$p(x_1, x_2, \ldots, x_n) = \sum_{i=1}^{k} c_i x_1^{m_{1,i}} x_2^{m_{2,i}} x_3^{m_{3,i}} \cdots x_n^{m_{n,i}},$$

where $m_{1,i}, \ldots m_{n,i}$, $i = 1, 2, \ldots, k$ *are nonnegative integers and* c_1, c_2, $\ldots$, c_k *are real numbers), then* f *is continuous at each point* $\mathbf{x} \in \mathbb{R}^n$.

2. *If* $f(\mathbf{x})$ *is a* **rational function** *(i.e. one of the form* $f(\mathbf{x}) = p(\mathbf{x})/q(\mathbf{x})$ *where* p *and* q *are polynomials), then* f *is continuous at each point in its domain. [If* $p(\mathbf{x})$ *and* $q(\mathbf{x})$ *have no common factors, then these are the points where* $q(\mathbf{x}) \neq 0$.*]*

Example 3.2.2 The function $f : \mathbb{R}^4 \to \mathbb{R}$ given by

$$f(x_1, x_2, x_3, x_4) = \frac{x_1 - 2x_2 + 3x_4}{(x_1^2 + x_2^2)(x_3 + x_4)}$$

is a rational function. By Theorem 3.2.4, it is continuous for $x_1^2 + x_2^2 \neq 0$ and $x_3 + x_4 \neq 0$. Thus f is continuous on the set $\{\mathbf{x} \in \mathbb{R}^4 \mid x_1^2 + x_2^2 \neq 0,\ x_3 + x_4 \neq 0\}$. ♦

A discontinuity in a rational function may actually not be too severe. In other words, it may be that $\lim_{\mathbf{x}\to\mathbf{a}} f(\mathbf{x})$ exists but f is not defined at $\mathbf{a}$. In this case we say that the discontinuity is **removable**.

Example 3.2.3 Consider

$$f(x,y) = \frac{x^4}{x^2+y^2}.$$

This function is discontinuous only at $(0,0)$, simply because $(0,0)$ is not in its domain. Notice, however, that

$$0 \le \frac{x^4}{x^2+y^2} = x^2 \frac{x^2}{x^2+y^2} \le x^2 \frac{x^2+y^2}{x^2+y^2} = x^2 \tag{3.2.5}$$

for all choices of (x,y) away from $(0,0)$. As (x,y) is chosen in disks of small radius about $(0,0)$, x^2 gets arbitrarily close to zero. By (3.2.5), this forces $x^4/(x^2+y^2)$ to get arbitrarily close to zero. Thus $\lim_{\mathbf{x}\to\mathbf{0}} f(\mathbf{x}) = 0$. The "patched-together" function

$$g(x,y) = \begin{cases} x^4/(x^2+y^2), & (x,y) \ne (0,0) \\ 0, & (x,y) = (0,0) \end{cases}$$

is continuous at $(0,0)$. Thus the discontinuity at $(0,0)$ is removable. ♦

To determine whether a discontinuity in a given function is removable generally requires analysis. The following examples indicate some of the possibilities.

Example 3.2.4 The function $f(x,y,z) = e^{-1/(x^2+y^2+z^2)}$ is discontinuous at $(0,0,0)$ because this point is not in its domain. To tell whether this is a removable discontinuity, we examine

$$\lim_{(x,y,z)\to(0,0,0)} e^{-1/(x^2+y^2+z^2)}. \tag{3.2.6}$$

Allowing $(x,y,z) \to (0,0,0)$ just means that

$$\|(x,y,z) - (0,0,0)\| = \sqrt{x^2+y^2+z^2} \to 0.$$

With $u = \sqrt{x^2+y^2+z^2}$ in (3.2.6), we have

$$\lim_{u\to 0^+} e^{-1/u^2} = \lim_{u\to 0^+} \frac{1}{e^{1/u^2}}.$$

Since $1/u^2 \to \infty$ as $u \to 0^+$, we see that $e^{1/u^2} \to \infty$, so $1/e^{1/u^2} \to 0$. Thus

$$\lim_{(x,y,z)\to(0,0,0)} e^{-1/(x^2+y^2+z^2)} = 0.$$

Because the limit exists, the discontinuity at the point (0, 0, 0) is removable. We could remove it by redefining f to have the value 0 at (0, 0, 0). ♦

Example 3.2.5 The function $f(x, y) = (x^2 - y^2)/(x^2 + y^2)$ has a discontinuity at $(0,0)$. If it is removable, then

$$\lim_{(x,y)\to(0,0)} \frac{x^2 - y^2}{x^2 + y^2} \tag{3.2.7}$$

exists. Consider what happens if (x,y) approaches $(0,0)$ along the x-axis. At such points, $y = 0$. We see that

$$\lim_{\substack{(x,y)\to(0,0)\\ y=0}} \frac{x^2 - y^2}{x^2 + y^2} = \lim_{x\to 0} \frac{x^2 - 0}{x^2 + 0} = \lim_{x\to 0} 1 = 1,$$

so if the limit in (3.2.7) exists, then it must be 1. Now consider what happens as (x,y) approaches $(0,0)$ along the y-axis:

$$\lim_{\substack{(x,y)\to(0,0)\\ x=0}} \frac{x^2 - y^2}{x^2 + y^2} = \lim_{y\to 0} \frac{0 - y^2}{0 + y^2} = \lim_{y\to 0}(-1) = -1.$$

Since this does not agree with 1, the limit in (3.2.7) does not exist: The discontinuity at $(0,0)$ is not removable. ♦

Example 3.2.6 The function $f(x,y) = (x+y)^2/(x-y)$ has a discontinuity at every point on the line $y = x$. None of these is removable: Let (a,a) be such a point and consider (x,y) approaching (a,a) along the vertical line $x = a$. If $y \to a^+$, then we have

$$\lim_{\substack{(x,y)\to(a,a^+)\\ x=a}} \frac{(x+y)^2}{x-y} = \lim_{y\to a^+} \frac{(a+y)^2}{a-y} = -\infty,$$

and if $y \to a^-$, then

$$\lim_{\substack{(x,y)\to(a,a^-)\\ x=a}} \frac{(x+y)^2}{x-y} = \lim_{y\to a^-} \frac{(a+y)^2}{a-y} = +\infty.$$

Since these limits are infinite (and also because they don't agree),

$$\lim_{(x,y)\to(a,a)} \frac{(x+y)^2}{x-y}$$

does not exist for any choice of (a, a). The discontinuities are therefore not removable. ♦

Example 3.2.7 The function $f(x, y) = \dfrac{x^2 y}{x^4 + y^2}$ has $(0, 0)$ as its only discontinuity. The trick we used in Example 3.2.5 gives a limit of 0 for (x, y) approaching $(0, 0)$ along both the x- and y-axes. Along a different approach, though, the limit might be different. Consider allowing (x, y) to approach $(0, 0)$ along the line $y = ax$:

$$\lim_{\substack{(x,y)\to(0,0) \\ y=ax}} \frac{x^2 y}{x^4 + y^2} = \lim_{x\to 0} \frac{x^2 \cdot ax}{x^4 + a^2 x^2} = \lim_{x\to 0} \frac{ax}{x^2 + a^2} = 0.$$

Thus along any line through the origin, the function's values approach 0.

This, unfortunately, still does not settle the question, for it may be that approaching $(0, 0)$ along some other curve through the origin will yield a limit different from 0. Observe that if (x, y) approaches $(0, 0)$ along the parabola $y = x^2$ then

$$\lim_{\substack{(x,y)\to(0,0) \\ y=x^2}} \frac{x^2 y}{x^4 + y^2} = \lim_{x\to 0} \frac{x^2 \cdot x^2}{x^4 + x^4} = \lim_{x\to 0} \frac{1}{2} = \frac{1}{2} \neq 0.$$

Thus when we approach $(0, 0)$ along the parabola $y = x^2$ we get a limit of $1/2$, so $\lim_{(x,y)\to(0,0)} f(x, y)$ does not exist. ♦

To complete our discussion of continuity, we give a definition for continuity of a vector-valued function of several variables.

Definition 3.2.3 Let $\mathbf{f}: D \subset \mathbb{R}^n \to \mathbb{R}^m$ and let $\mathbf{a}$ be a point in $\mathbb{R}^n$.

1. We say that $\lim_{\mathbf{x}\to\mathbf{a}} \mathbf{f}(\mathbf{x}) = \mathbf{L}$ if $\mathbf{f}$ is defined in some deleted open ball about $\mathbf{a}$ and for each positive number ε there exists a positive number δ such that $\|\mathbf{f}(\mathbf{x}) - \mathbf{L}\| < \varepsilon$ whenever $0 < \|\mathbf{x} - \mathbf{a}\| < \delta$.

2. We say that $\mathbf{f}$ is **continuous** at $\mathbf{a}$ if $\mathbf{a} \in D$ and $\lim_{x\to\mathbf{a}} \mathbf{f}(\mathbf{x}) = \mathbf{f}(\mathbf{a})$.

Any function $\mathbf{f} : D \subset \mathbb{R}^n \to \mathbb{R}^m$ has m real-valued component functions: $\mathbf{f}(\mathbf{x}) = (f_1(\mathbf{x}), f_2(\mathbf{x}), \ldots, f_m(\mathbf{x}))$. We would expect the continuity of $\mathbf{f}$ to have some connection with that of $f_1, f_2, \ldots, f_m$. Indeed, the connection is the strongest possible: They are equivalent. In the exercises, you are asked to prove the following theorem.

Theorem 3.2.5 *If* $\mathbf{f}: D \subset \mathbb{R}^n \to \mathbb{R}^m$ *is given by*

$$\mathbf{f}(\mathbf{x}) = (f_1(\mathbf{x}), f_2(\mathbf{x}), \ldots, f_m(\mathbf{x})),$$

then $\mathbf{f}$ *is continuous at a point* $\mathbf{a} \in D$ *if and only if* f_1, f_2, ..., f_m *are all continuous at* $\mathbf{a}$.

Example 3.2.8 By Theorem 3.2.5 the function

$$\mathbf{f}(x, y) = \left(\frac{xy - y}{xy - x}, \frac{1}{x^2 + y^2 - 9} \right)$$

is continuous precisely where both of its component functions

$$f_1(x, y) = \frac{xy - y}{xy - x} \quad \text{and} \quad f_2(x, y) = \frac{1}{x^2 + y^2 - 9}$$

are continuous. By Theorem 3.2.4, f_1 and f_2 are continuous on their respective domains. So $\mathbf{f}$ is continuous at each point $(x, y) \in \mathbb{R}^2$ such that $x \neq 0$, $y \neq 1$ and $x^2 + y^2 - 9 \neq 0$. These are all the points that lie off the x- and y-axes and off the circle $x^2 + y^2 = 9$. ♦

In elementary calculus, we freely use the fact that the composition of two continuous functions is continuous. Generally, the same result holds for functions of several variables; the following theorem makes this precise.

Theorem 3.2.6 *Let* $\mathbf{f}: D \subset \mathbb{R}^n \to E$, *where* $E \subset \mathbb{R}^m$, *and let* $\mathbf{g}: E \to \mathbb{R}^p$. *If* $\mathbf{f}$ *is continuous at* $\mathbf{a}$ *and* $\mathbf{g}$ *is continuous at* $\mathbf{f}(\mathbf{a})$, *then the composition function* $\mathbf{g} \circ \mathbf{f}$ *defined by* $(\mathbf{g} \circ \mathbf{f})(\mathbf{x}) = \mathbf{g}(\mathbf{f}(\mathbf{x}))$ *is continuous at* $\mathbf{a}$.

Proof Since $\mathbf{g}$ is continuous at $\mathbf{f}(\mathbf{a})$, $\mathbf{g}$ is defined in an open ball about $\mathbf{f}(\mathbf{a})$. For a given positive number ε there exists a positive number δ_g such that

$$\|\mathbf{g}(\mathbf{y}) - \mathbf{g}(\mathbf{f}(\mathbf{a}))\| < \varepsilon \tag{3.2.8}$$

whenever $\|\mathbf{y} - f(\mathbf{a})\| < \delta_g$. Also, because $\mathbf{f}$ is continuous at $\mathbf{a}$, there exists a positive number δ_f such that

$$\|\mathbf{f}(\mathbf{x}) - \mathbf{f}(\mathbf{a})\| < \delta_g$$

whenever $\|\mathbf{x} - \mathbf{a}\| < \delta_f$. So if $\mathbf{x} \in D$ satisfies $\|\mathbf{x} - \mathbf{a}\| < \delta_f$, then, with $\mathbf{y} = f(\mathbf{x})$ in (3.2.8), we see that

$$\|\mathbf{g}(\mathbf{f}(\mathbf{x})) - \mathbf{g}(\mathbf{f}(\mathbf{a}))\| < \varepsilon.$$

It follows that $\lim_{\mathbf{x} \to \mathbf{a}} \mathbf{g}(\mathbf{f}(\mathbf{x})) = \mathbf{g}(\mathbf{f}(\mathbf{a}))$; thus $\mathbf{g} \circ \mathbf{f}$ is continuous at $\mathbf{a}$. ♦

Theorem 3.2.6 gives us a way of determining the continuity of most elementary functions at a glance. In the next example we illustrate the line of reasoning.

Example 3.2.9 The function $h(x, y) = e^{x^2-y^2}$ is continuous at all points in the xy-plane. We can see this by noting first that h is a composition of the functions $g(u) = e^u$ and $f(x, y) = x^2 - y^2$. By Theorem 3.2.4, f is continuous everywhere, and by results in single-variable calculus, g is continuous everywhere. Then by Theorem 3.2.6, h is continuous for all choices of (x, y). ♦

Exercises 3.2

Evaluate the limits in exercises 1 through 11.

1. $\displaystyle\lim_{(x,y)\to(1,-1)} \frac{x+4y}{1-x+y}$

2. $\displaystyle\lim_{(x,y)\to(0,\pi/2)} \cos x \sin y$

3. $\displaystyle\lim_{(x,y,z)\to(1,1,-1/2)} \arctan\left(\frac{y+z}{x}\right)$

4. $\displaystyle\lim_{(x,y,z)\to(0,1,0)} \ln\left((x+2)y(z+10)\right)$

5. $\displaystyle\lim_{(x,y)\to(0,0)} \frac{\sin(x^2+y^2)}{x^2+y^2}$

6. $\displaystyle\lim_{(x,y)\to(0,0)} \frac{1-e^{x^2+y^2}}{x^2+y^2}$

7. $\displaystyle\lim_{(x,y)\to(0,0)} \frac{y^2}{\sqrt{x^2+y^2}}$

8. $\displaystyle\lim_{(x,y)\to(0,0)} \frac{x^2y}{x^2+y^2}$

9. $\displaystyle\lim_{(x,y)\to(0,0)} \frac{x^2(y+1)+y^2}{x^2+y^2}$ (Hint: Use the result in Exercise 8.)

10. $\displaystyle\lim_{(x,y,z)\to(0,0,0)} \frac{x^3}{x^2+y^2+z^2}$

11. $\displaystyle\lim_{(x,y)\to(1,1)} g(x,y)$, where $g(x,y) = \begin{cases} 1-x-y, & x+y<1 \\ -1+x+y, & x+y\geq 1 \end{cases}$

In each of the Exercises from 12 through 17, explain why the given limit does not exist.

12. $\displaystyle \lim_{(x,y)\to(0,0)} \frac{xy}{x^2+y^2}$

13. $\displaystyle \lim_{(x,y)\to(0,0)} \frac{x}{x^2+y^2}$

14. $\displaystyle \lim_{(x,y)\to(0,0)} \frac{1-\cos xy}{xy}$

15. $\displaystyle \lim_{(x,y)\to(0,0)} \frac{x^3y}{x^6+y^2}$

16. $\displaystyle \lim_{(x,y)\to(0,0)} \frac{xy^2-x^2y}{(x^2+y^2)^{3/2}}$

17. $\displaystyle \lim_{(x,y)\to(1,0)} h(x,y)$, where $h(x,y) = \begin{cases} 1, & x^2+y^2 \le 1 \\ 0, & x^2+y^2 > 1 \end{cases}$

In Exercises 18 through 21, determine the points of discontinuity in the given function. Are the discontinuities removable?

18. $\mathbf{f}(x,y) = \left(\dfrac{1}{x+y}, \dfrac{1}{x^2+y^2-1}\right)$

19. $g(x,y) = \dfrac{1-\cos(x^2+y^2)}{x^2+y^2}$

20. $h(x,y,z) = \tan(xyz)$

21. $k(x_1, x_2, x_3, x_4) = \ln\left((x_1-1)^2+(x_2-2)^2+(x_3-3)^2+(x_4-4)^2\right)$

Explain, by citing appropriate theorems, why each the functions in Exercises 22 through 26 is continuous at the point indicated.

22. $A(x,y) = \sin\left(\dfrac{x}{y}\right)$; $(\pi, 1)$.

23. $B(x,y) = x - y + \ln(xy)$; $(1, e)$.

24. $C(x,y,z) = z\sin(y\cos x)$; $(0, \pi/2, 1)$.

25. $\mathbf{D}(x,y) = \left(\dfrac{e^x}{y}, xe^{-y}\right)$; $(1, -1)$.

26. $\mathbf{E}(x,y) = \left(\sqrt{x^2+y^2}, \dfrac{1}{\sqrt{x^2+y^2}}\right)$; $(0, 4)$.

27. Prove that if $f : D \subset \mathbb{R}^n \to \mathbb{R}$ is continuous at $\mathbf{a}$ and $f(\mathbf{a}) \neq b$ for some number b, then there exists $r > 0$ such that $f(\mathbf{x}) \neq b$ for all $\mathbf{x} \in B_r(\mathbf{a})$.

28. Prove that if $f : D \subset \mathbb{R}^n \to \mathbb{R}$ and $\mathbf{a}$ is such that $\lim_{\mathbf{x}\to\mathbf{a}} f(\mathbf{x}) \neq b$ for some number b, then there exists $r > 0$ such that $f(\mathbf{x}) \neq b$ for all $\mathbf{x}$ in the deleted open ball $B_r^0(\mathbf{a})$.

29. Prove part 1 of Theorem 3.2.2.

30. Prove part 3 of Theorem 3.2.2. (*Hint*: Use the trick in the proof of the second part of the theorem along with Exercise 28.)

31. Prove Theorem 3.2.3.

32. Prove Theorem 3.2.4.

33. Prove Theorem 3.2.5.

34. Let f be the function of two variables defined in the following way: for $x = .x_1x_2x_3x_4\ldots$ and $y = .y_1y_2y_3\ldots$ in the interval $[0, 1)$ let $f(x, y) = .z_1z_2z_3z_4\ldots$, where $z_i = (x_i + y_i) \bmod 10$ (i.e., z_i is the remainder when $x_i + y_i$ is divided by 10). For example, $f(0.3258, 0.2564) = 0.5712$.

(**a**) Show that f is discontinuous at $(0.5, 0.5)$.

(**b**) Show that f is discontinuous at any point (x, y), where x and y are both nonzero and have terminating decimal expansions.

3.3 Open and Closed Sets and Continuity

In applications, it is important to determine when a function attains its largest or smallest value. To make such determinations, we must phrase the question precisely enough so that it has an answer. For instance, the question, "What are the maximum and minimum values of $f(x) = x^2$ for x in the interval $(1, \infty)$?" does not have an answer: The function does not attain a largest or smallest value on the interval. On the other hand, the question, "What are the maximum and minimum values of $f(x) = x^2$ on $[1, 4]$?" does have an answer: $f(1) = 1$ is the minimum value and $f(4) = 16$ is the maximum. This follows by the Extreme Value Theorem since f is continuous and the interval $[1, 4]$ is closed.

The objective in this section is to develop the n-dimensional analog of a closed interval and to state a corresponding Extreme Value Theorem for functions of several variables.

Definition 3.3.1 A set S in $\mathbb{R}^n$ is **open** if for each point $\mathbf{x} \in S$ there is a positive number δ which is small enough so that $B_\delta(\mathbf{x})$ is contained in S. A set S in $\mathbb{R}^n$ is **closed** if its **complement** $\mathbb{R}^n \backslash S = \{\mathbf{x} \in \mathbb{R}^n \mid \mathbf{x} \notin S\}$ is open.

These definitions make precise the notion that an open set is a "fat" or "fluffy" set that does not contain its edges and that a closed set has "sharp edges."

Example 3.3.1 Let $S = \{(x, y) \in \mathbb{R}^2 | x^2 + y^2 < 1\}$. Then S is open: For any point $(x, y) \in S$, if δ is any number less than $1 - \sqrt{x^2 + y^2}$, then $B_\delta((x, y)) \subset S$. See Figure 3.3.1. By definition, the complement of S, $\mathbb{R}^2 \backslash S = \{(x, y) | x^2 + y^2 \geq 1\}$ is closed. ♦

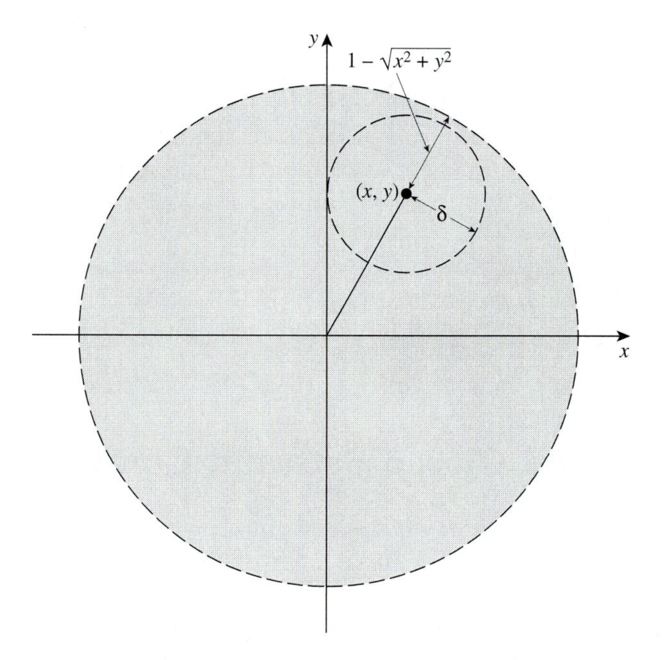

Figure 3.3.1 The set S of points (x, y) such that $x^2 + y^2 < 1$ is open since any disk about (x, y) of radius less than $1 - (x^2 + y^2)^{1/2}$ lies in S.

Definition 3.3.2 The **boundary** of a set S in $\mathbb{R}^n$ is the set $\partial S = \{\mathbf{x} \in \mathbb{R}^n \mid B_\delta(\mathbf{x})$ contains points in S and $\mathbb{R}^n \backslash S$ for all $\delta > 0\}$. The **interior** of a set S, denoted int (S), is the portion of the set remaining after its boundary has been removed. A set S in $\mathbb{R}^n$ is **bounded** if for any $\mathbf{x} \in S$ there exists $R > 0$ such that $S \subset B_R(\mathbf{x})$.

Example 3.3.2 Let S be the set in Example 3.3.1. Then the boundary of S consists of the circle $x^2 + y^2 = 1$; in set notation, $\partial S = \{(x, y) | x^2 + y^2 = 1\}$.

This holds because any disk about a point on this circle contains points in S and in $\mathbb{R}^2 \backslash S$. Since S does not contain any points of its boundary, there is nothing to removed from it to make the interior. Thus, in this case, the interior of S is simply S itself: $\text{int}(S) = S$. This set is bounded since for any point in S, the entire set is contained in $B_r((x, y))$ where r is any number bigger than $1 + \sqrt{x^2 + y^2}$. See Figure 3.3.2. ♦

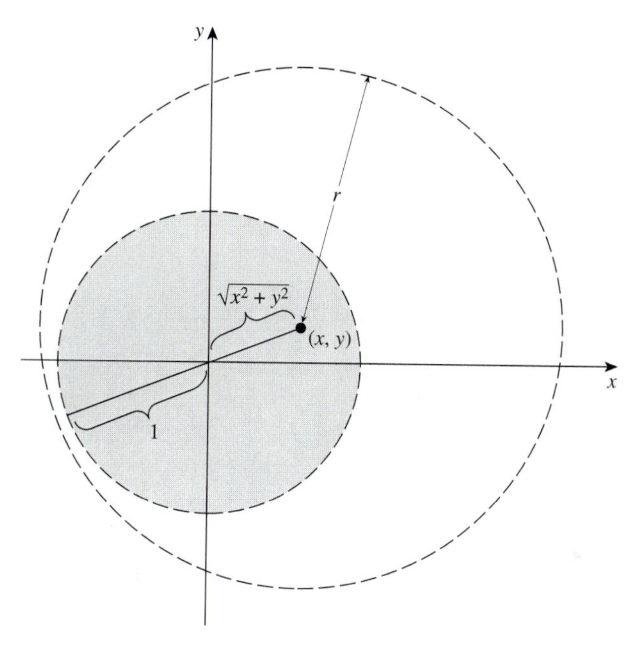

Figure 3.3.2 The set S of points (x, y) such that $x^2+y^2 < 1$ is bounded since it is contained in a ball of radius more than $1 + (x^2 + y^2)^{1/2}$ for any point (x, y) in S.

Example 3.3.3 The set $S = \{(x, y) | 0 \leq x < 1, 0 \leq y < 1\}$ is neither open nor closed. For instance, the point $(0, 0)$ is in S, but any disk $B_\delta((0, 0))$ about it contains points in S and in $\mathbb{R}^2 \backslash S$. See Figure 3.3.3. Thus S is not open. Also, the point $(1, 1)$, for example,is in $\mathbb{R}^2 \backslash S$, and any disk $B_\delta((1, 1))$ about it contains points in both S and $\mathbb{R}^2 \backslash S$. Thus $\mathbb{R}^2 \backslash S$ is not open, so $S = \mathbb{R}^2 \backslash (\mathbb{R}^2 \backslash S)$ is not closed. The boundary of S is clearly the union of the four sides of the square: $\partial S = S_1 \cup S_2 \cup S_3 \cup S_4$ where $S_1 = \{(x, y) | y = 0, 0 \leq x \leq 1\}$, $S_2 = \{(x, y) | x = 1, 0 \leq y \leq 1\}$, $S_3 = \{(x, y) | y = 1, 0 \leq x \leq 1\}$, and $S_4 = \{(x, y) | x = 0, 0 \leq y \leq 1\}$. The set is bounded since for any $(x, y) \in S$ it is contained in a ball about (x, y) with radius more than $\sqrt{2}$. ♦

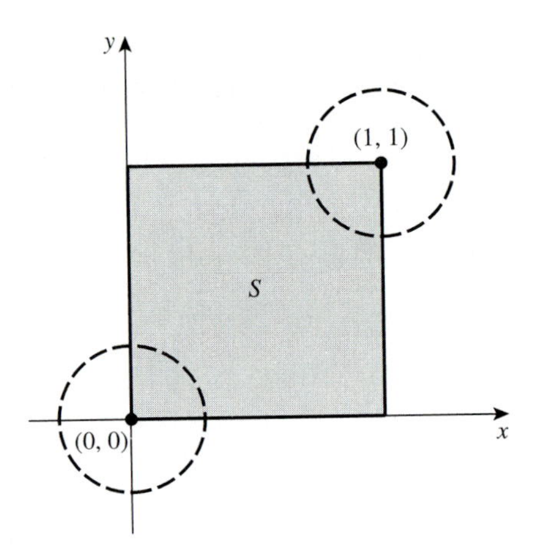

Figure 3.3.3 The set $S = \{(x, y)|0 \le x < 1, 0 \le y < 1\}$ is neither open nor closed.

Example 3.3.4 Show that the line parametrized by $\mathbf{x} = (3t + 1, t, -t + 1)$ is a closed subset of $\mathbb{R}^3$.

Solution The set in question is

$$L = \{(x, y, z) \mid (x, y, z) = t(3, 1, -1) + (1, 0, 1), -\infty < t < \infty\}.$$

We want to show that its complement is open. For any point $\mathbf{x}_0 = (x_0, y_0, z_0)$ in the complement of L, by Example 1.7.6, the distance to this line is

$$\begin{aligned} d &= \frac{\|((x_0, y_0, z_0) - (1, 0, 1)) \times (3, 1, -1)\|}{\|(3, 1, -1)\|} \\ &= \frac{1}{\sqrt{11}} \|(x_0 - 1, y_0, z_0 - 1) \times (3, 1, -1)\| \\ &= \frac{1}{\sqrt{11}} \|(1 - y_0 - z_0, -4 + x_0 + 3z_0, -1 + x_0 - 3y_0)\| \\ &= \frac{\sqrt{(1 - y_0 - z_0)^2 + (x_0 + 3z_0 - 4)^2 + (x_0 - 3y_0 - 1)^2}}{\sqrt{11}}. \end{aligned}$$

So if δ is any positive number smaller than d, then $B_\delta((x_0, y_0, z_0))$ misses L; that is, $B_\delta((x_0, y_0, z_0))$ lies entirely in $\mathbb{R}^3 \backslash L$. Thus $\mathbb{R}^3 \backslash L$ is open, and therefore L is closed.

You should convince yourself that $\partial L = L$. Also, the line is not bounded: The point $(1, 0, 1)$ is in L, and for any $r > 0$, $B_r((1, 0, 1))$ misses at least one point in L [e.g., $(r + 1)(3, 1, -1) + (1, 0, 1)$]. ♦

You have probably conjectured that a set's being closed has something to do with its boundary. Indeed, a set that contains its boundary is closed.

Theorem 3.3.1 *A set S in $\mathbb{R}^n$ is closed if and only if it contains its boundary.*

Proof Suppose that S is a closed set and let $\mathbf{x}$ be a point in ∂S. If $\mathbf{x}$ is not in S, then, since $\mathbb{R}^n \backslash S$ is open, there is a number $\delta > 0$ such that $B_\delta(\mathbf{x})$ lies entirely in $\mathbb{R}^n \backslash S$. But since $\mathbf{x} \in \partial S$, $B_\delta(\mathbf{x})$ must contain points in both S and $\mathbb{R}^n \backslash S$. This contradiction shows that it cannot be that $\mathbf{x}$ is not in S; thus $\mathbf{x}$ is in S. This reasoning applies to any point $\mathbf{x}$ in ∂S so S contains ∂S.

Conversely, suppose that S is a set that contains its boundary. We want to show that $\mathbb{R}^n \backslash S$ is open. Let $\mathbf{x}$ be a point in $\mathbb{R}^n \backslash S$. There exists a positive number δ such that $B_\delta(\mathbf{x})$ is in $\mathbb{R}^n \backslash S$; otherwise, $B_\delta(x)$ would contain points in both $\mathbb{R}^n \backslash S$ and S for each $\delta > 0$, thus implying that $\mathbf{x} \in \partial S \subset S$. This is true for each point in $\mathbb{R}^n \backslash S$, so $\mathbb{R}^n \backslash S$ is open and therefore S is closed. ♦

The main theorem of this section is the following.

Theorem 3.3.2 *If $f : D \subset \mathbb{R}^n \to \mathbb{R}$ is continuous on a nonempty closed and bounded subset S of D then f attains a minimum and a maximum value on S. That is, there exists a point $\mathbf{x}_0 \in S$ such that $f(\mathbf{x}_0) \leq f(\mathbf{x})$ for all $\mathbf{x} \in S$ and there exists $\mathbf{x}_1 \in S$ such that $f(\mathbf{x}_1) \geq f(\mathbf{x})$ for all $\mathbf{x} \in S$. We write $f(\mathbf{x}_0) = \min_{\mathbf{x} \in S} f(\mathbf{x})$ and $f(\mathbf{x}_1) = \max_{\mathbf{x} \in S} f(\mathbf{x})$ for short.*

A proof of this theorem is beyond the scope of this book, but its validity can be appreciated by studying examples.

Example 3.3.5 Find the maximum and minimum values of $f(x, y) = 2x + 3y$ on the disk $S = \{(x, y) | x^2 + y^2 \leq 1\}$.

Solution This problem can be conceived geometrically as simply asking for the highest and lowest points on the portion of the plane $z = 2x + 3y$ that lies within the cylinder $x^2 + y^2 = 1$. See Figure 3.3.4. Indeed the figure suggests that the maximum and minimum values occur on the boundary of S. To prove this, suppose in fact that, for example, the maximum value of f occurred at (a, b), where $a^2 + b^2 < 1$. Now consider a circle $(x-a)^2 + (y-b)^2 = r^2$ where $r < 1 - a^2 - b^2$. This circle lies entirely inside S. For any point (x, y) on this circle,

$$\begin{aligned} f(x, y) &= f((a, b) + (x - a, y - b)) \\ &= f(a, b) + f(x - a, y - b) \\ &= f(a, b) + (2(x - a) + 3(y - b)) \end{aligned}$$

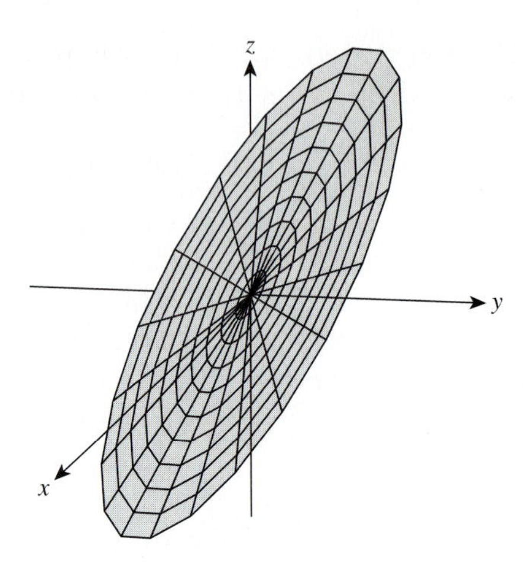

Figure 3.3.4 The portion of the plane $f(x, y) = 2x + 3y$ lying inside the cylinder $x^2 + y^2 = 1$

by the linearity of f. Now, choose $x = c$ and $y = d$, where $c > a$ and $d > b$. See Figure 3.3.5. Then $f(c, d) = f(a, b)+2(c-a)+3(d-b) > f(a, b)$, which

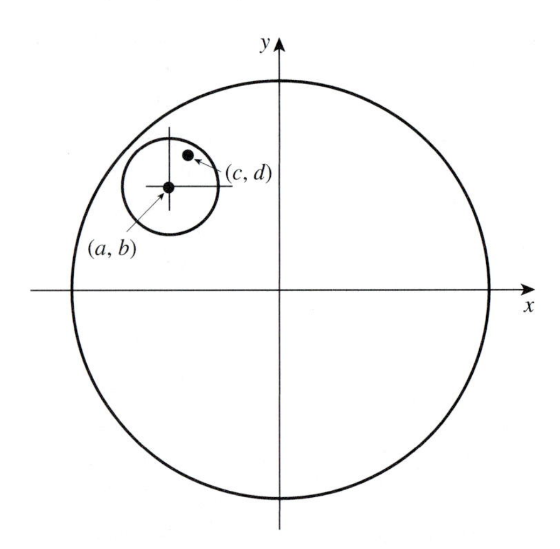

Figure 3.3.5 The maximum value of $f(x, y) = 2x+3y$ cannot occur at a point (a, b) in the interior of the disk because we can find another point (c, d) such that $c > a$ and $d > b$ inside the disk.

contracticts our assumption that $f(a, b)$ is the maximum value of f on S. So the point (a, b) at which the maximum occurs must be on the boundary. Similarly, the point at which the minimum occurs is on the boundary.

When (x, y) is on the boundary, $y = \pm\sqrt{1 - x^2}$, so we want the maximum and minimum values of

$$\begin{aligned} g(x) &= f\left(x, \sqrt{1 - x^2}\right) \\ &= 2x + 3\sqrt{1 - x^2}, \quad -1 \le x \le 1. \end{aligned}$$

and

$$\begin{aligned} h(x) &= f\left(x, -\sqrt{1 - x^2}\right) \\ &= 2x - 3\sqrt{1 - x^2}, \quad -1 \le x \le 1. \end{aligned}$$

These are ordinary single-variable optimization problems. For g we find the critical numbers in $[-1, 1]$ and calculate the values of g at these values and at the endpoints. The largest will be the maximum of g, and the smallest will be its minimum. The same procedure works for h.

First,

$$g'(x) = 2 - \frac{3x}{\sqrt{1 - x^2}}.$$

Setting this equal to zero and solving for x, we obtain the critical numbers of g:

$$\begin{aligned} 2\sqrt{1 - x^2} &= 3x \\ 4(1 - x^2) &= 9x^2 \\ 13x^2 &= 4 \\ x &= \pm\frac{2}{\sqrt{13}}. \end{aligned}$$

Then $g(1) = 2$, $g(-1) = -2$, $g(2/\sqrt{13}) = \sqrt{13}$, and $g(-2/\sqrt{13}) = 5/\sqrt{13}$. The largest value is $g(2/\sqrt{13}) = \sqrt{13}$ and the smallest is $g(-1) = -2$.

Similarly, we find the same critical numbers for h. Since $h(1) = 2$, $h(-1) = -2$, $h(2/\sqrt{13}) = -5/\sqrt{13}$, and $h(-2/\sqrt{13}) = -\sqrt{13}$, the largest value of h is $h(1) = 2$ and the smallest is $h(-2/\sqrt{13}) = -\sqrt{13}$.

Combining the results from g and h, we see that the maximum value of f on the boundary is $g(2/\sqrt{13}) = \sqrt{13}$ and the minimum is $h(-2/\sqrt{13}) = -\sqrt{13}$. These extreme values occur, respectively, at $(2/\sqrt{13}, 3/\sqrt{13})$ and $(-2/\sqrt{13}, -3/\sqrt{13})$. ♦

Although we do not prove it here, the maximum and minimum values of any linear function $f : \mathbb{R}^n \to \mathbb{R}$ on a closed and bounded set S occur on the boundary of S. This means, in particular, that if f is a linear function of two variables on a closed and bounded subset of the xy-plane, then we need look only at the values of the function on the curve that forms the boundary of the subset. In the exercises, you are asked to make use of this fact to find the maximum and minimum values of specific linear functions on given sets.

Exercises 3.3

In Exercises 1 throug 4, identify the given subset of $\mathbb{R}^2$ as open, closed, or neither. Also, find the boundary and complement.

1. $A = \{(x, y)|\ 1 < x^2 + y^2 \leq 4\}$

2. $B = \{(x, y)|\ 2x + 9y = 1\}$

3. $C = \{(x, y)|\ -5 < x < 5\}$

4. $D = \{(x, y)|\ x + y > 1,\ \ y - 2x < 1,\ \ y + 2x < 2\}$

Find the complement, interior, and boundary of the sets in Exercises 5 through 8.

5. $A = \{(x_1, x_2, \ldots, x_{10})|0 \leq x_3 < 1,\ 0 < x_6 \leq 2\}$

6. $B = \{\mathbf{x} \in \mathbb{R}^5|\mathbf{x} = (t+1,\ t,\ t-1, t-2,\ -t),\ t \in \mathbb{R}\}$

7. $C = \{\mathbf{x} \in \mathbb{R}^3|x_2 > -1\}$

8. $D = \{x \in \mathbb{R}|x^2 + 5x - 14 = 0\}$

Identify the subsets of $\mathbb{R}^4$ given in Exercises 9 through 11 as open or closed.

9. $\{(x_1,\ x_2,\ x_3,\ x_4)|\ -1 \leq x_1 \leq 1,\ \ 0 \leq x_2 \leq 1,\ \ -4 \leq x_3 \leq 4,\ \ 0 \leq x_4 \leq 1\}$

10. $\{\mathbf{x} \in \mathbb{R}^4|\ ||\mathbf{x} - (9,\ 3,\ 2,\ 7)|| > 6\}$

11. $\{\mathbf{x} \in \mathbb{R}^4|\ (4,\ 9,\ 0,\ 1) \cdot (\mathbf{x} - (2,\ 1,\ 5,\ -4)) = 0\}$

In Exercises 12 through 15, find the maximum and minimum values of the given linear function on the indicated set.

12. $f(x, y) = 1 + 2x + 5y;\ A = \{(x,\ y)|x \geq 0,\ y \geq 0,\ x + y \leq 1\}$

13. $f(x, y) = -4x + y;\ B = \{(x,\ y)|x^2 + y^2 \leq 81\}$

14. $f(x, y) = x/2 + y/3;\ C = \{(x,\ y)|x^2 \leq y \leq 16\}$

15. $f(x, y) = 6 - x + 4y;\ D = \{(x,\ y)|1 \leq x \leq 3,\ y + x = 1\}$

From your knowledge about the graphs of functions, find (approximately) the maximum and minimum values of the functions in Exercises 16 through 20 on the indicated sets, and find the approximate location of these extrema.

16. $g(x) = \sin x,\quad S = (0,\, 2\pi),\quad T = [-\pi/4, \pi]$

17. $h(x,\, y) = \sqrt{x^2 + y^2};\ S = \{(x,\, y)|\ x^2 + y^2 \leq 4\}$

18. $h(x,\, y) = \sqrt{x^2 + y^2};\ T = \{(x,\, y)|\ x + y = 0,\ -1 \leq x \leq 1\}$

19. $k(x,\, y) = 1 - x^2 - y^2;\ S = \{(x, y)|\ x + y = 0\}$

20. $k(x,\, y) = 1 - x^2 - y^2;\ T = \{(x,\, y)|\ 0 \leq x \leq 1,\ 0 \leq y \leq 1\}$

If S is a closed and bounded subset of $\mathbb{R}^3$ whose boundary is a **polyhedron** (i.e., the boundary consists of polygons), then a linear function f of three variables attains its extreme values on the boundary of S. But since each of these faces is the graph of a linear function, it follows that f restricted to one of the faces is a linear function of two variables. This two variable function, then, attains its extreme values on the boundary of the polygon. In other words, the extreme values of f occur on the edges of the polyhedron S. Use this fact to find the maximum and minimum values of the following functions on the indicated sets.

21. $f(x, y, z) = 3x+8y-z,\ S = \{(x, y, z)\,|\,0 \leq x \leq 1,\ 0 \leq y \leq 1,\ 0 \leq z \leq 1\}$

22. $f(x,\, y,\, z) = x - y + z;\ A = \{(x,\, y,\, z)|x \geq 0, y \geq 0\ z \geq 0,\ x + y + z \leq 1\}$

3.4 Partial Derivatives

A **partial derivative** of a function of several variables is a measure of the rate of change in that function with respect to one of its variables. When we take this measurement, we understand that the other variables are independent of the one of interest. For instance, if $T(x, y, z)$ represents the temperature of the air in a room at position (x, y, z), then the partial derivative of T with respect to x at, for example, the point $(0,\, 0,\, 0)$ is the derivative of the single-variable function $T(x, 0, 0)$ when x is 0.

The general notion of partial derivative is central in science. When a scientist studies a physical system, he or she identifies a numerical quantity that depends on other "inputs" to the system. By holding all but one of the inputs fixed, the scientist changes the remaining one and studies the resulting rate of change in the system. In effect, the scientist is determining a partial derivative.

Definition 3.4.1 For a function $f : D \subset \mathbb{R}^n \to \mathbb{R}$, the **partial derivative** of f with respect to the variable x_i is the function $\partial f/\partial x_i$ defined by

$$\begin{aligned}
&\frac{\partial f}{\partial x_i}(x_1, x_2, \dots, x_n) \\
&\quad = \lim_{h\to 0} \frac{f(x_1, \dots, x_{i-1}, x_i + h, x_{i+1}, \dots x_n) - f(x_1, \dots, x_{i-1}, x_i, x_{i+1}, \dots x_n)}{h}
\end{aligned}$$

whenever this limit exists.

Example 3.4.1 Let $f : \mathbb{R}^3 \to \mathbb{R}$ be given by $f(x, y, z) = x^2y/z$. Then, by Definition 3.4.1,

$$\begin{aligned}
\frac{\partial f}{\partial x}(x, y, z) &= \lim_{h\to 0} \frac{f(x+h, y, z) - f(x, y, z)}{h} \\
&= \lim_{h\to 0} \frac{(x+h)^2y/z - x^2y/z}{h} \\
&= \frac{y}{z} \lim_{h\to 0} \frac{x^2 + 2xh + h^2 - x^2}{h} \\
&= \frac{y}{z} \lim_{h\to 0} \frac{h(2x+h)}{h} = \frac{y}{z} \lim_{h\to 0}(2x+h) = \frac{2xy}{z}.
\end{aligned}$$

Similarly,

$$\begin{aligned}
\frac{\partial f}{\partial y}(x, y, z) &= \lim_{h\to 0} \frac{f(x, y+h, z) - f(x, y, z)}{h} \\
&= \lim_{h\to 0} \frac{x^2(y+h)/z - x^2y/z}{h} \\
&= \frac{x^2}{z} \lim_{h\to 0} \frac{y+h-y}{h} \\
&= \frac{x^2}{z} \lim_{h\to 0} \frac{h}{h} = \frac{x^2}{z} \lim_{h\to 0} 1 = \frac{x^2}{z}.
\end{aligned}$$

You should verify in the same way that

$$\frac{\partial f}{\partial z}(x, y, z) = \frac{-x^2y}{z^2}. \quad \blacklozenge$$

Example 3.4.1 suggests that partial derivatives can be evaluated using all the rules of ordinary differentiation (e.g., product, quotient, and chain rules) by regarding all the variables, except the one of interest, as constant. This is in fact the way we calculate partials most of the time.

Example 3.4.2 Calculate the partial derivatives of $f(x,y) = x\sin xy + ye^{-x^2}$.

Solution To calculate $\dfrac{\partial f}{\partial x}(x,y)$ we treat y as constant and use the product, chain, and sum rules as necessary:

$$\begin{aligned}\frac{\partial f}{\partial x}(x,y) &= \sin xy + x\cos xy \cdot y + ye^{-x^2}(-2x) \\ &= \sin xy + xy\cos xy - 2xye^{-x^2}.\end{aligned}$$

For $\dfrac{\partial f}{\partial y}(x,y)$, we proceed similarly: $\dfrac{\partial f}{\partial y}(x,y) = x^2\cos xy + e^{-x^2}$. ♦

Notation for partial derivatives appears in various forms. Instead of writing $\dfrac{\partial f}{\partial x_i}(x_1,\ldots,x_n)$, we may suppress the variables and just write $\dfrac{\partial f}{\partial x_i}$, bearing in mind that the function still depends on $x_1,\ldots,x_n$. Very often we write f_{x_i} instead of $\dfrac{\partial f}{\partial x_i}$. There is a risk of mistaking the subscript x_i for an index, but the use should be clear from the context. To conform to typographical conventions, we may write the partial as $\partial f/\partial x_i$. Also, when we write $\dfrac{\partial}{\partial x_i}$ (or $\partial/\partial x_i$) with an expression to its right, we mean take the partial of the expression with respect to x_i.

Partial derivatives of functions of two variables have natural geometric interpretations. The partial of $f(x,y)$ with respect to x at a point (a,b) is, by definition,

$$\frac{\partial f}{\partial x}(a,b) = \lim_{h\to 0}\frac{f(a+h,b)-f(a,b)}{h}.$$

This is just the *derivative* of the function of one variable $f(x,b)$ when $x=a$. Since the derivative can be interpreted as the slope of a line tangent to the curve, the *partial* derivative is the slope of the line tangent to the curve in which the surface $z=f(x,y)$ and the plane $y=b$ intersect. See Figure 3.4.1.

Similarly, we see that $\dfrac{\partial f}{\partial y}(a,b)$ is the derivative of the function $f(a,y)$ when $y=b$, so this partial is the slope of the line tangent to the curve in which $z=f(x,y)$ and $x=a$ intersect.

Example 3.4.3 Suppose that the elevation of points on a landscape is given by the function $f(x,\, y) = x^2(y-2x^2)+5$ where x and y are measured,

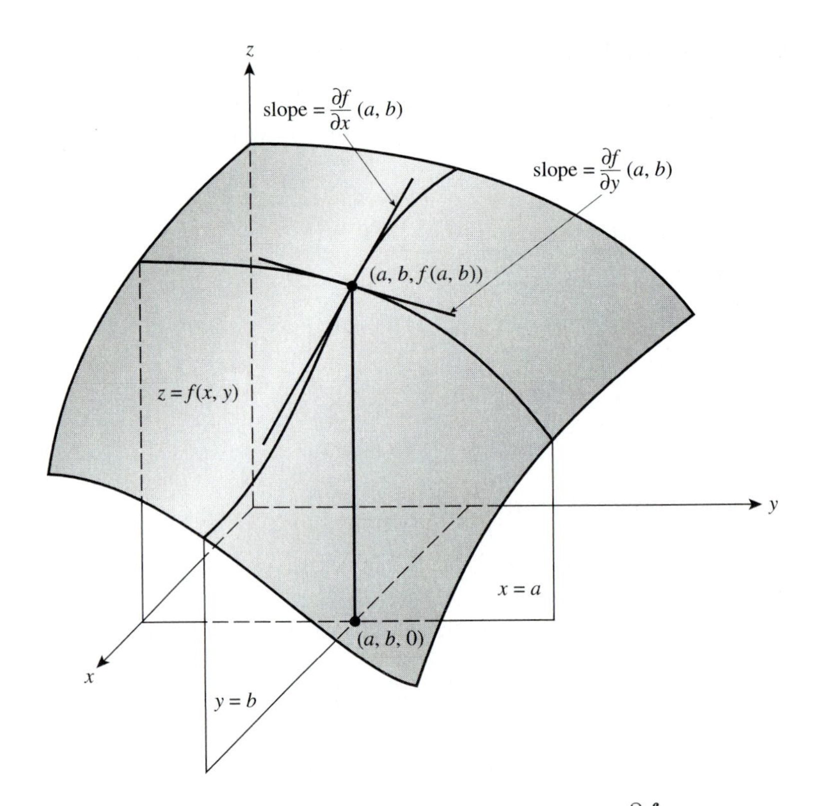

Figure 3.4.1 Geometric interpretation of $\dfrac{\partial f}{\partial x}(a, b)$ and $\dfrac{\partial f}{\partial y}(a, b)$

respectively, in thousands of feet east and north of some reference point. If a hiker is currently at the point on the landscape corresponding to the point $(1, 3)$ in the xy-plane and headed east, will she be going uphill or downhill?

Solution The question can be phrased in this way: At what rate is the hiker's elevation changing as the x-coordinate of her position increases from 1? In other words, what is the rate of change in $f(x, 3)$ when $x = 1$? By our preceding discussion, this is just $\dfrac{\partial f}{\partial x}(1, 3)$. Since $\dfrac{\partial f}{\partial x}(x, y) = 2xy - 8x^3$, we get $\dfrac{\partial f}{\partial x}(1, 3) = -2$. This value is negative, so it must be that the hiker is going downhill. ♦

Example 3.4.4 Let f be defined by

$$f(x,y) = \begin{cases} 1-|y|, & |y| < |x| \\ 1-|x|, & |y| \geq |x| \end{cases}.$$

Find $\dfrac{\partial f}{\partial x}(1,0)$ and $\dfrac{\partial f}{\partial y}(1,0)$ if they exist.

Solution We cannot use differentiation formulas directly, so we must resort to the definition. First, $\dfrac{\partial f}{\partial x}(1,0) = \dfrac{d}{dx}f(x,0)\Big|_{x=1}$. Now

$$f(x,0) = \left\{ \begin{array}{cc} 1, & 0 < |x| \\ 1-|x|, & 0 \geq |x| \end{array} \right\} = 1,$$

so $\dfrac{\partial f}{\partial x}(1,0) = \dfrac{d}{dx}(1)\Big|_{x=1} = 0$. Second,

$$f(1,y) = \left\{ \begin{array}{cc} 1-|y|, & |y| < 1 \\ 0, & |y| \geq 1 \end{array} \right\},$$

so for y near 0, $f(1,y) = 1-|y|$. For $\dfrac{d}{dy}f(1,y)\Big|_{y=0}$ to exist, we must have

$$\begin{aligned} \lim_{h\to 0^+} \frac{f(1,0+h)-f(1,0)}{h} &= \lim_{h\to 0^+} \frac{1-|h|-1}{h} \\ &= \lim_{h\to 0^+} \frac{-h}{h} = \lim_{h\to 0^+}(-1) = -1 \end{aligned}$$

equal to

$$\lim_{h\to 0^-} \frac{f(1,0+h)-f(1,0)}{h} = \lim_{h\to 0^-} \frac{+h}{h} = 1.$$

Since these two limits are different, we see that $\dfrac{\partial f}{\partial y}(1,0)$ does not exist. How can it be that one partial exists but the other does not? Refer back to Figure 1.2.11(a) in Section 1.2 to see a graph of f. The point (1, 0) corresponds to a point on the "crest" of the "rooftop." In the x-direction we walk the crest neither rising nor falling. In the y-direction, the slope of the tangent line is not uniquely defined. ♦

Higher-order partial derivatives are analogous to higher-order ordinary derivatives. For instance, the **second-order partial derivatives** of f are defined by

$$\frac{\partial^2 f}{\partial x_j\, \partial x_i} = \frac{\partial}{\partial x_j}\left(\frac{\partial f}{\partial x_i}\right), \quad 1 \leq i,\ j \leq n. \tag{3.4.1}$$

When $i \neq j$, we call these **mixed** partials. With the subscript notation, we write $f_{x_i x_j} = \dfrac{\partial^2 f}{\partial x_j\, \partial x_i}$ or $\partial^2 f/\partial x_j\, \partial x_i$. Notice that with the subscript notation, the partials are taken in the order, left to right, in which the subscripts appear; with the "∂" notation, the partials are taken in right-to-left order. Also, if $i = j$ in (3.4.1), we write $\partial^2 f/\partial x_i^2$ for the second partial.

Example 3.4.5 Let f be defined by

$$f(x, y) = 3x^2 + 4xy - 7y^2.$$

Then $\partial f/\partial x = 6x + 4y$ and $\partial f/\partial y = 4x - 14y$. By (3.4.1),

$$\frac{\partial^2 f}{\partial y\, \partial x} = \frac{\partial}{\partial y}\left(\frac{\partial f}{\partial x}\right) = 4, \quad \frac{\partial^2 f}{\partial x\, \partial y} = \frac{\partial}{\partial x}\left(\frac{\partial f}{\partial y}\right) = 4,$$

$$\frac{\partial^2 f}{\partial x^2} = \frac{\partial}{\partial x}\left(\frac{\partial f}{\partial x}\right) = 6, \quad \frac{\partial^2 f}{\partial y^2} = \frac{\partial}{\partial y}\left(\frac{\partial f}{\partial y}\right) = -14. \quad \blacklozenge$$

Partial derivatives of order higher than two are defined as you would expect. For instance, the third-order partials are

$$\frac{\partial}{\partial x_k}\left(\frac{\partial^2 f}{\partial x_j \partial x_i}\right), \qquad 1 \leq i, j, k \leq n.$$

How many different third-order partials are there?

In Example 3.4.5, you may have noticed that $\partial^2 f/\partial y\, \partial x = \partial^2 f/\partial x\, \partial y$. Coincidence? No. For almost all functions you encounter in this book, the mixed partials are equal at almost every point. The general principle at work here is expressed in the following theorem, which plays a surprisingly important role in our study of vector calculus.

Theorem 3.4.1 *(Clairaut's[3] Theorem) Suppose that the mixed partials of $f : D \subset \mathbb{R}^2 \to \mathbb{R}$ are defined in an open disk B about a point $\mathbf{a}$ in D. If $\partial^2 f/\partial y\, \partial x$ and $\partial^2 f/\partial x\, \partial y$ are continuous at $\mathbf{a}$ then*

$$\frac{\partial^2 f}{\partial y\, \partial x}(\mathbf{a}) = \frac{\partial^2 f}{\partial x\, \partial y}(\mathbf{a}). \tag{3.4.2}$$

[3] A. C. Clairaut, an eighteenth-century French mathematician, was interested in determining whether the flattening of the earth at the poles predicted by Isaac Newton was a real phenomenon. He took part in an expedition to Lapland to make measurements confirming the flattening, and he also formulated a mathematical model for it. In his model, there is a potential field related to gravitational and rotational forces. This field satisfies the identity in this theorem, which bears his name. See [4] for more details.

Proof (optional) Let $\mathbf{a} = (a, b)$. For small values of h,

$$\frac{\partial f}{\partial x}(x, y) \approx \frac{f(x+h, y) - f(x, y)}{h},$$

so it is plausible that

$$\begin{aligned}
&\frac{\partial^2 f}{\partial y\, \partial x}(a, b) \\
&\approx \frac{\frac{\partial f}{\partial x}(a, b+h) - \frac{\partial f}{\partial x}(a, b)}{h} \\
&\approx \frac{(f(a+h, b+h) - f(a, b+h)) - (f(a+h, b) - f(a, b))}{h^2} \\
&= \frac{1}{h^2}\left[(f(a+h, b+h) - f(a+h, b)) - (f(a, b+h) - f(a, b))\right].
\end{aligned}$$

This motivates us to examine the last quantity

$$\frac{1}{h^2}\left[(f(a+h, b+h) - f(a+h, b)) - (f(a, b+h) - f(a, b))\right]. \tag{3.4.3}$$

We will show that, as $h \to 0$, the quantity in (3.4.3) approaches both $\frac{\partial^2 f}{\partial x\, \partial y}(a, b)$ and $\frac{\partial^2 f}{\partial y\, \partial x}(a, b)$. From this it will follow that the mixed partials are equal at the point (a, b).

By the Mean Value Theorem applied to

$$g(x) = f(x, b+h) - f(x, b)$$

on the interval $[a, a+h]$, there exists $x^* \in [a, a+h]$ such that $g(a+h) - g(a) = g'(x^*)h$. Thus the bracketed quantity in (3.4.3) is

$$\left(\frac{\partial f}{\partial x}(x^*, b+h) - \frac{\partial f}{\partial x}(x^*, b)\right) \cdot h. \tag{3.4.4}$$

Now the difference in (3.4.4) is $k(b+h) - k(b)$ where $k(y) = \frac{\partial f}{\partial x}(x^*, y)$, so applying the Mean Value Theorem to k on the interval $[b, b+h]$, we see that there is a y^* such that the difference in (3.4.4) can be written as

$$k(b+h) - k(b) = k'(y^*) \cdot h = \frac{\partial}{\partial y}\left(\frac{\partial f}{\partial x}(x^*, y)\right)\Bigg|_{y=y^*} \cdot h = \frac{\partial^2 f}{\partial y\, \partial x}(x^*, y^*)h.$$

So (3.4.3) is

$$\frac{1}{h^2} \cdot \frac{\partial^2 f}{\partial y\, \partial x}(x^*, y^*)h \cdot h = \frac{\partial^2 f}{\partial y\, \partial x}(x^*, y^*).$$

By exactly the same reasoning, there exists $\tilde{x}$ between a and $a + h$ and $\tilde{y}$ between b and $b + h$ such that (3.4.3) is also equal to

$$\frac{\partial^2 f}{\partial x\, \partial y}(\tilde{x}, \tilde{y}).$$

Thus $\frac{\partial^2 f}{\partial x\, \partial y}(\tilde{x}, \tilde{y}) = \frac{\partial^2 f}{\partial y\, \partial x}(x^*, y^*)$. Since $(\tilde{x}, \tilde{y}) \to (a, b)$ and $(x^*, y^*) \to (a, b)$ as $h \to 0$, and since the mixed partials are continuous at (a, b), it follows that

$$\frac{\partial^2 f}{\partial x\, \partial y}(a, b) = \frac{\partial^2 f}{\partial y\, \partial x}(a, b). \quad \blacklozenge$$

Many physical systems are readily described by **partial differential equations** in which partial derivatives of an unknown function are related to one another and to the function itself. For example,

$$\frac{\partial^2 u}{\partial x^2} + \frac{\partial^2 u}{\partial y^2} = 0 \tag{3.4.5}$$

is a partial differential equation commonly called **Laplace's**[4] **equation** or the **potential equation**. This equation describes, for instance, the shape of a soap film suspended on a wire frame. It also may describe the equilibrium temperature distribution in a planar region where the temperature on the boundary of the region is specified.

Example 3.4.6 Show that $u = \ln(x^2 + y^2)$ is a solution to Laplace's equation.

Solution First, compute $\partial u/\partial x = 2x/(x^2 + y^2)$ and

$$\frac{\partial^2 u}{\partial x^2} = \frac{2(x^2 + y^2) - 2x(2x)}{(x^2 + y^2)^2} = \frac{2(y^2 - x^2)}{(x^2 + y^2)^2}. \tag{3.4.6}$$

Since the function is **symmetric** in the variables (i.e. interchanging the roles of x and y yields the same function), we can just write down

[4]Pierre-Simon Laplace (1749-1827) brought this equation to general attention in his work on celestial mechanics. According to [1], Euler had discovered the same equation in connection with fluid mechanics. The equation arises also in the theory of heat flow and electromagnetism.

$$\frac{\partial^2 u}{\partial y^2} = \frac{2(x^2 - y^2)}{(y^2 + x^2)^2}. \qquad (3.4.7)$$

Adding (3.4.6) and (3.4.7), we have

$$\frac{\partial^2 u}{\partial x^2} + \frac{\partial^2 u}{\partial y^2} = \frac{2(y^2 - x^2)}{(x^2 + y^2)^2} + \frac{2(x^2 - y^2)}{(x^2 + y^2)^2} = 0.$$

Thus the given function is a solution to Laplace's equation. ♦

Exercises 3.4

Calculate the first partial derivatives of each function with respect to the independent variables. If a point is given, evaluate the partials at that point.

1. $g(x, y) = x^2y^3 - x^3y^2$; $(-2, 3)$

2. $f(x, y) = (x - y)\cos(x + y)$; $(\pi/2, 3\pi/2)$

3. $f(x, y) = e^{xy} - ye^x$

4. $h(x, y) = 1/\sqrt{x^2 + y^2}$

5. $g(x, y, z) = x\sin(2y + z)$; $(1, \pi/4, \pi/4)$

6. $h(r, \theta) = r\cos\theta$

7. $B(\rho, \theta, \phi) = \rho\sin\phi\cos\theta$

8. $P(h, k) = e^{a-h}\sin(h - k)$

9. $F(x_1, x_2, x_3, x_4) = \dfrac{x_1 - x_2}{x_3 + x_4}$

10. $k(x_1, x_2, x_3, x_4) = x_2x_3x_4 - x_1x_3x_4 + x_1x_2x_4 - x_1x_2x_3$

11. $f(x_1, x_2, x_3, x_4) = \ln\dfrac{x_1 + x_2}{x_3 - x_4}$; $(5, 3, 2, -2)$

12. $f(x_1, x_2, x_3, x_4, x_5) = x_1\arccos(x_2x_3) + \arctan(x_4/x_5)$

13. $M(x_1, x_2, \ldots, x_n) = \displaystyle\sum_{i=1}^{n} a_ix_i$

14. $G(x_1, x_2, \ldots, x_n) = \sqrt{\displaystyle\sum_{i=1}^{n} x_i^2}$

15. In Example 3.4.3, suppose that the hiker is instead going north. Will she then be going uphill or downhill? At what rate will she be ascending or descending? If she is going south, will she go uphill or downhill? At what rate?

16. An empirical formula for the surface area of a child's body is

$$A = 0.0177h^{0.72}w^{0.42},$$

where h is the child's height in inches and w is the child's weight in pounds (see [2]). If a child is 40 inches tall and weighs 50 pounds, which will cause a larger increase in surface area, an increase of 1 inch in height or 1 pound in weight?

17. The **ideal gas law** states that for a gas confined to a container with volume V at temperature T, the pressure P exerted on the container wall is given by $P = RT/V$ where R is an empirically determined constant. (In this equation, volume is measured in cubic centimeters and temperature in Kelvin.) Suppose that an ideal gas is in a vessel with volume 1000 cm^3 at 600 K. Which will have a bigger effect on pressure, a small decrease in volume or a small increase in temperature?

18. For three resistors, with resistances r_1, r_2, and r_3, wired in parallel, the effective resistance R is given by $R = 1/(1/r_1 + 1/r_2 + 1/r_3)$. Suppose that the resistances are 10, 15, and 20 ohms, respectively. To which resistor is the effective resistance most sensitive?

19. In Figure 3.4.2 are shown some level curves of a function $f(x, y)$. Use this to give an approximation for the following partial derivatives

(a) $\dfrac{\partial f}{\partial x}(1, 0)$ (b) $\dfrac{\partial f}{\partial y}(0, -2.8)$

In Exercises 20 through 24, compute the indicated partial derivatives by using Definition 3.4.1.

20. $f(x, y) = 2x^2 + x - y;\quad \dfrac{\partial f}{\partial x}(x, y),\ \dfrac{\partial f}{\partial y}(x, y)$

21. $g(x, y) = xe^y;\quad \dfrac{\partial g}{\partial x}(x, y),\ \dfrac{\partial g}{\partial y}(x, y)$

22. $h(x, y) = \sin x;\quad \dfrac{\partial h}{\partial x}(x, y),\ \dfrac{\partial h}{\partial y}(x, y)$

23. $F(x, y) = \begin{cases} e^{-1/(x^2+y^2)}, & x^2 + y^2 \neq 0; \\ 0, & x^2 + y^2 = 0 \end{cases};\quad \dfrac{\partial F}{\partial x}(0, 0),\ \dfrac{\partial F}{\partial y}(0, 0)$

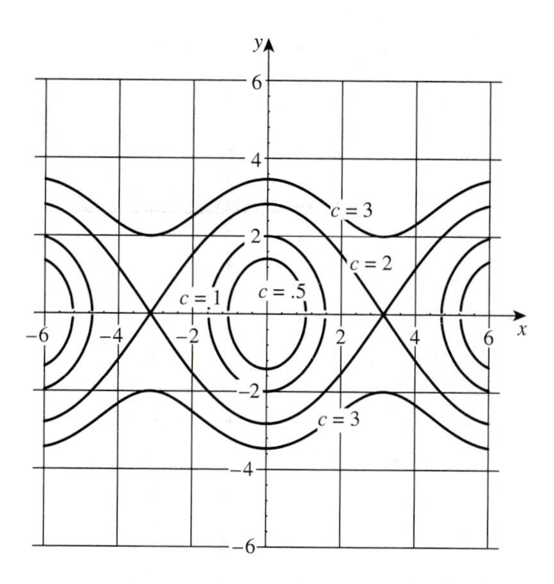

Figure 3.4.2 Figure for Exercise 19

24. $G(x,y) = \begin{cases} \sin(x^2+y^2)/x^2+y^2, & x^2+y^2 \neq 0; \\ 1, & x^2+y^2 = 0 \end{cases}$; $\dfrac{\partial G}{\partial x}(0,0), \dfrac{\partial G}{\partial y}(0,0)$

In Exercises 25 through 30, calculate all of the second-order partials of the following functions.

25. $u(x,y) = 4x^3y^4 + x^2y^5 + xy$

26. $v(x,y) = \sin x \cos y$

27. $w(x,y) = e^{xy}$

28. $f(x,y,z) = x\ln(y \sin z)$

29. $r(v,\theta) = \dfrac{v^2 \sin 2\theta}{g}$

30. $A(P,r,t) = P\left(1 + \dfrac{r}{n}\right)^{nt}$

31. Show that $f(x,y) = e^{x-y}$ is a solution to the partial differential equation $\dfrac{\partial u}{\partial x} + \dfrac{\partial u}{\partial y} = 0$.

32. Show that the partial differential equation $\dfrac{\partial u}{\partial x} + k\dfrac{\partial u}{\partial y} = 0$ has $u = f(y - kx)$ as a solution where f is any differentiable function.

33. A function u of two variables is called **harmonic** if it satisfies Laplace's equation,

$$\frac{\partial^2 u}{\partial x^2} + \frac{\partial^2 u}{\partial y^2} = 0.$$

Show that the following functions are harmonic:

(a) $u = \arctan \dfrac{y}{x}$

(b) $u = \ln(x^2 + y^2)$

34. Find conditions on the coefficients in $q(x, y) = ax^2 + bxy + cy^2$ that ensure that q is harmonic. Show that when a, b, and c, satisfy these conditions, q is an indefinite quadratic form.

35. Prove that if u and v have continuous mixed partials and satisfy the **Cauchy-Riemann**[5] **equation**

$$\frac{\partial u}{\partial x} = \frac{\partial v}{\partial y} \quad \text{and} \quad \frac{\partial u}{\partial y} = -\frac{\partial v}{\partial x} \tag{3.4.8}$$

then both u and v are harmonic.

36. The partial differential equation

$$\frac{\partial^4 u}{\partial x^4} = -\frac{1}{c^2}\frac{\partial^2 u}{\partial t^2},$$

where c is a constant, comes up in the study of deflections of a thin beam. Show that

$$u(x, t) = \sin(\lambda \pi x)\cos(\lambda^2 \pi^2 ct)$$

is a solution for any choice of the parameter λ.

37. The **Kortweg-DeVries equation**

$$\frac{\partial u}{\partial t} + u\frac{\partial u}{\partial x} + \frac{\partial^3 u}{\partial x^3} = 0,$$

[5] Augustin-Louis Cauchy (1789-1857), a Frenchman, contributed to such areas as geometry, analysis, number theory, and mechanics. His textbooks on analysis presented calculus in substantially the form that it is taught in present-day introductory courses. Bernhard Riemann (1826-1866) contributed to mathematics in general through his analytic perspective. In particular, he used analysis to elucidate problems in number theory and geometry. The process of integration presented in introductory calculus texts, called **Riemann integration** in his honor, was but one small part of his work.

arises in modeling shallow water waves (called **solitons**) . Show that

$$u(x,t) = 12a^2 \ \operatorname{sech}^2(ax - 4a^3t)$$

is a solution to the Kortweg-DeVries equation.

38. Show that $g(x,\, y,\, z) = 1/(x^2 + y^2 + z^2)^{1/2}$ is a solution to the three-dimensional Laplace equation $\dfrac{\partial^2 u}{\partial x^2} + \dfrac{\partial^2 u}{\partial y^2} + \dfrac{\partial^2 u}{\partial z^2} = 0$.

39. Show directly (i.e., without Clairaut's Theorem) that if $f(x,\, y) = x^n y^m$, where m and n are nonnegative integers, then $\partial^2 f/\partial y\, \partial x = \partial^2 f/\partial x\, \partial y$. Use this to show that if $f(x,\, y)$ is any polynomial then $\partial^2 f/\partial y\, \partial x = \partial^2 f/\partial x\, \partial y$.

40. Show that $u = \exp\left(\sum_{i=1}^{n} a_i x_i\right)$, where the coefficients satisfy $\sum_{i=1}^{n} a_i^2 = 1$, is a solution to

$$\sum_{i=1}^{n} \frac{\partial^2 u}{\partial x_i^2} = u.$$

41. In economics, the **Cobb-Douglas production function**

$$P(x,\, y) = Kx^\alpha y^{1-\alpha},$$

where $0 < \alpha < 1$, is supposed to measure the productive capacity of an economic entity (such as the United States) as a function of capital x and labor y. Show that P is a solution to the differential equation

$$x\frac{\partial u}{\partial x} + y\frac{\partial u}{\partial y} = u. \qquad (3.4.9)$$

Equation (3.4.9) is often called **Euler's[6] equation**. The function P has the desirable property (from an economist's point of view) that it exhibits constant *returns to scale*: If x and y are replaced by tx and ty for some constant t, then the value of P is scaled up by the factor t.

42. The **Sharpe-Lotka partial differential equation**

$$\frac{\partial u}{\partial t} + \frac{\partial u}{\partial a} = -ku$$

[6]Leonhard Euler (1707-1783), was born in Switzerland and spent most of his life at the Academies of Sciences in St. Petersburg and Berlin. He produced mathematics at a prodigious rate throughout his life, as evidenced by the fact that his papers constituted at least half of the journals of both academies for many years after his death. Many mathematical notions, equations, and theorems bear his name.

arises in modeling an age-structured population. Here $u = u(a, t)$ represents the number of individuals of age a at a time t; k is a given constant mortality rate. Find a relationship between constants c_1 and c_2 such that $u = \exp(c_1 a + c_2 t)$ is a solution to the Sharpe-Lotka equation.

43. Let f be the function defined in Example 3.4.4. For each of the following, determine whether the partial derivative exists; if it does, find its value.

(a) $\dfrac{\partial f}{\partial x}(0, 0)$ (b) $\dfrac{\partial f}{\partial y}(0, 0)$

(c) $\dfrac{\partial f}{\partial x}(1, 2)$ (d) $\dfrac{\partial f}{\partial y}(1, 2)$

(e) $\dfrac{\partial f}{\partial x}(1, 1)$ (f) $\dfrac{\partial f}{\partial y}(1, 1)$

[*Suggestion*: Consult Figure 1.2.11(a).]

44. In Exercise 12 in Section 3.2 you showed that the limit of the function $xy/x^2 + y^2$ as $(x, y) \to (0, 0)$ does not exist. Thus the function

$$f(x, y) = \begin{cases} xy/(x^2 + y^2), & x^2 + y^2 \neq 0 \\ 0, & x^2 + y^2 = 0 \end{cases}$$

is not continuous at $(0, 0)$. Show, however, that both $\dfrac{\partial f}{\partial x}(0, 0)$ and $\dfrac{\partial f}{\partial y}(0, 0)$ exist and are equal to zero. How can this happen? If you have a computer graphing program, examine the graph of f near $(0, 0)$. What are the values of f along the lines $x = 0$ and $y = 0$?

3.5 Differentiability and the Total Derivative

In elementary calculus you learned that the graph of a function f has a tangent line precisely where the function is differentiable (i.e., at numbers a where the limit $\lim_{x\to a}(f(x) - f(a))/(x - a)$ exists). Another way to say this is that a function f is differentiable at a if, upon continued magnification of the graph about the point $(a, f(a))$, the graph is indistinguishable from a line. See Figure 3.5.1. On the other hand, f is not differentiable at a point where continued magnification of the graph fails to produce a limiting line. See Figure 3.5.2.

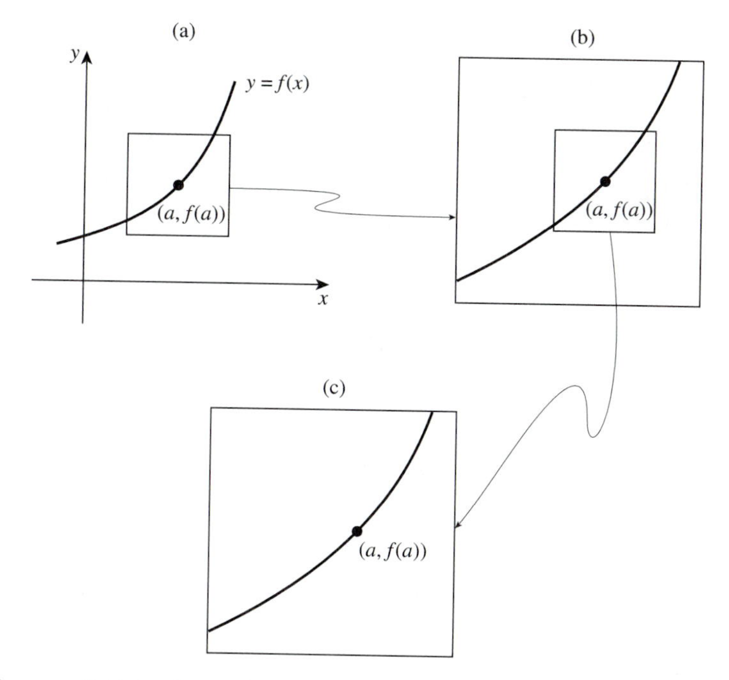

Figure 3.5.1 Zooming in on a point on the graph of $y = f(x)$ at which f is differentiable, we see that the graph looks more and more like a line.

We generalize the latter view of differentiability to functions of several variables. For instance, we will say that a function of two variables $f(x, y)$ is differentiable at a point (a, b) if the graph of $z = f(x, y)$ near that point is indistinguishable from a plane. Since a plane through the point $(a, b, f(a, b))$ has an equation of the form

$$z - f(a,b) = a_1(x - a) + a_2(y - b)$$

(if it is not parallel to the z-axis), this means that the nonlinear function $f(x, y) - f(a, b)$ must be indistinguishable from a linear transformation $T(x - a, y - b) = a_1(x - a) + a_2(y - b)$ for (x, y) near (a, b).

Definition 3.5.1 A function $\mathbf{f} : U \subset \mathbb{R}^n \to \mathbb{R}^m$ is **differentiable** at a point $\mathbf{a} \in U$ if there exists a linear transformation $\mathbf{T} : \mathbb{R}^n \to \mathbb{R}^m$ such that

$$\lim_{\mathbf{x} \to \mathbf{a}} \frac{||\mathbf{f}(\mathbf{x}) - \mathbf{f}(\mathbf{a}) - \mathbf{T}(\mathbf{x} - \mathbf{a})||}{||\mathbf{x} - \mathbf{a}||} = 0. \tag{3.5.1}$$

We call $\mathbf{T}$ the **derivative** (or **total derivative**) **of f at a**. Often we use the slightly cumbersome but expressive notation $\mathbf{Df}(\mathbf{a})$ for the total derivative.

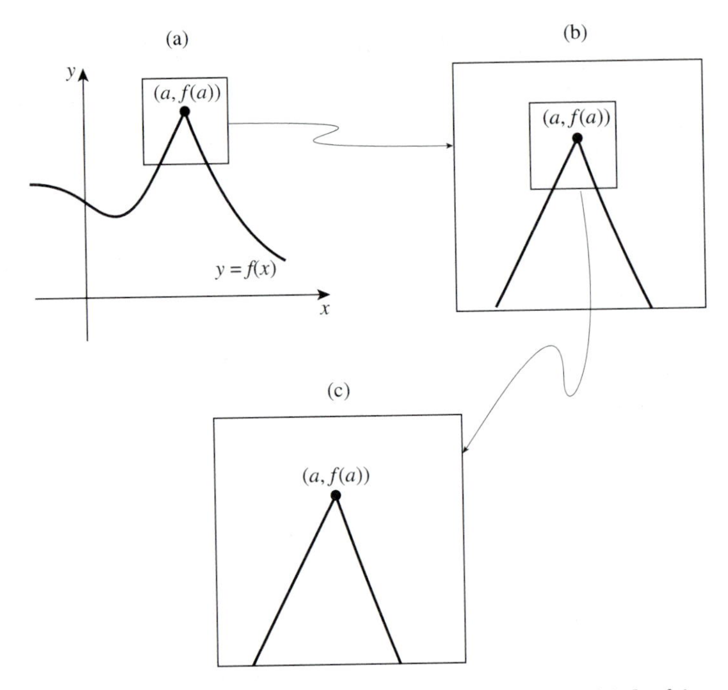

Figure 3.5.2 When we zoom in on a point at which f is not differentiable, the graph does not come to resemble a line.

If the function f is scalar-valued, then, in this book, we will print the total derivative of f at $\mathbf{a}$ in italics: $Df(\mathbf{a})$. This is consistent with the way we distinguish between vector- and scalar-valued functions elsewhere.

By what we know about linear transformations, calculating the derivative of $\mathbf{f}$ at $\mathbf{a}$ amounts to finding the matrix for $\mathbf{Df}(\mathbf{a})$. So what are the entries of this matrix? In some calculus books, the following theorem is called the Fundamental Lemma. It gives an answer to this question in the special case of a scalar valued function.

Theorem 3.5.1 *If $f : U \subset \mathbb{R}^n \to \mathbb{R}$ has partial derivatives that are defined on an open ball about a point $\mathbf{a} \in U$ and if $\partial f/\partial x_1$, $\partial f/\partial x_2$, ..., $\partial f/\partial x_n$ are continuous at $\mathbf{a}$, then f is differentiable at $\mathbf{a}$ and its derivative is given by*

$$(Df(\mathbf{a}))(\mathbf{x}) = \begin{bmatrix} \frac{\partial f}{\partial x_1}(\mathbf{a}) & \frac{\partial f}{\partial x_2}(\mathbf{a}) & \cdots & \frac{\partial f}{\partial x_n}(\mathbf{a}) \end{bmatrix} \mathbf{x}. \tag{3.5.2}$$

Proof For the sake of clarity, we prove the theorem for $n = 2$. (You have the opportunity in the exercises to write a proof of the general form.)

Let $\mathbf{a} = \begin{bmatrix} a \\ b \end{bmatrix}$, and let $\mathbf{x} = \begin{bmatrix} x \\ y \end{bmatrix}$ be the variables for f. The basic line of reasoning is to write the difference $f(\mathbf{x}) - f(\mathbf{a})$ as a telescoping sum and apply the Mean Value Theorem to each of its terms. First, we can write

$$f(\mathbf{x}) - f(\mathbf{a}) = (f(x, y) - f(a, y)) + (f(a, y) - f(a, b)). \qquad (3.5.3)$$

For the moment, we regard x and y as fixed and consider the function g of one variable given by

$$g(t) = f(t, y),$$

where t varies over the interval between a and x. Since $\partial f/\partial x$ is defined on the interval, g is continuous and differentiable, so by the Mean Value Theorem, there exists a number $\overline{x}$ between x and a (which depends on the choice of y) such that

$$g(x) - g(a) = g'(\overline{x})(x - a).$$

In terms of f, this says that

$$f(x, y) - f(a, y) = \frac{\partial f}{\partial x}(\overline{x}, y)(x - a). \qquad (3.5.4)$$

We apply the same reasoning to the function $h(t) = f(a, t)$ for t varying over the interval between b and y, and the conclusion is similar: There exists $\overline{y}$ between b and y such that

$$f(a, y) - f(a, b) = \frac{\partial f}{\partial y}(a, \overline{y})(y - b). \qquad (3.5.5)$$

Let $Jf(\mathbf{a}) = \left[\frac{\partial f}{\partial x}(a,\, b) \quad \frac{\partial f}{\partial y}(a,\, b)\right]$. Using (3.5.4) and (3.5.5) in (3.5.3), we can now write

$$\begin{aligned}
&f(\mathbf{x}) - f(\mathbf{a}) - Jf(\mathbf{a})(\mathbf{x} - \mathbf{a}) \\
&\quad = \frac{\partial f}{\partial x}(\overline{x}, y)(x - a) + \frac{\partial f}{\partial y}(a, \overline{y})(y - b) - \left[\frac{\partial f}{\partial x}(a,\, b) \quad \frac{\partial f}{\partial y}(a,\, b)\right]\begin{bmatrix} x - a \\ y - b \end{bmatrix} \\
&\quad = \left(\frac{\partial f}{\partial x}(\overline{x}, y) - \frac{\partial f}{\partial x}(a, b)\right)(x - a) + \left(\frac{\partial f}{\partial y}(a, \overline{y}) - \frac{\partial f}{\partial y}(a, b)\right)(y - b) \\
&\quad = \left(\frac{\partial f}{\partial x}(\overline{x}, y) - \frac{\partial f}{\partial x}(a, b),\, \frac{\partial f}{\partial y}(a, \overline{y}) - \frac{\partial f}{\partial y}(a, b)\right) \cdot (x - a,\, y - b) \\
&\quad = \left(\frac{\partial f}{\partial x}(\overline{x}, y) - \frac{\partial f}{\partial x}(a, b),\, \frac{\partial f}{\partial y}(a, \overline{y}) - \frac{\partial f}{\partial y}(a, b)\right) \cdot (\mathbf{x} - \mathbf{a}).
\end{aligned}$$

By the Cauchy-Schwarz inequality, the absolute value of this dot product is no more than the product of the magnitudes of the vectors. So we see that

$$\begin{aligned} &\frac{|f(\mathbf{x}) - f(\mathbf{a}) - Jf(\mathbf{a})(\mathbf{x} - \mathbf{a})|}{||\mathbf{x} - \mathbf{a}||} \\ &\leq \frac{\left|\left|\left(\frac{\partial f}{\partial x}(\overline{x}, y) - \frac{\partial f}{\partial x}(a, b),\ \frac{\partial f}{\partial y}(a, \overline{y}) - \frac{\partial f}{\partial y}(a, b)\right)\right|\right| \, ||(\mathbf{x} - \mathbf{a})||}{||(\mathbf{x} - \mathbf{a})||} \\ &= \left|\left|\left(\frac{\partial f}{\partial x}(\overline{x}, y) - \frac{\partial f}{\partial x}(a, b),\ \frac{\partial f}{\partial y}(a, \overline{y}) - \frac{\partial f}{\partial y}(a, b)\right)\right|\right|. \end{aligned} \tag{3.5.6}$$

As $\mathbf{x} \to \mathbf{a}$, we must have $(\overline{x}, a) \to \mathbf{a}$ and $(a, \overline{y}) \to \mathbf{a}$. Since $\partial f/\partial x$ and $\partial f/\partial y$ are continuous at $\mathbf{a}$, the right-hand side of (3.5.6) goes to zero as $\mathbf{x} \to \mathbf{a}$. Thus the quotient on the left-hand side of (3.5.6) is squeezed between 0 and a quantity approaching zero:

$$\lim_{\mathbf{x} \to \mathbf{a}} \frac{||f(\mathbf{x}) - f(\mathbf{a}) - Jf(\mathbf{a})(\mathbf{x} - \mathbf{a})||}{||\mathbf{x} - \mathbf{a}||} = 0.$$

By Definition 3.5.1, the matrix of partials $Jf(\mathbf{a})$ represents the total derivative of f at $\mathbf{a}$. ♦

The matrix $Jf(\mathbf{a})$ defined in the proof of Theorem 3.5.1 is called the **Jacobian[7] matrix of f at $\mathbf{a}$**. In general, if $f : D \subset \mathbb{R}^n \to \mathbb{R}$ then the Jacobian matrix is

$$Jf = \begin{bmatrix} \frac{\partial f}{\partial x_1} & \frac{\partial f}{\partial x_2} & \cdots & \frac{\partial f}{\partial x_n} \end{bmatrix}.$$

Example 3.5.1 Calculate the Jacobian matrix of $f(x, y) = 2x^3 + 5xy^2 + y^3$ at the point $\mathbf{a} = (-3, 1)$.

Solution By Theorem 3.5.1, we just need the partials of f at $\mathbf{a}$:

$$\begin{aligned} \frac{\partial f}{\partial x}(-3, 1) &= (6x^2 + 5y^2)\Big|_{(-3,1)} = 59 \\ \frac{\partial f}{\partial y}(-3, 1) &= (10xy + 3y^2)\Big|_{(-3,1)} = -27. \end{aligned}$$

Thus $Jf(-3, 1) = \begin{bmatrix} 59 & -27 \end{bmatrix}$ and

$$(\mathbf{Df}(-3, 1))\,(\mathbf{x}) = \begin{bmatrix} 59 & -27 \end{bmatrix} \begin{bmatrix} x \\ y \end{bmatrix} = 59x - 27y. \quad ♦$$

[7] This matrix is named after Carl Gustav Jacob Jacobi (1804-1851) who introduced it in a paper in 1841.

For a real-valued function f, it is traditional to write the total derivative in an alternative way using **differential notation**:

$$df = \frac{\partial f}{\partial x_1} dx_1 + \frac{\partial f}{\partial x_2} dx_2 + \cdots + \frac{\partial f}{\partial x_n} dx_n. \tag{3.5.7}$$

In this notation, df is called the **total differential** of f. It is understood that if the partials are all evaluated at $\mathbf{a} = (a_1, a_2, \ldots, a_n)$, then the variables $dx_1, dx_2, \ldots, dx_n$ stand, respectively, for $x_1 - a_1, x_2 - a_2, \ldots, x_n - a_n$. You can think of (3.5.7) as (3.5.2) "multiplied out." The quantities $dx_1, dx_2, \ldots, dx_n$ are read as "differential in x_1," ..., "differential in x_n," or sometimes as "increment in x_1," ..., "increment in x_n."

Example 3.5.2 The total differential of $f(x, y) = xe^{-y} + ye^{2x}$ is

$$df = \frac{\partial f}{\partial x} dx + \frac{\partial f}{\partial y} dy = \left(e^{-y} + 2ye^{2x}\right) dx + \left(-xe^{-y} + e^{2x}\right) dy. \quad \blacklozenge$$

For a vector-valued function $\mathbf{f}$ from $\mathbb{R}^n$ to $\mathbb{R}^m$, the total derivative $\mathbf{Df(a)}$ is a linear transformation from $\mathbb{R}^n$ to $\mathbb{R}^m$, so the Jacobian matrix $J\mathbf{f}(\mathbf{a})$ is an $m \times n$ matrix whose rows are the Jacobian matrices of the component functions. This is the content of the following theorem.

Theorem 3.5.2 *Let* $\mathbf{f} : U \subset \mathbb{R}^n \to \mathbb{R}^m$ *be given by the formula*

$$\mathbf{f}(\mathbf{x}) = (f_1(\mathbf{x}), f_2(\mathbf{x}), \ldots, f_m(\mathbf{x})).$$

If $\partial f_i/\partial x_j$ *is continuous at* $\mathbf{a}$ *for* $i = 1, \ldots, m$ *and* $j = 1, \ldots, n$, *then* $\mathbf{f}$ *is differentiable at* $\mathbf{a}$ *and the derivative of* $\mathbf{f}$ *at* $\mathbf{a}$ *is the linear transformation* $\mathbf{Df(a)} : \mathbb{R}^n \to \mathbb{R}^m$ *given by*

$$(\mathbf{Df(a)})(\mathbf{x}) = \begin{bmatrix} \partial f_1/\partial x_1(\mathbf{a}) & \partial f_1/\partial x_2(\mathbf{a}) & \cdots & \partial f_1/\partial x_n(\mathbf{a}) \\ \partial f_2/\partial x_1(\mathbf{a}) & \partial f_2/\partial x_2(\mathbf{a}) & \cdots & \partial f_2/\partial x_n(\mathbf{a}) \\ \vdots & \vdots & & \vdots \\ \partial f_m/\partial x_1(\mathbf{a}) & \partial f_m/\partial x_2(\mathbf{a}) & \cdots & \partial f_m/\partial x_n(\mathbf{a}) \end{bmatrix} \mathbf{x}. \tag{3.5.8}$$

Example 3.5.3 Find the derivative of

$$\mathbf{f}(x_1, x_2) = (x_1 x_2 + x_1^3 x_2^2,\ x_1^2 + x_2^2,\ x_1/x_2)$$

at the point $(1, -1)$.

Solution Here the component functions are

$$\begin{aligned} f_1(x_1, x_2) &= x_1x_2 + x_1^3x_2^2 \\ f_2(x_1, x_2) &= x_1^2 + x_2^2 \\ f_3(x_1, x_2) &= \frac{x_1}{x_2}. \end{aligned}$$

So the Jacobian matrix is

$$\begin{aligned} J\mathbf{f}(1, -1) &= \begin{bmatrix} \partial f_1/\partial x_1(1,-1) & \partial f_1/\partial x_2(1,-1) \\ \partial f_2/\partial x_1(1,-1) & \partial f_2/\partial x_2(1,-1) \\ \partial f_3/\partial x_1(1,-1) & \partial f_3/\partial x_2(1,-1) \end{bmatrix} \\ &= \begin{bmatrix} x_2 + 3x_1^2x_2^2 & x_1 + 2x_1^3x_2 \\ 2x_1 & 2x_2 \\ 1/x_2 & -x_1/x_2^2 \end{bmatrix}\Bigg|_{(1,-1)} = \begin{bmatrix} 2 & -1 \\ 2 & -2 \\ -1 & -1 \end{bmatrix}, \end{aligned}$$

and the derivative is

$$(\mathbf{Df}(1,-1))(\mathbf{x}) = \begin{bmatrix} 2 & -1 \\ 2 & -2 \\ -1 & -1 \end{bmatrix} \mathbf{x}. \quad \blacklozenge$$

The definition of differentiability says that the difference

$$(f(\mathbf{x}) - f(\mathbf{a})) - (\mathbf{Df}(\mathbf{a}))(\mathbf{x} - \mathbf{a})$$

is less than a constant multiple of $||\mathbf{x} - \mathbf{a}||$ for all choices of $\mathbf{x}$ sufficiently close to $\mathbf{a}$. So for $\mathbf{x}$ near $\mathbf{a}$, this difference is approximately 0. Rearranging the terms, we can write this as

$$\mathbf{f}(\mathbf{x}) \approx \mathbf{f}(\mathbf{a}) + \mathbf{Df}(\mathbf{a})(\mathbf{x} - \mathbf{a}) \tag{3.5.9}$$

for $\mathbf{x}$ near $\mathbf{a}$. We call the right-hand side of (3.5.9) the **differential approximation of f near a**. When the function is real-valued, (3.5.9) is often written in differential notation:

$$f(\mathbf{x}) \approx f(\mathbf{a}) + df.$$

With $\Delta f = f(\mathbf{x}) - f(\mathbf{a})$, this is written succinctly as

$$\Delta f \approx df. \tag{3.5.10}$$

Example 3.5.4 Suppose that $\mathbf{f}$ is differentiable at the point $(1,2)$, satisfies $\mathbf{f}(1,2) = (2,\ 6,\ -4)$, and has Jacobian matrix

$$J\mathbf{f}(1,2) = \begin{bmatrix} 1 & -1 \\ 0 & 6 \\ -4 & 3 \end{bmatrix}.$$

Find an approximation for $\mathbf{f}(1.2,\ 1.9)$.

Solution We simply apply (3.5.9) with $\mathbf{a} = (1,\ 2)$ and $\mathbf{x} = (1.2,\ 1.9)$:

$$\begin{aligned} \mathbf{f}(1.2,\ 1.9) &\approx \mathbf{f}(1,2) + J\mathbf{f}(\mathbf{a})\left((1.2,\ 1.9) - (1,\ 2)\right) \\ &= (2,\ 6,\ -4) + \begin{bmatrix} 1 & -1 \\ 0 & 6 \\ -4 & 3 \end{bmatrix} \begin{bmatrix} 0.2 \\ -0.1 \end{bmatrix} \\ &= \begin{bmatrix} 2 \\ 6 \\ -4 \end{bmatrix} + \begin{bmatrix} 0.3 \\ -0.6 \\ -1.1 \end{bmatrix} = \begin{bmatrix} 2.3 \\ 5.4 \\ -5.1 \end{bmatrix}. \quad \blacklozenge \end{aligned}$$

Example 3.5.5 The earth's surface is not actually spherical; it is "flattened" at the poles so that cross sections through the north and south poles are ellipses with semimajor radius 6370 km and semiminor radius 6350 km. Approximately what volume of water is needed to provide 1 mm of rainfall over the entire earth's surface?

Solution The description of the earth's surface means that it is an ellipsoid. Since the volume inside an ellipsoid $x^2/a^2+y^2/b^2+z^2/c^2 = 1$ is $V = 4\pi abc/3$, we could use this formula to calculate the volume of the earth (with $a = b = 6370$ and $c = 6350$) and subtract that from the volume inside the surface parallel to this ellipsoid that is 1 mm $= 0.000001$ km larger. It turns out that this parallel surface is not an ellipsoid, and we do not have a ready formula for the volume it encloses.

A second possibility is to assume that the parallel surface is well approximated by the ellipsoid with $a = b = 6370.000001$ and $c = 6350.000001$. Then the volume of water will be the difference between the volume enclosed by the larger ellipsoid and that enclosed by the smaller. (You should try this.)

A third approach is to use a differential approximation: The volume between the surfaces is an increment in the volume function which, by (3.5.10), is approximated by the differential in the volume dV corresponding to increments $da = db = dc = 0.000001$ km in the independent variables. To compute dV, we first compute

$$\left.\frac{\partial V}{\partial a}\right|_{(6370,6370,6350)} = \left.\frac{4\pi}{3}bc\right|_{(6370,6370,6350)} = 1.6943 \times 10^8.$$

Similarly, the partials with respect to b and c are

$$\frac{\partial V}{\partial b} = 1.6943 \times 10^8, \qquad \frac{\partial V}{\partial c} = 1.6997 \times 10^8.$$

So the volume of water is approximately

$$\begin{aligned} dV &= \frac{\partial V}{\partial a}da + \frac{\partial V}{\partial b}db + \frac{\partial V}{\partial c}dc \\ &= 1.6943 \times 10^8 \times 10^{-6} + 1.6943 \times 10^8 \times 10^{-6} + 1.6997 \times 10^8 \times 10^{-6} \\ &= 509\,\text{km}^3. \quad \blacklozenge \end{aligned}$$

Example 3.5.6 A function may have partials at a point but fail to be differentiable at that point. Consider, for instance,

$$f(x,y) = \begin{cases} \dfrac{x^3 - y^3}{x^2 + y^2}, & x^2 + y^2 \neq 0 \\ 0, & x^2 + y^2 = 0. \end{cases}$$

We can easily calculate $\frac{\partial f}{\partial x}(0,0)$ and $\frac{\partial f}{\partial y}(0,0)$ by Definition 3.4.1:

$$\begin{aligned} \frac{\partial f}{\partial x}(0,0) &= \lim_{x\to 0}\frac{f(x,0) - f(0,0)}{x - 0} = \lim_{x\to 0}\frac{x^3/x^2 - 0}{x} \\ &= \lim_{x\to 0}\frac{x^3}{x^3} = \lim_{x\to 0} 1 = 1 \end{aligned}$$

and

$$\begin{aligned} \frac{\partial f}{\partial y}(0,0) &= \lim_{y\to 0}\frac{f(0,y) - f(0,0)}{y - 0} = \lim_{y\to 0}\frac{-y^3/y^2 - 0}{y} \\ &= \lim_{y\to 0}\frac{-y^3}{y^3} = \lim_{y\to 0} -1 = -1. \end{aligned}$$

If f is differentiable at (0, 0), then the total derivative is the linear transformation $T(x, y) = 1 \cdot x - 1 \cdot y = x - y$. But

$$\lim_{(x,y)\to(0,0)} \frac{|f(x,y) - f(0,0) - (x-y)|}{\sqrt{x^2+y^2}} = \lim_{(x,y)\to(0,0)} \frac{x^2y - y^2x}{(x^2+y^2)^{3/2}}$$

does not exist (as you showed in Exercise 16 in Section 3.2). ♦

This example illustrates that it is not enough simply for a function to have partial derivatives defined at a point in order for it to be differentiable there. For a function to be differentiable at a point $\mathbf{a}$, the partial derivatives must be "well-behaved" in an open ball about $\mathbf{a}$. In particular, if the partials are continuous in an open ball about $\mathbf{a}$, then f *is* differentiable at $\mathbf{a}$.

Exercises 3.5

In Exercises 1 through 16, calculate the Jacobian matrix for the given function at the indicated point. Then write a formula for the total derivative.

1. $e(x, y) = 8x - 7y + 2$; $\mathbf{a} = (-4, -5)$

2. $f(x, y) = 8x - 7y + 2$; $\mathbf{a} = (1, -3)$

3. $g(x, y) = \dfrac{x - y}{x + y}$; $\mathbf{a} = (1/2, 3/2)$

4. $k(r, \theta) = r \sin\theta$; $\mathbf{a} = (5, \pi/6)$

5. $m(x, y, z) = \sin x \cos y \sin z$; $\mathbf{a} = (0, \pi/4, \pi/3)$

6. $p(x_1, x_2, x_3, x_4, x_5, x_6) = \dfrac{x_1^2 + x_2^2 + x_3^2}{x_4^2 + x_5^2 + x_6^2}$; $\mathbf{a} = (3, 4, 5, 1, 1, -1)$

7. $q(x_1, x_2, x_3, \ldots, x_{30}) = \ln(x_1^2 + x_2^2 + x_3^2 + \ldots + x_{30}^2)$;
$\mathbf{a} = (-1, 1, -1, 1, \ldots, -1, 1)$

8. $\mathbf{E}(x, y) = (2x + 3y - 9, 9x - y + 4)$; $\mathbf{a} = (6, 12)$

9. $\mathbf{F}(x, y) = (x^2 - y^2, 2xy)$; $\mathbf{a} = (3, 1)$

10. $\mathbf{G}(x, y) = \left(\arctan\dfrac{y}{x}, \dfrac{x}{x^2 + y^2}, \dfrac{y}{x^2 + y^2}\right)$; $\mathbf{a} = (1, 1)$

11. $\mathbf{H}(r, \theta) = (r\cos\theta, r\sin\theta)$; $\mathbf{a} = (\sqrt{2}, 3\pi/2)$

12. $\mathbf{J}(x, y, z) = (x + 8y - 7z + 2, 2x + z + 8, -x - y + z + 5)$;
$\mathbf{a} = (-2, 0, 5)$

13. $\mathbf{K}(x, y, z) = (z^2 - y^2, x^2 - z^2, y^2 - x^2)$; $\mathbf{a} = \left(\dfrac{1}{2}, \dfrac{1}{3}, \dfrac{1}{6}\right)$

14. $\mathbf{M}(x_1, x_2, x_3, x_4) = \left(\dfrac{x_1 + x_2}{x_3 + x_4}, \dfrac{x_3 + x_4}{x_1 + x_2}\right)$; $\mathbf{a} = (1, 1, 1, 1)$

15. $\mathbf{N}(x_1, x_2, x_3, x_4, x_5) = (x_1 e^{x_2}, x_3 e^{-x_4}, x_5 e^{x_1})$;
$\mathbf{a} = (0, -\ln 6, 4, \ln 2, 11)$

16. $\mathbf{f}(t) = (5t + 1, 2\cos 3t, 3\sin 3t)$; $a = \pi/6$

In Exercises 17 through 20, use the given values of $\mathbf{f}$ and $J\mathbf{f}$ at the indicated point to find an approximation for $\mathbf{f}(\mathbf{x})$.

17. $\mathbf{f}(3, -1) = (2, 6)$; $J\mathbf{f}(3, -1) = \begin{bmatrix} 2 & 8/3 \\ -3 & 2 \end{bmatrix}$; $\mathbf{x} = (3.1, -0.8)$

18. $f(0, 0, 1, 0) = 14$; $Jf(0, 0, 1, 0) = [-\frac{1}{3} \quad -\frac{1}{2} \quad \frac{2}{3} \quad 1]$; $\mathbf{x} = (-0.01, -0.03, 1.02, 0.04)$

19. $\mathbf{f}(0, 0, 0) = (1, -1)$; $J\mathbf{f}(0, 0, 0) = \begin{bmatrix} 0 & 2 & 1 \\ 4 & 3 & 5 \end{bmatrix}$; $\mathbf{x} = (0.0125, -0.1, 0.067)$

20. $\mathbf{f}(4) = (3.21, -5.05, 4.8)$; $J\mathbf{f}(4) = \begin{bmatrix} 12 \\ 0 \\ -1 \end{bmatrix}$; $x = 4.13$

In Exercises 21 through 26, calculate df.

21. $f(x, y) = \dfrac{x}{2} + \dfrac{y}{4} + 1$

22. $f(x, y) = 2x^2 + xy + y^2 - 10$

23. $f(x, y) = \ln \sqrt{e^x - y}$

24. $f(x, y, z) = x \sec y - y \csc z$

25. $f(x_1, x_2, x_3, x_4, x_5) = \dfrac{x_1 x_2 x_3}{x_4 x_5}$

26. $f(x_1, x_2, x_3, \ldots, x_n) = \sum_{i=1}^{n} i x_i$

27. A metal box with dimensions 1 meter by 1.5 meter by 2 meters is to be covered with gold leaf that is 0.001 mm thick. Approximately what volume of gold will be needed to do the job?

28. A machinist has measured the height and radius of a cylindrical piece of metal to be 5.8 cm and 3.1 cm, respectively. If her measurements are in error by at most 0.002 and 0.001 cm, respectively, by approximately how much may the volume she calculates for the cylinder differ from its actual volume?

29. A conical machine part with base radius 8 cm and height 4 cm is to be coated with an oil film that is 0.001 mm thick. Approximately what volume of oil will be used?

30. In Example 3.5.5, approximately how much water would be needed for 0.1 mm rainfall over the earth's surface?

31. Show that if $\mathbf{f} : U \subset \mathbb{R}^n \to \mathbb{R}^m$ is differentiable at $\mathbf{a}$, then

$$(\mathbf{Df}(\mathbf{a}))(\mathbf{x}) = \lim_{h \to 0} \frac{\mathbf{f}(\mathbf{a} + h\mathbf{x}) - \mathbf{f}(\mathbf{a})}{h} \tag{3.5.11}$$

for all $\mathbf{x} \in \mathbb{R}^n$. (*Suggestion*: Show that

$$\left\| \frac{\mathbf{f}(\mathbf{a} + h\mathbf{x}) - \mathbf{f}(\mathbf{a})}{h} - \mathbf{Df}(\mathbf{a})(\mathbf{x}) \right\|$$

goes to 0 as $h \to 0$.)

In Exercises 32 through 35, use (3.5.11) to calculate $\mathbf{Df}(\mathbf{a})$.

32. $f(x, y) = 5x - 8y + 6$; $\mathbf{a} = (3, -4)$

33. $f(x, y) = x^2y$; $\mathbf{a} = (1, 2)$

34. $\mathbf{f}(x, y) = (6x + y, x - y/2)$; $\mathbf{a} = (0, 0)$

35. $\mathbf{f}(x, y, z) = (xy, x - z, z^2)$; $\mathbf{a} = (0, 0, 1)$

36. Prove that if $\mathbf{f} : U \subset \mathbb{R}^n \to \mathbb{R}^m$ is differentiable at $\mathbf{a} \in \mathbb{R}^n$ then the derivative is unique, that is,that there is only *one* linear transformation $\mathbf{T}$ such that

$$\lim_{\mathbf{x} \to \mathbf{a}} \frac{\|\mathbf{f}(\mathbf{x}) - \mathbf{f}(\mathbf{a}) - \mathbf{T}(\mathbf{x} - \mathbf{a})\|}{\|\mathbf{x} - \mathbf{a}\|} = 0.$$

[*Suggestion*: Suppose, for a contradiction, that $\mathbf{S}$ and $\mathbf{T}$ are both derivatives of $\mathbf{f}$ at $\mathbf{a}$ and they differ at $\mathbf{y}$. Then, on the one hand,

$$0 < \|\mathbf{T}(\mathbf{y}) - \mathbf{S}(\mathbf{y})\| = \frac{\|\mathbf{T}(h\mathbf{y}) - \mathbf{S}(h\mathbf{y})\|}{\|h\mathbf{y}\|} \|\mathbf{y}\|,$$

but on the other,

$$\begin{aligned} &\frac{\|\mathbf{T}(h\mathbf{y}) - \mathbf{S}(h\mathbf{y})\|}{\|h\mathbf{y}\|} \\ &\leq \frac{\|\mathbf{T}(h\mathbf{y}) - \mathbf{f}(\mathbf{a} + h\mathbf{y}) - \mathbf{f}(\mathbf{a})\|}{\|h\mathbf{y}\|} + \frac{\|\mathbf{f}(\mathbf{a} + h\mathbf{y}) - \mathbf{f}(\mathbf{a}) - \mathbf{S}(h\mathbf{y})\|}{\|h\mathbf{y}\|}. \end{aligned}$$]

37. Show that if f has continuous second-order partials and $\mathbf{F}(\mathbf{x}) = (Jf(\mathbf{x}))^T$, then $J\mathbf{F}(\mathbf{x})$ is a symmetric matrix.

A sort of converse to Theorem 3.5.1 (which we do not prove in this book) is this: If $f : U \subset \mathbb{R}^n \to \mathbb{R}$ is differentiable at $\mathbf{a}$ then $\partial f/\partial x_1$, $\partial f/\partial x_2$, $\ldots$, $\dfrac{\partial f}{\partial x_n}$ are all defined at $\mathbf{a}$. Use this result to show that the functions in Exercises 38 through 40 are *not* differentiable at the indicated point.

38. $f(x, y) = \sqrt{x^2 + y^2}$; $(0, 0)$

39. $g(x, y) = |x - y|$; $(3, 3)$

40. $h(x, y) = \left\{ \begin{array}{cc} \sqrt{x^2 + y^2} \sin\left(1/\sqrt{x^2 + y^2}\right), & x^2 + y^2 \neq 0 \\ 0, & x^2 + y^2 = 0 \end{array} \right\}$; $(0, 0)$

41. Prove the general form of (3.5.2). [*Hint*:

$$\begin{aligned} & f(\mathbf{x}) - f(\mathbf{a}) - Jf(\mathbf{a})(\mathbf{x} - \mathbf{a}) \\ = \; & (f(\mathbf{x}) - f(x_1, \ldots, x_{n-1}, a_n)) \\ + \; & (f(x_1, \ldots, x_{n-1}, a_n) - f(x_1, \ldots, x_{n-2}, a_{n-1}, a_n)) \\ + \; & (f(x_1, \ldots, x_{n-2}, a_{n-1}, a_n) - f(x_1, \ldots, x_{n-3}, a_{n-2}, a_{n-1}, a_n)) \\ + \; & \ldots + (f(x_1, a_2, a_3, \ldots, a_n) - f(\mathbf{a})) \\ - \; & \sum_{i=1}^{n} \frac{\partial f}{\partial x_i}(\mathbf{a})(x_i - a_i). \end{aligned}$$

Combine into a single sum, use the Mean Value Theorem, and use the Cauchy-Schwarz inequality.]

3.6 The Chain Rule

In many applications where a function of several variables arises, the independent variables may themselves depend upon other variables. For instance, suppose that the temperature at points (x, y, z) is $T(x, y, z)$. Now if $x = f(t)$, $y = g(t)$, and $z = h(t)$ is a parametrization of a path through a region of space where T is the temperature, a natural question to ask is, "How fast is the temperature changing with respect to motion along the path?" In other words, if $U = T(f(t), g(t), h(t))$, what is dU/dt? Of course, if we have actual formulas for f, g, and h, then we would just plug them into the formula for T and then differentiate. However, we may not have actual formulas for f, g, h, or T. In this case we may still need a formula for dU/dt into which we could substitute for f, g, h, or T. The Chain Rule provides a general way of producing such formulas.

With the notation which we have developed in Section 3.5, the Chain Rule for functions of several variables looks virtually identical to that for functions of a single variable.

First we set down a bit of notation. If $\mathbf{f} : U \subset \mathbb{R}^m \to \mathbb{R}^p$ and $\mathbf{g} : \mathbf{f}(U) \to \mathbb{R}^m$, the **composition** of $\mathbf{g}$ with $\mathbf{f}$ is the function $\mathbf{g} \circ \mathbf{f}$ defined by

$$(\mathbf{g} \circ \mathbf{f})(\mathbf{x}) = \mathbf{g}(\mathbf{f}(\mathbf{x})).$$

Think of composition in this way: Start with $\mathbf{x}$, then apply the first function ($\mathbf{f}$ in this case), and then apply the second function ($\mathbf{g}$).

Theorem 3.6.1 *(Chain Rule) Let $\mathbf{f} : U \subset \mathbb{R}^n \to \mathbb{R}^p$ and $\mathbf{g} : \mathbf{f}(U) \to \mathbb{R}^m$. Suppose that $\mathbf{f}$ is differentiable at $\mathbf{a}$ with derivative $\mathbf{Df}(\mathbf{a})$ and $\mathbf{g}$ is differentiable at $\mathbf{f}(\mathbf{a})$ with derivative $\mathbf{Dg}(\mathbf{f}(\mathbf{a}))$. Then the composite function $\mathbf{g} \circ \mathbf{f} : \mathbb{R}^n \to \mathbb{R}^m$ is differentiable at $\mathbf{a}$ and the derivative of $\mathbf{g} \circ \mathbf{f}$ at $\mathbf{a}$ is the linear transformation $\mathbf{D}(\mathbf{g} \circ \mathbf{f})(\mathbf{a})$ given by*

$$(\mathbf{D}(\mathbf{g} \circ \mathbf{f})(\mathbf{a}))(\mathbf{x}) = \mathbf{Dg}(\mathbf{f}(\mathbf{a}))((\mathbf{Df}(\mathbf{a}))(\mathbf{x})). \tag{3.6.1}$$

In terms of the Jacobian matrix,

$$J(\mathbf{g} \circ \mathbf{f})(\mathbf{a}) = J\mathbf{g}(\mathbf{f}(\mathbf{a}))J\mathbf{f}(\mathbf{a}).$$

Proof The proof is somewhat technical and we omit it. See, for example, [3] for details. ♦

Example 3.6.1 Let

$$\mathbf{f}(x_1, x_2) = (x_1^2 + x_2^2, x_1 x_2)$$

and

$$\mathbf{g}(x_1, x_2) = (3x_1 x_2, x_1 - x_2, 7x_2^2).$$

Then $\mathbf{f}$ is differentiable at $(-6, 1)$ and $\mathbf{g}$ is differentiable at $\mathbf{f}(-6, 1) = (37, -6)$, so by the Chain Rule

$$\begin{aligned} J(\mathbf{g} \circ \mathbf{f})(-6, 1) &= J\mathbf{g}(\mathbf{f}(-6, 1))J\mathbf{f}(-6, 1) \\ &= J\mathbf{g}(37, -6)J\mathbf{f}(-6, 1). \end{aligned}$$

Now

$$J\mathbf{g}(37, -6) = \left.\begin{bmatrix} 3x_2 & 3x_1 \\ 1 & -1 \\ 0 & 14x_2 \end{bmatrix}\right|_{(37,-6)} = \begin{bmatrix} -18 & 111 \\ 1 & -1 \\ 0 & -84 \end{bmatrix}$$

and

$$J\mathbf{f}(-6, 1)) = \left.\begin{bmatrix} 2x_1 & 2x_2 \\ x_2 & x_1 \end{bmatrix}\right|_{(-6,1)} = \begin{bmatrix} -12 & 2 \\ 1 & -6 \end{bmatrix},$$

so

$$J(\mathbf{g} \circ \mathbf{f})(-6, 1) = \begin{bmatrix} -18 & 111 \\ 1 & -1 \\ 0 & -84 \end{bmatrix} \begin{bmatrix} -12 & 2 \\ 1 & -6 \end{bmatrix} = \begin{bmatrix} 327 & 702 \\ -13 & 8 \\ -84 & 504 \end{bmatrix}. \quad ♦$$

Example 3.6.2 Let $\mathbf{f} : \mathbb{R} \to \mathbb{R}^3$ and $g : \mathbb{R}^3 \to \mathbb{R}$. With $u = g(\mathbf{f}(t))$, find a formula for du/dt.

Solution Write $\mathbf{f}(t) = \begin{bmatrix} f_1(t) \\ f_2(t) \\ f_3(t) \end{bmatrix}$ and $g(x_1, x_2, x_3) = g\left(\begin{bmatrix} x_1 \\ x_2 \\ x_3 \end{bmatrix}\right)$. Then $J\mathbf{f}(t) = \begin{bmatrix} f_1'(t) \\ f_2'(t) \\ f_3'(t) \end{bmatrix}$ and

$$Jg(\mathbf{f}(t)) = \left[\frac{\partial g}{\partial x_1}(\mathbf{f}(t)) \quad \frac{\partial g}{\partial x_2}(\mathbf{f}(t)) \quad \frac{\partial g}{\partial x_3}(\mathbf{f}(t))\right],$$

so by the Chain Rule,

$$\begin{aligned}\frac{du}{dt} = J(g \circ \mathbf{f})(t) &= Jg(\mathbf{f}(t))J\mathbf{f}(t) \\ &= \frac{\partial g}{\partial x_1}(\mathbf{f}(t))f_1'(t) + \frac{\partial g}{\partial x_2}(\mathbf{f}(t))f_2'(t) + \frac{\partial g}{\partial x_3}(\mathbf{f}(t))f_3'(t).\end{aligned}$$

Since $x_1 = f_1(t)$, $x_2 = f_2(t)$, and $x_3 = f_3(t)$, this can be written more simply as

$$\frac{du}{dt} = \frac{\partial u}{\partial x_1}\frac{\partial x_1}{\partial t} + \frac{\partial u}{\partial x_2}\frac{\partial x_2}{\partial t} + \frac{\partial u}{\partial x_3}\frac{\partial x_3}{\partial t}. \quad \blacklozenge \tag{3.6.2}$$

Example 3.6.3 In Example 3.6.2, if the functions g and $\mathbf{f}$ are defined by $g(x_1, x_2, x_3) = 3x_1^2 + x_2^2x_3^3$ and $\mathbf{f}(t) = (1/t,\ t^3,\ \ln\ t)$, and $u = (g \circ \mathbf{f})(t)$, then $\partial g/\partial x_1 = 6x_1$, $\partial g/\partial x_2 = 2x_2x_3^3$, $\partial g/\partial x_3 = 3x_2^2x_3^2$, and $J\mathbf{f}(t) = \begin{bmatrix} -1/t^2 \\ 3t^2 \\ 1/t \end{bmatrix}$,

so substituting into (3.6.2), we obtain

$$\frac{du}{dt} = 6x_1\frac{-1}{t^2} + 2x_2x_3^3 \cdot 3t^2 + 3x_2^2x_3^2 \cdot \frac{1}{t},$$

in terms of the variables x_1, x_2, x_3, and t.

Of course, when, as in this case, the functions are given explicitly by formulas, then it would be as simple to substitute the formulas for $\mathbf{g}$ into that for f and compute the derivative directly. The cost paid by doing this is that substituting effectively destroys information about the action of the respective functions. ♦

The Chain Rule is useful for finding formulas for the partial derivatives of functions which are themselves the composition of two (or more) intermediate functions. For example, suppose that u is a function of two variables; for short we write this as $u = u(x, y)$. In turn suppose that x and y are functions of two other variables s and t: $x = x(s, t)$ and $y = y(s, t)$. Then

$$u = u(\mathbf{f}(s, t))$$

where $\mathbf{f}(s, t) = \begin{bmatrix} x(s,t) \\ y(s,t) \end{bmatrix}$. By the Chain Rule,

$$Ju = Ju(\mathbf{f}(s, t))J\mathbf{f}(s, t). \tag{3.6.3}$$

Restated in components, (3.6.3) is

$$\begin{bmatrix} \partial u/\partial s & \partial u/\partial t \end{bmatrix} = \begin{bmatrix} \partial u/\partial x & \partial u/\partial y \end{bmatrix} \begin{bmatrix} \partial x/\partial s & \partial x/\partial t \\ \partial y/\partial s & \partial y/\partial t \end{bmatrix}, \tag{3.6.4}$$

Multiplying (3.6.4) out, we obtain

$$\frac{\partial u}{\partial s} = \frac{\partial u}{\partial x}\frac{\partial x}{\partial s} + \frac{\partial u}{\partial y}\frac{\partial y}{\partial s} \tag{3.6.5}$$

$$\frac{\partial u}{\partial t} = \frac{\partial u}{\partial x}\frac{\partial x}{\partial t} + \frac{\partial u}{\partial y}\frac{\partial y}{\partial t}. \tag{3.6.6}$$

The pattern becomes clearer after a little practice so that it becomes unnecessary for us to make the substitutions that led to (3.6.5). For instance, if $u = u(x, y, z)$, $x = x(r, s, t)$, $y = y(r, s, t)$, and $z = z(r, s, t)$, then by remembering that the entries in the product of two matrices are dot products of rows and columns, we can easily write

$$\frac{\partial u}{\partial r} = \frac{\partial u}{\partial x}\frac{\partial x}{\partial r} + \frac{\partial u}{\partial y}\frac{\partial y}{\partial r} + \frac{\partial u}{\partial z}\frac{\partial z}{\partial r}, \tag{3.6.7}$$

$$\frac{\partial u}{\partial s} = \frac{\partial u}{\partial x}\frac{\partial x}{\partial s} + \frac{\partial u}{\partial y}\frac{\partial y}{\partial s} + \frac{\partial u}{\partial z}\frac{\partial z}{\partial s}, \tag{3.6.8}$$

and

$$\frac{\partial u}{\partial t} = \frac{\partial u}{\partial x}\frac{\partial x}{\partial t} + \frac{\partial u}{\partial y}\frac{\partial y}{\partial t} + \frac{\partial u}{\partial z}\frac{\partial z}{\partial t}. \tag{3.6.9}$$

Many physical problems are described by partial differential equations. For instance if $u = u(x, y)$ is the temperature of a metal plate at position (x, y) on the plate and the metal is in thermal equilibrium, then under several other simplifying assumptions, the function u satisfies Laplace's equation,

$$\frac{\partial^2 u}{\partial x^2} + \frac{\partial^2 u}{\partial y^2} = 0.$$

If the plate is circular, then the temperature on the boundary, and hence elsewhere on the plate, is more readily described using polar coordinates (r, θ). In such a situation it is reasonable to transform the problem, equation and all, to polar coordinates. In particular, it would be necessary to write $\partial^2 u/\partial x^2$ and $\partial^2 u/\partial y^2$ in terms of r, θ, and the partials of u with respect to r and θ. We do this in the following example.

Example 3.6.4 Represent Laplace's equation in polar coordinates.

Solution Let $u = u(x,\ y)$ be a function of x and y, where x and y in turn depend on r and θ by

$$\begin{bmatrix} x \\ y \end{bmatrix} = \mathbf{f}(r, \theta) = \begin{bmatrix} r\cos\theta \\ r\sin\theta \end{bmatrix}.$$

Then u depends on r and θ through the formula $u(r, \theta) = u(\mathbf{f}(r, \theta))$, and by the Chain Rule, $Ju(r, \theta) = Ju(\mathbf{f}(r,\ \theta))J\mathbf{f}(r, \theta)$.

This equation really means

$$\begin{bmatrix} \frac{\partial u}{\partial r} & \frac{\partial u}{\partial \theta} \end{bmatrix} = \begin{bmatrix} \frac{\partial u}{\partial x} & \frac{\partial u}{\partial y} \end{bmatrix} \begin{bmatrix} \cos\theta & -r\sin\theta \\ \sin\theta & r\cos\theta \end{bmatrix}, \tag{3.6.10}$$

where we are writing $\partial u/\partial x$ and $\partial u/\partial y$ as shorthand for $\dfrac{\partial u}{\partial x}(\mathbf{f}(r, \theta))$ and $\dfrac{\partial u}{\partial y}(\mathbf{f}(r, \theta))$. We want to solve for $\partial u/\partial x$ and $\partial u/\partial y$.

The rightmost matrix in (3.6.10) has determinant r, which is nonzero except when $r = 0$. Its inverse is $\begin{bmatrix} \cos\theta & \sin\theta \\ -\sin\theta/r & \cos\theta/r \end{bmatrix}$, so multiplying both sides of (3.6.10) by this matrix gives

$$\begin{bmatrix} \frac{\partial u}{\partial r} & \frac{\partial u}{\partial \theta} \end{bmatrix} \begin{bmatrix} \cos\theta & \sin\theta \\ -\sin\theta/r & \cos\theta/r \end{bmatrix} = \begin{bmatrix} \frac{\partial u}{\partial x} & \frac{\partial u}{\partial y} \end{bmatrix}. \tag{3.6.11}$$

Multiplying the left member of (3.6.11) out, we obtain formulas for $\partial u/\partial x$ and $\partial u/\partial y$:

$$\frac{\partial u}{\partial x} = \cos\theta\frac{\partial u}{\partial r} - \frac{\sin\theta}{r}\frac{\partial u}{\partial \theta} \tag{3.6.12}$$

$$\frac{\partial u}{\partial y} = \sin\theta\frac{\partial u}{\partial r} + \frac{\cos\theta}{r}\frac{\partial u}{\partial \theta}. \tag{3.6.13}$$

We can now use (3.6.12) and (3.6.13) to obtain $\partial^2 u/\partial x^2$ and $\partial^2 u/\partial y^2$ directly:

$$\begin{aligned}
\frac{\partial^2 u}{\partial x^2} &= \frac{\partial}{\partial x}\left(\frac{\partial u}{\partial x}\right) = \cos\theta\frac{\partial}{\partial r}\left(\frac{\partial u}{\partial x}\right) - \frac{\sin\theta}{r}\frac{\partial}{\partial \theta}\left(\frac{\partial u}{\partial x}\right) \qquad (3.6.14)\\
&= \cos\theta\frac{\partial}{\partial r}\left(\cos\theta\frac{\partial u}{\partial r} - \frac{\sin\theta}{r}\frac{\partial u}{\partial \theta}\right) - \frac{\sin\theta}{r}\frac{\partial}{\partial \theta}\left(\cos\theta\frac{\partial u}{\partial r} - \frac{\sin\theta}{r}\frac{\partial u}{\partial \theta}\right)\\
&= \cos\theta\left(\cos\theta\frac{\partial^2 u}{\partial r^2} - \sin\theta\left(-\frac{1}{r^2}\frac{\partial u}{\partial \theta} + \frac{1}{r}\frac{\partial^2 u}{\partial r\,\partial \theta}\right)\right)\\
&\quad -\frac{\sin\theta}{r}\left(-\sin\theta\frac{\partial u}{\partial r} + \cos\theta\frac{\partial^2 u}{\partial \theta\,\partial r} - \frac{1}{r}\left(\cos\theta\frac{\partial u}{\partial \theta} + \sin\theta\frac{\partial^2 u}{\partial \theta^2}\right)\right)\\
&= \cos^2\theta\frac{\partial^2 u}{\partial r^2} - 2\frac{\sin\theta\cos\theta}{r}\frac{\partial^2 u}{\partial r\,\partial \theta} + \frac{\sin^2\theta}{r^2}\frac{\partial^2 u}{\partial \theta^2}\\
&\quad +2\frac{\sin\theta\cos\theta}{r^2}\frac{\partial u}{\partial \theta} + \frac{\sin^2\theta}{r}\frac{\partial u}{\partial r}.
\end{aligned}$$

Notice that to simplify this last expression, we have assumed the continuity of the mixed partials so that, by Clairaut's Theorem, they are equal. Similarly (you should verify),

$$\begin{aligned}
\frac{\partial^2 u}{\partial y^2} &= \sin^2\theta\frac{\partial^2 u}{\partial r^2} + 2\frac{\cos\theta\sin\theta}{r}\frac{\partial^2 u}{\partial r \partial \theta} + \frac{\cos^2\theta}{r^2}\frac{\partial^2 u}{\partial \theta^2}\\
&\quad -2\frac{\cos\theta\sin\theta}{r^2}\frac{\partial u}{\partial \theta} + \frac{\cos^2\theta}{r}\frac{\partial y}{\partial r}.
\end{aligned} \tag{3.6.15}$$

Adding (3.6.14) and (3.6.15) and simplifying the right-hand side, we obtain

$$\frac{\partial^2 u}{\partial x^2} + \frac{\partial^2 u}{\partial y^2} = \frac{\partial^2 u}{\partial r^2} + \frac{1}{r}\frac{\partial u}{\partial r} + \frac{1}{r^2}\frac{\partial^2 u}{\partial \theta^2}. \tag{3.6.16}$$

Thus Laplace's Equation in polar coordinates is

$$\frac{\partial^2 u}{\partial r^2} + \frac{1}{r}\frac{\partial u}{\partial r} + \frac{1}{r^2}\frac{\partial^2 u}{\partial \theta^2} = 0. \quad \blacklozenge \tag{3.6.17}$$

This formidable-looking derivation is not uncommon when **differential operators** such as the **Laplacian** ∇^2, defined by

$$\nabla^2 u = \frac{\partial^2 u}{\partial x^2} + \frac{\partial^2 u}{\partial y^2},$$

are represented in other coordinate systems. It involves nothing more than applications of the Chain Rule and a few other derivative formulas.

Implicit differentiation is another important application that demands the Chain Rule. Consider, for instance, the equation $e^x y + 2e^y z - e^z = 0$. Values can be chosen for any two of the variables, and this determines the value of the third. Thus the equation defines z as a function of x and y. It then makes sense to ask, "Is this function differentiable?" and "What are its partial derivatives?"

Consider an equation of the form

$$f(x, y, z) = 0$$

where f is a differentiable function of x, y, and z. Suppose that the equation defines $z = z(x, y)$ as a differentiable function of x and y. Then, to emphasize the dependence, we write

$$f(x, y, z(x, y)) = 0 \tag{3.6.18}$$

to see that the left-hand side is a differentiable function of x and y. Differentiating both sides with respect to x and using the Chain Rule, we obtain

$$\frac{\partial f}{\partial x}(x, y, z(x, y))\frac{\partial x}{\partial x} + \frac{\partial f}{\partial y}(x, y, z(x, y))\frac{\partial y}{\partial x} + \frac{\partial f}{\partial z}(x, y, z(x, y))\frac{\partial z}{\partial x}(x, y) = 0.$$

Now, $\partial x/\partial x = 1$, and $\partial y/\partial x = 0$ since y is independent of x. So this simplifies to

$$\frac{\partial f}{\partial x} + \frac{\partial f}{\partial z}\frac{\partial z}{\partial x} = 0$$

when we suppress the variables. Solving for $\partial z/\partial x$ we get

$$\frac{\partial z}{\partial x} = -\frac{\partial f/\partial x}{\partial f/\partial z}. \tag{3.6.19}$$

By differentiating (3.6.18) with respect to y, we have

$$\frac{\partial z}{\partial y} = -\frac{\partial f/\partial y}{\partial f/\partial z}. \tag{3.6.20}$$

Example 3.6.5 Find a formula for $\partial z/\partial x$ and $\partial z/\partial y$ if $e^x y + 2e^y z - e^z = 0$ defines z implicitly as a function of x and y.

Solution By (3.6.19) and (3.6.20),

$$\frac{\partial z}{\partial x} = \frac{-ye^x}{2e^y - e^z} \quad \text{and} \quad \frac{\partial z}{\partial y} = \frac{-e^x - 2e^y z}{2e^y - e^z}. \quad \blacklozenge$$

Exercises 3.6

In Exercises 1 through 9, obtain formulas for du/dt.

1. $u = x^4 - y^4$, $x = \cos t$, $y = \sin(t^2)$
2. $u = e^{x_1+4x_2-x_3}$, $x_1 = \ln t$, $x_2 = \ln((t+1)/t)$, $x_3 = 1/t$
3. $u = \arctan(y/x)$, $x = \sqrt{t}$, $y = t^{3/2}$
4. $u = f(\mathbf{g}(t))$, $\mathbf{g}(t) = (\sin t,\ \cos t,\ t)$
5. $u = \dfrac{x+y}{x-y}$, $x = f(t)$, $y = g(t)$
6. $u = f(\cos t,\ \sin t)$
7. $u = g(3t+1,\ 4t-5,\ -t)$
8. $u = \tan(a(t)/b(t))$
9. $u = x_1^4 + x_2^3 - x_3^2 + x_4$, $x_i = f_i(t)$, $i = 1, 2, 3, 4.$

In Exercises 10 through 18, obtain formulas for $\partial u/\partial s$ and $\partial u/\partial t$.

10. $u = x^2y^3 + x - 3y$; $x = t^2 - s$, $y = t + s^2$
11. $u = x + y - z$; $x = 3s + 2t$, $y = -s + t$, $z = s + t$
12. $u = \dfrac{x_1 + 1}{x_1^2 + x_2^2}$, $x_1 = t_1^2 + t_2$, $x_2 = t_1 - t_2^2$
13. $u = \ln(x_1x_2x_3)$, $x_1 = t_1 + t_2$, $x_2 = t_1 - 3t_2$, $x_3 = 1/(t_1 - t_2)$
14. $u = h(s+t, 2s-9t)$
15. $u = g(s^2 - t^2,\ 2st)$
16. $u = s/(x_1(t)^2 + x_2(t)^2 + x_3(t)^2)$
17. $u = tf(s+t,\ 2s+3t) - sf(2s+3t,\ s+t)$

18. $u = e^t g(e^2 + e^{-s},\, \ln t,\, \ln(e^{t+s} + 1))$

In Exercises 19 through 22, obtain a formula for the partial derivatives indicated. (Assume that the first variable of f is x and the second is y.)

19. $u = f(tr,\, ts);\ \dfrac{\partial u}{\partial t}, \dfrac{\partial u}{\partial r}, \dfrac{\partial u}{\partial s}.$

20. $u = f(tx,\, ty);\ \dfrac{\partial u}{\partial t}, \dfrac{\partial u}{\partial x}, \dfrac{\partial u}{\partial y}$ (*Note*: Here we are allowing x and y to stand not only for the variables of f, but we are also letting them appear in the formula.)

21. $u = f(-x,\, -y)\ ;\ \dfrac{\partial u}{\partial x}, \dfrac{\partial u}{\partial y}$

22. $u = f(y,\, x);\ \dfrac{\partial u}{\partial x}, \dfrac{\partial u}{\partial y}$

In Exercises 23 through 26, use the Chain Rule to find the derivative of $\mathbf{g} \circ \mathbf{f}$ at the indicated point $\mathbf{a}$.

23. $\mathbf{g}(x, y) = (x^2y^3, 3x - y^2)$, $\mathbf{f}(x, y) = (-y, x)$, $\mathbf{a} = (3, 2)$

24. $\mathbf{g}$ as in Exercise 23, $\mathbf{f}(x_1,\, x_2,\, x_3) = (x_1x_3,\, x_2/(x_1x_3))$, $\mathbf{a} = (3,\, 1,\, -1)$

25. $\mathbf{g}(x_1,\, x_2,\, x_3) = (x_2^2,\, x_3^2,\, x_1^2)$, $\mathbf{f}(x_1,\, x_2,\, x_3) = (\sin x_1, \cos x_2,\, \sin(x_1 + x_2 + x_3))$ $\mathbf{a} = (\pi/4,\, -\pi/4,\, \pi/6)$

26. $\mathbf{g}(x_1,\, x_2) = (x_1^3,\, x_2)$, $\mathbf{f}(x_1,\, x_2,\, x_3) = (4x_1 + x_2 + x_3^2,\, x_1x_3)$, $\mathbf{a} = (0,\, 1,\, 0)$

27. Suppose that the temperature in space is given by

$$T(x,\, y,\, z)) = \frac{1}{x^2 + y^2 + z^2}$$

and let $\mathbf{x} = (3t^2 - t,\, t^2,\, t^3)$ be a parametrization for a path. What is the rate of change in temperature along the path when $t = 1$?

28. Suppose that in a steady-state fluid flow,

$$\mathbf{v}(x,\, y,\, z) = \left(-y,\, x,\, \sqrt{x^2 + z^2}\right)$$

is the velocity of the fluid at position $(x,\, y,\, z)$. At what rate will $\mathbf{v}$ change as we move from the point $(1,\, 0,\, -4)$ along the path $\mathbf{x} = (2t + 1,\, t^3,\, 6t - 4)$?

29. Suppose that an ideal gas (see Exercise 17 in Section 3.4) occupies a container with volume 900 cm^3 at temperature 400 K. If the volume is increasing at 10 cm^3/min and the temperature is increasing at 15 K/min, at what rate is the pressure in the container changing?

30. When a sound with frequency f is produced by an object traveling along a straight line with speed u and a listener is traveling along the same line in the opposite direction with speed v, the listener *hears* the frequency

$$\phi = \left(\frac{c - v}{c + u}\right) f$$

where c is the speed of sound in air, about 330 m/s. (This is known as the **Doppler effect**.) Suppose that the source is traveling at 30 m/s, accelerating at 1 m/s^2, and emitting a tone of 440 Hz. If the listener is traveling at 15 m/s. and accelerating at 1.5 m/sec^2, how fast is the perceived frequency changing?

31. Economists attempt to measure how useful or satisfying people find goods or services with **utility functions**. Suppose that the utility a person derives from consuming x ounces of beer per week and watching y minutes of videotaped movies per week is

$$u(x, y) = 1 - e^{-0.0003x^2 - 0.0000009y^2}.$$

Further suppose that she currently drinks 70 ounces of beer per week and watches 300 minutes of movies per week. If she is increasing her consumption of beer by 4 ounces per week and cutting back on her movie watching by 10 minutes per week, is the utility she derives from these activities increasing or decreasing? At what rate?

32. A function $f(x, y)$ is **homogeneous of degree** n if it satisfies the equation

$$f(tx, ty) = t^n f(x, y) \tag{3.6.21}$$

for all x, y, and t.

(a) Show that $g(x, y) = (xy - x^2)^2/(x^2y + y^3)$ is homogeneous of degree 1.

(b) Show that if f is any homogeneous of degree n the it satisfies **Euler's formula**

$$x\frac{\partial f}{\partial x}(x, y) + y\frac{\partial f}{\partial y}(x, y) = nf(x, y). \tag{3.6.22}$$

(Hint: Treating each side of (3.6.21) as a function of the three variables x, y, and t, use the Chain Rule to compute the partials with respect to t. Then set t equal to 1.)

(c) Verify that the function in part (a) satisfies Euler's formula.

33. Show that any function of the form $f(x, y) = Kx^{\alpha}y^{1-\alpha}$, where K and α are constants, is homogeneous.

In Exercises 34 through 37, find the indicated partial derivatives, assuming that the given equation implicitly defines the appropriate differentiable function.

34. $x^3y^2z + xy - z^3 = 0;\ \dfrac{\partial z}{\partial x}, \dfrac{\partial z}{\partial y}.$

35. $x \sin \dfrac{y}{z} + z \cos \dfrac{x}{y} = y;\ \dfrac{\partial z}{\partial x}, \dfrac{\partial z}{\partial y}.$

36. $x_1^2 + 3x_2^2 + x_3^2 + 7x_4^2 = 9;\ \dfrac{\partial x_4}{\partial x_1}, \dfrac{\partial x_4}{\partial x_2}, \dfrac{\partial x_4}{\partial x_3}.$

37. $ze^{x+2y} + z^2 - x - y = 0;\ \dfrac{\partial x}{\partial z}, \dfrac{\partial x}{\partial y}.$

38. Show that if $f(x, y, z) = 0$ defines each variable implicitly as a differentiable function of the other two, then

$$\frac{\partial z}{\partial x}\frac{\partial x}{\partial y}\frac{\partial y}{\partial z} = -1.$$

39. Show that if $f(x, y, z, w) = 0$ defines each variable implicitly as a function of the other variables, then

$$\frac{\partial w}{\partial x}\frac{\partial x}{\partial y}\frac{\partial y}{\partial z}\frac{\partial z}{\partial w} = 1.$$

The **Laplacian** of u in three dimensions is defined by

$$\nabla^2 u = \frac{\partial^2 u}{\partial x^2} + \frac{\partial^2 u}{\partial y^2} + \frac{\partial^2 u}{\partial z^2}.$$

40. Show that in cylindrical coordinates (r, θ, w),

$$\nabla^2 u = \frac{\partial^2 u}{\partial r^2} + \frac{1}{r}\frac{\partial u}{\partial r} + \frac{1}{r^2}\frac{\partial^2 u}{\partial \theta^2} + \frac{\partial^2 u}{\partial w^2}.$$

41. Show that in spherical coordinates (ρ, θ, ϕ), the Laplacian is

$$\nabla^2 u = \frac{\partial^2 u}{\partial \rho^2} + \frac{2}{\rho}\frac{\partial u}{\partial \rho} + \frac{1}{\rho^2}\frac{\partial^2 u}{\partial \phi^2} + \frac{\cot\phi}{\rho^2}\frac{\partial u}{\partial \phi} + \frac{1}{\rho^2 \sin^2\phi}\frac{\partial^2 u}{\partial \theta^2}.$$

3.7 The Gradient and Directional Derivative

For a function $f : D \subset \mathbb{R}^n \to \mathbb{R}$, the entries of the Jacobian matrix $J\mathbf{f}$ are often used to construct a vector that has considerable geometric significance. For instance, if f represents the temperature at various points in a region of space, then this vector, called the gradient of f, points in the direction in which the temperature is increasing most rapidly.

Definition 3.7.1 The **gradient** of $f : D \subset \mathbb{R}^n \to \mathbb{R}$ is the vector-valued function given by

$$\vec{\nabla} f(\mathbf{x}) = \left(\frac{\partial f}{\partial x_1}(\mathbf{x}), \frac{\partial f}{\partial x_2}(\mathbf{x}), \ldots, \frac{\partial f}{\partial x_n}(\mathbf{x}) \right).$$

Sometimes we write $\operatorname{grad} f$ for the gradient.

Example 3.7.1 Let $f(x, y, z) = yz - xz + xy$. Then $\partial f/\partial x = y - z$, $\dfrac{\partial f}{\partial y} = z + x$, and $\partial f/\partial z = y - x$, so

$$\vec{\nabla} f(x, y, z) = (y - z, z + x, y - x). \quad \blacklozenge$$

Notice that with our convention of writing vectors as column matrices, $\nabla f = (Jf)^T$. Using this observation you should convince yourself that the total derivative of f can be written as

$$(Df(\mathbf{a}))(\mathbf{x}) = \vec{\nabla} f(\mathbf{a}) \cdot \mathbf{x}. \tag{3.7.1}$$

Imagine yourself standing on a hillside and deciding to walk in a south-eastern direction. Will you be walking uphill? downhill? how steeply? The answers, of course, depend on the orientation of the hill. If you think of the hill's surface as the graph of a function f of two variables, and your chosen direction as a unit vector $\mathbf{u}$ in $\mathbb{R}^2$, then the answer to these questions is, "The rate of change in f in the direction of $\mathbf{u}$." This notion is captured precisely in the definition of the directional derivative.

Definition 3.7.2 Let $f : U \subset \mathbb{R}^n \to \mathbb{R}$, let $\mathbf{a} \in U$, and let $\mathbf{u}$ be a unit vector in $\mathbb{R}^n$. The **directional derivative** f **at a in the direction u** is

$$D_{\mathbf{u}} f(\mathbf{a}) = \lim_{h \to 0} \frac{f(\mathbf{a} + h\mathbf{u}) - f(\mathbf{a})}{h}$$

provided that the limit exists.

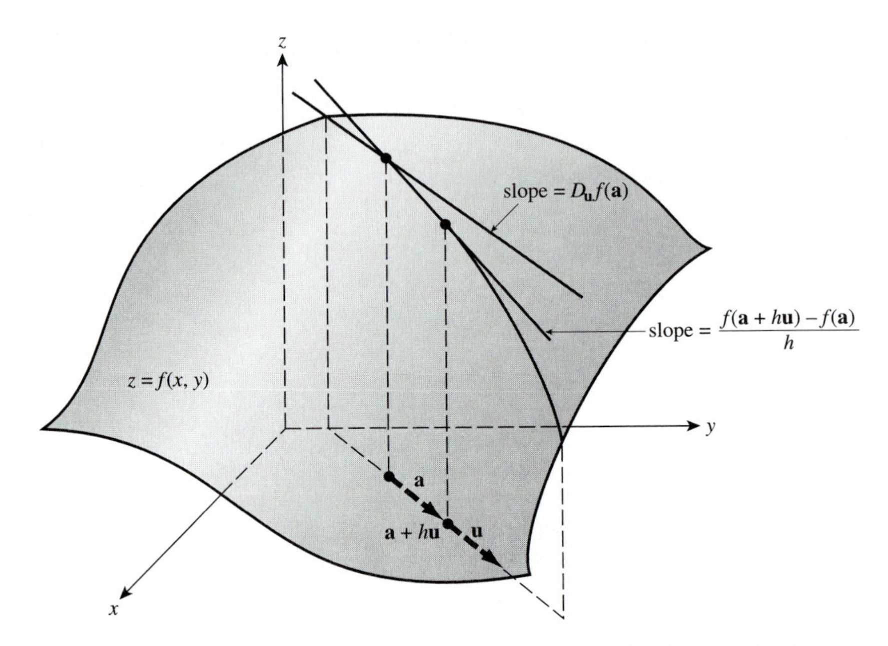

Figure 3.7.1 The directional derivative of f at $\mathbf{a}$ in the direction $\mathbf{u}$

From this definition we see that the directional derivative is a generalization of a partial derivative: whenever the unit vector $\mathbf{u}$ is chosen to be $(0, \ldots, 0, 1, 0, \ldots, 0)$, where the 1 appears in the ith position, then $D_{\mathbf{u}}f(\mathbf{a}) = \dfrac{\partial f}{\partial x_i}(\mathbf{a})$ (Verify this!).

Also, in the particular case where $f : U \subset \mathbb{R}^2 \to \mathbb{R}$, the directional derivative is easily visualized as the limit of slopes of secant lines as shown in Figure 3.7.1.

Theorem 3.7.1 *If $f : U \subset \mathbb{R}^n \to \mathbb{R}$ is differentiable at $\mathbf{a}$ and $\mathbf{u}$ is a unit vector then $D_{\mathbf{u}}f(\mathbf{a})$ exists and*

$$D_{\mathbf{u}}f(\mathbf{a}) = \vec{\nabla} f(\mathbf{a}) \cdot \mathbf{u}. \tag{3.7.2}$$

Proof This is just an application of the Chain Rule. Let $g(h) = f(\mathbf{a} + h\mathbf{u})$. Then $g(0) = f(\mathbf{a})$, and the directional derivative of f at $\mathbf{a}$ is just the derivative of g when $h = 0$:

$$D_{\mathbf{u}}f(\mathbf{a}) \quad = \quad \lim_{h\to 0} \frac{f(\mathbf{a}+h\mathbf{u}) - f(\mathbf{a})}{h}$$

$$\begin{aligned} &= \lim_{h \to 0} \frac{g(h) - g(0)}{h} \\ &= g'(0). \end{aligned}$$

Using the Chain Rule and (3.7.1), we have

$$g'(h) = Dg(h) \quad = \quad \vec{\nabla} f(\mathbf{a} + h\mathbf{u}) \cdot \frac{d}{dh}(\mathbf{a} + h\mathbf{u}).$$

In particular, when $h = 0$, we get $g'(0) = \vec{\nabla} f(\mathbf{a} + 0\mathbf{u}) \cdot \mathbf{u} = \vec{\nabla} f(\mathbf{a}) \cdot \mathbf{u}$. That is, $D_{\mathbf{u}} f(\mathbf{a}) = \vec{\nabla} f(\mathbf{a}) \cdot \mathbf{u}$. ♦

Example 3.7.2 Find the directional derivative of $f(x_1, x_2, x_3) = 2x_1^3 x_2^2 x_3$ at the point $(-1, 1, -1)$ in the direction $(6, 11, 7)$.

Solution A unit vector in the prescribed direction is

$$\mathbf{u} = \frac{(6, 11, 7)}{\sqrt{36 + 121 + 49}} = \frac{(6, 11, 7)}{\sqrt{206}}.$$

By Theorem 3.7.1,

$$\begin{aligned} D_{\mathbf{u}} f(-1, 1, -1) &= \left.(6x_1^2 x_2^2 x_3, 4x_1^3 x_2 x_3, 2x_1^3 x_2^2)\right|_{(-1, 1, -1)} \cdot \frac{(6, 11, 7)}{\sqrt{206}} \\ &= \frac{(-6, 4, -2) \cdot (6, 11, 7)}{\sqrt{206}} = \frac{-6}{\sqrt{206}}. \end{aligned}$$ ♦

In what direction is the directional derivative largest? Using Theorem 3.7.1, the definition of the dot product, and the fact that $\mathbf{u}$ is a unit vector, we see that

$$\begin{aligned} D_{\mathbf{u}} f(x_0) &= ||\vec{\nabla} f(\mathbf{a})||\, ||\mathbf{u}|| \cos\theta \\ &= ||\vec{\nabla} f(\mathbf{a})|| \cos\theta, \end{aligned} \tag{3.7.3}$$

where θ is the angle between $\vec{\nabla} f(\mathbf{a})$ and $\mathbf{u}$. This is largest when $\theta = 0$ (i.e., when $\mathbf{u}$ points in the direction of $\vec{\nabla} f(\mathbf{a})$). Moreover, the directional derivative in the direction of $\vec{\nabla} f(\mathbf{x}_0)$ is $||\vec{\nabla} f(\mathbf{x}_0)||$.

Example 3.7.3 Olga Olaafsen is the world's toughest mountain climber; she always takes the steepest route up the mountainside. Suppose that she finds herself at the point $(1, 3, 67.38)$ on the surface of Mount Gauss, whose elevation in feet at each point is given by $10,000e^{-x^2-(y-1)^2}$. In what direction should she head to maintain her reputation?

Solution We are really asking, "In what direction does the gradient at (1, 3) point?" Here the function is elevation, given by $z = 10,000e^{-x^2-(y-1)^2}$. So

$$\begin{aligned}\vec{\nabla} z &= 10,000\left(-2xe^{-x^2-(y-1)^2}, -2(y-1)e^{-x^2-(y-1)^2}\right)\\ &= -20,000e^{-x^2-(y-1)^2}(x, y-1).\end{aligned}$$

At the point (1, 3),

$$\vec{\nabla} f(1, 3) = -20,000e^{-5}(1, 2),$$

so she should head in the direction of the vector

$$\mathbf{u} = \frac{\vec{\nabla} f(1, 3)}{||\vec{\nabla} f(1, 3)||} = \left(\frac{-1}{\sqrt{5}}, -\frac{2}{\sqrt{5}}\right). \quad \blacklozenge$$

There is a close relationship between the gradient of a function of two variables and the level curves of the function. Let f be differentiable at $\mathbf{a} \in \mathbb{R}^2$ and let

$$f(x, y) = c \tag{3.7.4}$$

be the equation for the level curve of f that passes through $\mathbf{a}$. (What must c be?) We can think of this level curve as parametrized by $\mathbf{x} = \mathbf{r}(t)$, so that $\mathbf{r}(0) = \mathbf{a}$. See Figure 3.7.2(a). Substituting $\mathbf{r}$ into (3.7.4) for (x, y), we obtain

$$f(\mathbf{r}(t)) = c. \tag{3.7.5}$$

Differentiating both sides of (3.7.5) and using the Chain Rule on the left, we have

$$\begin{aligned}\frac{d}{dt} f(\mathbf{r}(t)) &= \frac{d}{dt} c;\\ \vec{\nabla} f(\mathbf{r}(t)) \cdot \mathbf{r}'(t) &= 0.\end{aligned}$$

In particular, when $t = 0$, this gives

$$\vec{\nabla} f(\mathbf{a}) \cdot \mathbf{r}'(0) = 0.$$

If $\mathbf{r}'(0) \neq \mathbf{0}$, this means that $\vec{\nabla} f(\mathbf{a})$ is perpendicular to a vector that is tangent to the level curve through $\mathbf{a}$. See Figure 3.7.2(b). Since this reasoning works for any point on the level curve, we have proven the following theorem.

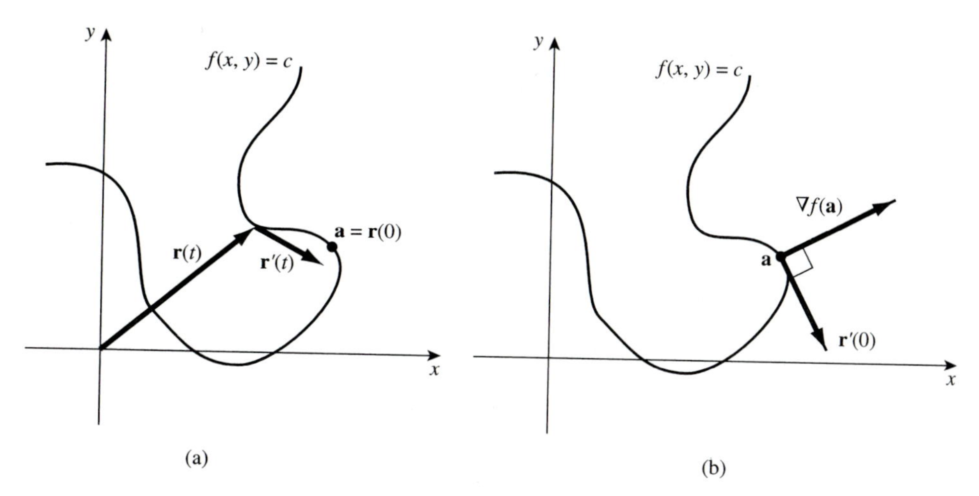

Figure 3.7.2 (a) A level curve of f parametrized. (b) $\vec{\nabla} f(\mathbf{a})$ is perpendicular to the line tangent to the level curve $f(x, y) = c$ at point $\mathbf{a}$.

Theorem 3.7.2 *If $\vec{\nabla} f(\mathbf{a}) \neq \mathbf{0}$ and the level curve of f through $\mathbf{a}$ has a tangent vector $\mathbf{T}$ at $\mathbf{a}$, then $\vec{\nabla} f(\mathbf{a})$ is perpendicular to $\mathbf{T}$.*

Example 3.7.4 Find the equation for the line tangent to the curve $3x^2y^4 - x^2 = 8$ at the point $(-2, 1)$.

Solution If we have a point $\mathbf{a}$ on the line and a vector $\mathbf{n}$ that is perpendicular to it, then the equation for the line is $((x, y) - \mathbf{a}) \cdot \mathbf{n} = 0$, exactly analogous to the equation for a plane in $\mathbb{R}^3$.

We already have $\mathbf{a} = (-2, 1)$ given. By Theorem 3.7.2, we can take the perpendicular vector to be

$$\begin{aligned} \vec{\nabla} f(-2, 1) &= \left(6xy^4 - 2x,\ 12x^2y^3\right)\Big|_{(-2,1)} \\ &= (-8, 48). \end{aligned}$$

The equation of the tangent line is, then,

$$\begin{aligned} ((x, y) - (-2, 1)) \cdot (-8, 48) &= 0 \\ (x + 2,\ y - 1) \cdot (-8, 48) &= 0 \\ -8(x + 2) + 48(y - 1) &= 0 \\ -x + 6y - 8 &= 0. \quad \blacklozenge \end{aligned}$$

Note too that since $\vec{\nabla} f(\mathbf{a})$ points in the direction of greatest increase in f, if $c' > c$ and the two constants are close, then $\vec{\nabla} f(\mathbf{a})$ points toward the side of the level curve $f(x, y) = c$ on which the level curve $f(x, y) = c'$ lies. See Figure 3.7.3.

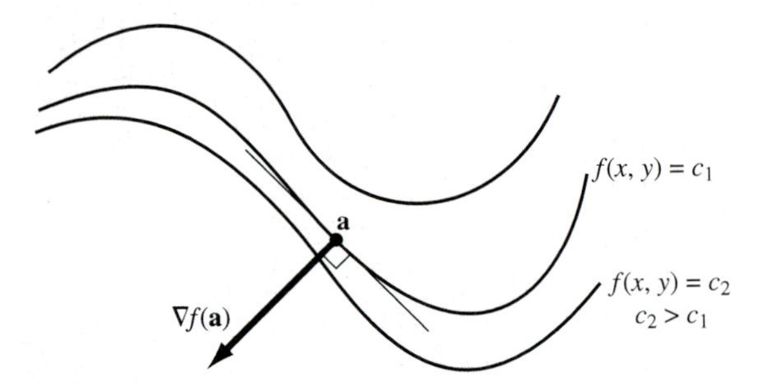

Figure 3.7.3 The gradient $\vec{\nabla} f(\mathbf{a})$ points in the direction of greatest increase in f, perpendicular to the level curve of f passing through $\mathbf{a}$.

The same ideas carry over to three dimensions: The gradient of a function f at a point $\mathbf{a} \in \mathbb{R}^3$ is perpendicular to a plane that is tangent to the level surface

$$f(x, y, z) = c \tag{3.7.6}$$

that passes through $\mathbf{a}$. A line of reasoning similar to that leading up to Theorem 3.7.2 justifies this claim. Let $\mathbf{x} = \mathbf{r}(t)$ be a parametrization of a curve that lies in the level surface such that $\mathbf{r}(0) = \mathbf{a}$ and $\mathbf{r}'(0) \neq \mathbf{0}$. By (3.7.6),

$$f(\mathbf{r}(t)) = c,$$

so when we differentiate we obtain

$$\vec{\nabla} f(\mathbf{r}(t)) \cdot \mathbf{r}'(t) = 0.$$

In particular, when $t = 0$ this yields

$$\vec{\nabla} f(\mathbf{a}) \cdot \mathbf{r}'(0) = 0.$$

Thus $\vec{\nabla} f(\mathbf{a})$ is perpendicular to the curve at $\mathbf{a}$. See Figure 3.7.4.

Since the choice of curve was arbitrary, $\vec{\nabla} f(\mathbf{a})$ is perpendicular to each vector in the "bundle" of possible tangents at $\mathbf{a}$. These vectors span a plane through $\mathbf{a}$ which we call the **plane tangent** to $f(x, y, z) = 0$ at $\mathbf{a}$. Its equation is

$$(\mathbf{x} - \mathbf{a}) \cdot \vec{\nabla} f(\mathbf{a}) = 0,$$

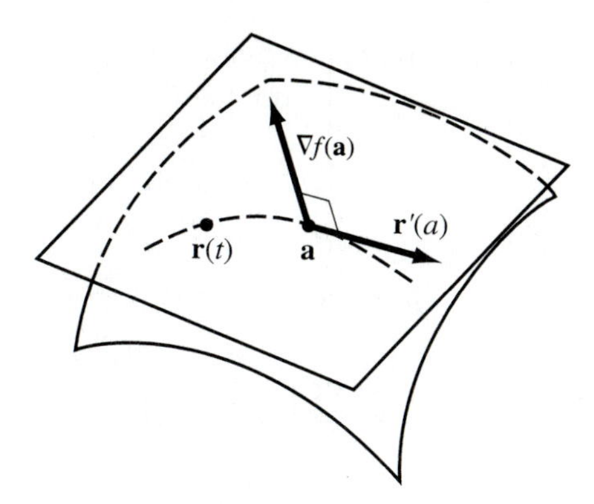

Figure 3.7.4 The gradient of f at $\mathbf{a}$ is perpendicular to each vector in the "bundle" of tangent vectors at $\mathbf{a}$.

or in terms of components,

$$\frac{\partial f}{\partial x}(\mathbf{a})(x - a_1) + \frac{\partial f}{\partial y}(\mathbf{a})(y - a_2) + \frac{\partial f}{\partial z}(\mathbf{a})(z - a_3) = 0. \qquad (3.7.7)$$

Also, at any point, $\mathbf{a} \in \mathbb{R}^3$, the function $f(x, y, z)$ increases most rapidly in the direction of $\vec{\nabla} f(\mathbf{a})$, and that rate of increase is $||\vec{\nabla} f(\mathbf{a})||$.

Example 3.7.5 Find an equation for the plane tangent to the surface $x^2 - y^2 + z^2 = 0$ at $(\sqrt{5}, 3, 2)$.

Solution We want the plane tangent to the level surface of $f(x, y, z) = x^2 - y^2 + z^2$ at the point $(\sqrt{5}, 3, 2)$. A normal to the surface at $(\sqrt{5}, 3, 2)$ is

$$\vec{\nabla} f(\sqrt{5}, 3, 2) = (2x, -2y, 2z)|_{(\sqrt{5}, 3, 2)} = (2\sqrt{5}, -6, 4).$$

Thus an equation for the tangent plane is

$$2\sqrt{5}(x - \sqrt{5}) - 6(y - 3) + 4(z - 2) = 0. \quad \blacklozenge$$

Example 3.7.6 Show that the ellipsoid $x^2/4 + y^2/9 + z^2 = 1$ and the sphere $(x - 10)^2 + (y - 5)^2 + \left(z - 37\sqrt{23}/6\right)^2 = 925$ are tangent at their point of intersection $\left(1, 1, \sqrt{23}/6\right)$.

Solution You should verify that the point $\left(1, 1, \sqrt{23}/6\right)$ is common to both surfaces. To show that they are tangent at this point it is sufficient to show that the normals to the respective tangent planes are parallel. These normals are just the gradients of the functions

$$f(x, y, z) = \frac{x^2}{4} + \frac{y^2}{9} + z^2$$

and $g(x, y, z) = (x-10)^2 + (y-5)^2 + \left(z - 37\sqrt{23}/6\right)^2$. Now

$$\begin{aligned}\vec{\nabla} f\left(1, 1, \frac{\sqrt{23}}{6}\right) &= \left.\left(\frac{x}{2}, \frac{2y}{9}, 2z\right)\right|_{(1,1,\sqrt{23}/6)} \\ &= \left(\frac{1}{2}, \frac{2}{9}, \frac{\sqrt{23}}{3}\right)\end{aligned}$$

and

$$\begin{aligned}\vec{\nabla} g\left(1, 1, \frac{\sqrt{23}}{6}\right) &= \left.\left(2(x-10), 2(y-5), 2z - \frac{37}{3}\sqrt{23}\right)\right|_{(1,1,\sqrt{23}/6)} \\ &= (-18, -8, -12\sqrt{23}) \\ &= -36\left(\frac{1}{2}, \frac{2}{9}, \frac{\sqrt{23}}{3}\right).\end{aligned}$$

So $\vec{\nabla} g\left(1, 1, \sqrt{23}/6\right) = -36\vec{\nabla} f\left(1, 1, \sqrt{23}/6\right)$; the normals are parallel, so the two surfaces have a common tangent plane at $\left(1, 1, \sqrt{23}/6\right)$. ♦

The idea of level set makes sense for any function $f : D \subset \mathbb{R}^n \to \mathbb{R}$. These are the points $(x_1, x_2, \ldots, x_n) \in \mathbb{R}^n$ such that

$$f(x_1, x_2, \ldots, x_n) = c, \tag{3.7.8}$$

for some constant c. By considering curves in $\mathbb{R}^n$ parametrized by $\mathbf{x} = \mathbf{r}(t)$ that lie in the level surface and pass through a point $\mathbf{a}$ in the set, we are once again drawn to the conclusion that $\vec{\nabla} f(\mathbf{a})$ is orthogonal to the tangent vectors at $\mathbf{a}$. This "tangent bundle" orthogonal to $\vec{\nabla} f(\mathbf{a})$ we define to be the **hyperplane tangent** to the level set at $\mathbf{a}$. Its equation is formally the same as in dimensions 2 and 3:

$$(\mathbf{x} - \mathbf{a}) \cdot \vec{\nabla} f(\mathbf{a}) = 0. \tag{3.7.9}$$

Example 3.7.7 Find the equation for the hyperplane in $\mathbb{R}^5$ that is tangent to the level set of $f(x_1, x_2, x_3, x_4, x_5) = x_1 \sin x_2 + x_3 - x_4 \cos x_5$ at the point $(1, \pi/4, -2, 1, 3\pi/4)$.

Solution We need only calculate $\vec{\nabla} f$ at the point indicated and use (3.7.9):

$$\begin{aligned}
&\vec{\nabla} f\left(1, \frac{\pi}{4}, -2, 1, \frac{3\pi}{4}\right) \\
&= (\sin x_2,\ x_1 \cos x_2,\ 1,\ -\cos x_5,\ x_4 \sin x_5)|_{(1, \frac{\pi}{4}, -2, 1, \frac{3\pi}{4})} \\
&= \left(\frac{1}{\sqrt{2}}, \frac{1}{\sqrt{2}}, 1, \frac{1}{\sqrt{2}}, \frac{1}{\sqrt{2}}\right).
\end{aligned}$$

Then the tangent hyperplane is

$$\left((x_1, x_2, x_3, x_4, x_5) - \left(1, \frac{\pi}{4}, -2, 1, \frac{3\pi}{4}\right)\right) \cdot \left(\frac{1}{\sqrt{2}}, \frac{1}{\sqrt{2}}, 1, \frac{1}{\sqrt{2}}, \frac{1}{\sqrt{2}}\right) = 0.$$

Multiplying this out and simplifying, we obtain

$$x_1 - 1 + x_2 - \frac{\pi}{4} + \sqrt{2}(x_3 + 2) + x_4 - 1 + x_5 - \frac{3\pi}{4} = 0$$

or

$$x_1 + x_2 + \sqrt{2}x_3 + x_4 + x_5 = \pi + 2 - 2^{3/2}. \quad \blacklozenge$$

Exercises 3.7

In Exercises 1 through 10, calculate the gradient.

1. $f(x, y) = \dfrac{3}{4}x^2y^4 - \dfrac{5}{4}xy^6$

2. $g(x, y) = x \sin y + y \cos^2 x$

3. $h(x, y, z) = xe^{yz}$

4. $F(x, y, z) = \dfrac{x + yz}{xy - z}$

5. $G(x_1, x_2, x_3, x_4, x_5, x_6) = \dfrac{1}{(x_1 - x_4)^2 + (x_2 - x_5)^2 + (x_3 - x_6)^2}$

6. $H(x_1, x_2, \cdots, x_n) = \ln\left(\displaystyle\sum_{i=1}^{n} x_i^2\right)$

7. $P(x_1, x_2, \cdots, x_n) = \exp\left(\displaystyle\sum_{i=1}^{n} x_i^2\right)$

8. $f(\mathbf{x}) = ||\mathbf{x}||^2$, $\mathbf{x} \in \mathbb{R}^n$

9. $f(\mathbf{x}) = \dfrac{1}{||\mathbf{x}||^2}$, $\mathbf{x} \in \mathbb{R}^n$

10. $q(\mathbf{x}) = \mathbf{a} \cdot \mathbf{x}$, $\mathbf{x} \in \mathbb{R}^n$, $\mathbf{a} \in \mathbb{R}^n$ constant

In Exercises 11 through 21, calculate the directional derivative of f at the indicated point in the direction given.

11. $f(x, y) = 7x^2 + 4xy - 3y^2$; $\mathbf{a} = \left(\dfrac{2}{3}, \dfrac{3}{2}\right)$; $\mathbf{u} = \left(\dfrac{1}{\sqrt{5}}, \dfrac{2}{\sqrt{5}}\right)$

12. $f(x, y) = x^{2/3} + y^{2/3}$; $\mathbf{a} = (1, 2)$; $\mathbf{u} = \left(\dfrac{-3}{\sqrt{34}}, \dfrac{5}{\sqrt{34}}\right)$

13. $f(x, y) = \ln(x + e^y)$; $\mathbf{a} = (e, 1)$; $\mathbf{u} = \left(\dfrac{e}{\sqrt{e^2+1}}, \dfrac{1}{\sqrt{e^2+1}}\right)$

14. $f(x, y) = \arcsin xy$; $\mathbf{a} = \left(\dfrac{1}{5}, 3\right)$, $\mathbf{u} = \left(\dfrac{1}{\sqrt{2}}, -\dfrac{1}{\sqrt{2}}\right)$

15. $f(x, y) = x^3 + y^2$; $\mathbf{a} = (1, 5)$; $\mathbf{u}$ points from $\mathbf{a}$ toward $(3, 6)$

16. $f(x, y, z) = \dfrac{x+y}{z}$; $\mathbf{a} = (1, 2, 5)$; $\mathbf{u} = \left(\dfrac{1}{\sqrt{3}}, -\dfrac{1}{\sqrt{3}}, \dfrac{1}{\sqrt{3}}\right)$

17. $f(x, y, z) = x \cos y \sin z$; $\mathbf{a} = \left(1, \dfrac{\pi}{4}, \dfrac{5\pi}{6}\right)$; $\mathbf{u} = \left(\dfrac{3}{\sqrt{10}}, 0, -\dfrac{1}{\sqrt{10}}\right)$

18. $f(x, y, z) = \dfrac{\sqrt{x^2+y^2}}{z}$; $\mathbf{a} = (3, -4, 1)$; $\mathbf{u}$ points from $\mathbf{a}$ toward the origin.

19. $f(x, y, z) = \ln(xyz+1)$; $\mathbf{a} = (2, e, \frac{1}{e})$; $\mathbf{u}$ points from $\mathbf{a}$ toward $(3, 6, e)$.

20. $F(x_1, x_2, x_3) = \sqrt{x_1^2 + x_2^2 + x_3^2}$; $\mathbf{a} = (1, 0, -5)$, $\mathbf{u} = \left(\dfrac{1}{2}, \dfrac{1}{3}, -\dfrac{\sqrt{23}}{6}\right)$

21. $f(x_1, x_2, x_3, x_4) = e^{x_1 x_2} \cos(x_3 x_4)$; $\mathbf{a} = (4, 3, 1, \frac{\pi}{2})$; $\mathbf{u}$ points in the direction from $\mathbf{a}$ to $(5, -2, 0, \pi/4)$

22. Suppose that the elevation (in miles) of points on a mountain is given by $z = 2e^{-x^2-y^2} + e^{-(x-2)^2-(y-2)^2}$, where x and y are, respectively, the distances east and north of a reference point. If a mountain climber finds herself 3 miles east and 1 mile south of the reference point and she is headed southwest, is she going uphill or downhill? In what direction(s) should she head if she wants to stay at the same elevation?

23. Suppose that the temperature in a piece of metal at position $(x,\, y,\, z)$ is given by $T(x,\, y,\, z) = 100/(1 + x^2 + y^2 + z^2) + 500/(1 + (y - 2)^2)$. At the point $(1,\, 0,\, 1)$, in what direction is the temperature increasing most rapidly? decreasing most rapidly?

24. Suppose that the concentration in mg/cm^3 of a chemical at postion $(x,\, y,\, z)$ is given by

$$C(x,\, y,\, z) = 50 + z \cos 2\pi x \sin 2\pi y.$$

In what direction is the concentration increasing most rapidly at the point $\left(\frac{1}{6},\, \frac{1}{8},\, 3\right)$? What is the rate of change in concentration in this direction at this point?

25. Prove that $\vec{\nabla}(kf) = k\vec{\nabla} f$ for any constant k.

26. Prove that $\vec{\nabla}(fg) = f\vec{\nabla} g + g\vec{\nabla} f$.

27. Prove that $\vec{\nabla}(\mathbf{f} \cdot \mathbf{g}) = \sum_{i=1}^{n} (f_i \vec{\nabla} g_i + g_i \vec{\nabla} f_i)$.

In Exercises 28 through 31, find the equation for the line tangent to the curve at the indicated point.

28. $\dfrac{x^2}{25} - \dfrac{y^2}{81} = 1;\ (-5,\, 0)$

29. $ye^x - x = 4;\ (0,\, 4)$

30. $x^3 + y^3 = -\dfrac{7}{2}xy;\ (2,\, -1)$

31. $\sqrt{\sin x + y} = 2 - y;\ (\pi,\, 1)$

In Exercises 32 through 34, find the equation for the line tangent to the level curve of the given function at the point indicated.

32. $f(x, y) = \ln(1 + x^2 + y^2)$, $\mathbf{a} = \left(-\frac{2}{3}, \frac{1}{3}\right)$.

33. $f(x, y) = e^{-x^2+2xy-y^2}$, $\mathbf{a} = (1, 0)$.

34. $f(x, y) = x \tan y$, $\mathbf{a} = (-1, \pi/3)$.

In Exercises 35 through 41, find the equation for the plane tangent to the surface at the indicated point.

35. $x^3y^2 + xy - z - 3x = 14,\ (1,\, 3,\, -5)$

36. $x^2 + \dfrac{y^2}{4} + \dfrac{z^2}{9} = 1,\ (1/2,\, -1,\, 3/\sqrt{2})$

37. $z = \dfrac{x+y}{x-2y}$, $(3,\, 1,\, 4)$

38. $6x + y + 8z = -2$, $(1,\, 0,\, -1)$

39. $x^2 + y^2 + z^2 = 27$, $(1,\, 5,\, -1)$

40. $\left(\sqrt{x^2+y^2} - 3\right)^2 + z^2 = 4$, $\left(1,\, 2,\, \sqrt{6\sqrt{5} - 10}\right)$

41. $x^2 + y^2 + z^2 = 6$, $(1,\, 2,\, -1)$

42. Find the point on the paraboloid $2x - z^2 - 3y^2 = 0$ at which the tangent plane is parallel to the plane $5x + y + z = 1$.

43. Find the points on the surface $x^4 + y^4 + z^4 = 1$ where the tangent plane is parallel to the plane $x + y + z = 1$.

44. Find the point at which the ellipsoid $x^2/4 + y^2 + z^2 = 1$ is tangent to one of the hyperboloids in the family $x^2 + y^2 - (z+1)^2 = c^2$.

45. Find the points $(a,\, b,\, c)$ on the paraboloid $z = x^2 + y^2$ such that the tangent plane at $(a,\, b,\, c)$ passes through the points $\mathbf{p}_1 = (1,\, 2,\, 1)$ and $\mathbf{p}_2 = (0,\, 3,\, -2)$. [*Hint*: At such a point, $((a,\, b,\, c) - \mathbf{p}_1) \times ((a,\, b,\, c) - \mathbf{p}_2)$ is parallel to $\vec{\nabla} f(a,\, b,\, c)$; you may need to use a computer or calculator to obtain numerical approximations to the resulting equations.]

46. Show that any line tangent to the circle $x^2 + y^2 = 1$ has the equation

$$(\cos\theta)x + (\sin\theta)y = 1$$

for some θ in the interval $[0,\, 2\pi)$.

47. Show that if $(x_0,\, y_0)$ is a point on the ellipse

$$\frac{x^2}{a^2} + \frac{y^2}{b^2} = 1,$$

then the line tangent to the ellipse at $(x_0,\, y_0)$ is

$$\frac{x_0 x}{a^2} + \frac{y_0 y}{b^2} = 1.$$

48. Show that if $(x_0,\, y_0,\, z_0)$ is a point on the ellipsoid

$$\frac{x^2}{a^2} + \frac{y^2}{b^2} + \frac{z^2}{c^2} = 1,$$

then the plane tangent to the surface at $(x_0,\, y_0,\, z_0)$ is

$$\frac{x_0 x}{a^2} + \frac{y_0 y}{b^2} + \frac{z_0 z}{c^2} = 1.$$

49. Show, by using Euler's formula (3.6.22), that if F is homogeneous of degree n and (x_0, y_0, z_0) is a point on a level surface of F, then the equation for the plane tangent to the surface $F(x, y, z) = F(x_0, y_0, z_0)$ at (x_0, y_0, z_0) is

$$\frac{\partial F}{\partial x}(x_0, y_0, z_0)x + \frac{\partial F}{\partial y}(x_0, y_0, z_0)y + \frac{\partial F}{\partial z}(x_0, y_0, z_0)z = nF(x_0, y_0, z_0).$$

50. Find two points on the circle $x^2 + (y-1)^2 = 1$ at which a level curve of $f(x,y) = x^2 - y^2$ is tangent to the circle. [*Suggestion*: The first equation's graph is a level curve of $g(x,y) = x^2 + (y-1)^2 - 1$; at a point of tangency the gradients of f and g will be parallel ($\vec{\nabla} f = \lambda \vec{\nabla} g$ for some number λ). Solve the resulting system of three equations in the three unknowns x, y, and λ. Make a sketch!]

51. Find eight points on the sphere $x^2 + y^2 + z^2 = 1$ at which a level surface to $f(x, y, z) = xyz$ is tangent to the sphere. (*Hint*: Again, $\vec{\nabla} f = \lambda \vec{\nabla} g$.)

52. Find a function $f : \mathbb{R}^3 \to \mathbb{R}$ such that

$$\vec{\nabla} f(\mathbf{x}) = \begin{bmatrix} 3 & 5 & 0 \\ 5 & -1 & 2 \\ 0 & 2 & 6 \end{bmatrix} \mathbf{x}.$$

3.8 Divergence and Curl

In physical applications such as mechanics, fluid mechanics, and electromagnetic theory, other notions of the derivative of a vector field besides the total derivative are useful. In particular, for a given vector field $\mathbf{F}$ on a subset of $\mathbb{R}^3$, we can calculate the **divergence** of $\mathbf{F}$ and the **curl** of $\mathbf{F}$. For instance, if $\mathbf{F}$ is the velocity field of a fluid, then the divergence measures the instantaneous rate of fluid flow outward at each point. The curl of $\mathbf{F}$ in this case measures the amount of rotation undergone by microscopic parts of the fluid.

We will first give definitions of these two differential operators, and then we will examine some examples to see why these measure what we want.

Definition 3.8.1 Let $\mathbf{F} : D \subset \mathbb{R}^3 \to \mathbb{R}^3$ be a differentiable vector field.

1. The **divergence** of $\mathbf{F}$ is the scalar-valued function div $(\mathbf{F})$, defined by

$$\operatorname{div}(\mathbf{F}) = \frac{\partial F_1}{\partial x} + \frac{\partial F_2}{\partial y} + \frac{\partial F_3}{\partial z}.$$

2. The **curl** of $\mathbf{F}$ is the vector field curl ($\mathbf{F}$), defined by

$$\operatorname{curl}(\mathbf{F}) = \left(\frac{\partial F_3}{\partial y} - \frac{\partial F_2}{\partial z}, \frac{\partial F_1}{\partial z} - \frac{\partial F_3}{\partial x}, \frac{\partial F_2}{\partial x} - \frac{\partial F_1}{\partial y}\right).$$

A handy mnemonic for calculating div and curl is to regard the symbol $\vec{\nabla}$ (read this as "del") as the "vector" $(\partial/\partial x, \partial/\partial y, \partial/\partial z)$ so that $\operatorname{div}\mathbf{F} = \vec{\nabla}\cdot\mathbf{F}$ (read "del dot $\mathbf{F}$") and $\operatorname{curl}\mathbf{F} = \vec{\nabla}\times\mathbf{F}$ (read "del cross $\mathbf{F}$").

Example 3.8.1 Let $\mathbf{F} = (3xy + z, \sqrt{z} + y, z/y)$. Then

$$\begin{aligned}\operatorname{div}\mathbf{F} &= \frac{\partial}{\partial x}(3xy + z) + \frac{\partial}{\partial y}(\sqrt{z} + y) + \frac{\partial}{\partial z}\left(\frac{z}{y}\right) \\ &= 3y + 1 + \frac{1}{y}. \quad \blacklozenge\end{aligned}$$

Example 3.8.2 Let $\mathbf{f}(\mathbf{x}) = \dfrac{\mathbf{x}}{||\mathbf{x}||}$. Then

$$\begin{aligned}\operatorname{div}\mathbf{f} &= \operatorname{div}\left(\frac{x}{\sqrt{x^2+y^2+z^2}}, \frac{y}{\sqrt{x^2+y^2+z^2}}, \frac{z}{\sqrt{x^2+y^2+z^2}}\right) \\ &= \frac{y^2+z^2}{(x^2+y^2+z^2)^{3/2}} + \frac{x^2+z^2}{(x^2+y^2+z^2)^{3/2}} + \frac{x^2+y^2}{(x^2+y^2+z^2)^{3/2}} \\ &= \frac{2}{\sqrt{x^2+y^2+z^2}} = \frac{2}{||\mathbf{x}||}. \quad \blacklozenge\end{aligned}$$

Example 3.8.3 Find a formula for div $(f\mathbf{F})$.

Solution Because we are taking the "derivative" of a product, we would expect a formula that is at least reminiscent of the product formula. Thus a reasonable guess is

$$\operatorname{div}(f\mathbf{F}) = \vec{\nabla}f\cdot\mathbf{F} + f\operatorname{div}\mathbf{F}. \tag{3.8.1}$$

To check this, we first let $\mathbf{F} = (P, Q, R)$. Then

$$\begin{aligned}\operatorname{div}(f\mathbf{F}) &= \frac{\partial}{\partial x}(fP) + \frac{\partial}{\partial y}(fQ) + \frac{\partial}{\partial z}(fR) \\ &= f_xP + fP_x + f_yQ + fQ_y + f_zR + fR_z \\ &= (f_x, f_y, f_z)\cdot(P, Q, R) + f(P_x + Q_y + R_z) \\ &= \vec{\nabla}f\cdot\mathbf{F} + f\operatorname{div}\mathbf{F}.\end{aligned}$$

We find that our guess for the differentiation formula was correct! ♦

Example 3.8.4 The curl of $\mathbf{F} = (3xy + z,\ \sqrt{z} + y,\ z/y)$ is

$$\begin{aligned}
\operatorname{curl}\mathbf{F} &= \vec{\nabla} \times \mathbf{F} \\
&= \begin{vmatrix} \mathbf{i} & \mathbf{j} & \mathbf{k} \\ \partial/\partial x & \partial/\partial y & \partial/\partial z \\ 3xy + z & \sqrt{z} + y & z/y \end{vmatrix} \\
&= \left(-\frac{z}{y^2} - \frac{1}{2\sqrt{z}}\right)\mathbf{i} + (1-0)\mathbf{j} + (0 - 3x)\mathbf{k} \\
&= -\frac{2z^{3/2} + y^2}{2y^2\sqrt{z}}\,\mathbf{i} + \mathbf{j} + 3x\,\mathbf{k} \\
&= \left(-\frac{2z^{3/2} + y^2}{2y^2\sqrt{z}},\ 1,\ 3x\right).
\end{aligned}$$ ♦

Example 3.8.5 Determine a formula for curl $(f\mathbf{F})$ where f is scalar-valued and $\mathbf{F}$ is a vector field.

Solution As in Example 3.8.3, we expect the derivative of this product to be a "derivative" of f "times" $\mathbf{F}$ plus f "times" a "derivative" of $\mathbf{F}$. The two products must each be vectors, so we might guess that

$$\operatorname{curl}(f\mathbf{F}) = \vec{\nabla} \times (f\mathbf{F}) = \vec{\nabla} f \times \mathbf{F} + f\vec{\nabla} \times \mathbf{F}. \tag{3.8.2}$$

To verify this, we just calculate the curl of the product in terms of the component functions of $\mathbf{F} = (P,\ Q,\ R)$:

$$\begin{aligned}
\operatorname{curl}(f\mathbf{F}) &= (f_yR + fR_y - f_zQ - fQ_z)\mathbf{i} \\
&\quad + (f_zP + fP_z - f_xR - fR_x)\mathbf{j} \\
&\quad + (f_xQ + fQ_x - f_yP - fP_y)\mathbf{k} \\
&= ((f_yR - f_zQ)\mathbf{i} + (f_zP - f_xR)\mathbf{j} + (f_xQ - f_yP)\mathbf{k}) \\
&\quad + (f(R_y - Q_z)\mathbf{i} + f(P_z - R_x)\mathbf{j} + f(Q_x - P_y)\mathbf{k}) \\
&= \vec{\nabla} f \times \mathbf{F} + f\vec{\nabla} \times \mathbf{F}.
\end{aligned}$$ ♦

In Section 2.4 we examined the way in which linear transformations act geometrically. To obtain a concept of the divergence of a vector field, we return to some of those considerations.

Example 3.8.6 Suppose that a cloud of dust particles is moving so that the velocity of a particle passing through a point (x, y, z) is given by

$$\mathbf{v}(x, y, z) = (ax, by, cz)$$

where a, b, and c are given constants. We want to measure the time rate of change in the volume of a region about the origin occupied by a portion of the dust. For convenience we will take that portion of dust to be a small cube centered at the origin. By exactly the same reasoning as we used in Example 2.4.5, we see that a particle initially at (x, y, z) will be at position

$$\mathbf{P}(x, y, z) = \left(xe^{at}, ye^{bt}, ze^{ct}\right) = \begin{bmatrix} e^{at} & 0 & 0 \\ 0 & e^{bt} & 0 \\ 0 & 0 & e^{ct} \end{bmatrix} \begin{bmatrix} x \\ y \\ z \end{bmatrix} \tag{3.8.3}$$

after t time units. In particular, consider all the dust particles that initially occupy a cube of side length s that is centered at the origin. Since the transformation $\mathbf{P}$ in Equation 3.8.3 is linear, after t time units, the dust will occupy a rectangular solid with sides of length se^{at}, se^{bt}, and se^{ct}. The volume of this solid is

$$se^{at} \cdot se^{bt} \cdot se^{ct} = s^3 e^{(a+b+c)t},$$

so the average rate of change in the volume occupied by the dust over the time 0 to t is

$$\frac{s^3 e^{(a+b+c)t} - s^3}{t} = \frac{s^3 \left(e^{(a+b+c)t} - 1\right)}{t}.$$

The instantaneous rate of change in this volume will be the limit as $t \to 0$ of this expression:

$$\lim_{t \to 0} \frac{s^3 \left(e^{(a+b+c)t} - 1\right)}{t} = s^3(a + b + c). \tag{3.8.4}$$

From this we can obtain the **instantaneous rate of change** in the **volume, per unit volume** by dividing (3.8.4) by the volume s^3. This is simply

$$a + b + c.$$

Notice that this depends only on the original velocity field and that it is the divergence of $\mathbf{v}$. When the sum $a + b + c$ is positive we see that the cube is expanding, so *on average*, the dust particles are moving outward, or diverging, from the origin. (A positive divergence in this case does not necessarily mean that if we sit at the origin and watch all the nearby dust that we will see particles going away from us in all directions. See Exercise 27.)

For a simple linear vector field such as this, we also see that the divergence is the same at each point in space. ♦

In general, if $\mathbf{F}$ on $\mathbb{R}^3$ is the velocity field for a fluid then $\operatorname{div}\mathbf{F}$ evaluated at a point $\mathbf{a}$ measures the instantaneous rate of expansion (or compression) per unit volume in a microscopic portion of fluid centered at $\mathbf{a}$. For this reason, if $\mathbf{F}$ satisfies $\operatorname{div}\mathbf{F} = 0$, then $\mathbf{F}$ is sometimes called **incompressible**. In Chapter 5 we will be able to gain further insight to this claim by means of the Divergence Theorem.

Example 3.8.7 Consider a cloud of particles each of which follows a circular path centered on the z-axis that lies in a plane perpendicular to the z-axis. Suppose the angular speed is a positive constant ω so that rotation is counterclockwise as viewed from the positive z-axis. Then (see Example 2.4.3 in Section 2.4) the position of a particle initially at $(x,\, y,\, z)$ is

$$\mathbf{p}(t) = ((\cos\omega t)x - (\sin\omega t)y,\ (\sin\omega t)x + (\cos\omega t)y,\ z)\,,$$

so the velocity field at $(x,\, y,\, z)$ is

$$\begin{aligned}\mathbf{v} &= \mathbf{p}'(0)\\ &= \left.((-\omega\sin\omega t)x - (\omega\cos\omega t)y,\ (\omega\cos\omega t)x - (\omega\sin\omega t)y,\ 0)\right|_{t=0}\\ &= (-\omega y,\ \omega x,\ 0).\end{aligned}$$

Now notice that

$$\operatorname{curl}\mathbf{v} = (0,\ 0,\ 2\omega)$$

so that

$$\begin{aligned}(\operatorname{curl}\mathbf{v}) \times \mathbf{x} &= (-2\omega y,\ 2\omega x,\ 0)\\ &= 2(-\omega y,\ \omega x,\ 0)\\ &= 2\mathbf{v}.\end{aligned}$$

Thus, in the case of rigid rotation, the velocity field is given by

$$\mathbf{v}(\mathbf{x}) = \left(\frac{1}{2}\operatorname{curl}\mathbf{v}(\mathbf{0})\right) \times \mathbf{x}$$

or, using matrix notation,

$$\mathbf{v}(\mathbf{x}) = \frac{1}{2}\begin{bmatrix} 0 & \omega & 0 \\ -\omega & 0 & 0 \\ 0 & 0 & 0 \end{bmatrix}\mathbf{x}.$$

In this case, the magnitude of curl $\mathbf{v}$ is twice the angular speed and the vector points along the axis of rotation. ♦

In general, if $\mathbf{F}$ is the velocity field of a fluid, curl $\mathbf{F}$ evaluated at a point $\mathbf{a}$ provides a measure of the angular velocity of a microscopic portion of the fluid centered at $\mathbf{a}$; its direction is along the axis of this rotation. For this reason, a vector field whose curl is zero is called **irrotational**. In Chapter 5 we will be able to make this idea clearer by means of Stokes' Theorem. If $\mathbf{F}$ is instead a force field and curl $\mathbf{F} = \mathbf{0}$, then $\mathbf{F}$ is, for reasons we will discover in Section 5.1, called **path-independent** or **conservative**.

Given a twice-differentiable scalar-valued function f on a subset D of $\mathbb{R}^3$, we can calculate grad $f = \vec{\nabla} f$ to obtain a vector field and then we can calculate div $(\vec{\nabla} f)$ to obtain a scalar-valued function again. This last function is the **Laplacian** of f. See Exercise 33 in Section 3.4 and Exercises 40 and 41 in Section 3.6. We can remember how to calculate $\vec{\nabla}^2 f$ by the formula

$$\vec{\nabla}^2 f = \frac{\partial^2 f}{\partial x^2} + \frac{\partial^2 f}{\partial y^2} + \frac{\partial^2 f}{\partial z^2}. \tag{3.8.5}$$

A function f whose Laplacian is zero is called **harmonic**. For example the temperature distribution throughout a solid that is in thermal equilibrium is a harmonic function.

There are many differentiation formulas involving gradient, divergence, curl, and Laplacian which resemble formulas from single-variable calculus. We encountered some of these in the earlier examples in this section. See, in particular, (3.8.1) and (3.8.2). Because these will arise in this book from time to time and because they are generally helpful in calculations, we summarize them in Table 3.8.1. In the exercises, you are asked to prove some of these formulas.

- “Powers of $\mathbf{x}$” formulas
  1. $\vec{\nabla}\|\mathbf{x}\| = \dfrac{\mathbf{x}}{\|\mathbf{x}\|}$
  2. $\vec{\nabla}\|\mathbf{x}\|^k = k\|\mathbf{x}\|^{k-2}\mathbf{x}; \quad k \in \mathbb{R}$
  3. $\operatorname{div}\left(\|\mathbf{x}\|^k\mathbf{x}\right) = (k+3)\|x\|^k; \quad k \in \mathbb{R}$
  4. $\operatorname{curl}\left(\|\mathbf{x}\|^k\mathbf{x}\right) = \mathbf{0}$

- “Constant multiple” rules
  5. $\vec{\nabla}(cF) = c\vec{\nabla}f$
  6. $\operatorname{div}(c\mathbf{F}) = c\operatorname{div}\mathbf{F}$
  7. $\operatorname{curl}(c\mathbf{F}) = c\operatorname{curl}\mathbf{F}$

- Sum rules
  8. $\vec{\nabla}(f+g) = \vec{\nabla}f + \vec{\nabla}g$
  9. $\operatorname{div}(\mathbf{F}+\mathbf{G}) = \operatorname{div}\mathbf{F} + \operatorname{div}\mathbf{G}$
  10. $\operatorname{curl}(\mathbf{F}+\mathbf{G}) = \operatorname{curl}\mathbf{F} + \operatorname{curl}\mathbf{G}$

- Product rules
  11. $\vec{\nabla}(fg) = g\vec{\nabla}f + f\vec{\nabla}\mathbf{g}$
  12. $\vec{\nabla}(\mathbf{F}\cdot\mathbf{G}) = \sum_{i=1}^{3}(F_i\vec{\nabla}G_i + G_i\vec{\nabla}F_i)$
  13. $\operatorname{div}(f\mathbf{G}) = f\operatorname{div}\mathbf{G} + \mathbf{G}\cdot\vec{\nabla}f$
  14. $\operatorname{div}(\mathbf{F}\times\mathbf{G}) = \mathbf{G}\cdot\operatorname{curl}\mathbf{F} - \mathbf{F}\cdot\operatorname{curl}\mathbf{G}$
  15. $\operatorname{curl}(f\mathbf{G}) = f\operatorname{curl}\mathbf{G} + \vec{\nabla}f\times\mathbf{G}$

- Composition of differential operators
  16. $\operatorname{div}\vec{\nabla}f = \vec{\nabla}^2 f$
  17. $\operatorname{curl}\vec{\nabla}f = \mathbf{0}$
  18. $\operatorname{div}\operatorname{curl}\mathbf{F} = 0$

Table 3.8.1 Differentiation formulas for grad , div , and curl

Exercises 3.8

In Exercises 1 through 7, calculate the divergence and curl of the given vector field.

1. $\mathbf{B} = \left(\frac{1}{2}x - \frac{1}{3}y + \frac{1}{4}z,\ -\frac{1}{3}x + y + \frac{1}{5}z,\ \frac{1}{4}x + \frac{1}{5}y - \frac{1}{6}z\right)$
2. $\mathbf{C} = (-3y + z,\ 3x - 9z,\ -x + 9y)$
3. $\mathbf{E} = (4x + y,\ 7x - 3y - 5z,\ 8x)$
4. $\mathbf{F} = \left(\frac{1}{4}x^4 y,\ xz^2 + y,\ xyz + 2\right)$
5. $\mathbf{G} = \left(\frac{x}{y},\ \frac{y}{z},\ \frac{z}{x}\right)$
6. $\mathbf{H} = \left(\frac{x}{\sqrt{x^2+y^2+z^2}},\ \frac{y}{\sqrt{x^2+y^2+z^2}},\ \frac{z}{\sqrt{x^2+y^2+z^2}}\right)$
7. $\mathbf{J} = (z \sin xy,\ x \cos xy,\ y \tan xy)$
8. Prove formula 3 in Table 3.8.1.
9. Prove formula 4 in Table 3.8.1.
10. Prove formula 7 in Table 3.8.1.
11. Prove formula 12 in Table 3.8.1.
12. Prove formula 14 in Table 3.8.1.
13. Prove formula 17 in Table 3.8.1.
14. Prove formula 18 in Table 3.8.1.
15. Show that $\vec{\nabla} \ln \|\mathbf{x}\| = \mathbf{x}/\|\mathbf{x}\|^2$.
16. Show that $\operatorname{curl}(\mathbf{c} \times \mathbf{F}) = (\operatorname{div}\mathbf{F})\mathbf{c} - (\mathbf{c} \cdot \nabla)\mathbf{F}$ for any constant vector $\mathbf{c}$.
17. Show that $\operatorname{div}(\vec{\nabla} f \times \vec{\nabla} g) = 0$.
18. Find a formula for $\operatorname{div}(\mathbf{F}/f)$.
19. Find a formula for $\operatorname{curl}(\mathbf{F}/f)$.
20. Show that if A is a 3×3 matrix and

$$\mathbf{F}(\mathbf{x}) = A\mathbf{x},$$

then

$$(\operatorname{curl}\mathbf{F}) \times \mathbf{x} = (A - A^T)\mathbf{x}.$$

21. Show that if A is symmetric, then

$$\operatorname{curl}(A\mathbf{x}) = \mathbf{0}.$$

22. Show that if A is skew-symmetric, then

$$\operatorname{curl}(A\mathbf{x}) \times \mathbf{x} = 2A\mathbf{x}$$

and

$$\operatorname{div}(A\mathbf{x}) = 0.$$

23. Show that if $J\mathbf{F}(\mathbf{a})$ is symmetric, then $\operatorname{curl}\mathbf{F}(\mathbf{a}) = \mathbf{0}$.

24. Show that if $J\mathbf{F}(\mathbf{a})$ is skew-symmetric, then

$$\mathbf{DF}(\mathbf{a})(\mathbf{x}) = \frac{1}{2}\operatorname{curl}\mathbf{F}(\mathbf{a}) \times \mathbf{x}.$$

25. Show that if $\mathbf{F}$ is a vector field on $\mathbb{R}^3$, then

$$\mathbf{DF}(\mathbf{a})(\mathbf{x}) = A\mathbf{x} + \frac{1}{2}\operatorname{curl}\mathbf{F}(\mathbf{a}) \times \mathbf{x},$$

where A is a 3×3 matrix such that $\operatorname{tr}(A) = \operatorname{div}\mathbf{F}(\mathbf{a})$. (*Hint*: See Exercise 32 in Section 2.2.)

26. Recall (see Exercise 38 in Section 2.2) the definition of the norm of a square matrix. Show that if $\mathbf{F} : \mathbb{R}^3 \to \mathbb{R}^3$, then the value of r that minimizes $||rI - J\mathbf{f}(\mathbf{a})||$ is

$$r = \frac{\operatorname{div}\mathbf{F}(\mathbf{a})}{3}.$$

This result says that the linear transformation $\mathbf{T}(\mathbf{x}) = \frac{1}{3}\operatorname{div}\mathbf{F}(\mathbf{a})I\mathbf{x}$ is the "closest" dilation approximation to $\mathbf{DF}(\mathbf{a})$.

27. Suppose that each of the following is a velocity field for a cloud of dust particles. Explain geometrically what happens to a cubical region of dust centered at the origin over a short time interval. In each case, calculate the divergence at the origin and explain the result geometrically.

(a) $\mathbf{v} = (x,\ y,\ z)$

(b) $\mathbf{v} = (2x,\ -y,\ 0)$

(c) $\mathbf{v} = (3x,\ -y,\ -2z)$

(d) $\mathbf{v} = (x,\ -4y,\ z)$

(e) $\mathbf{v} = (-x,\ -y,\ -z)$

28. Suppose that $\mathbf{F} = (0,\ kz,\ 0)$ is the velocity field of a fluid moving in space.

(a) Describe the motion of particles moving according to this field.

(b) Calculate $\operatorname{div}\mathbf{F}$ and interpret the result.

(c) Calculate $\operatorname{curl}\mathbf{F}$ and interpret the result.

3.9 Mean Value Theorems: Taylor's Theorem

The role of the Mean Value Theorem in elementary calculus is absolutely central to that subject. Similarly, its generalization, Taylor's Theorem, provides not only a means of effectively approximating the values of algebraic and transcendental functions such as $(1 + x)^r$, $\sin x$, $\arcsin x$, e^x, and $\ln x$ but also a way of estimating how good these approximations are.

In the context of functions of several variables, the Mean Value Theorem and Taylor's Theorem have their analogs. Here, too, Taylor's Theorem is important in estimating errors in polynomial approximations, and it also provides a "second derivative" test for identifying local extrema.

Besides purely theoretical uses, Taylor's Theorem is important in applications. Many of the problems in physics and chemistry involving heat conduction, diffusion, wave propagation, and the like are posed in terms of partial differential equations (PDEs), in which the unknown (e.g., temperature or displacement) is a function of three spatial variables and a time variable. Such PDEs cannot, except for very special geometries, be solved explicitly in terms of a formula involving the independent variables, so one resorts to a discretized version of the problem, that is, a **difference equation** whose solution at discrete points is supposed to approximate the solution to the original PDE at those points. Taylor's Theorem for functions of several variables provides the means to estimate the errors incurred by the discretization.

Since the version of Taylor's with which you are familiar involves derivatives of various orders, it is natural to expect that a theorem in vector calculus that bears the same name will also involve derivatives of order greater than one. This is indeed the case. We begin by defining what amounts to the second derivative of a function of several variables.

Definition 3.9.1 Given a function $f : D \subset \mathbb{R}^n \to \mathbb{R}$ that has second-order partials at $\mathbf{a} \in D$, the **Hessian**[8] **matrix** for f at $\mathbf{a}$ is

$$Hf(\mathbf{a}) = \left[\frac{\partial^2 f}{\partial x_j \partial x_i}(\mathbf{a}) \right].$$

The **Hessian form for** f **at a** is the quadratic form defined by

$$h(\mathbf{x}) = \mathbf{x}^T Hf(\mathbf{a})\mathbf{x}.$$

Example 3.9.1 Find the Hessian matrix and Hessian form for $f(x, y) = 1/\sqrt{x^2 + y^2}$ at the point (3, 4).

Solution First we have

$$\frac{\partial f}{\partial x} = \frac{-x}{(x^2 + y^2)^{3/2}} \quad \text{and} \quad \frac{\partial f}{\partial y} = \frac{-y}{(x^2 + y^2)^{3/2}},$$

so

$$\begin{aligned} \frac{\partial^2 f}{\partial x^2} &= \frac{2x^2 - y^2}{(x^2 + y^2)^{5/2}}, \\ \frac{\partial^2 f}{\partial y^2} &= \frac{2y^2 - x^2}{(x^2 + y^2)^{5/2}} \end{aligned}$$

and

$$\frac{\partial^2 f}{\partial y \partial x} = \frac{\partial^2 f}{\partial x \partial y} = \frac{3xy}{(x^2 + y^2)^{5/2}}.$$

By Definition 3.9.1,

$$\begin{aligned} H\mathbf{f}(3, 4) &= \begin{bmatrix} \dfrac{\partial^2 f}{\partial x^2}(3, 4) & \dfrac{\partial^2 f}{\partial y \partial x}(3, 4) \\ \dfrac{\partial^2 f}{\partial x \partial y}(3, 4) & \dfrac{\partial^2 f}{\partial y^2}(3, 4) \end{bmatrix} = \begin{bmatrix} 2/3125 & 36/3125 \\ 36/3125 & 23/3125 \end{bmatrix} \\ &= \frac{1}{3125} \begin{bmatrix} 2 & 36 \\ 36 & 23 \end{bmatrix}. \end{aligned}$$

---

[8] This matrix is named for Ludwig Otto Hesse (1811-1874) who applied it to solving systems of algebraic equations. According to [1], the same quantity was used many years earlier by Lagrange.

Then the Hessian form is

$$\begin{aligned} h(x,y) &= \frac{1}{3125}\begin{bmatrix} x & y \end{bmatrix}\begin{bmatrix} 2 & 36 \\ 36 & 23 \end{bmatrix}\begin{bmatrix} x \\ y \end{bmatrix} \\ &= \frac{1}{3125}\begin{bmatrix} x & y \end{bmatrix}\begin{bmatrix} 2x+36y \\ 36x+23y \end{bmatrix} \\ &= \frac{1}{3125}(2x^2+72xy+23y^2). \quad \blacklozenge \end{aligned}$$

You can think of the Hessian matrix as representing the second derivative of f at $\mathbf{a}$. Indeed, the Hessian matrix is just the Jacobian matrix for $\vec{\nabla} f(\mathbf{x})$.

We can now state (and prove!) a theorem that will be very useful to us in Section 3.10, when we discuss extrema of functions of several variables.

Theorem 3.9.1 *(Taylor's Theorem of order 1) Suppose that $f : D \subset \mathbb{R}^n \to \mathbb{R}$ has second-order partials that are continuous in an open ball B centered at a point $\mathbf{a} \in D$. Then for each point $\mathbf{x} \in B$ there exists a point $\mathbf{x}_0$ on the segment from $\mathbf{a}$ to $\mathbf{x}$ such that*

$$f(\mathbf{x}) = f(\mathbf{a}) + \vec{\nabla} f(\mathbf{a}) \cdot (\mathbf{x}-\mathbf{a}) + \frac{1}{2}(\mathbf{x}-\mathbf{a})^T Hf(\mathbf{x}_0)(\mathbf{x}-\mathbf{a}). \qquad (3.9.1)$$

Proof (optional) Recall Taylor's Theorem for a function of one variable: If $g(t)$ has continuous second derivative on an open interval I about a number α, then for each $t \in I$ there exists a number β between t and α such that

$$g(t) = g(\alpha) + g'(\alpha)(t-\alpha) + \frac{1}{2}g''(\beta)(t-\alpha)^2.$$

The idea of our proof here is simply to reduce the multivariable question to a single variable one. Toward that end, pick a point $\mathbf{x}$ in the ball B and define g by $g(t) = f(\mathbf{a}+t(\mathbf{x}-\mathbf{a}))$. Since g is the composition of twice continuously differentiable functions, its second derivative is continuous over the interval of t-values for which $\mathbf{a}+t(\mathbf{x}-\mathbf{a}) \in B$. Now by the Chain Rule,

$$g'(t) = \vec{\nabla} f(\mathbf{a}+t(\mathbf{x}-\mathbf{a})) \cdot (\mathbf{x}-\mathbf{a}),$$

and in particular

$$g'(0) = \vec{\nabla} f(\mathbf{a}) \cdot (\mathbf{x}-\mathbf{a}).$$

Applying the Chain Rule again, and then the Product Rule (see Exercise 27 in Section 3.7), we obtain

$$
\begin{aligned}
g''(t) & \\
&= \frac{d}{dt}\left(\vec{\nabla} f(\mathbf{a}+t(\mathbf{x}-\mathbf{a}))\cdot(\mathbf{x}-\mathbf{a})\right)\\
&= \frac{d}{dt}\sum_{i=1}^{n}\frac{\partial f}{\partial x_i}(\mathbf{a}+t(\mathbf{x}-\mathbf{a}))(x_i-a_i)\\
&= \sum_{i=1}^{n}\frac{d}{dt}\left(\frac{\partial f}{\partial x_i}(\mathbf{a}+t(\mathbf{x}-\mathbf{a}))(x_i-a_i)\right)\\
&= \sum_{i=1}^{n}\left(\frac{d}{dt}\left(\frac{\partial f}{\partial x_i}(\mathbf{a}+t(\mathbf{x}-\mathbf{a}))\right)(x_i-a_i)+\frac{\partial f}{\partial x_i}(\mathbf{a}+t(\mathbf{x}-\mathbf{a}))\frac{d}{dt}(x_i-a_i)\right)\\
&= \sum_{i=1}^{n}(x_i-a_i)\vec{\nabla}\frac{\partial f}{\partial x_i}(\mathbf{a}+t(\mathbf{x}-\mathbf{a}))\cdot(\mathbf{x}-\mathbf{a})\\
&= \sum_{i=1}^{n}(x_i-a_i)\left[\frac{\partial^2 f}{\partial x_1\,\partial x_i}(\mathbf{a}+t(\mathbf{x}-\mathbf{a}))\quad\cdots\quad\frac{\partial^2 f}{\partial x_n\,\partial x_i}(\mathbf{a}+t(\mathbf{x}-\mathbf{a}))\right](\mathbf{x}-\mathbf{a})\\
&= (\mathbf{x}-\mathbf{a})^T Hf(\mathbf{a}+t(\mathbf{x}-\mathbf{a}))(\mathbf{x}-\mathbf{a}).
\end{aligned}
$$

By Taylor's Theorem applied to g when $t=1$, there exists t_0 between 0 and 1 such that

$$g(1)=g(0)+g'(0)\cdot 1+\frac{1}{2}g''(t_0)(1-0)^2.$$

In terms of f, this says exactly that

$$f(\mathbf{x})=f(\mathbf{a})+\vec{\nabla}f(\mathbf{a})\cdot(\mathbf{x}-\mathbf{a})+\frac{1}{2}(\mathbf{x}-\mathbf{a})^T Hf(\mathbf{a}+t_0(\mathbf{x}-\mathbf{a}))(\mathbf{x}-\mathbf{a}).$$

Since $\mathbf{x}_0=\mathbf{a}+t_0(\mathbf{x}-\mathbf{a})$ is between $\mathbf{x}$ and $\mathbf{a}$, we have shown that there exists $\mathbf{x}_0$ between $\mathbf{x}$ and $\mathbf{a}$ such that

$$f(\mathbf{x})=f(\mathbf{a})+\vec{\nabla}f(\mathbf{a})\cdot(\mathbf{x}-\mathbf{a})+\frac{1}{2}(\mathbf{x}-\mathbf{a})^T Hf(\mathbf{x}_0)(\mathbf{x}-\mathbf{a}).\quad\blacklozenge$$

The function $p_1(\mathbf{x})=f(\mathbf{a})+\vec{\nabla}f(\mathbf{a})\cdot(\mathbf{x}-\mathbf{a})$ is called the **first-degree Taylor polynomial for f centered at a.**

A consequence of Taylor's Theorem is this: If the **remainder** term $\frac{1}{2}(\mathbf{x}-\mathbf{a})^T Hf(\mathbf{x}_0)(\mathbf{x}-\mathbf{a})$ in (3.9.1) is small for all $\mathbf{x}$ in a ball surrounding $\mathbf{a}$, then $f(\mathbf{x})$ is well-approximated by the first degree Taylor polynomial near $\mathbf{a}$.

Example 3.9.2 Find the first-degree Taylor polynomial and remainder for $f(x, y) = e^{x+y}$ at $\mathbf{a} = (0, 0)$.

Solution We have $f(0, 0) = e^0 = 1$ and

$$\vec{\nabla} f(0, 0) = (e^{x+y}, e^{x+y})\big|_{(0,0)} = (1, 1),$$

so

$$\begin{aligned} p_1(x, y) &= f(0, 0) + \vec{\nabla} f(0, 0) \cdot ((x, y) - (0, 0)) \\ &= 1 + (1, 1) \cdot (x, y) = 1 + x + y. \end{aligned}$$

Also,

$$Hf(x, y) = \begin{bmatrix} e^{x+y} & e^{x+y} \\ e^{x+y} & e^{x+y} \end{bmatrix} = e^{x+y} \begin{bmatrix} 1 & 1 \\ 1 & 1 \end{bmatrix},$$

and so by Theorem 3.9.1 there exists (x_0, y_0) between (x, y) and $(0, 0)$ such that the remainder is

$$\begin{aligned} &\frac{1}{2}\left(\begin{bmatrix} x & y \end{bmatrix} - \begin{bmatrix} 0 & 0 \end{bmatrix}\right) e^{x_0+y_0} \begin{bmatrix} 1 & 1 \\ 1 & 1 \end{bmatrix} \left(\begin{bmatrix} x \\ y \end{bmatrix} - \begin{bmatrix} 0 \\ 0 \end{bmatrix}\right) \\ = \;& \frac{e^{x_0+y_0}}{2} \begin{bmatrix} x & y \end{bmatrix} \begin{bmatrix} 1 & 1 \\ 1 & 1 \end{bmatrix} \begin{bmatrix} x \\ y \end{bmatrix} \\ = \;& \frac{e^{x_0+y_0}}{2}(x^2 + 2xy + y^2). \end{aligned}$$

In this situation we can visualize the graph of f along with that of p_1. See Figure 3.9.1. Of course, the graph of $z = p_1(x, y)$ is just the plane tangent to $z = e^{x+y}$ at the point $(0, 0, 1)$. ♦

To express Taylor's Theorem in general, it is helpful to invent some notation. By the symbol $\vec{\nabla}$ we mean the differential **operator** (function), which acts on a scalar-valued function f to yield the gradient function. We "abuse" our notation by writing $\vec{\nabla} = (\partial/\partial x_1, \partial/\partial x_x, \ldots, \partial/\partial x_n)$ so that putting $\vec{\nabla}$ to the left of a function symbol can be thought of as "multiplication" of $\vec{\nabla}$ by f. Incidentally, this symbol is often called the "del" operator. Regarding $\vec{\nabla}$ as a vector, we can symbolize many expressions in a compact form. For instance, if $\mathbf{g}$ is a vector-valued function we can write $\mathbf{g} \cdot \vec{\nabla}$ for the operator that acts on a scalar function f in this way:

$$(\mathbf{g} \cdot \vec{\nabla})f = g_1 \frac{\partial f}{\partial x_1} + g_2 \frac{\partial f}{\partial x_2} + \cdots + g_n \frac{\partial f}{\partial x_n},$$

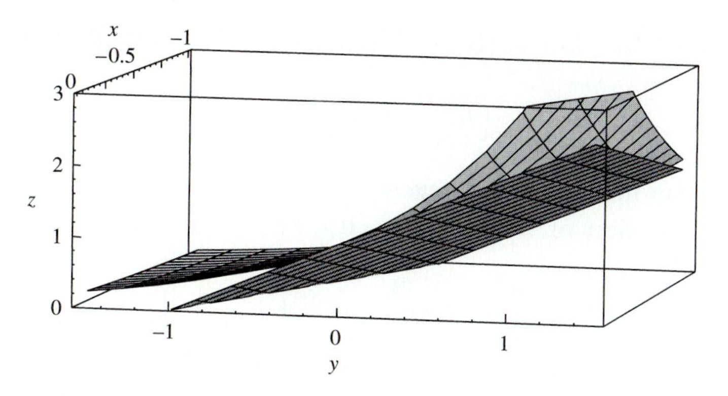

Figure 3.9.1 The graph of $f(x, y) = e^{x+y}$ along with that of its first degree Taylor polynomial $p(x, y) = 1+x+y$ centered at (0, 0)

[i.e., $\mathbf{g} \cdot \vec{\nabla} = g_1\partial/\partial x_1 + \cdots + g_n\partial/\partial x_n$]. In particular, if $\mathbf{g(x)} = \mathbf{h}$, a constant vector, let us examine what happens when we apply $\mathbf{h} \cdot \vec{\nabla}$ over and over, at least in the case where f depends only on $n = 2$ variables and has continuous mixed partials of all orders. With

$$\mathbf{h} \cdot \vec{\nabla} = h_1\frac{\partial}{\partial x} + h_2\frac{\partial}{\partial y},$$

we have

$$(\mathbf{h} \cdot \vec{\nabla})f = h_1\frac{\partial f}{\partial x} + h_2\frac{\partial f}{\partial y}$$

and

$$\begin{aligned}
(\mathbf{h} \cdot \vec{\nabla})^2 f &= (\mathbf{h} \cdot \vec{\nabla})(\mathbf{h} \cdot \vec{\nabla})f \\
&= h_1\frac{\partial}{\partial x}\left(h_1\frac{\partial f}{\partial x} + h_2\frac{\partial f}{\partial y}\right) + h_2\frac{\partial}{\partial y}\left(h_1\frac{\partial f}{\partial x} + h_2\frac{\partial f}{\partial y}\right) \\
&= h_1^2\frac{\partial^2 f}{\partial x^2} + 2h_1h_2\frac{\partial^2 f}{\partial x\,\partial y} + h_2^2\frac{\partial^2 f}{\partial y^2}.
\end{aligned}$$

Similarly, you can verify that

$$\begin{aligned}
(\mathbf{h} \cdot \vec{\nabla})^3 f &= (\mathbf{h} \cdot \vec{\nabla})(\mathbf{h} \cdot \vec{\nabla})^2 f \\
&= h_1^3\frac{\partial^3 f}{\partial x^3} + 3h_1^2h_2\frac{\partial^3 f}{\partial x^2\,\partial y} + 3h_1h_2^2\frac{\partial^3 f}{\partial x\,\partial y^2} + h_2^3\frac{\partial^3 f}{\partial y^3}.
\end{aligned}$$

Here, as above, we are relying on Clairaut's Theorem about the equality of mixed partials. In general, $(\mathbf{h}\cdot\vec{\nabla})^m$ represents a "binomial expansion" of

$$\left(h_1\frac{\partial}{\partial x}+h_2\frac{\partial}{\partial y}\right)^m f,$$

where m is any nonnegative integer.

With a function of n variables, $(\mathbf{h}\cdot\vec{\nabla})^m$ represents the "multinomial expansion" of

$$\left(h_1\frac{\partial}{\partial x_1}+h_2\frac{\partial}{\partial x_2}+\ldots+h_n\frac{\partial}{\partial x_n}\right)^m.$$

With this notation, we can now state Taylor's Theorem.

Theorem 3.9.2 *(Taylor's Theorem of order m) Suppose that $f : D \in \mathbb{R}^n \to \mathbb{R}$ has $m+1$ order partials that are continuous in an open ball B centered at a point $\mathbf{a}$ in D. Then for each point $\mathbf{x} \subset B$ there exists a point $\mathbf{x}_0$ on the segment from $\mathbf{a}$ to $\mathbf{x}$ such that*

$$f(\mathbf{x}) \tag{3.9.2}$$

$$= \sum_{k=0}^{m}\frac{1}{k!}(\mathbf{h}\cdot\vec{\nabla})^k f(\mathbf{y})\bigg|_{\substack{\mathbf{y}=\mathbf{a}\\ \mathbf{h}=\mathbf{x}-\mathbf{a}}} + \frac{1}{(m+1)!}(\mathbf{h}\cdot\vec{\nabla})^{m+1} f(\mathbf{y})\bigg|_{\substack{\mathbf{y}=\mathbf{x}_0\\ \mathbf{h}=\mathbf{x}-\mathbf{a}}}.$$

A proof of this theorem is beyond the scope of this book.

The sum in (3.9.2) is called the **Taylor polynomial of degree m for f centered at a**. The second term on the right-hand side of Equation 3.9.2 is the **remainder**. If the remainder goes to zero as $m \to \infty$ for all $\mathbf{x}$ in some open ball, then f is equal to its **Taylor series** centered at $\mathbf{a}$:

$$f(\mathbf{x}) = \sum_{k=0}^{\infty}\frac{1}{k!}(\mathbf{h}\cdot\vec{\nabla})^k f(\mathbf{y})\bigg|_{\substack{\mathbf{y}=\mathbf{a}\\ \mathbf{h}=\mathbf{x}-\mathbf{a}}}.$$

Example 3.9.3 Calculate the second-degree Taylor polynomial of $f(x, y) = e^{-(x^2+y^2)}$ at the point $(0, 0)$.

Solution We just calculate the quantities required in (3.9.2):

$$(\mathbf{h}\cdot\vec{\nabla})^0 f(\mathbf{y})\bigg|_{\substack{\mathbf{y}=\mathbf{0}\\ \mathbf{h}=(x,\,y)}} = f(\mathbf{y})|_{\mathbf{y}=\mathbf{0}} = f(\mathbf{0}) = 1$$

$$(\mathbf{h}\cdot\vec{\nabla})^1 f(\mathbf{y})\bigg|_{\substack{\mathbf{y}=\mathbf{0}\\ \mathbf{h}=(x,\,y)}} = \left(h_1\frac{\partial f}{\partial x}(\mathbf{y})+h_2\frac{\partial f}{\partial y}(\mathbf{y})\right)\bigg|_{\substack{\mathbf{y}=\mathbf{0}\\ \mathbf{h}=(x,\,y)}}$$

$$= x\frac{\partial f}{\partial x}(\mathbf{0})+y\frac{\partial f}{\partial y}(\mathbf{0}).$$

Now

$$\frac{\partial f}{\partial x}(\mathbf{0}) = -2xe^{-x^2-y^2}\Big|_{(0,0)} = 0$$

and

$$\frac{\partial f}{\partial y}(\mathbf{0}) = -2ye^{-x^2-y^2}\Big|_{(0,0)} = 0,$$

so

$$(\mathbf{h}\cdot\vec{\nabla})f(\mathbf{y})\Big|_{\substack{\mathbf{y}=\mathbf{0}\\ \mathbf{h}=(x,\,y)}} = 0.$$

Finally,

$$\begin{aligned}(\mathbf{h}\cdot\vec{\nabla})^2 f(\mathbf{y})\Big|_{\substack{\mathbf{y}=\mathbf{0}\\ \mathbf{h}=(x,\,y)}} &= \left(h_1^2\frac{\partial^2 f}{\partial x^2}(\mathbf{0}) + 2h_1h_2\frac{\partial^2 f}{\partial x\,\partial y}(\mathbf{0}) + h_2^2\frac{\partial^2 f}{\partial y^2}(\mathbf{0})\right)\Bigg|_{\mathbf{h}=(x,\,y)} \\ &= x^2\frac{\partial^2 f}{\partial x^2}(0,\,0) + 2xy\frac{\partial^2 f}{\partial x\,\partial y}(0,\,0) + y^2\frac{\partial^2 f}{\partial y^2}(0,\,0).\end{aligned}$$

You should verify that

$$\frac{\partial^2 f}{\partial x^2}(0,\,0) = -2,\quad \frac{\partial^2 f}{\partial x\partial y}(0,\,0) = 0,\quad \text{and}\quad \frac{\partial^2 f}{\partial y^2}(0,\,0) = -2.$$

So

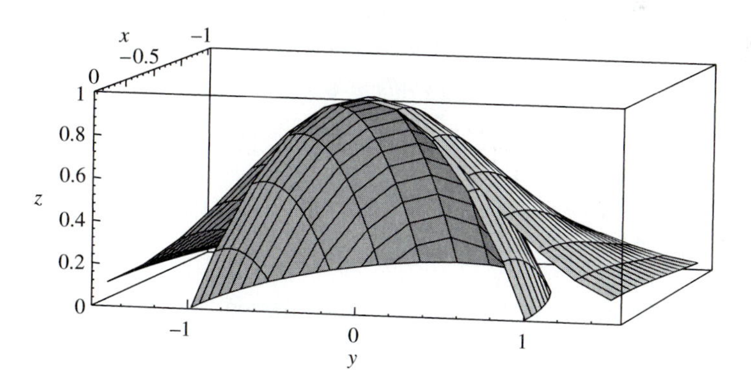

Figure 3.9.2 The graph of $f(x,y) = e^{-x^2-y^2}$ with that of its second-degree Taylor polynomial $p_2(x,\,y) = 1 - x^2 - y^2$ centered at $(0,\,0)$

$$(\mathbf{h}\cdot\vec{\nabla})^2 f(\mathbf{y})\Big|_{\substack{\mathbf{y}=\mathbf{0}\\ \mathbf{h}=(x,\,y)}} = -2x^2 - 2y^2.$$

Thus the second-degree Taylor polynomial is

$$\begin{aligned} p_2(x,\, y) &= 1 + \frac{1}{1!}\cdot 0 + \frac{1}{2!}(-2x^2 - 2y^2) \\ &= 1 - x^2 - y^2. \end{aligned}$$

See Figure 3.9.2 for graphs of the two functions superimposed. Note that p_2 is a very close approximation for f near (0, 0). This is typical. ♦

In the special case when we want the second-degree ($m = 2$) Taylor polynomial, it is convenient to write the sum in (3.9.2) in terms of the gradient and Hessian. You should verify that

$$p_2(\mathbf{x}) = f(\mathbf{a}) + \vec{\nabla} f(\mathbf{a}) \cdot (\mathbf{x} - \mathbf{a}) + \frac{1}{2}(\mathbf{x} - \mathbf{a})^T H f(\mathbf{a})(\mathbf{x} - \mathbf{a}). \tag{3.9.3}$$

Exercises 3.9

In Exercises 1 through 7, compute the Hessian matrix and Hessian form of the function at the point indicated.

1. $f(x,\, y) = 2x^2 - 3xy + 14y^2$; $\mathbf{a} = (1,\, 4)$

2. $f(x,\, y) = \ln(x + y)$; $\mathbf{a} = (1,\, 1)$

3. $f(x,\, y,\, z) = \arctan x + yz$; $\mathbf{a} = (0,\, 3,\, 1)$

4. $f(x_1,\, x_2,\, x_3,\, x_4) = x_1^2 - x_2x_3 + x_4$; $\mathbf{a} = (0,\, 0,\, 0,\, 0)$

5. $f(x_1,\, x_2,\, x_3,\, x_4) = \dfrac{x_1 + x_2}{x_3 + x_4}$; $\mathbf{a} = (1,\, 1,\, 1,\, 1)$

6. $f(x_1, x_2, \ldots, x_n) = x_1 + x_2 + \ldots + x_n$; $\mathbf{a}$ arbitrary

7. $f(\mathbf{x}) = ||\mathbf{x}||^2$; $\mathbf{a}$ arbitrary

In Exercises 8 through 16, find the second-degree Taylor polynomial for the given function at the point indicated.

8. $f(x,\, y) = 4x^2 + 3y^2$; $\mathbf{a} = (1,\, -1)$

9. $f(x,\, y) = 2x^2 + 3y^2 - 4x + 42y + 145$; $\mathbf{a} = (1,\, 7)$

10. $f(x,\, y) = x^3 + y^3$; $\mathbf{a} = (2,\, 3)$

11. $f(x,\, y) = \sin(x^2 + y^2)$; $\mathbf{a} = (0,\, 0)$

12. $f(x,\, y,\, z) = x + ye^z$; $\mathbf{a} = (1,\, 1,\, 0)$

13. $f(x, y, z) = x \cos yz$; $\mathbf{a} = (-4, \pi/4, 1)$

14. $f(x, y, z) = z \tan^{-1}(y/x)$; $\mathbf{a} = (1, 1, 1)$

15. $f(x, y, z, w) = 3x + 2y - 42 + w$; $\mathbf{a} = (1, -1, 2, 0)$

16. $f(\mathbf{x}) = ||\mathbf{x}||^2$; $\mathbf{a} = \mathbf{0}$

17. Show that the Hessian of a linear transformation is identically zero.

18. Show that the Hessian of a quadratic form is two times the symmetric matrix that represents the form.

19. Show that if $f(x, y) = g(ax + by)$ and p is a Taylor polynomial of g centered at 0, then the corresponding Taylor polynomial of f centered at $(0, 0)$ is given by

$$P(x, y) = p(ax + by).$$

For Exercises 20 through 26, use Exercise 19 and your knowledge of Taylor polynomials for elementary functions to write down the Taylor polynomial of the indicated degree centered at $(0, 0)$.

20. $f(x, y) = e^{x+y}$; $m = 4$

21. $s(x, y) = e^{(x+y)^2}$; $m = 4$

22. $t(x, y) = (x - y)e^{(x-y)/4}$; $m = 4$

23. $g(x, y) = \sin(\pi(x - y))$; $m = 5$

24. $k(x, y) = \cos\left(\dfrac{x + y}{6}\right)$; $m = 2$

25. $r(x, y) = \sqrt{1 + 9(x + y)^2}$; $m = 2$

26. $u(x, y) = \dfrac{1}{1 - x^2 - 2xy - y^2}$; $m = 6$

3.10 Local Extrema

One of the principal applications of calculus is to solve optimization problems. Many scientific questions can be posed using the language of (multivariable) calculus in such a way that the answer to the question is, "the point at which this function attains its minimum value" or "the maximum value of this function." For instance, in designing an apparatus to make a chemical product, an engineer may have to adjust the size of the vessel where reactions take place, the pressure inside the vessel, and the amount of catalyst in order to maximize the amount of chemical produced per unit time. A physicist may use a minimum energy principle to understand the trajectory of a spacecraft as it moves among the planets. An economist attempting to predict the unemployment rate a year from now may pose that question as a minimization problem in which, for example,the present prime interest rate, inflation rate, unemployment rate, and Dow Jones Industrial average are all independent variables.

In this section we are concerned with the general problem of finding and identifying local extrema of functions of several variables. The main tool here is a second derivative test, analogous to a test with the same name for functions of one variable. Indeed, recall that if f is a real-valued function of one variable such that $f'(a) = 0$, then $f''(a) < 0$ implies that $f(a)$ is a local maximum value, and $f''(a) > 0$ implies that $f(a)$ is a local minimum value. In geometric terms, if the line tangent to the graph of f at a has slope zero, then concavity downward at a implies a local maximum at a, and concavity upward implies a local minimum.

For a function of several variables, there is an analogous result: If $\vec{\nabla} f(\mathbf{a}) = \mathbf{0}$ then a negative definite Hessian form $h(\mathbf{x}) = \mathbf{x}^T Hf(\mathbf{a})\mathbf{x}$ implies that f has a local maximum value at $\mathbf{a}$, and a positive definite Hessian form implies that $f(\mathbf{a})$ is a local minimum value. In the two-dimensional case, this means, for instance, that if the plane tangent to $z = f(x, y)$ at the point $\mathbf{a}$ is horizontal and the graph of the Hessian form h is concave downward, then f has a local maximum at $\mathbf{a}$; if the Hessian form is concave upward, then f has a local minimum.

Definition 3.10.1 A function $f : D \subset \mathbb{R}^n \to \mathbb{R}$ has a **local minimum** at $\mathbf{a} \in D$ if there is an open ball U about $\mathbf{a}$ such that $f(\mathbf{x}) \geq f(\mathbf{a})$ for all $\mathbf{x} \in U$. Similarly, f has a **local maximum** at $\mathbf{a} \in D$ if $f(\mathbf{x}) \leq f(\mathbf{a})$ for all $\mathbf{x}$ in some open ball about $\mathbf{a}$. In either of these cases we say that f has a **local extremum** at $\mathbf{a}$. We say that f has a **saddle** at $\mathbf{a}$ if $f(\mathbf{x})$ takes on both positive and hegative values in every open ball about $\mathbf{a}$.

Example 3.10.1 The graph of the function $f(x, y) = x^2 + (y - 2)^2$ is an elliptic paraboloid that is centered at $(0, 2)$ and opens upward. Because we are familiar with the graph, we can tell immediately that the function has a local minimum at $(0, 2)$, corresponding to the lowest point on the graph. To show that $f(0, 2)$ is a local minimum value without referring to a picture, we can simply observe that $f(0, 2) = 0 \leq x^2 + (y - 2)^2 = f(x, y)$ for any choice of $(x, y) \in \mathbb{R}^2$.

Note, too, that the plane tangent to the graph of f is horizontal at $(0, 2) : \vec{\nabla} f(0, 2) = \mathbf{0}$. ♦

Example 3.10.2 The graph of $g(x, y) = 1 - \sqrt{x^2 + y^2}$ is a cone with vertex at (0, 0, 1) that opens downward. The function g attains a local maximum value at this peak on the graph. We can show this without reference to the graph simply by noting that $g(0,\,0) = 1 \geq 1 - \sqrt{x^2 + y^2} = g(x,\,y)$ for any choice of $(x,\,y) \in \mathbb{R}^2$. Notice that $\vec{\nabla}\mathbf{g}$ is undefined at (0, 0). ♦

These two examples illustrate the general behavior of functions at local extrema.

Theorem 3.10.1 *If $f : D \subset \mathbb{R}^n \to \mathbb{R}$ has a local extremum at a point $\mathbf{a} \in D$, then either $\vec{\nabla} f(\mathbf{a}) = \mathbf{0}$ or the gradient at $\mathbf{a}$ is undefined.*

Proof First, suppose that f has a local maximum at $\mathbf{a} = (a_1, a_2, \ldots, a_n)$. Then there exists an open ball B about $\mathbf{a}$ such that $f(\mathbf{a}) \geq f(\mathbf{x})$ for all $\mathbf{x} \in B$. Now the gradient is either defined at $\mathbf{a}$ or else it isn't. If $\vec{\nabla} f(\mathbf{a})$ is defined, then $\dfrac{\partial f}{\partial x_i}(\mathbf{a})$ is defined for $i = 1, 2, \ldots, n$. What is the numerical value of $\dfrac{\partial f}{\partial x_i}(\mathbf{a})$? By definition,

$$\begin{aligned}\frac{\partial f}{\partial x_i}(\mathbf{a}) & \qquad\qquad (3.10.1) \\ = \lim_{x_i \to a_i} & \frac{f(a_1, \ldots, a_{i-1}, x_i, a_{i+1}, \ldots, a_n) - f(a_1, \ldots, a_{i-1}, a_i, a_{i+1}, \ldots, a_n)}{x_i - a_i}\end{aligned}$$

for $i = 1, 2, \ldots, n$.

Since $f(\mathbf{a})$ is a local maximum value, the numerator of the fraction in (3.10.1) is nonnegative, so the limit is also nonnegative: $\dfrac{\partial f}{\partial x_i}(\mathbf{a}) \geq 0$. For $x_i < a_i$, the quotient in Equation 3.10.1 is nonpositive, so the limit must also be nonpositive: $\dfrac{\partial f}{\partial x_i}(\mathbf{a}) \leq 0$. Thus $\dfrac{\partial f}{\partial x_i}(\mathbf{a}) = 0$, $i = 1, 2, \ldots, n$, so $\vec{\nabla} f(\mathbf{a}) = \mathbf{0}$. The alternative in this case is that the gradient is undefined at $\mathbf{a}$.

In the case that f has a local minimum at $\mathbf{a}$, the same reasoning again leads to one of two alternatives: $\vec{\nabla} f(\mathbf{a}) = \mathbf{0}$ or the gradient is undefined.

Definition 3.10.2 A point $\mathbf{a}$ in the domain of $f : D \subset \mathbb{R}^n \to \mathbb{R}$ is called a **critical point** of f if either $\vec{\nabla} f(\mathbf{a}) = \mathbf{0}$ or the gradient of f is undefined at $\mathbf{a}$.

With this definition we can restate Theorem 3.10.1 in a useful form.

Corollary 3.10.1 *The only points in the domain of a function f at which it may have a local extremum are the critical points.*

Example 3.10.3 Find and identify the local extrema of $f(x, y) = 4x^2 + 3y^2 - 24x + 6y + 39$.

Solution The gradient is

$$\vec{\nabla} f(x, y) = (8x - 24, 6y + 6),$$

and it equals zero when $8x - 24 = 0$ and $6y + 6 = 0$. Thus the only critical point of f is $(3, -1)$. Completing the square, we can write $f(x, y) = 4(x-3)^2 + 3(y+1)^2$ to see that $f(3, -1) = 0$ is the smallest value f attains. Thus f has the local (and global) minimum 0 at $(3, -1)$. ♦

Example 3.10.4 Find and identify the local extrema of

$$f(x, y) = (x^2 + y^2)^{2/3}.$$

Solution The gradient of f is

$$\vec{\nabla} f = \left(\frac{4x}{3(x^2+y^2)^{1/3}}, \frac{4y}{3(x^2+y^2)^{1/3}} \right) = \frac{4}{3(x^2+y^2)^{1/3}}(x, y),$$

which is undefined at the point $(0, 0)$. There is no point in f's domain where $\vec{\nabla} f$ is zero, so $(0, 0)$ is the only critical point. Since f is nonnegative and $f(0, 0) = 0$, we see that $f(0, 0)$ is a local minimum value. ♦

Example 3.10.5 Find and identify the local extrema of $f(x, y) = x^{2/3}(y - 1)^{2/3}$.

Solution The gradient is $\vec{\nabla} f(x, y) = \left(\dfrac{2}{3} \dfrac{(y-1)^{2/3}}{x^{1/3}}, \dfrac{2}{3} \dfrac{x^{2/3}}{(y-1)^{1/3}} \right)$.

It is undefined whenever the denominator of either component is zero. Thus any point on the line $x = 0$ or on the line $y = 1$ is a critical point. Note that $f(0, y) = f(x, 1) = 0$ for all $x, y \in \mathbb{R}$ and that $f(x, y) \geq 0$ for all $(x, y) \in \mathbb{R}^2$. Therefore *each point* on these lines is a local minimum. See Figure 3.10.1. ♦

Example 3.10.6 Find the local extrema of $f(x, y) = x^3 + x^2 y - y^2 - 4y$.

Solution The gradient is

$$\vec{\nabla} f(x, y) = (3x^2 + 2xy, \quad x^2 - 2y - 4),$$

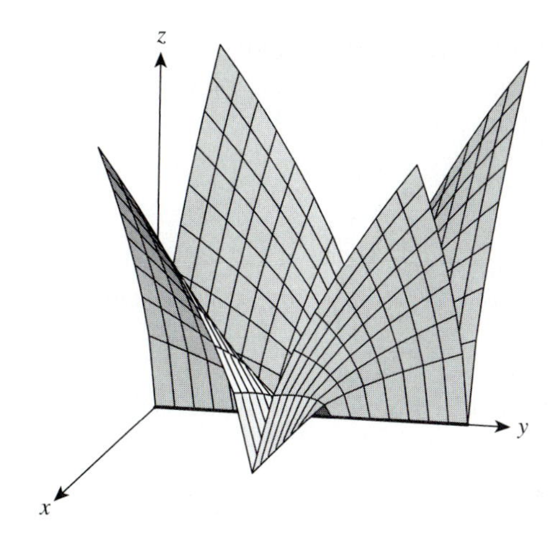

Figure 3.10.1 The graph of $z = x^{2/3}(y-1)^{2/3}$. The function has local minima and critical points whenever $x = 0$ or $y = 1$.

and it is zero precisely when both components are simultaneously zero. Such points are those where x and y satisfy the equations

$$x(3x + 2y) = 0 \qquad (3.10.2)$$

$$x^2 - 2y - 4 = 0. \qquad (3.10.3)$$

We attempt to treat such systems of nonlinear equations so that we find all the solutions. For (3.10.2), $x = 0$ is a solution. Substituting this into (3.10.3) gives us $2y + 4 = 0$, so $y = -2$ and thus $(0, -2)$ is a critical point. Another way for (3.10.2) to be satisfied is that $3x + 2y = 0$. Then

$$y = -\frac{3}{2}x, \qquad (3.10.4)$$

so substituting into (3.10.3), we have

$$\begin{aligned} x^2 + 3x - 4 &= 0 \\ (x-1)(x+4) &= 0 \\ x &= 1, -4. \end{aligned}$$

Putting $x = 1$ into (3.10.4), we see that $\left(1, -\frac{3}{2}\right)$ is a critical point. Also, $x = -4$ gives $(-4, 6)$ as a critical point.

Unfortunately, it is not simply a matter of rewriting $f(x, y)$ as we did in Example 3.10.3 to see whether f has extrema at these points. The next theorem and its corollary, however, will give us a tool for making this determination. ♦

Theorem 3.10.2 *Suppose that $f : D \subset \mathbb{R}^n \to \mathbb{R}$ has continuous second order partials in an open ball B about a critical point $\mathbf{a} \in D$ where $\vec{\nabla} f(\mathbf{a}) = 0$.*

1. *If the Hessian form $h(\mathbf{x}) = \mathbf{x}^T Hf(\mathbf{a})\mathbf{x}$ for f at $\mathbf{a}$ is positive definite, then f has a local minimum at $\mathbf{a}$.*
2. *If the Hessian form for f at $\mathbf{a}$ is negative definite, then f has a local maximum at $\mathbf{a}$.*
3. *If the Hessian form for f at $\mathbf{a}$ is indefinite, then f has a saddle at $\mathbf{a}$.*

Proof (optional) Since the second order partials of f are continuous, (3.9.1) is applicable, so for each $\mathbf{x} \in B$ there exists $\mathbf{x}_0$ between $\mathbf{a}$ and $\mathbf{x}$ such that

$$f(\mathbf{x}) = f(\mathbf{a}) + \frac{1}{2}(\mathbf{x} - \mathbf{a})^T Hf(\mathbf{x}_0)(\mathbf{x} - \mathbf{a}). \qquad (3.10.5)$$

Note that since $\vec{\nabla} f(\mathbf{a}) = \mathbf{0}$, the second term in (3.9.1) is zero. Thus, whether f has a local extremum at $\mathbf{a}$ is linked directly to whether the quadratic form $\mathbf{z}^T Hf(\mathbf{x}_0)\mathbf{z}$ is positive definite, negative definite, or indefinite.

1. Suppose that the quadratic form $\mathbf{z}^T Hf(\mathbf{a})\mathbf{z}$ is positive definite. Because the second order partials of f, namely the entries in the Hessian matrix, are continuous, it follows[9] that there exists a ball C about $\mathbf{a}$ such that $\mathbf{z}^T Hf(\mathbf{x})\mathbf{z}$ is positive definite for all $\mathbf{x} \in C$. In particular, the point $\mathbf{x}_0$ associated with $\mathbf{x}$ by Taylor's Theorem also lies in C and so the quadratic form $\mathbf{z}^T Hf(\mathbf{x}_0)\mathbf{z}$ is positive definite. Thus for all $\mathbf{x} \in C$, (substituting $\mathbf{x} - \mathbf{a}$ for $\mathbf{z}$)

$$(\mathbf{x} - \mathbf{a})^T Hf(\mathbf{x}_0)(\mathbf{x} - \mathbf{a}) > 0.$$

By (3.10.5), then

$$\begin{aligned} f(\mathbf{x}) &= f(\mathbf{a}) + \frac{1}{2}(\mathbf{x} - \mathbf{a})^T Hf(\mathbf{x}_0)(\mathbf{x} - \mathbf{a}) \\ &\geq f(\mathbf{a}) + 0 = f(\mathbf{a}), \end{aligned}$$

[9]There is a somewhat technical argument to show this rigorously. It makes use of the fact that if a continuous function is positive at a point $\mathbf{a}$, then there is a ball about $\mathbf{a}$ such that $f(\mathbf{a})$ is positive for all $\mathbf{x}$ in that ball. See Theorem 3.2.1.

for all $\mathbf{x}$ in the open ball C. This means that f has a local minimum at $\mathbf{a}$.
2. Suppose that the quadratic form $\mathbf{z}^T Hf(\mathbf{a})\mathbf{z}$ is negative definite. By the same reasoning as in the proof of part 1, there exists an open ball C about $\mathbf{a}$ such that $\mathbf{z}^T Hf(\mathbf{x})\mathbf{z}$ is negative definite for all $\mathbf{x} \in C$. In particular, the associated point $\mathbf{x}_0$ also lies in C, so by (3.10.5),

$$\begin{aligned} f(\mathbf{x}) &= f(\mathbf{a}) + \frac{1}{2}(\mathbf{x}-\mathbf{a})^T Hf(\mathbf{x}_0)(\mathbf{x}-\mathbf{a}) \\ &\le f(\mathbf{a}) + 0 = f(\mathbf{a}) \end{aligned}$$

for all $\mathbf{x} \in C$. Thus f has a local maximum at $\mathbf{a}$.
3. Suppose that the quadratic form $\mathbf{z}^T Hf(\mathbf{a})\mathbf{z}$ is indefinite. It follows[10] that there exists an open ball C about $\mathbf{a}$ such that $\mathbf{z}^T Hf(\mathbf{x})\mathbf{z}$ is indefinite for all $\mathbf{x} \in C$. In particular, $\mathbf{x}_0$ is in C, so $(\mathbf{x}-\mathbf{a})^T Hf(\mathbf{x}_0)(\mathbf{x}-\mathbf{a})$ assumes both positive and negative values for values of $\mathbf{x}$ arbitrarily close to $\mathbf{a}$. By (3.10.5), $f(\mathbf{x})$ takes on values both greater than and less than $f(\mathbf{a})$ for values of $\mathbf{x}$ arbitrarily close to $\mathbf{a}$. Thus f does not have a local extremum at $\mathbf{a}$; f has a saddle at $\mathbf{a}$. ♦

Corollary 3.10.2 *(Second Derivative Test) Suppose that $f : D \subset \mathbb{R}^n \to \mathbb{R}$ has continuous second order partials in an open ball B about a critical point $\mathbf{a} \in D$ where $\vec{\nabla} f(\mathbf{a}) = \mathbf{0}$. Let $[h_{ij}] = Hf(\mathbf{a})$ and let*

$$\begin{aligned} H_1 = [h_{11}],\ H_2 &= \begin{bmatrix} h_{11} & h_{12} \\ h_{21} & h_{22} \end{bmatrix}, \\ H_3 &= \begin{bmatrix} h_{11} & h_{12} & h_{13} \\ h_{21} & h_{22} & h_{23} \\ h_{31} & h_{32} & h_{33} \end{bmatrix}, \ldots, \\ H_n &= \begin{bmatrix} h_{11} & h_{12} & h_{13} & \cdots & h_{1n} \\ h_{21} & h_{22} & h_{23} & \cdots & h_{2n} \\ h_{31} & h_{32} & h_{33} & \cdots & h_{3n} \\ \vdots & \vdots & \vdots & & \vdots \\ h_{n1} & h_{n2} & h_{n3} & \cdots & h_{nn} \end{bmatrix}. \end{aligned}$$

1. If

$$\det(H_1) > 0,\ \det(H_2) > 0,\ \det(H_3) > 0,\ \ldots,\ \det(H_n) > 0 \qquad (3.10.6)$$

then f has a local minimum at $\mathbf{a}$.

[10] Again, by an argument that is somewhat technical.

2. *If*

$$\det(H_1) < 0,\ \det(H_2) > 0,\ \det(H_3) < 0,\ \det(H_4) > 0,\ \dots \quad (3.10.7)$$

[i.e., if the signs of the determinants alternate starting with $\det(H_1)$ *negative], then* f *has a local maximum at* $\mathbf{a}$.

3. *If* $\det(H_i) \neq 0$ *for* $i = 1, \dots, n$, *but the signs are neither all positive nor alternating (starting with* H_1 *negative), then* f *has a saddle at* $\mathbf{a}$.

4. *If* $\det(H_i) = 0$ *for some* i, *then this test provides no information.*

Proof 1. By Theorem 2.5.2, the quadratic Hessian form $h(\mathbf{x}) = \mathbf{x}^T Hf(\mathbf{a})\mathbf{x}$ is positive definite if the determinants of the upper left-hand square submatrices $H_1, H_2, \dots, H_n$ of $Hf(\mathbf{a})$ are all positive. So if (3.10.6) holds, then h is positive definite, and by Theorem 3.10.2, f has a local minimum at $\mathbf{a}$.
2. See the exercises.
3. See the exercises. ♦

Example 3.10.7 Returning to Example 3.10.6, we try to use the second derivative test:

$$Hf(\mathbf{x}) = \begin{bmatrix} 6x + 2y & 2x \\ 2x & -2 \end{bmatrix}.$$

For the critical point $(0, -2)$,

$$Hf(0, -2) = \begin{bmatrix} -4 & 0 \\ 0 & -2 \end{bmatrix}.$$

Since $\det(H_1) = -4 < 0$ and $\det(H_2) = 8 > 0$, we know by Corollary 3.10.2 that f has a local maximum at $(0, -2)$. At the critical point $\left(1, -\frac{3}{2}\right)$,

$$Hf\left(1, -\frac{3}{2}\right) = \begin{bmatrix} 3 & 2 \\ 2 & -2 \end{bmatrix}.$$

Since $\det(H_1) = 3 > 0$ and $\det(H_2) = -6 - 4 = -10 < 0$, by Corollary 3.10.2, f has a saddle at $\left(1, -\frac{3}{2}\right)$.

Finally,

$$Hf(-4, 6) = \begin{bmatrix} -12 & -8 \\ -8 & -2 \end{bmatrix},$$

so $\det(H_1) = -12 < 0$ and $\det(H_2) = 24 - 64 = -40 < 0$. By Corollary 3.10.2, f has a saddle at $(-4, 6)$. ♦

Example 3.10.8 Find the second degree polynomial $p(t) = x + yt + zt^2$ that minimizes $\int_0^1 \left(p(t) - \frac{1}{t+1}\right)^2 dt$. (This function is the best **least-squares fit** of a second degree polynomial to the rational function $1/(t+1)$ on the interval $[0, 1]$.)

Solution By working the integral out longhand, or by using a computer algebra system, we can determine how the integral depends on x, y, and z. By substituting the the formula for $p(t)$ and integrating, we find that the integral evaluates as

$$\begin{aligned} f(x, y, z) &= x^2 + \frac{1}{3}y^2 + \frac{1}{5}z^2 + xy + \frac{1}{2}yz + \frac{2}{3}xz - (2\ln 2)x \\ &\quad + 2(\ln 2 - 1)y + (1 - 2\ln 2)z + \frac{1}{2}. \end{aligned}$$

It is this function we want to minimize. First we calculate

$$\begin{aligned} &\vec{\nabla} f(\mathbf{x}) \\ &= \left(2x + y + \frac{2}{3}z - 2\ln 2,\ x + \frac{2}{3}y + \frac{1}{2}z + 2(\ln 2 - 1)\ ,\right. \\ &\quad \left.\frac{2}{3}x + \frac{1}{2}y + \frac{2}{5}z + (1 - 2\ln 2)\right) \end{aligned}$$

and then solve $\vec{\nabla} f(\mathbf{x}) = \mathbf{0}$. This is a system of three linear equations in three unknowns, and the only solution is

$$\begin{aligned} \mathbf{a} &= (-51 + 75\ln 2,\ 282 - 408\ln 2,\ -270 + 390\ln 2) \\ &\approx (0.9860,\ -0.8041,\ 0.3274). \end{aligned}$$

Now

$$Hf(\mathbf{x}) = \begin{bmatrix} 2 & 1 & 2/3 \\ 1 & 2/3 & 1/2 \\ 2/3 & 1/2 & 2/5 \end{bmatrix}.$$

Since this is constant, $Hf(\mathbf{a}) = Hf(\mathbf{x})$. We check that $\det(H_1) = 2 > 0$, $\det(H_2) = \begin{vmatrix} 2 & 1 \\ 1 & 2/3 \end{vmatrix} = \frac{1}{3} > 0$, and $\det(H_3) = \det(Hf(\mathbf{a})) = \frac{1}{270} > 0$. By Corollary 3.10.2, $f(\mathbf{a}) = 0.0000186$ is a local minimum. The graphs

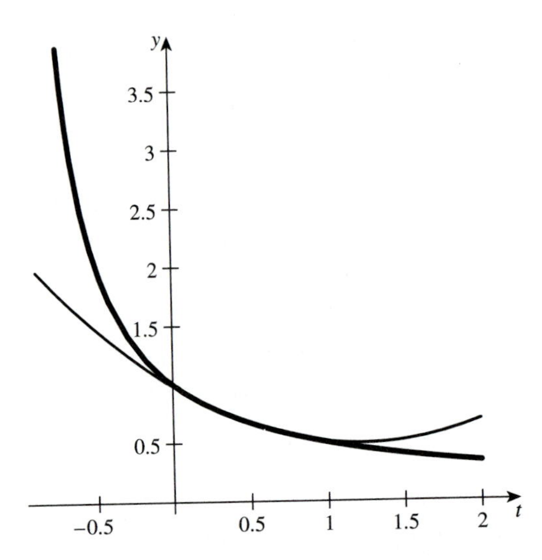

Figure 3.10.2 The graph of $y = 1/(t+1)$ (thick curve) along with that of the second-degree polynomial $y = 0.9860-0.8041t+0.3274t^2$ (thin curve), which is the best least squares fit on the interval $[0, 1]$

of $y = 1/(t+1)$ and $y = p(t) = 0.9860 - 0.8041t + 0.3274t^2$ are shown superimposed in Figure 3.10.2.

As you can see, the agreement between the two is quite close over the range $[0, 1]$. Outside this interval the graphs quickly diverge from one another. ♦

When the Hessian form is neither positive definite nor negative definite nor indefinite at a critical point, generally anything can happen. In the next example we explore some of the possibilities.

Example 3.10.9 The function $f(x, y) = x^4 + y^4$ has, you should verify, a single local minimum at $(0, 0)$. Now

$$Hf(\mathbf{x}) = \begin{bmatrix} 12x^2 & 0 \\ 0 & 12y^2 \end{bmatrix} \quad \text{and} \quad Hf(0, 0) = \begin{bmatrix} 0 & 0 \\ 0 & 0 \end{bmatrix}.$$

Thus $\det(H_1) = \det(H_2) = 0$. This shows that in the situation when two of the determinants of the submatrices are 0, there may be a local minimum.

The function $g(x, y) = x^3 + y^3$ does not (again, verify this) have a local minimum at its critical point $(0, 0)$. Here

$$Hg(\mathbf{x}) = \begin{bmatrix} 6x & 0 \\ 0 & 6x \end{bmatrix} \quad \text{and} \quad Hg(0, 0) = \begin{bmatrix} 0 & 0 \\ 0 & 0 \end{bmatrix}.$$

Again $\det(H_1) = \det(H_2) = 0$, so this shows that when the determinants of the submatrices are zero, there may be no local extremum. ♦

In single-variable calculus, you learned the following fact: If a function f has first- and second-order derivatives that are continuous at a number a, and if $f'(a) = 0$ and $f''(a) > 0$, then f has a local minimum at a. With this in mind, consider a function f of two variables with first- and second-order partials that are continuous at a point (a_1, a_2). Suppose that $\vec{\nabla} f(a_1, a_2) = \mathbf{0}$ and that every vertical plane through the point $(a_1, a_2, f(a_1, a_2))$ intersects the graph of f in a curve that is concave upward. We would say with virtual certainty that f has a local minimum at (a_1, a_2). Yet this is not necessarily true! Consider the following example.

Example 3.10.10 Rebecca has been hiking in the mountains and valleys of eastern Tennessee, and she stops to rest. Looking straight ahead, she notes that the ground slopes upward. Rotating her head just a bit, she sees that the same is true in that direction. Slowly, she turns all the way around at her resting spot and she sees that in *every* direction, the ground slopes upward. Rebecca must be at the bottom of a valley, right? Wrong.

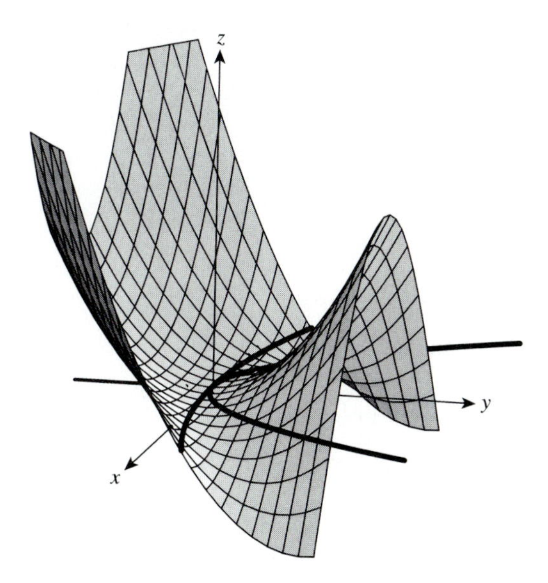

Figure 3.10.3 The surface $z = (y - 8x^2)(y - x^2)$ has a horizontal tangent plane and is concave upward in every direction at the point $(0, 0, 0)$, yet there is not a local minimum at this point.

Suppose that the landscape is described by the graph of

$$z = (y - 8x^2)(y - x^2) \tag{3.10.8}$$

and that she sits at the point (0, 0, 0). Along any line $y = mx$, $(m \neq 0)$ through the origin in the xy-plane,

$$\begin{aligned} z &= (mx - 8x^2)(mx - x^2) \\ &= m^2x^2 - 9mx^3 + 8x^4, \end{aligned}$$

so

$$\begin{aligned} \frac{dz}{dx} &= 2m^2x - 27mx^2 + 32x^3 \\ \text{and } \frac{d^2z}{dx^2} &= 2m^2 - 54mx + 96x^2. \end{aligned}$$

When $x = 0$, we have $dz/dx = 0$ and $d^2z/dx^2 = 2m^2 > 0$, so along the line $y = mx$, the graph has a strict local minimum. (You should also verify that even along the lines $y = 0$ and $x = 0$, the function has a local minimum.) This function, then, exhibits exactly the shape that Rebecca is observing.

But notice what happens if she walks away from (0, 0, 0) along a curve on the surface corresponding to $y = 5x^2$. Her altitude is given by $z = (5x^2 - 8x^2)(5x^2 - x^2) = -3x^2 \cdot 4x^2 = -12x^2$; she will go downhill!

How can this be? Take a look at Figure 3.10.3 where the graph of this equation is shown. The heavy curves are the parabolas $y = 8x^2$ and $y = x^2$ in which the surface intersects the xy-plane. ♦

Exercises 3.10

In Exercises 1 through 11, find all of the critical points of the given function.

1. $f(x, y) = 2x^2 + xy - 4y^2$

2. $f(x, y) = x^2 + 3y^2 - 2x + 12y + 13$

3. $f(x, y) = (5x - 2y)^2$

4. $f(x, y) = \sqrt{x^2 + y^2}$

5. $f(x, y) = xy^{2/3}$

6. $f(x, y) = 15 + 17x + 24y$

7. $f(x, y) = \cos x \sin y$

8. $f(x, y) = e^{-x^2-y^2}$

9. $f(x, y, z) = 4x + 16y + 256z$

10. $f(x, y, z) = x^{4/3} + y^{2/3} + z^{1/3}$

11. $f(x, y, z) = \dfrac{1}{x^2 + y^2 + z^2 + 1}$

Use the Second Derivative Test to identify the local extrema of the functions in Exercises 12 through 21.

12. $f(x, y) = x^2 + y^2 + 2x - y + 3$

13. $f(x, y) = -3x^2 + 6xy + 2y^2 + 12x - 12y$

14. $f(x, y) = x^3 + xy^2 - y^2 - x$

15. $f(x, y) = x^4 - 3x^2y^2 + y^4$

16. $f(x, y) = x + y - \ln(xy)$

17. $f(x, y) = (x^2 + y^2)e^{-y}$

18. $f(x, y) = \cos x \sin y$

19. $f(x_1, x_2, x_3) = x_1^2 + x_2^2 + x_3^2 + x_2x_3.$

20. $f(x_1, x_2, x_3) = \dfrac{1}{1 + x_1^2 + x_2^2 + x_3^2}$

21. $f(x_1, x_2, x_3, x_4) = x_1 + x_2 + x_3 + 2x_4 - \ln|x_1x_2x_3x_4|$

22. Find the points on the ellipsoid $x^2 + y^2/4 + z^2/9 = 1$ that are closest to the origin.

23. A rectangular box is to be inscribed in the tetrahedron whose faces are the coordinate planes and the plane $x/a + y/b + z/c = 1$ (Here a, b, and c are given positive constants.). One corner of the box touches the plane, the opposite corner is at the origin, and the faces of the box are parallel to the coordinate planes. See Figure 3.10.4. Find the dimensions of the largest such box.

24. A closed rectangular box is to contain 1000 cm^3. If the material for the top and bottom costs 2 cents per square centimeter and the material for the sides costs 4 cents per square centimeter, find the dimensions of the box that minimize its cost.

25. For a set of data (x_1, y_1), (x_2, y_2),..., (x_n, y_n), the **least-squares regression line** is the line $y = mx + b$, where m and b are chosen to

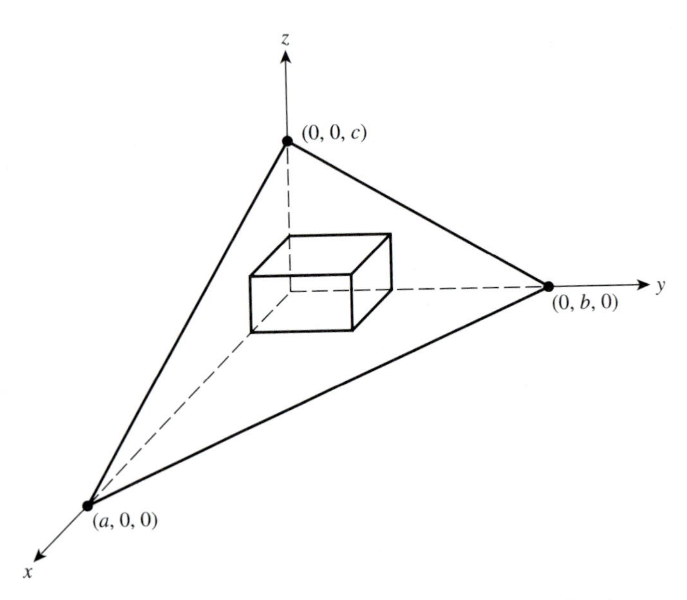

Figure 3.10.4 A box inscribed in a tetrahedron

minimize the sum

$$S(m, b) = \sum_{i=1}^{n} (y_i - mx_i - b)^2.$$

Show that the values of m and b that minimize this sum are

$$m = \frac{\sum_{i=1}^{n} x_i y_i - n\,\overline{x}\,\overline{y}}{\sum_{i=1}^{n} x_i^2 - n\overline{x}^2} \tag{3.10.9}$$

and

$$b = \overline{y} - m\overline{x}, \tag{3.10.10}$$

where $\overline{x} = \frac{1}{n}\sum_{i=1}^{n} x_i$ and $\overline{y} = \frac{1}{n}\sum_{i=1}^{n} y_i$ are the averages of the x's and y's.

26. Use Exercise 25 to find the regression line for the data (0, 10), (2, 6), (3, 7), (4, 6), (5, 3), (8, 1).

27. Just how "good" is the fit of the regression line in Exercise 26? (*Hint*: See Example 2.1.5 in Section 2.1.)

28. Find conditions on a, b, and c so that the function $f(x, y) = ax^3 + by^3 + cxy$ has a local maximum.

In Exercises 29 through 33, find the best least-squares fit of the polynomial to the given function on the interval indicated. You may wish to use a computer algebra system to aid with your calculations.

29. $p(t) = x + yt;\quad f(t) = 1/t;\quad [1,\, 2]$

30. $p(t) = x + yt;\quad f(t) = \sin t;\quad [-\pi/2,\, \pi/2]$

31. $p(t) = x + yt + zt^2;\quad f(t) = \sin t;\quad [-\pi/1,\, \pi/2]$

32. $p(t) = x + yt + zt^2;\quad f(t) = t^3;\quad [-1,\, 1]$

33. $p(t) = x + yt + zt^2;\quad f(t) = e^t;\quad [-1,\, 1]$

34. Determine the behavior of $f(x, y) = x^2 + y^4$ at its critical point. What is the Hessian form at the critical point? What are $\det(H_1)$ and $\det(H_2)$?

35. Determine the behavior of $g(x, y) = x^2 + y^3$ at its critical point. What is the Hessian form? What are $\det(H_1)$ and $\det(H_2)$?

36. Determine the behavior of $h(x, y, z) = x^2 + y^2 - z^2$ at its critical point. Can you use the Second Derivative Test?

37. In Example 3.10.10, verify that the sections of the surface cut by the planes $x = 0$ and $y = 0$ have local minima at the origin.

38. Verify that the gradient of the function in Example 3.10.10 is zero at $\mathbf{0}$. What can you conclude about the behavior of the function at $\mathbf{0}$ by examining the Hessian matrix?

Chapter 4

Integration

In this chapter we explore the concept of integrating functions over various types of sets. We begin with **line** (or **path**) **integrals**. The general idea is to extend ordinary integration of functions over straight-line segments in $\mathbb{R}$ (intervals) to integration of functions along curvy lines that lie in $\mathbb{R}^2$, $\mathbb{R}^3$, $\mathbb{R}^4$, ... Such integrals may represent length, mass, or area when the integrand is scalar-valued. We also develop the notion of integrating a vector field along a path. When the vector field is a force, the path integral represents work (or energy); when the vector field is the velocity of a fluid and the path is closed, the path integral represents the fluid's net "circulation" around the path.

Next, we motivate, define, calculate, and apply **double integrals** of real-valued functions of two variables over closed and bounded sets in $\mathbb{R}^2$. By way of analogy, double integrals are to volume what ordinary definite integrals are to area. Such integrals can also be interpreted as area or mass.

Triple integrals are three-dimensional analogs to double integrals: To define them, we apply a limiting process to real-valued functions of three variables on closed and bounded subsets of $\mathbb{R}^3$. The resulting number can represent such physical quantities as volume or mass. After that, we study **surface integrals** of scalar-valued functions of three variables over two-dimensional subsets of $\mathbb{R}^3$ (surfaces). These integrals can represent such physical quantities as area or mass. We also define the surface integral of a vector field so that when the field is the velocity of a fluid, the surface integral of the field represents the per-unit-time rate of material flow through the surface.

The various notions of integration that we examine here are in many ways independent of one another, but they are all analogous. To develop a conceptual framework, these analogies can be quite helpful. In Chapter 5 we will see that there are concrete connections among the types of integrals; these connections provide us with a theoretical framework that can make our thinking and our ability to apply them more complete.

In the last two sections of this chapter, we examine the process of changing variables in double and triple integrals. This is the direct analog of the substitution method which you learned in elementary calculus. In particular, we obtain formulas for changing triple integrals from rectangular to cylindrical coordinates and from rectangular to spherical coordinates.

4.1 Paths, Curves, and Length

In Section 1.8 we saw that a continuous vector-valued function $\mathbf{f}$ from $\mathbb{R}$ into $\mathbb{R}^2$ or $\mathbb{R}^3$ can be thought of as tracing a curve as t varies over an interval in $\mathbb{R}$. This description is not quite refined enough to be a definition of the term *curve*. For example, the "curve" traced by $\mathbf{f}(t) = \cos t\mathbf{i} + \sin t\mathbf{j}$ as t ranges from 0 to 4π actually represents a trip twice around the unit circle in a counterclockwise direction. This should be distinguished from the "curve" consisting simply of the points on the unit circle."

To reflect the dynamical interpretation of vector-valued functions as the changing position of a particle, and to distinguish between this and the more purely geometrical notion of a curve, we adopt the following terminology. The **path** parametrized by a continuous vector-valued function $\mathbf{f} : I \to \mathbb{R}^n$, where I is an interval in $\mathbb{R}$, is the ordered set of points $\mathbf{f}(t)$ in $\mathbb{R}^n$ that are traced as t varies from smaller to larger values in I. With this terminology, then, we phrase the example above as, "$\mathbf{x} = \cos t\mathbf{i} + \sin t\mathbf{j}, 0 \leq t \leq 4\pi$ is a parametrization of the path that starts at the point $(1, 0)$ and goes twice counterclockwise around the unit circle. We will say that a path parametrized by $\mathbf{f} : [a, b] \to \mathbb{R}^n$ is **simple** if it has no points of self-intersection except possibly at the endpoints. That is, $\mathbf{f}(t_1) \neq \mathbf{f}(t_2)$ for all choices of t_1 and t_2 in (a, b) such that $t_1 \neq t_2$. We say that the path is **closed** if $\mathbf{f}(b) = \mathbf{f}(a)$. See Figure 4.1.1.

We will call a set of points C in $\mathbb{R}^n$ a **curve** if it is the image of a function $\mathbf{f} : I \subset \mathbb{R} \to \mathbb{R}^n$, and in this case we say that the curve is **parametrized** by $\mathbf{f}$. We say that the curve is **simple** if $\mathbf{f}$ can be chosen so that the path it parametrizes is simple.

In order to develop calculus on paths and curves, we must restrict our attention to those that do not have "kinks" or other pathologies. We therefore want to have a well-defined nonzero tangent at each point that varies continuously with the parameter. Since the derivative of a vector-valued function is tangent to the path parametrized by the function, it is natural

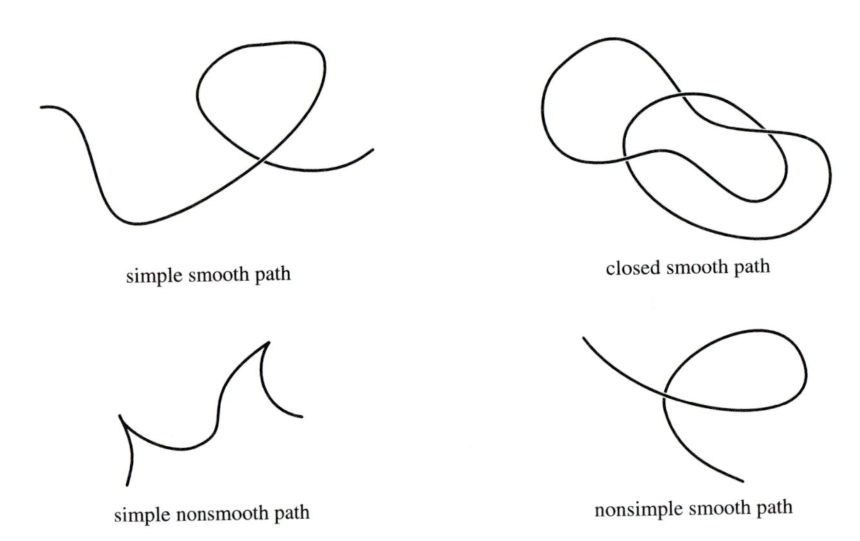

Figure 4.1.1 Paths that are smooth, piecewise smooth, simple, nonsimple, and closed.

to define smoothness in terms of the derivative. A path parametrized by $\mathbf{f} : I \to \mathbb{R}^n$ is **smooth** if $\mathbf{f}'$ is continuous on I and $\mathbf{f}'(t) \neq \mathbf{0}$ for all $t \in I$. A curve is **smooth** if it has a parametrization for which the corresponding path is smooth. A path is **piecewise smooth** if I consists of finitely many subintervals on which $\mathbf{f}$ is smooth. A piecewise smooth curve is defined accordingly. Again, see Figure 4.1.1.

Example 4.1.1 The curve (and path) parametrized by $\mathbf{f}(t) = (\sin 2t, \sin t)$, $0 \leq t \leq 2\pi$ is closed and smooth, but not simple. To see that the curve is closed, simply note that $\mathbf{f}(0) = (0, 0) = \mathbf{f}(2\pi)$. To see that the curve is smooth, note first that $\mathbf{f}'(t) = (2\cos 2t, \cos t)$, which is continuous for all t. If $\mathbf{f}'(t) = 0$ for some t then $2\cos 2t = 0$ and $\cos t = 0$. By the latter equation, t is either $\pi/2$ or $3\pi/2$. But $\cos(2 \cdot \pi/2) = -1 \neq 0$ and $\cos(2 \cdot (3\pi/2)) = \cos 6\pi = 1 \neq 0$, so $\mathbf{f}'(t)$ is never $\mathbf{0}$. Thus the curve is smooth.

To see that it is not simple, we must show that there are two different values of t, say t_1 and t_2, such that $\mathbf{f}(t_1) = \mathbf{f}(t_2)$. Any such numbers must be solutions to the vector equation $(\sin 2t_1, \sin t_1) = (\sin 2t_2, \sin t_2)$. Equating components, we have

$$\sin 2t_1 = \sin 2t_2 \tag{4.1.1}$$

and

$$\sin t_1 = \sin t_2. \tag{4.1.2}$$

From (4.1.2), we see that t_1 and t_2 are related by the equation $t_1 + t_2 = \pi$

(check this!). Substituting for t_1 in (4.1.1), we obtain

$$\begin{aligned}\sin(2\pi - 2t_2) &= \sin 2t_2\\ \sin(-2t_2) &= \sin 2t_2.\end{aligned}$$

In the interval $[0, 2\pi]$, the only solution to this equation is $t_2 = 0$, so $t_1 = \pi - 0 = \pi$, and the only point of self-intersection is $(\sin 0, \sin 0) = (\sin(2\pi), \sin\pi) = (0,0)$. See Figure 4.1.2 for a computer-generated plot of the curve. ♦

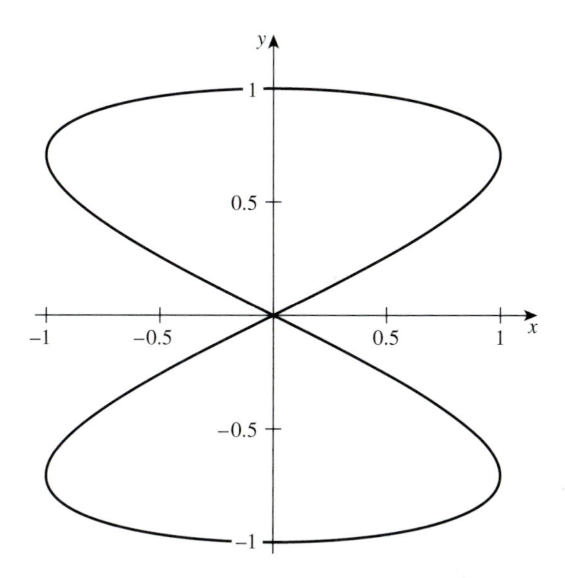

Figure 4.1.2 The curve parametrized by the function $\mathbf{f}(t) = (\sin 2t, \sin t)$, $0 \le t \le 2\pi$

Typically we regard a path as **oriented**, (i.e., as being traced in a particular direction). If $\mathbf{f}$ is defined on an interval $[a, b]$ parametrizes a path with endpoints $\mathbf{a}$ and $\mathbf{b}$, we will say that the path is **oriented** from $\mathbf{a}$ to $\mathbf{b}$ if $\mathbf{f}$ satisfies $\mathbf{f}(a) = \mathbf{a}$ and $\mathbf{f}(b) = \mathbf{b}$. Generally, if C is a path, then $-C$ represents the path with the opposite orientation.

Example 4.1.2 The segment C from the origin in $\mathbb{R}^3$ to the point $(3, 2, -1)$ is parametrized by $\mathbf{x} = (3t, 2t, -t)$, $0 \le t \le 1$. On the other hand, the segment from $(3, 2, -1)$ to the origin, denoted by $-C$, is parametrized by $\mathbf{x} = (3 - 3t, 2 - 2t, -1 + t)$, $0 \le t \le 1$. ♦

Given a smooth path parametrized by $\mathbf{f} : [a, b] \to \mathbb{R}^n$, it is natural for us to ask, "How long is it?" That is, "How far does a particle with $\mathbf{f}(t)$ as its position travel as t ranges from a to b?" Partition $[a, b]$ into subintervals with endpoints $a = t_0 < t_1 < t_2 \cdots < t_m = b$, and consider the line segments that join the pairs of points $\mathbf{f}(t_{i-1})$ and $\mathbf{f}(t_i)$, $i = 1, \ldots, m$. The sum of the lengths of these segments [or equivalently, the sum of the magnitudes of the vectors $\mathbf{f}(t_i) - \mathbf{f}(t_{i-1})$] should approximate what we would like to call the length of the path. See Figure 4.1.3 for the case $n = 3$.

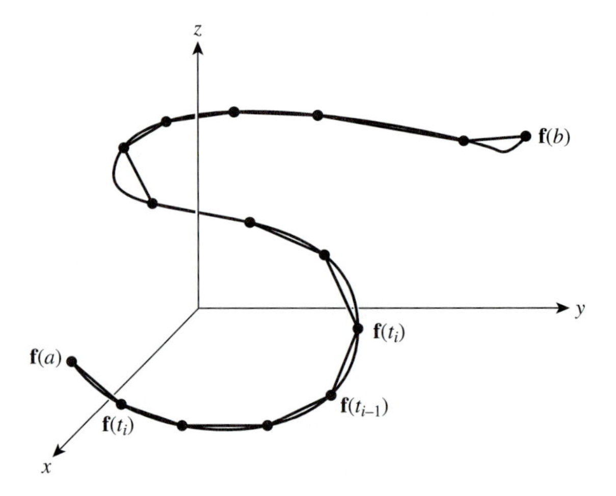

Figure 4.1.3 The sum of the lengths of the segments approximates what we want to call the length of the curve.

Suppose that for finer partitions the approximate lengths calculated in this way converge to a single number. In other words, suppose that the limit of the sum of these lengths as the partition norms are refined exists. This number we *define* to be the length of the path.

To obtain a formula for length defined in this way, we consider an approximating sum:

$$\sum_{i=1}^{m} ||\mathbf{f}(t_i) - \mathbf{f}(t_{i-1})|| = \sum_{i=1}^{m} \sqrt{\sum_{j=1}^{n} (f_j(t_i) - f_j(t_{i-1}))^2}. \tag{4.1.3}$$

By the Mean Value Theorem applied to each component f_j on the intervals $[t_{i-1}, t_i]$, there exist points $t_{ji} \in (t_{i-1}, t_i)$ such that

$$f_j(t_i) - f_j(t_{i-1}) = f'_j(t_{ji})(t_i - t_{i-1})$$

for $i = 1, \ldots, m$ and $j = 1, \ldots, n$. So, writing Δt_i for $t_i - t_{i-1}$, we have

$$\sum_{i=1}^{m} ||\mathbf{f}(t_i) - \mathbf{f}(t_{i-1})|| = \sum_{i=1}^{m} \sqrt{\sum_{j=1}^{n} f'_j(t_{ji})^2}\, \Delta t_i.$$

It is plausible (and can be proved rigorously) that the sums under the square roots can be made arbitrarily close to the sums

$$\sum_{j=1}^{n} f_j'(t_i)^2$$

for a sufficiently fine partition, under the assumption that the derivatives f_j' are continuous. Thus $\sum_{i=1}^{m} ||\mathbf{f}(t_i) - \mathbf{f}(t_{i-1})||$ can be made arbitrarily close to $\sum_{i=1}^{m} \sqrt{\sum_{j=1}^{n} f_j'(t_i)^2 \, \Delta t_i} = \sum_{i=1}^{m} ||\mathbf{f}'(t_i)|| \, \Delta t_i$, which is a Riemann sum for the integral $\int_a^b ||\mathbf{f}'(t)|| \, dt$. So under the condition that $\mathbf{f}'(t)$ is continuous, the approximating sums (4.1.3) can be made arbitrarily close to $\int_a^b ||\mathbf{f}'(t)|| \, dt$.

Definition 4.1.1 The length of a smooth path C in $\mathbb{R}^n$ parametrized by $\mathbf{f}(t)$, $a \le t \le b$, is

$$L(C) = \int_a^b ||\mathbf{f}'(t)|| \, dt.$$

The length of a smooth curve parametrized by $\mathbf{f}$ is the length of the corresponding path.

Suppose that $\mathbf{f}(t)$, $a \le t \le b$ and $\mathbf{g}(s)$, $c \le s \le d$ are two different parametrizations of a smooth path C. It *should* be the case that

$$\int_a^b ||\mathbf{f}'(t)|| \, dt \quad \text{and} \quad \int_c^d ||\mathbf{g}'(s)|| \, ds \tag{4.1.4}$$

are the same. In fact, they are, and readers for whom this statement is sufficient may safely skip the next paragraph. Those who are intrigued should read on.

Since $\mathbf{g}$ is smooth, there is a smooth function $G : C \to [c, d]$ that exactly "undoes" $\mathbf{g}$ [i.e., as a point $\mathbf{x}$ moves along the path parametrized by $\mathbf{g}$, $G(\mathbf{x})$ goes monotonically from c to d]. See Figure 4.1.4. In particular, for each $t \in [a, b]$, we have $\mathbf{f}(t) = \mathbf{g}(G(\mathbf{f}(t))) = \mathbf{g}(H(t))$, where $H(t) = G(\mathbf{f}(t))$ is smooth and increasing (why?). Then

$$\begin{aligned} \int_a^b ||\mathbf{f}'(t)|| \, dt &= \int_a^b ||\mathbf{g}'(H(t))H'(t)|| \, dt \\ &= \int_a^b ||\mathbf{g}'(H(t))|| \, |H'(t)| \, dt. \end{aligned} \tag{4.1.5}$$

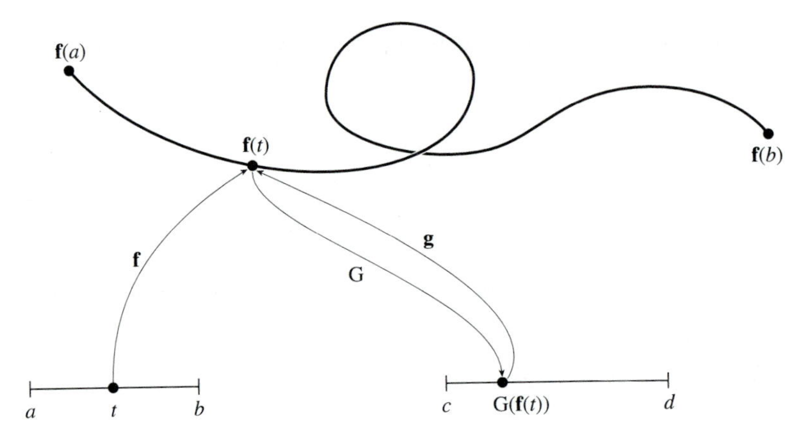

Figure 4.1.4 The definition of the length of a path is independent of the particular parametrization.

Since H is increasing, $H(a) = c$, $H(b) = d$, and $H'(t) > 0$, so by the substitution $s = H(t)$, this integral becomes

$$\int_a^b ||\mathbf{g}'(H(t))||H'(t)\,dt = \int_c^d ||\mathbf{g}'(s)||ds.$$

The integrals in (4.1.4) are the same, so the length of a path is independent of the particular parametrization used to calculate it.

Example 4.1.3 Find the length of 3 turns of the helix C parametrized by $\mathbf{f}(t) = (\cos t, \sin t, t)$.

Solution Three turns correspond to t varying from 0 to 6π. Since

$$\mathbf{f}'(t) = (-\sin t, \cos t, 1),$$

we have

$$||\mathbf{f}'(t)|| = \sqrt{\sin^2 t + \cos^2 t + 1} = \sqrt{2},$$

so the length is $L(C) = \displaystyle\int_0^{6\pi} \sqrt{2}\,dt = 6\pi\sqrt{2}$. ♦

Example 4.1.4 Find the length of the hypocycloid parametrized by $\mathbf{f}(t) = (a\cos^3 t, a\sin^3 t)$. See Figure 4.1.5.

Solution The entire curve is traced as t varies from 0 to 2π. Since the curve is symmetric about the origin, it suffices to find the length of the part from $t = 0$ to $t = \pi/2$ and multiply by 4. First,

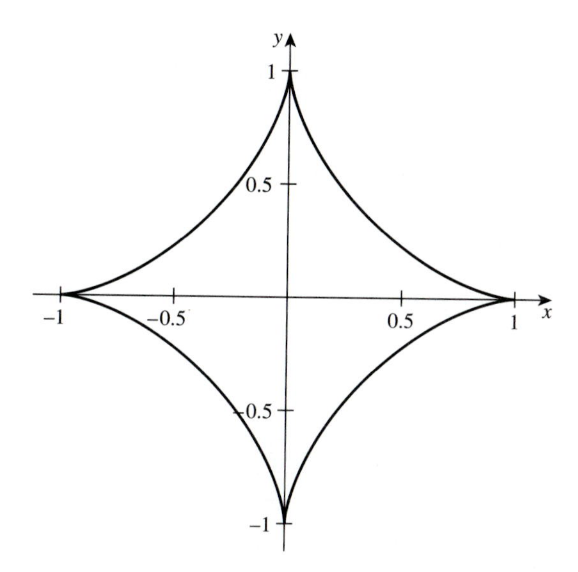

Figure 4.1.5 The hypocycloid of four cusps parametrized by $\mathbf{f}(t) = (a\cos^3 t, a\sin^3 t)$, $0 \le t \le 2\pi$ $(a = 1)$.

$$\begin{aligned}\mathbf{f}'(t) &= (-3a\cos^2 t\sin t,\ 3a\sin^2 t\cos t) \\ &= 3a\cos t\sin t(-\cos t,\ \sin t),\end{aligned}$$

so for $0 \le t \le \pi/2$,

$$||\mathbf{f}'(t)|| = 3a\cos t\sin t.$$

Thus the length of the hypocycloid H is

$$L(H) = 4\int_0^{\pi/2} 3a\cos t\sin t\,dt = 12a\frac{1}{2}\sin^2 t\Big|_0^{\pi/2} = 6a. \quad \blacklozenge$$

Example 4.1.5 Determine the length of the groove in a typical LP (long-playing) record [1] To do this, assume that the record plays for 25 minutes,

[1] In constructing this example, I am aware that college students nowadays are not likely to be so familiar with this form of sound recording; compact disks (CDs) have almost completely replaced long-playing records. It is my understanding that a compact disk player's laser and read head follow a "pitted path" on the CD to read the digitally encoded sound there, but I have never seen the apparatus at work. On the other hand, I have seen a stylus follow a record groove. If you have never seen one, ask your parents to show you that old turntable in the attic.

that it rotates at 33 revolutions per minute, and that the inner and outer radii of the grooved portion of the record are, respectively, 6 cm and 14.5 cm.

Solution In 25 minutes, the turntable revolves $33 \times 25 = 825$ times. If we assume that the spacing between successive turns of the grooves is constant, a reasonable formula for the groove in polar coordinates is $r = a\theta$, $\theta_1 \leq \theta \leq \theta_2$ where a, θ_1, and θ_2 are to be determined. See Figure 4.1.6 for a schematic picture. The difference between θ_2 and θ_1 is $\theta_2 - \theta_1 = 825 \cdot 2\pi$, and we

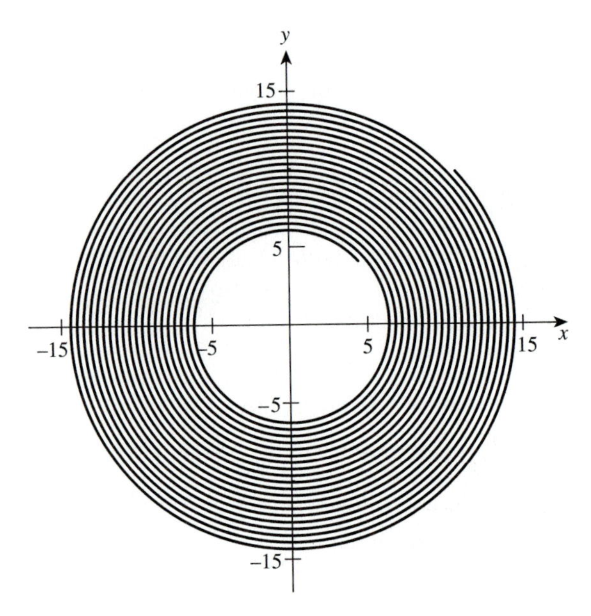

Figure 4.1.6 Parametrization of a record groove

must also have $6 = a\theta_1$ and $14.5 = a\theta_2$. Solving these three equations simultaneously, we obtain $a = 0.00164$, $\theta_1 = 6/a = 3658.5\,\text{rad}$, and $\theta_2 = 14.5/a = 8841.5\,\text{rad}$. Using the transformations from polar to rectangular coordinates, we have

$$\mathbf{f}(\theta) = 0.00164\theta(\cos\theta, \sin\theta), \quad 3658.5 \leq \theta \leq 8841.5$$

as a parametrization for the groove. After differentiating and simplifying, we have

$$||\mathbf{f}'(\theta)|| = 0.00164\sqrt{1+\theta^2},$$

so the length of the groove is

$$L(G) = 0.00164 \int_{3658.5}^{8841.5} \sqrt{1+\theta^2}\, d\theta \approx 53125.8 \text{ cm } \approx 0.3301 \text{ mile.}$$

Does this value agree with your intuition? ♦

Example 4.1.6 Find an expression for the length of the curve in which the cylinders $x^2 + z^2 = 1$ and $y^2 + z^2 = 1$ intersect.

Solution Inspection of Figure 4.1.7 reveals that this curve consists of four pieces of equal length, all meeting at the points $(0, 0, -1)$ and $(0, 0, 1)$. Since each of these pieces is symmetric with respect to the xy-plane, it suffices to obtain the length of the piece in the first octant joining (1, 1, 0) to (0, 0, 1) and multiply by 8. We must produce a parametrization of this curve, and by

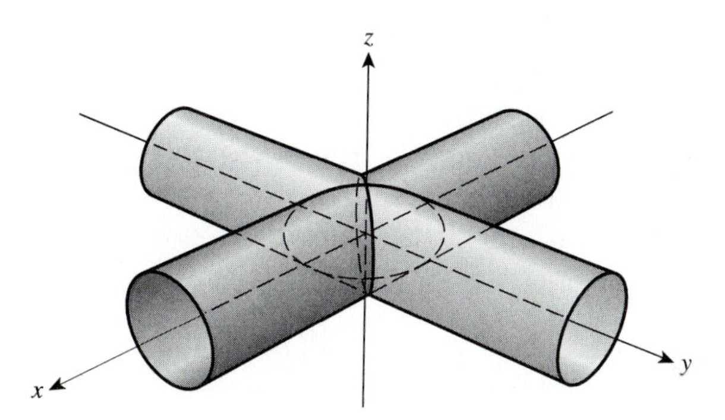

Figure 4.1.7 The cylinders $x^2+z^2 = 1$ and $y^2+z^2 = 1$ intersect in a curve parametrized by $(x, y, z) = (\cos t, \cos t, \sin t)$, $0 \leq t \leq \pi/2$.

our earlier observation, any parametrization will do. Since x, y, and z will be functions of t that satisfy the two given equations, we first try $x = \cos t$ (why?). As t ranges from 0 to $\pi/2$, the x-coordinate will decrease from 1 to 0. We must also choose the functions for y and z so that the curve is oriented from the point $(1, 1, 0)$ to the point $(0, 0, 1)$. By the first equation, $z^2 = 1 - x^2$, so $z = \pm\sqrt{1 - \cos^2 t} = \pm \sin t$. Since z must increase from 0 to 1, we choose the positive sign. By the second equation, $y = \pm\sqrt{1 - z^2} = \pm\sqrt{1 - \sin^2 t} = \pm \cos t$. Again we use the positive sign so that y goes from 1 to 0. Thus the arc is parametrized by

$$(x, y, z) = (\cos t, \cos t, \sin t), \quad 0 \leq t \leq \frac{\pi}{2}.$$

Thus the length of the curve in which the cylinders intersect is

$$\begin{aligned} 8\int_0^{\pi/2}\left\|\frac{d}{dt}(\cos t,\cos t,\sin t)\right\|\,dt &= 8\int_0^{\pi/2}\sqrt{\sin^2 t+\sin^2 t+\cos^2 t\,dt} \\ &= 8\int_0^{\pi/2}\sqrt{\sin^2 t+1\,dt}. \end{aligned}$$

An antiderivative of the integral in terms of elementary functions cannot be found (this is an elliptic integral), but using a numerical technique such as Simpson's rule, we get 15.2808 as the approximate length. ♦

For each smooth path C there is a special parametrization that is occasionally important. Given $\mathbf{f}(t)$, $a \le t \le b$ for C, we define the **pathlength function** to be

$$L(t) = \int_a^t ||\mathbf{f}'(u)||\,du, \quad a \le t \le b. \tag{4.1.6}$$

Since $\mathbf{f}$ is smooth, the integral is continuous, and by the Fundamental Theorem of Calculus, $L'(t) = ||\mathbf{f}'(t)||$. By this observation, $L'(t) > 0$, so L increases from 0 to B (the length of the path) as t goes from a to b. This means that L is one-to-one, so it has a differentiable inverse L^{-1}. Thus $\mathbf{g}(s) = \mathbf{f}(L^{-1}(s))$ for $0 \le s \le B$ is a parametrization of the path with the following important property:

$$\mathbf{g}'(s) = \mathbf{f}'(L^{-1}(s))(L^{-1})'(s) = \frac{\mathbf{f}'(L^{-1}(s))}{L'(L^{-1}(s))} = \frac{\mathbf{f}'(L^{-1}(s))}{||\mathbf{f}'(L^{-1}(s))||},$$

and $||\mathbf{g}'(s)|| = 1$ for all $s \in [0, B]$. This can be interpreted as saying that a particle whose position along its path is given by $\mathbf{g}(s)$ will move with constant speed 1.

Example 4.1.7 Parametrize the helix $(\cos t, \sin t, t)$ $0 \le t \le 2\pi$ in terms of its arclength.

Solution We have $L(t) = \int_0^t ||(-\sin u, \cos u, 1)||du = \int_0^t \sqrt{2}du = \sqrt{2}t$. Solving the equation $s = \sqrt{2}t$ gives $t = s/\sqrt{2}$, which represents L^{-1}. Thus $\mathbf{g}(s) = \left(\cos(s/\sqrt{2}), \sin(s/\sqrt{2}), s/\sqrt{2}\right)$, $0 \le s \le 2\sqrt{2}\pi$, is an arclength parametrization. ♦

The letter s is often used to represent the arclength function:

$$s = \int_a^t ||\mathbf{f}'(t)||\,dt.$$

This means that

$$\frac{ds}{dt} = ||\mathbf{f}'(t)||,$$

which can be interpreted physically as saying that the rate of change in the distance traveled by a particle with position function $\mathbf{f}$ is its speed.

Exercises 4.1

1. Explain why the path parametrized by $\mathbf{f}(t) = (t^2, t^3)$ is not smooth.

2. Explain why the path parametrized by $\mathbf{f}(t) = (t, |t-1|)$ is not smooth.

3. Show that $\mathbf{f}(t) = (\cos t, \sin t)$, $0 < t < 2\pi$ and $\mathbf{g}(t) = \left(\frac{t^2-1}{t^2+1}, \frac{-2t}{t^2+1}\right)$, $-\infty < t < \infty$ parametrize the same curve.

4. Show that $\mathbf{f}(t) = (e^{-t}, 1+e^t)$, $0 \le t < \infty$ and $\mathbf{g}(t) = \left(t^3 - 2, \frac{t^3-1}{t^3-2}\right)$, $\sqrt[3]{2} < t \le \sqrt[3]{3}$ parametrize the same curve.

5. Show that the path parametrized by $\mathbf{f}(t) = (t^2 - 1, t^3 - t)$ is smooth but not simple.

6. Show that the curve $\mathbf{f}(t) = (\sin t, \sin 4t)$, $0 \le t < 2\pi$ is smooth but not simple. (*Hint*: There are three points at which the curve intersects itself.)

7. Suppose that a path in $\mathbb{R}^n$ is parametrized by $\mathbf{f}(t)$ where each component of $\mathbf{f}$ is a periodic function of t. What can you say about the path? (*Hint*: Different components may have different periods.)

8. Write a parametrization for the straight-line path from the point $(1, 3, 2)$ to the point $(5, 7, -4)$. Write a parametrization for the reverse path.

9. Write two different parametrizations of the closed path that starts at the point $(1, 0)$ and goes around the ellipse $x^2 + y^2/4 = 1$ three times.

In Exercises 10 through 13, compute the length of the path with the given parametrization.

10. $\mathbf{f}(t) = (t^2, t^3)$, $1 \le t \le 2$

11. $\mathbf{f}(t) = (5t + 3, -13t + 1, t - 8)$, $1 \le t \le 3$

12. $\mathbf{f}(t) = (e^t \cos t, e^t \sin t, e^t)$, $0 \le t \le 3$

13. $\mathbf{f}(t) = \left(\frac{1}{\sqrt{2}}t + 1, \frac{1}{\sqrt{2}}t^2 + 1, -\frac{1}{3}t^3, \frac{1}{\sqrt{2}}t\right)$, $-1 \le t \le 2$

14. Show that the arclength formula yields the expected value for the circumference of a circle with radius a.

15. Find the length of one arch of the cycloid $\mathbf{f}(t) = \big(a(t-\sin t), a(1-\cos t)\big)$.

16. Obtain an expression for the length of the ellipse

$$\frac{x^2}{a^2} + \frac{y^2}{b^2} = 1.$$

(The integral will be in terms of a and b; moreover, it cannot be evaluated in terms of elementary functions, so do not try to use the Fundamental Theorem to evaluate it.)

17. Find the arclength function for the cycloid in Exercise 15.

18. Find the arclength function for the spiral parametrized by $\mathbf{f}(t) = (t\cos t, t\sin t)$, $0 \leq t$.

19. Find the arclength parametrization of the circle parametrized by $\mathbf{f}(t) = \big(a\sin(t^2), a\cos(t^2)\big)$, $0 \leq t \leq \sqrt{2\pi}$.

20. Find the arclength parametrization of the line segment joining the origin to the point (a, b, c).

4.2 Line Integrals

In this section we discuss two separate but related notions of integration over curves. First, we introduce the line integral of a scalar-valued function along a curve. Its definition is motivated by considering how we can reasonably define the mass of a filament (or wire) whose linear density varies from point to point. It will turn out that the integral we produce has other physical interpretations. Second, we introduce the line integral of a vector field. When a particle moves along a curve under the influence of a force field, it is reasonable to try to measure the amount of work done by the force on the particle during this motion. We define the line integral of a vector field in such a way that it gives a reasonable measure of this work. Here, too, the line integral can have other physical interpretations. For instance, if the vector field represents the velocity of a fluid, then the line integral around a closed curve measures the tendency of the fluid particles to "circulate" around the curve.

Line Integrals of Scalar-Valued Functions

Consider a piece of wire that is bent into some interesting shape and whose density varies from point to point. The density is then a function of the position in space of the points on the wire. Let $u(\mathbf{x})$ represent the density at point $\mathbf{x}$ on the wire, and let $\mathbf{x} = \mathbf{g}(s)$, $0 \leq s \leq B$, be the arclength parametrization for the curve on which the wire resides. Let $0 = s_0 < s_1 < \cdots < s_n = B$ be a partition of the interval $[0, B]$. If u is continuous, the density at each point of the short segment of curve from $\mathbf{g}(s_{i-1})$ to $\mathbf{g}(s_i)$ is approximately $u(\mathbf{g}(s_i))$. The length of this segment of curve is $s_i - s_{i-1}$, so the mass of the wire along this segment is approximately $u(\mathbf{g}(s_i))(s_i - s_{i-1})$. See Figure 4.2.1. Thus we approximate the mass of the wire with the sums

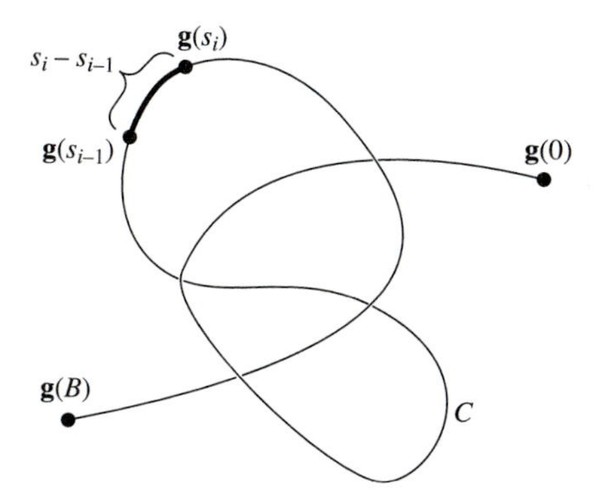

Figure 4.2.1 The mass of the piece of wire along the curve segment C_i is approximately the product of the density at one point on C_i with its length.

$$\sum_{i=1}^{n} u(\mathbf{g}(s_i))(s_i - s_{i-1}). \tag{4.2.1}$$

This is a Riemann sum for the integral $\int_0^B u(\mathbf{g}(s))ds$. It turns out that just as for the formula that defines pathlength, this integral is independent of the particular parametrization for the curve. The following definition is motivated by this discussion.

Definition 4.2.1 The **line integral** (or **path integral**) of a continuous function $u : \mathbb{R}^n \to \mathbb{R}$ over a smooth path C parametrized by $\mathbf{f}(t)$, $a \leq t \leq b$ is

$$\int_C u\,dL = \int_a^b u(\mathbf{f}(t))||\mathbf{f}'(t)||\,dt. \tag{4.2.2}$$

The notation $\int_C u\,dL$ reminds us that the path integral is independent of the specific parametrization.

Example 4.2.1 Find the line integral of $u(x,y) = x^2 + y^2$ over the upper half of the counterclockwise-oriented unit circle.

Solution A parametrization for the path C is

$$\mathbf{f}(t) = \cos t\,\mathbf{i} + \sin t\,\mathbf{j}, \quad 0 \le t \le \pi.$$

Then $u(\mathbf{f}(t)) = \cos^2 t + \sin^2 t = 1$ and $||\mathbf{f}'(t)|| = 1$, so $\displaystyle\int_C u\,dL = \int_0^\pi 1 \cdot 1\,dt = \pi$. ♦

Example 4.2.2 Consider the path C traced by $\mathbf{f}(t) = (t^3, t^2, t, 1)$, $0 \le t \le 2$ in $\mathbb{R}^4$ and the function $u : \mathbb{R}^4 \to \mathbb{R}$ given by $u(x_1, x_2, x_3, x_4) = 36x_3^3 + 8\sqrt{x_2} + x_4 - 1$. Find $\int_C u\,dL$.

Solution We have $u(\mathbf{f}(t)) = 36t^3 + 8t + 1 - 1 = 36t^3 + 8t$, and $\mathbf{f}'(t) = (3t^2, 2t, 1, 0)$ so that $||\mathbf{f}'(t)|| = \sqrt{9t^4 + 4t^2 + 1}$. Thus

$$\begin{aligned}
\int_C u\,dL &= \int_0^2 (36t^3 + 8t)\sqrt{9t^4 + 4t^2 + 1}\,dt \\
&= \frac{2}{3}\left(9t^4 + 4t^2 + 1\right)^{3/2}\Big|_0^2 = \frac{2}{3}\left(161^{3/2} - 1\right).
\end{aligned}$$

Since u is nonnegative along this path, we can interpret this number as the mass of a wire along C with linear density given by u. For instance, if u were in units of grams per centimeter, then the value of the integral would have units of grams. ♦

Example 4.2.3 A portion of a circular room with wall of height h and a spherical domed ceiling of radius a is to be concealed by a curtain that hangs from ceiling to floor along a circular arc of radius a that passes through the center of the room. See Figure 4.2.2. How much material will be required to make this curtain?

Solution The problem is to find the area of the curved surface shown in Figure 4.2.2. If the curve at the base of the curtain were composed of short line segments of length ΔL_i and we knew the height $u(x_i, y_i)$ of the curtain at points (x_i, y_i) on these segments, then the area would be approximated by $\sum u(x_i y_i)\,\Delta L_i$. This is essentially a Riemann sum for the line integral of the height function u over the circular arc C on the floor. Thus the area we want will be $\displaystyle\int_C u\,dL$. We need a formula for u and a parametrization for C.

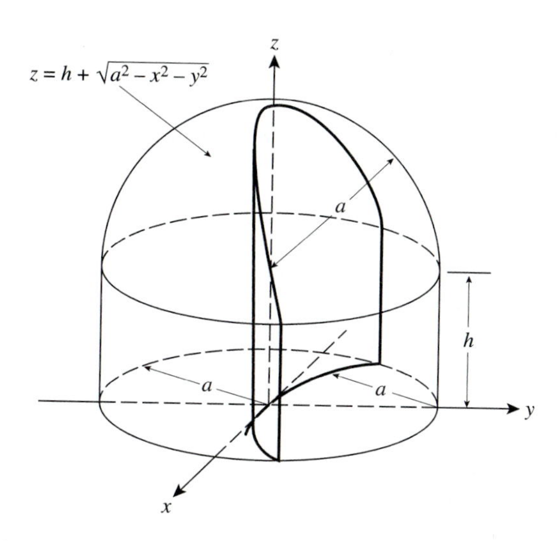

Figure 4.2.2 Finding the area of a curtain suspended from a domed ceiling

Place the room in a coordinate system with origin at the center of the floor, z-axis through the peak of the roof, and the center of the circular arc on the y-axis. Then the roof is the graph of the function

$$u(x, y) = h + \sqrt{a^2 - x^2 - y^2},$$

and the equation for the circular arc on the floor is $x^2 + (y - a)^2 = a^2$. The curtain and wall meet when x and y satisfy both the equations $x^2 + y^2 = a^2$ and $x^2 + (y - a)^2 = a^2$. See Figure 4.2.3. Solving the latter for x^2 and substituting into the former, we get

$$\begin{aligned} a^2 - (y - a)^2 + y^2 &= a^2 \\ y^2 &= (y - a)^2 \\ y &= \pm(y - a). \end{aligned}$$

Only the negative sign produces an equation with a solution: $y = a/2$. Thus $x^2 = a^2 - (a/2)^2 = 3a^2/4$, so $x = \pm\sqrt{3}/2a$, and the arc C joins the points $\left(-\sqrt{3}/2a, a/2\right)$ and $\left(\sqrt{3}/2a, a/2\right)$. In order to parametrize C, we set $x = a\cos t$ and $y = a\sin t + a$. This gives the point $\left(-\sqrt{3}/2a,\ a/2\right)$ when $t = 7\pi/6$, and it gives $\left(\sqrt{3}/2a,\ a/2\right)$ when $t = 11\pi/6$. Since the curtain is symmetric about the yz-plane, its area will be twice that of the portion behind this plane. Thus we will integrate with respect to t from $7\pi/6$ to

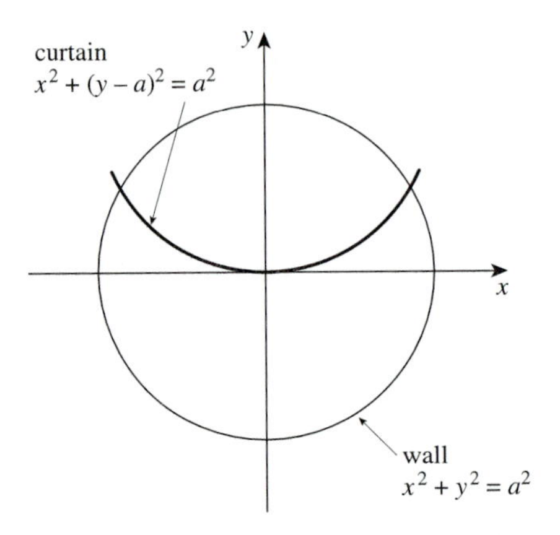

Figure 4.2.3 Top view of the room and and curtain

$3\pi/2$ and multiply by 2. The area of the curtain is, then,

$$\begin{aligned}
&\int_C u\,dL \\
&= \int_{7\pi/6}^{11\pi/6} \left(h + \sqrt{a^2 - a^2\cos^2 t - a^2(\sin t + 1)^2}\right) \cdot \sqrt{a^2\sin^2 t + a^2\cos^2 t}\,dt \\
&= 2a\int_{7\pi/6}^{3\pi/2} \left(h + a\sqrt{-1 - 2\sin t}\right)\,dt \\
&= 2a\left(\frac{\pi h}{3} + a\int_{7\pi/6}^{3\pi/2} \sqrt{-1 - 2\sin t}\,dt\right).
\end{aligned}$$

The integral in the last expression cannot be evaluated in terms of elementary functions, so an approximation is necessary. By Simpson's rule or some other numerical integration method we find that the integral is approximately 0.812598. Thus the area of the curtain is $2\pi ah/3 + 0.812598a^2$. ♦

This example illustrates that, in general, if $u(x, y)$ is a nonnegative continuous function and C is a curve in the xy-plane, then *the line integral $\int_C u\,dL$ represents the surface area of the portion of the cylinder lying above C between the xy-plane and the surface $z = u(x, y)$.*

Line Integrals of Vector Fields

We also define the line integral of a vector field. This definition is motivated by the physical concept of **work** or **energy**. Consider a continuous vector

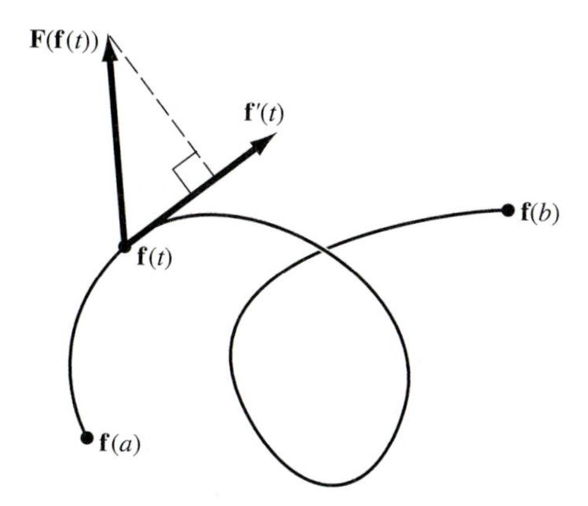

Figure 4.2.4 If **F** is a force field, the line integral of **F** along a path C measures the work done by **F** on a particle that moves along C.

field $\mathbf{F}:\mathbb{R}^n\to\mathbb{R}^n$ defined on some region in $\mathbb{R}^n$. (If $n=3$ we can think of **F** as assigning to each point in $\mathbb{R}^3$ a force vector, which in general will vary from point to point.) Let C be a smooth path parametrized by $\mathbf{x}=\mathbf{f}(t)$, $a\le t\le b$. If we partition $[a,b]$ into short subintervals $[t_{i-1},t_i]$, $i=1,2,\dots,m$, then motion along the short arc from $\mathbf{f}(t_{i-1})$ to $\mathbf{f}(t_i)$ is approximated by motion along the line segment between these points. See Figure 4.2.4. Also, since **F** is continuous, it has approximately the value $\mathbf{F}(\mathbf{f}(t_{i-1}))$ on the segment, and thus $\mathbf{F}(\mathbf{f}(t_{i-1})\cdot(\mathbf{f}(t_i)-\mathbf{f}(t_{i-1}))$ approximates the work done by **F** on a particle that moves along C from $\mathbf{f}(t_{i-1})$ to $\mathbf{f}(t_i)$. Summing these approximations yields an approximation for the work done by **F** on a particle that moves from one end of C to the other. Thus we consider

$$\sum_{i=1}^{m}\mathbf{F}(\mathbf{f}(t_{i-1}))\cdot(\mathbf{f}(t_i)-\mathbf{f}(t_{i-1}))=\sum_{i=1}^{m}\mathbf{F}(\mathbf{f}(t_{i-1}))\cdot\frac{\mathbf{f}(t_i)-\mathbf{f}(t_{i-1})}{t_i-t_{i-1}}\,(t_i-t_{i-1}).$$

This sum can be made arbitrarily close to $\displaystyle\sum_{i=1}^{m}\mathbf{F}(\mathbf{f}(t_{i-1}))\cdot\mathbf{f}'(t_{i-1})(t_i-t_{i-1})$ by taking $\max_i(t_i-t_{i-1})$ to be sufficiently small. This is a Riemann sum for $\displaystyle\int_a^b\mathbf{F}(\mathbf{f}(t))\cdot\mathbf{f}'(t)\,dt$.

Definition 4.2.2 The **line integral** (or **path integral** of a continuous vector field **F** over a smooth path C is

$$\int_C\mathbf{F}\cdot d\mathbf{x}=\int_a^b\mathbf{F}(\mathbf{f}(t))\cdot\mathbf{f}'(t)\,dt, \qquad (4.2.3)$$

where $\mathbf{f}(t)$, $a\le t\le b$, is a parametrization of C.

Since $\mathbf{f}'(t) = ||\mathbf{f}'(t)|| \cdot \mathbf{f}'(t)/||\mathbf{f}'(t)|| = ||\mathbf{f}'(t)||\mathbf{T}(t)$, where $\mathbf{T}$ given by $\mathbf{T}(t) = \mathbf{f}'(t)/||\mathbf{f}'(t)||$ is the **unit tangent vector**, we can write this integral as

$$\int_a^b \mathbf{F}(\mathbf{f}(t)) \cdot \mathbf{T}(t)||\mathbf{f}'(t)||\, dt = \int_C \mathbf{F} \cdot \mathbf{T}\, dL \tag{4.2.4}$$

to see that the line integral of a vector field $\mathbf{F}$ is actually the line integral of the scalar function $\mathbf{F} \cdot \mathbf{T}$. Thus we can think of the line integral of a force field as simply the *net* contribution of force in the direction of $\mathbf{T}$ as the curve is traversed.

Example 4.2.4 Find the work done by the vector field $\mathbf{F}(x, y, z) = (x + y, y^2, x - z)$ on a particle that moves along the line segment from $(1, 2, 1)$ to $(0, 3, -5)$.

Solution We need first a parametrization of the line segment C. By the methods of Chapter 1, this is

$$\begin{aligned} \mathbf{x} = \mathbf{f}(t) &= (1, 2, 1) + t((0, 3, -5) - (1, 2, 1)) \\ &= (1, 2, 1) + t(-1, 1, -6) \\ &= (-t + 1, t + 2, -6t + 1), \quad 0 \le t \le 1. \end{aligned}$$

Thus

$$\begin{aligned} &\int_C \mathbf{F} \cdot d\mathbf{x} \\ &= \int_0^1 ((-t + 1) + (t + 2), (t + 2)^2, (-t + 1) - (-6t + 1)) \cdot (-1, 1, -6)\, dt \\ &= \int_0^1 (-3 + (t + 2)^2 - 30t)\, dt \\ &= \left.(-3t + \frac{1}{3}(t + 2)^3 - 15t^2)\right|_0^1 \\ &= -9 - \frac{8}{3} = \frac{-35}{3}, \end{aligned}$$

in appropriate units (e.g., N·m if $\mathbf{F}$ is in newtons and distance is measured in meters). ♦

There are other forms of notation for line integrals, one of which is the **differential notation**. Remember that if $x = f(t)$ then the differential of x is $dx = f'(t)\, dt$, where we regard dt a variable that is independent of t. If $\mathbf{x} = \mathbf{f}(t)$, then the total differential of $\mathbf{x}$ is $d\mathbf{x} = \mathbf{f}'(t)\, dt = (f_1'(t)\, dt,\ f_2'(t)\, dt,\ \ldots,\ f_n'(t)\, dt) = (dx_1,\ dx_2,\ \ldots,\ dx_n)$, which is a vector that depends on the two independent variables t and dt. Then the line integral of $\mathbf{F}$ can be written as

$$\int_C \mathbf{F} \cdot d\mathbf{x} = \int_C F_1(\mathbf{x})\,dx_1 + F_2(\mathbf{x})\,dx_2 + l\,dots + F_n(\mathbf{x})\,dx_n. \qquad (4.2.5)$$

An expression such as $F_1\,dx_1 + \ldots + F_n\,dx_n$ is called a **differential form**, so our formula for a line integral gives a way of integrating a differential form over a curve. We can think of the right-hand side of 4.2.5 as a "multiplied-out" version of the left-hand side. One advantage of the differential notation is that it does not refer to a specific parametrization of the path and the values of the *field* along it.

Example 4.2.5 Evaluate $\int_C (x_1 - x_2)\,dx_1 + (x_2^2 - x_1)\,dx_2$ where C is the path in $\mathbb{R}^2$ given by $\mathbf{x} = (1 - t^2, 3t + 1)$, $0 \le t \le 1$.

Solution Here the components of position are $f_1(t) = 1 - t^2$ and $f_2(t) = 3t + 1$, so $f_1'(t) = -2t$ and $f_2'(t) = 3$. Thus

$$\begin{aligned}
&\int_C (x_1 - x_2)\,dx_1 + (x_2^2 - x_1)\,dx_2 \\
&= \int_0^1 [(1 - t^2 - (3t + 1))(-2t) + ((3t + 1)^2 - (1 - t^2))3]\,dt \\
&= \int_0^1 (2t^3 + 36t^2 + 18t)\,dt = \frac{1}{2} + 12 + 9 = \frac{43}{2}. \quad \blacklozenge
\end{aligned}$$

Recall that if C is a path, then $-C$ refers to the same path but with opposite orientation. It is an exercise to show that the line integral of a scalar-valued function is independent of the path's orientation, whereas the line integral of a vector field satisfies the relation $\int_{-C} \mathbf{F} \cdot d\mathbf{x} = -\int_C \mathbf{F} \cdot d\mathbf{x}$. Also, if a path C is piecewise smooth and $C_1, C_2, \ldots, C_m$ are smooth arcs whose union is C, then we define $\int_C \mathbf{F} \cdot d\mathbf{x} = \int_{C_1} \mathbf{F} \cdot d\mathbf{x} + \cdots + \int_{C_m} \mathbf{F} \cdot d\mathbf{x}$.

For curves composed of line segments parallel to the coordinate axes, the differential notation is very convenient for evaluating line integrals. This observation will be helpful later in connection with the Fundamental Theorem of Line Integrals.

Example 4.2.6 Evaluate $\int_C (x + 2y)\,dx + (x^2 - y^3)\,dy$, where C consists of the segments from $(1, 1)$ to $(3, 1)$ and $(3, 1)$ to $(3, -1)$.

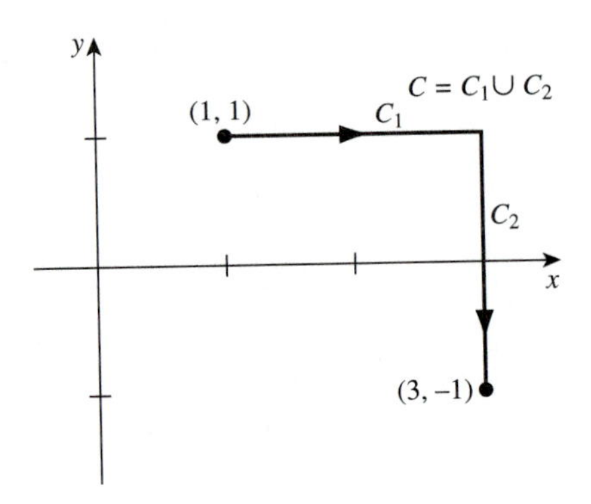

Figure 4.2.5 A piecewise smooth path $C = C_1 \cup C_2$ from (1, 1) to (3, −1)

Solution Let C_1 and C_2 denote the respective segments that make up C. See Figure 4.2.5. For any parametrization of C_1, the y-component has the constant value 1, so $dy = 0$ along C_1. For the x-component of the parametrization we simply let x be the parameter, for x ranging from 1 to 3. Similarly, $x = 3$ and $dx = 0$ along C_2, and we let the parameter y range from 1 to −1. Thus

$$\begin{aligned}
&\int_C (x+2y)\,dx + (x^2 - y^3)\,dy \\
&= \int_{C_1} (x+2y)\,dx + (x^2 - y^3)\,dy + \int_{C_2} (x+2y)\,dx + (x^2 - y^3)\,dy \\
&= \int_{C_1} (x + 2 \cdot 1)\,dx + \int_{C_2} (3^2 - y^3)\,dy \\
&= \int_1^3 (x+2)\,dx + \int_1^{-1} (9 - y^3)\,dy \\
&= 8 - 18 = -10. \quad \blacklozenge
\end{aligned}$$

Finally, we give an example of how line integrals of vector fields arise in physical applications.

Example 4.2.7 By Newton's Law of Gravitation, an object of mass m in the space surrounding the earth experiences a gravitational force directed toward the earth's center that is proportional to m and inversely proportional to the square of the distance from the center of the earth to the object. Thus if the center of the earth is at the origin of a coordinate system and $\mathbf{x} = (x_1, x_2, x_3)$ is the object's position, then the gravitational force is

$$\mathbf{F}(\mathbf{x}) = \left(-\frac{GmM}{||\mathbf{x}||^2}\right)\frac{\mathbf{x}}{||\mathbf{x}||},$$

where M is the mass of the earth and G is the universal gravitational constant ($M \approx 5.98 \times 10^{24}$ kg and $G \approx 6.67 \times 10^{-11}$ N·m^2/kg^2). Assuming that gravitation is the only force acting, how much energy is needed to propel a rocket of mass m from an altitude of R_1 to a higher one of R_2 along a straight line directed outward from the center of the earth? See Figure 4.2.6.

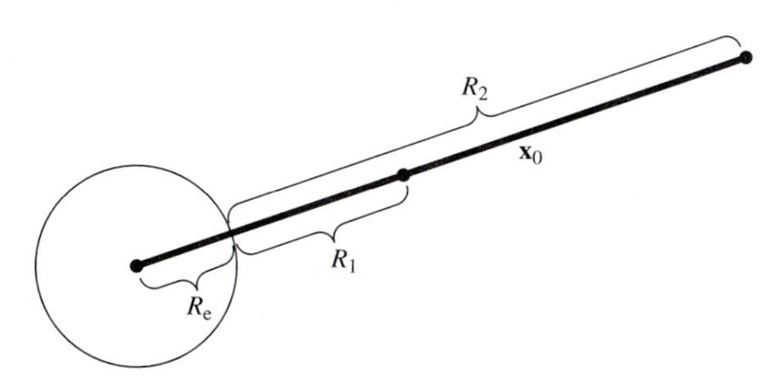

Figure 4.2.6 Calculating the energy to move a rocket of mass m from altitude R_1 to altitude R_2

Solution We'll try to work this problem without dealing in components of the vectors. Let $\mathbf{x}_0$ be a point at altitude R_1. Then $||\mathbf{x}_0|| = R_1 + R_e$, where R_e ($\approx 6.37 \times 10^6$ m) is the radius of the earth. Again, see Figure 4.2.6. The segment C from $\mathbf{x}_0$ straight out to an altitude of R_2 is parametrized by $\mathbf{x} = [t/(R_1 + R_e)]\mathbf{x}_0$ for $R_1 + R_e \le t \le R_2 + R_e$. Along this segment, the force field is

$$\begin{aligned}
\mathbf{F}\left(\frac{t}{R_1 + R_e}\mathbf{x}_0\right) &= -\frac{GmM}{||[t/(R_1 + R_e)]\mathbf{x}_0||^3} \cdot \frac{t}{R_1 + R_e}\mathbf{x}_0 \\
&= -\frac{GmMt}{t^3\,||\mathbf{x}_0/(R_1 + R_e)||^3\,(R_1 + R_e)}\mathbf{x}_0 \\
&= -\frac{GmM}{(R_1 + R_e)t^2}\mathbf{x}_0.
\end{aligned}$$

[The last step follows because $\mathbf{x}_0/(R_1 + R_e)$ is a unit vector.] Also the differential in $\mathbf{x}$ is

$$d\mathbf{x} = \frac{d}{dt}\left(\frac{t}{R_1 + R_e}\mathbf{x}_0\right)dt = \frac{\mathbf{x}_0}{R_1 + R_e}\,dt.$$

Thus the work done by the earth's gravitational field on the rocket as it moves on this path is

$$\begin{aligned}
\int_C \mathbf{F}\cdot d\mathbf{x} &= \int_{R_1+R_e}^{R_2+R_e}\left(-\frac{GmM}{(R_1+R_e)t^2}\mathbf{x}_0\right)\cdot\frac{\mathbf{x}_0}{R_1+R_e}\,dt \\
&= -GmM\int_{R_1+R_e}^{R_2+R_e}\frac{\|\mathbf{x}_0\|^2}{(r_1+R_e)^2}\cdot\frac{1}{t^2}\,dt \\
&= -GmM\int_{R_1+R_e}^{R_2+R_e}\frac{1}{t^2}\,dt \\
&= GmM\cdot\frac{1}{t}\bigg|_{R_1+R_e}^{R_2+R_e} = -\frac{GmM(R_2-R_1)}{(R_1+R_e)(R_2+R_e)}.
\end{aligned}$$

The "−" sign indicates that the force field opposes the motion along the path, and this was to be expected. Thus the work done by the force that propels the object to the higher altitude is $+\dfrac{GmM(R_2-R_1)}{(R_1+R_e)(R_2+R_e)}$. ♦

When a path C is closed (i.e., when the beginning and ending points coincide) we sometimes use the notation $\oint_C \mathbf{F}\cdot d\mathbf{x}$ to remind ourselves of the special nature of the path. Also, in the case of a closed path, there is a physical interpretation for $\oint_C \mathbf{F}\cdot d\mathbf{x}$ when $\mathbf{F}$ is the velocity field for a fluid.

Let C be a simple closed curve and think about the fluid particles that, at one instant, lie on C. A short time Δt later, these fluid particles will have moved so that they lie along a new curve $C_{\Delta t}$. See Figure 4.2.7. Think of a

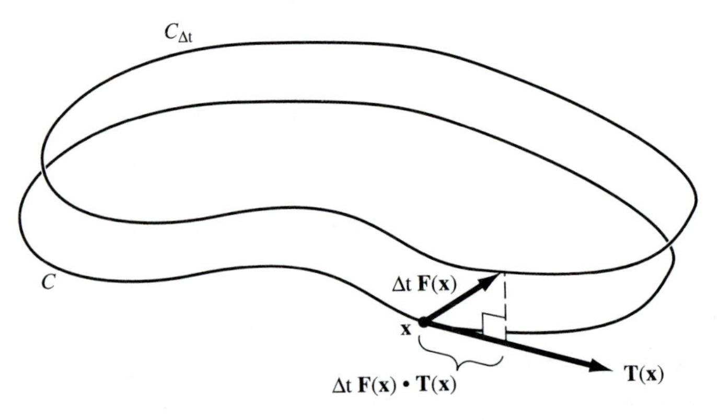

Figure 4.2.7 Fluid particles initially along C have moved (approximately) to positions along $C_{\Delta t}$ during time Δt.

thin smoke ring's motion over a very short time interval. During time Δt, a particle at position $\mathbf{x}$ will have moved to a new position that differs from $\mathbf{x}$ by approximately $\Delta t\mathbf{F}(\mathbf{x})$, and the component of this displacement in the direction of C is

$$\Delta t\mathbf{F}(\mathbf{x}) \cdot \mathbf{T}(\mathbf{x}),$$

where $\mathbf{T}(\mathbf{x})$ is the unit tangent vector to C at $\mathbf{x}$. If $\mathbf{x}_1, \mathbf{x}_2, \ldots, \mathbf{x}_n$ are closely spaced points along C ordered according to the orientation of C, then

$$\Delta t \sum_{i=1}^{n} \mathbf{F}(\mathbf{x}_i) \cdot \mathbf{T}(\mathbf{x}_i) \tag{4.2.6}$$

approximately measures the net component of displacement of particles around C during time Δt. The expression in (4.2.6) is a Riemann sum for $\Delta t \oint_C \mathbf{F} \cdot d\mathbf{x}$, which approximately measures the total amount of "circulation" the fluid particles originally along C have undergone in time Δt. The instantaneous rate of circulation about C is, then,

$$\lim_{\Delta t \to 0} \frac{\Delta t \oint_C \mathbf{F} \cdot d\mathbf{x}}{\Delta t} = \oint_C \mathbf{F} \cdot d\mathbf{x}.$$

For this reason, whenever $\mathbf{F}$ represents a velocity field, we call $\oint_C \mathbf{F} \cdot d\mathbf{x}$ the (net) **circulation of F around** C.

Exercises 4.2

In Exercises 1 through 4, evaluate the line integral over the indicated path.

1. $\int_C u\,dL$, where $u(x, y) = x^2 + y^2$ and C is parametrized by $\mathbf{f}(t) = (3t - 2,\ t + 1)$, $-1 \le t \le 0$.

2. $\int_C (16x - y^2 + 49)\,dL$, where C is parametrized by $\mathbf{f}(t) = (t^2 + 1,\ 4t + 7)$, $0 \le t \le 3$.

3. $\int_C (x + y + z)\,dL$, where C is the line segment joining the origin to the point $(-8, 2, 1/3)$.

4. $\int_C \left(\frac{x_1 + x_2}{x_3 - x_4}\right) dL$, where C is the segment joining the point $(6, 0, 3, 1)$ to the point $(5, 1, 5, 3)$.

5. Find the mass of a wire along the line segment from $(5, 5/3, 3/2)$ to $(10, 10/3, 3)$ if the density at each point is inversely proportional to the distance between that point and the origin.

6. Find the mass of a wire in the shape of the helix traced by $(\cos t, \sin t, t/\pi)$, $\pi \le t \le 3\pi$ if its density at each point is proportional to the distance from the point to the xy-plane.

In Exercises 7 through 10, use the remark after Example 4.2.3 to find the surface areas of the indicated cylinders.

7. The cylinder lying above the segment parametrized by $\mathbf{x} = (2t-1, 4t-3)$, $0 \le t \le 1$ and below the surface $z = x + y + 4$.

8. The cylinder lying above the curve parametrized by $\mathbf{x} = (t^2, t^3)$, $1 \le t \le 4$ and below the surface $z = y\sqrt{4+9x}$.

9. The cylinder lying above the cycloid parametrized by $\mathbf{x} = (t - \sin t, 1 - \cos t)$, $0 \le t \le 2\pi$ and below the surface $z = y$.

10. The cylinder lying above the unit circle $x^2 + y^2 = 1$ and below the surface $z = x + y + 3$.

In Exercises 11 through 16, evaluate the line integral of each vector field over the path indicated.

11. $\mathbf{F}(x, y, z) = (y - x, x - y - xz, -z + xy)$; C is the line segment from the origin to the point $(-3, 5, 2)$.

12. $\mathbf{F}(x, y) = \left(\dfrac{y}{\sqrt{x^2+y^2}}, \dfrac{-x}{\sqrt{x^2+y^2}}\right)$; C is the portion of the unit circle in the first quadrant that goes from $(0, 1)$ to $(1, 0)$.

13. $\mathbf{F}(x, y) = (y, y + 1 - x^2)$; C consists of the line segments from $(5, -1)$ to $(5, 2)$ and from $(5, 2)$ to $(0, 2)$.

14. $\mathbf{F}(x, y, z) = (z\cos x, y\sin x, y^2 + z^2)$; C consists of the line segments in $\mathbb{R}^3$ joining $(0, 0, 0)$ to $(\pi, 0, 0)$, $(\pi, 0, 0)$ to $(\pi, 1, 0)$, and $(\pi, 1, 0)$ to $(\pi, 1, -1)$.

15. $\mathbf{F}(\mathbf{x}) = \begin{bmatrix} 0 & 1/2 & 1/3 \\ 1/2 & 0 & -1/2 \\ 1/3 & -1/2 & 0 \end{bmatrix} \mathbf{x}$; C consists of line segments joining $(1, 1, 1)$ to $(1, 1, 2)$, $(1, 1, 2)$ to $(1, 3, 2)$, and $(1, 3, 2)$ to $(4, 3, 2)$.

16. $\mathbf{F}(\mathbf{x}) = \begin{bmatrix} 0 & 1 & 0 \\ 0 & 0 & 1 \\ 1 & 0 & 0 \end{bmatrix} \mathbf{x}$; C consists of the segments joining $(0, 0, 0)$ to $(\sqrt{2}, 0, 0)$, $(\sqrt{2}, 0, 0)$ to $(\sqrt{2}, 1, 0)$, and $(\sqrt{2}, 1, 0)$ to $(\sqrt{2}, 1, \sqrt{5})$.

In Exercises 17 through 20, evaluate the line integral, where C has the given parametrization.

17. $\int_C (2x - y)\,dx + (3x + 2y)\,dy$; $\mathbf{x} = (t + 1, e^{-t})$, $0 \leq t \leq 1$

18. $\int_C (x^2 + yz)\,dx + z\,dy + (y - x)\,dz$; $\mathbf{x} = (t,\ 2t - 1,\ -8t + 2)$, $0 \leq t \leq 1$

19. $\int_C x_2\,dx_1 + x_3\,dx_2 + x_4\,dx_3 + x_1\,dx_4$, $\mathbf{x} = (1, t^3, t^5, 1 + t)$, $-1 \leq t \leq 1$

20. $\int_C (x - y - x(x^2 + y^2))\,dx + (x + y - y(x^2 + y^2))\,dy$; $\mathbf{x} = (\cos t, \sin t)$, $0 \leq t \leq 2\pi$

21. Find an expression for the work done by the gravitational field $\mathbf{F} = (0, 0, -g)$ near the earth's surface on a particle of mass m that moves from the origin to position $\mathbf{x}_0 = (a, b, c)$ along

(a) the line segment from the origin to this point,

(b) the curve consisting of the segments from $(0, 0, 0)$ to $(a, 0, 0)$, $(a, 0, 0)$ to $(a, b, 0)$, and $(a, b, 0)$ to (a, b, c).

22. Show that $\int_C u\,dL = \int_{-C} u\,dL$. [*Hint*: If C is parametrized by $\mathbf{x} = \mathbf{f}(t)$, $a \leq t \leq b$ then $-C$ is parametrized by $\mathbf{x} = \mathbf{f}(a + b - t)$, $a \leq t \leq b$.]

23. Show that $\int_{-C} \mathbf{F} \cdot d\mathbf{x} = -\int_C \mathbf{F} \cdot d\mathbf{x}$.

24. In Example 4.2.7, suppose that the rocket is propelled straight up from the earth's surface to an altitude R_2 that is much larger than R_e (i.e., $R_2 \to \infty$). Approximately how much energy will be required?

25. Suppose that the Universal Law of Gravitation was actually inverse-first power instead of inverse-square power. In this case, write down a formula for the gravitational force discussed in Example 4.2.7 and calculate the work to move an object from altitude R_e to $R_e + R_2$. From this hypothetical model, what would a rocket designer conclude about the prospect of sending a spaceship away from the earth so that it would never come back?

In Exercises 26 through 30, find the circulation of the vector field about the closed path indicated.

26. $\oint_C (x - y)\,dx + (x + y)\,dy$; C is the counterclockwise oriented circle $x^2 + y^2 = 9$.

27. $\oint_C -y\,dx + x\,dy$; C consists of the segments from $(0,0)$ to $(-1,1)$, $(-1,1)$ to $(-1,-1)$, $(-1,-1)$ to $(3,0)$, and $(3,0)$ to $(0,0)$.

28. $\oint_C y\,dx + z\,dy - x\,dz$; C is parametrized by $\mathbf{x} = (-\sin t, \cos t, \sin t)$, $0 \le t \le 2\pi$.

29. $\oint_C dx_1 - dx_2 + dx_3 - dx_4$; C consists of the segments from $(0,0,0,0)$ to $(1,1,1,0)$, $(1,1,1,0)$ to $(-1,1,-1,1)$, and $(-1,1,-1,1)$ to $(0,0,0,0)$.

30. $\oint_C x_2x_3\,dx_1 - x_3x_4\,dx_2 + \frac{1}{2}\,dx_3 + x_3 dx_4$; C is parametrized by $\mathbf{x} = (t^2 - 1,\ 4, t^3 - t, 1 - t^2)$, $-1 \le t \le 1$.

4.3 Double Integrals

In this section we are concerned with the double integral of a scalar-valued function, the two-dimensional analog of an ordinary definite integral. A physical motivation for this concept is the problem of determining a formula for the volume of a solid T in $\mathbb{R}^3$. Here we think of T as bounded below by

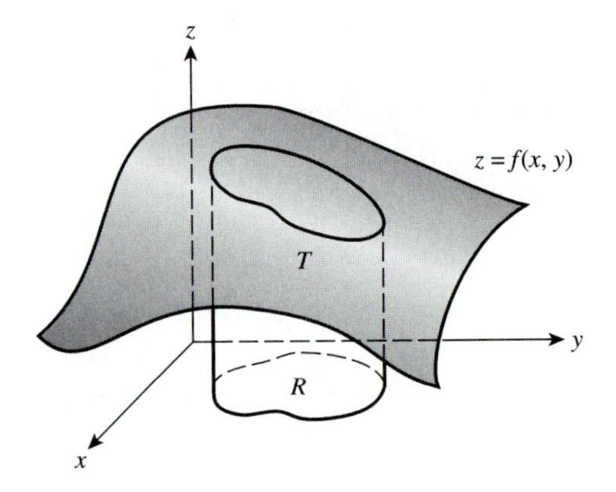

Figure 4.3.1 A solid lying above a region R and beneath the graph of $z = f(x, y)$

a region R in the xy-plane, above by the graph $z = f(x, y)$ of a function f that is nonnegative for $(x, y) \in R$, and on the sides by the cylinder lying above the boundary of R. See Figure 4.3.1.

We can obtain an approximation of the volume of T by effectively "digitizing" the solid T and adding up the volumes of the boxes in this digitized

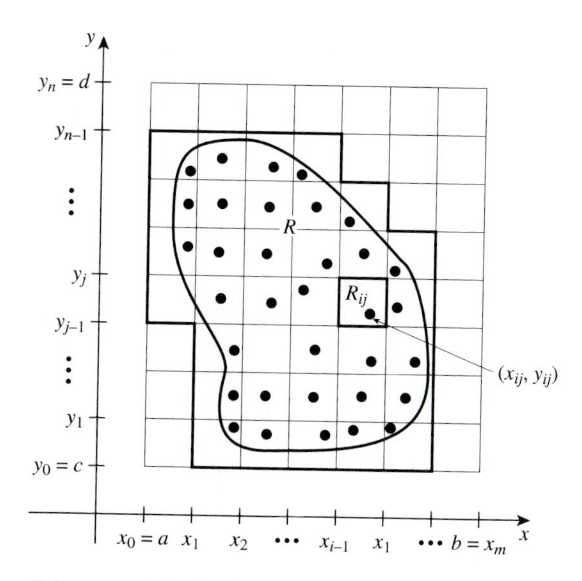

Figure 4.3.2 A partition of a region R

image. First, find a rectangle $B = [a, b] \times [c, d]$ in the xy-plane that contains R. With $a = x_0 < x_1 < x_2 < \cdots < x_m = b$ and $c = y_0 < y_1 < y_2 < \cdots < y_m = d$, partition B into mn rectangles $R_{ij} = [x_{i-1}, x_i] \times [y_{j-1}, y_j]$, $1 \le i \le m$, $1 \le j \le n$, and choose a point (x_{ij}, y_{ij}) in each R_{ij}. Let P represent the collection of subrectangles such that the chosen point lies in R. In symbols,

$$P = \{R_{ij} | (x_{ij}, y_{ij}) \in R\}.$$

See Figure 4.3.2. Then, over each rectangle in P, construct a—typically tall and slender—rectangular solid with height $f(x_{ij}, y_{ij})$. The sum of the volumes of these rectangular solids,

$$I(f, P) = \sum_{(x_{ij}, y_{ij}) \in R} f(x_{ij}, y_{ij}) \, \Delta x_i \, \Delta y_j, \tag{4.3.1}$$

is an approximation to the volume of T. (Here $\Delta x_i = x_i - x_{i-1}$, $\Delta y_j = y_j - y_{j-1}$, and the notation "$\sum_{(x_{ij}, y_{ij}) \in R}$" means "sum over all the rectangles in P.") See Figure 4.3.3. If we repeated this process with a finer partition of T, and hence of R, then the result would be another sum of the form in (4.3.1) with more terms but smaller Δx's and Δy's. Consider what may happen when we calculate Riemann sums for successively finer partitions: the resulting numbers could just bounce around randomly, *or* they might show a tendency toward just *one* number. This latter eventuality is what we expect, and it does indeed occur much of the time. To make this idea precise, it will be helpful to introduce a bit of terminology and notation.

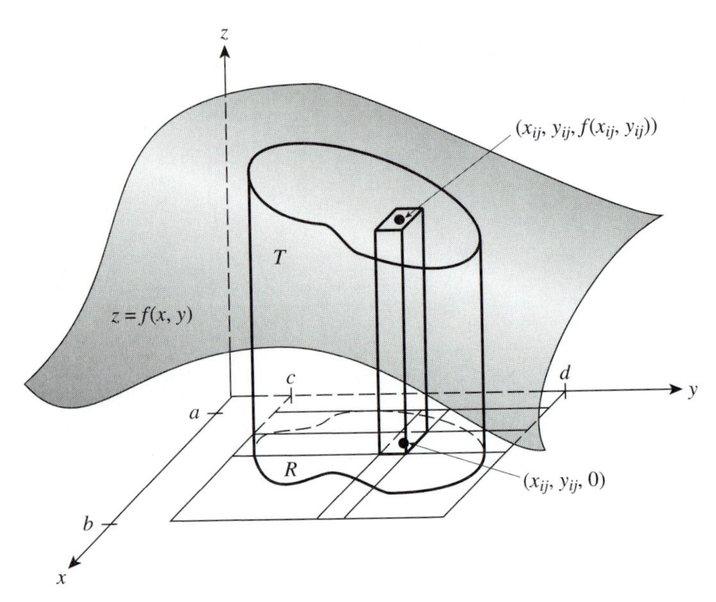

Figure 4.3.3 The volume of the box shown is one term of the sum in (4.3.1).

Definition 4.3.1 The **norm** of a partition P is the length of the longest diagonal of the rectangles in the partition. We denote the norm of P by $N(P)$.

We are now ready for our main definition.

Definition 4.3.2 Let $f : D \subset \mathbb{R}^2 \to \mathbb{R}$ and let R be a subset of D. We say that f is **integrable** over R if there is a number, denoted $\iint\limits_R f(x, y)\, dA$, that satisfies the condition: For any positive number ε there exists a positive number δ such that $\left| I(f, p) - \iint\limits_R f(x, y) dA \right| < \varepsilon$ whenever P is a partition of R such that $N(P) < \delta$. We write this in symbols as

$$\iint\limits_R f(x, y)\, dA = \lim_{\delta \to 0+} \sum_{\substack{(x_{ij}, y_{ij}) \in R \\ N(P) < \delta}} f(x_{ij}, y_{ij})\, \Delta x_i\, \Delta y_j. \tag{4.3.2}$$

This number is called the **double integral of f over R**. At times we may omit the variables and just write $\iint\limits_R f\, dA$ for the double integral.

A natural question is, "Which functions are integrable?" It is beyond the scope of this book to prove the following theorem, but we state it here to exhibit a simply checked criterion for integrability and to remind ourselves that this is an important question.

Theorem 4.3.1 *If $f : D \subset \mathbb{R}^2 \to \mathbb{R}$ is continuous and R is a region contained in D, then f is integrable on R.*

Example 4.3.1 Estimate the value of the integral $\iint\limits_R f\, dA$, where $f(x, y) = x^2 + y^2$ and R is the L-shaped region shown in Figure 4.3.4.

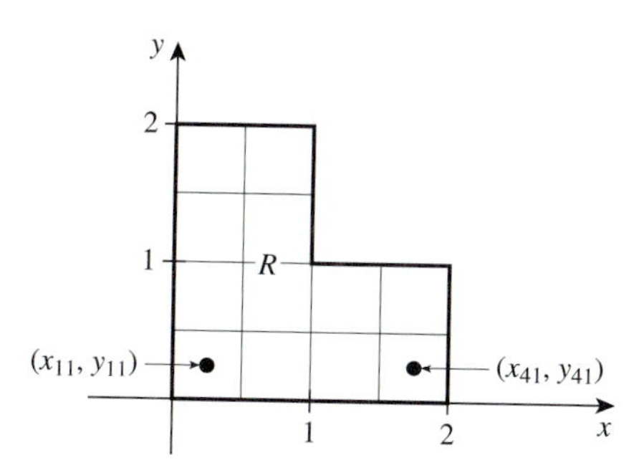

Figure 4.3.4 A partition of the L-shaped region R of Example 4.3.1

Solution The function is continuous everywhere, so by Theorem 4.3.1, it is integrable over R. We use the partition shown in Figure 4.3.4 consisting of 12 squares, each of side length 1/2. We choose the points at which to evaluate f to be the midpoints of these squares. Thus $\Delta x_i = 1/2$, $\Delta y_j = 1/2$, and the midpoints are (1/4, 1/4), (1/4, 3/4), (1/4, 5/4), (1/4, 7/4), (3/4, 1/4),

(3/4, 3/4), (3/4, 5/4), (3/4, 7/4), (5/4, 1/4), (5/4, 3/4), (7/4, 1/4), and (7/4, 3/4). Thus

$$\begin{aligned}
&I(f, P)\\
&= f(1/4, 1/4)\cdot 1/2\cdot 1/2 + f(1/4, 3/4)\cdot 1/2\cdot 1/2 + \cdots\\
&\qquad \cdots + f(7/4, 3/4)\cdot 1/2\cdot 1/2\\
&= \frac{1}{4}\left(\frac{2}{16}+\frac{10}{16}+\frac{26}{16}+\frac{50}{16}+\frac{10}{16}+\frac{18}{16}+\frac{34}{16}+\frac{58}{16}+\frac{26}{16}+\frac{34}{16}+\frac{50}{16}+\frac{58}{16}\right)\\
&= \frac{376}{64} = 5.875,
\end{aligned}$$

so the volume is approximately 5.875.

Just how good *is* this estimate? The smallest value of f on each little square occurs at the lower left corner and the largest at the upper right. So $\iint_R f\, dA$ is greater than or equal to

$$\frac{1}{4}(f(0, 0) + f(1/2, 0) + f(1, 0) + f(3/2, 0) + f(0, 1/2) + f(1/2, 1/2)$$
$$+f(1, 1/2) + f(3/2, 1/2) + f(0, 1) + f(1/2, 1) + f(0, 3/2)$$
$$+ f(1/2, 3/2)) = 15/4 = 3.75$$

and less than or equal to

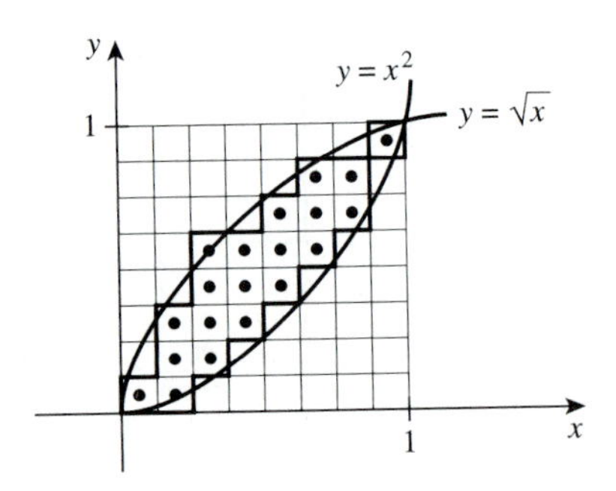

Figure 4.3.5 A partition of the region bounded by the curves $y = x^2$ and $y = x^{1/2}$

$$\frac{1}{4}(f(1/2, 1/2) + f(1, 1/2) + f(3/2, 1/2) + f(2, 1/2) + f(1/2, 1) + f(1, 1)$$
$$+f(3/2, 1) + f(2, 1) + f(1/2, 3/2) + f(1, 3/2) + f(1/2, 2)$$
$$+ f(1, 2)) = 140/16 = 8.75.$$

Thus the actual integral is between 3.75 and 8.75, so the difference between the integral and $I(f, P)$ is less than the difference $8.75 - 3.75 = 5$. Thus the average of these upper and lower estimates, $8.75 + 3.75/2 = 6.25$, is within 2.5 of the exact value. ♦

It is a simple matter to write a computer program to approximate double integrals. In the following example we indicate a method that is modified readily to handle other double integrals.

Example 4.3.2 Approximate $\iint_R \sin(x^2 + y^2)\, dA$, where R is the region bounded by the curves $y = x^2$ and $y = \sqrt{x}$.

```
a = 0
b = 0
c = 0
d = 1
f(x,y) = sin(x^2 + y^2)
sum = 0
input m, n
dx = (b - a) / m
dy = (d - c) / n
x = a - dx / 2
for i = 1 to m do
        x = x + dx
        y = c - dy / 2
        for j = 1 to n do
                y = y + dy
                if (x^2 <= y) and (y <= sqrt(x))
                        then sum = sum + f(x,y)
output dx * dy * sum
```

Figure 4.3.6 An algorithm for approximating $\iint_R \sin(x^2 + y^2)\, dA$ where $R = \{(x, y) \,|\, x^2 \le y \le x^{1/2}\}$

Solution We enclose $R = \left\{(x, y)|x^2 \le y \le \sqrt{x}\right\}$ in the square $[0, 1] \times [0, 1]$ and partition it into mn rectangles each of width $1/m$ and height $1/n$. See

Figure 4.3.5. Then

$$R_{ij} = \left[\frac{i-1}{m}, \frac{i}{m}\right] \times \left[\frac{j-1}{n}, \frac{j}{n}\right], \quad \Delta x_i = \frac{1}{m}, \quad \text{and} \quad \Delta y_j = \frac{1}{n}.$$

For convenience, take the midpoints of these rectangles as the points at which to evaluate the integrand:

$$(x_{ij}, y_{ij}) = \left(\frac{2i-1}{2m}, \frac{2j-1}{2n}\right), \quad i = 1, 2, \ldots, m, \; j = 1, 2, \ldots, n.$$

Now, initialize the approximating Riemann sum to zero, and then check whether (x_{11}, y_{11}) satisfies $x_{11}^2 \leq y_{11} \leq \sqrt{x_{11}}$; if so, then increment the sum by $f(x_{11}, y_{11})$, and otherwise don't increment. Then check whether (x_{12}, y_{12}) satisfies $x_{12}^2 \leq y_{12} \leq \sqrt{x_{12}}$; if so, then increment the sum by $f(x_{12}, y_{12})$. We continue this until we have checked $f(x_{1n}, y_{1n})$. Then we start over with $f(x_{21}, y_{21})$: Check whether (x_{2j}, y_{2j}) is in R and increment the sum by $f(x_{2j}, y_{2j})$ accordingly, for $j = 1, 2, \ldots, n$. We continue this until we have checked (x_{mj}, y_{mj}), $j = 1, 2, \ldots, n$. Finally, multiply this sum by $\Delta x_i \, \Delta y_j = 1/(mn)$; the resulting number is the Riemann sum. This algorithm is summarized in Figure 4.3.6. Choosing $m = 10$, $n = 10$, and applying this algorithm, we find the approximation to be 0.149083; choosing $m = 50$ and $n = 50$, we find that it is 0.146217; choosing $m = 500$ and $n = 500$ we find that it is 0.145373. ♦

It would, of course, be unfortunate if double integrals could only be approximated with the aid of a computer. In fact, double integrals can often be evaluated as **iterated integrals** provided that the region of integration is of a particularly simple type.

Definition 4.3.3 A region R^2 is **y-simple** if it is of the form

$$R = \{(x, y) \,|\, a \leq x \leq b, \; g_1(x) \leq y \leq g_2(x)\}$$

for constants a and b and functions g_1 and g_2. See Figure 4.3.7(a). A region R is **x-simple** if it is of the form $R = \{(x, y) | c \leq y \leq d, \; h_1(y) \leq x \leq h_2(y)\}$ for constants c and d and functions $h_1(x)$ and $h_2(x)$. See Figure 4.3.7(b).

Example 4.3.3 The region bounded by the curves $y = x^2$ and $y = 1 + x^2/3$ is y-simple since it is of the form

$$\left\{(x, y) \,\middle|\, -\sqrt{3/2} \leq x \leq \sqrt{3/2}, \;\; x^2 \leq y \leq 1 + x^2/3\right\}.$$

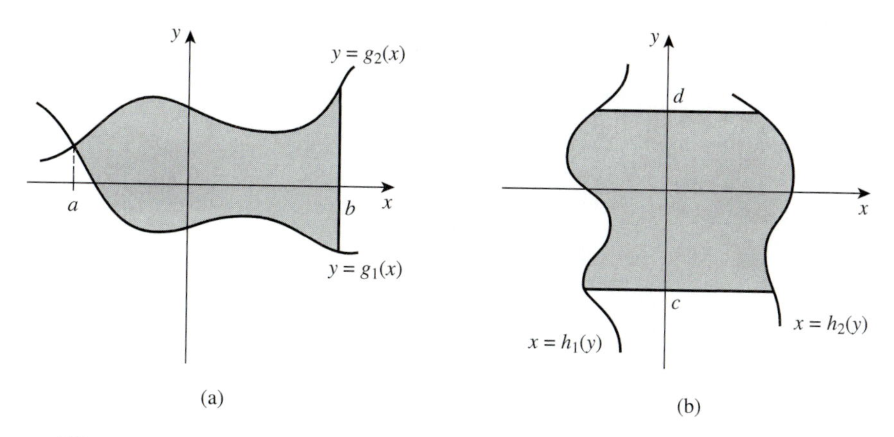

Figure 4.3.7 (a) Typical y-simple region. (b) Typical x-simple region.

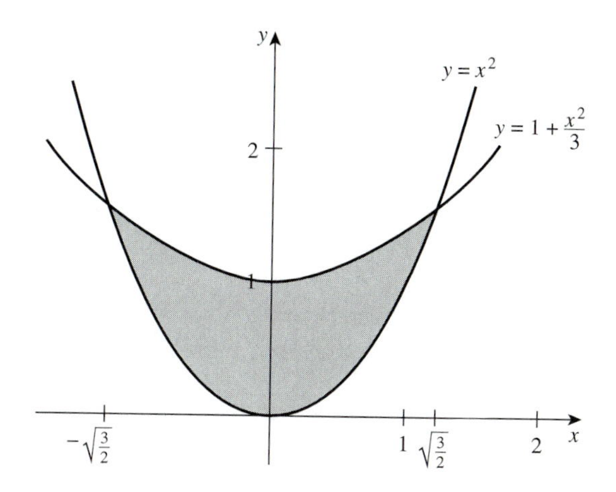

Figure 4.3.8 The region bounded by $y = x^2$ and $y = 1 + x^2/3$ is y-simple.

It is not x-simple. See Figure 4.3.8.

The region bounded by $y = 1$ $y = \sqrt{x}$ and the x- and y-axes are x-simple. It is of the form $\left\{(x,\, y)|0 \le y \le 1,\ \ 0 \le x \le y^2\right\}$ This region is also y-simple. See Figure 4.3.9. ♦

We can now obtain a formula for evaluating many double integrals over y-simple or x-simple regions. Let R be a y-simple region given by $a \le x \le b$ and $g_1(x) \le y \le g_2(x)$, and let f be continuous on R. Then by Theorem 4.3.1, f is integrable. That is, the limit in (4.3.2) exists. In particular, it equals

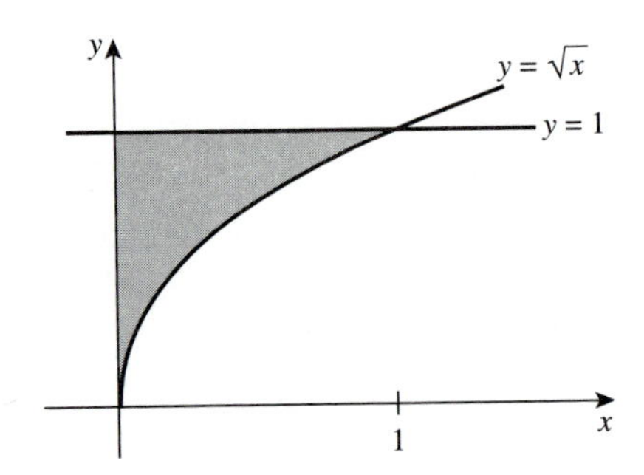

Figure 4.3.9 The region bounded by $y = 1$, $y = x^{1/2}$, and the x- and y-axes is both x-simple and y-simple.

what we get by taking the limit of Riemann sums relative to partitions P, where all of the x-increments are the same, all the y-increments are the same, and the point (x_{ij}, y_{ij}) at which f is evaluated is the upper right-hand corner of each subrectangle. With $\Delta x_i = (b - a)/m$, for $i = 1, 2, \ldots, m$, and $\Delta y_j = (d - c)/n$, for $j = 1, 2, \ldots, n$, we have $x_i = a + i\Delta x_i$, $y_j = c + j\Delta y_j$, and $(x_{ij}, y_{ij}) = (x_i, y_j)$. In this case, the Riemann sums are

$$I(f, P) = \sum_{(x_i, y_j) \in R} f(x_i, y_j)\Delta x_i \Delta y_j = \sum_{i=1}^{m} \left(\sum_{j=k(i)}^{K(i)} f(x_i, y_j)\Delta y_j \right) \Delta x_i, \tag{4.3.3}$$

where $k(i)$ is the smallest j-value such that $(x_i, y_j) \in R$ and $K(i)$ is the largest j-value such that $(x_i, y_j) \in R$. In Figure 4.3.10, for instance, $m = 8$, $n = 12$, $k(4) = 3$, and $K(4) = 11$. The sum in parentheses in (4.3.3) is a Riemann sum for $f(x_i, y)$, which is a function of y alone on the y-interval $[g_1(x_i), g_2(x_i)]$. Since f is continuous, $f(x_i, y)$ is a continuous function of y, and as $\max\limits_j \Delta y_j \to 0$ (equivalently, $n \to \infty$) these sums converge to $\displaystyle\int_{g_1(x_i)}^{g_2(x_i)} f(x_i, y)\, dy$. Thus

$$\lim_{n \to \infty} I(f, P) = \sum_{i=1}^{m} \left(\int_{g_1(x_i)}^{g_2(x_i)} f(x_i, y)\, dy \right) \Delta x_i, \tag{4.3.4}$$

and *this* is a Riemann sum for the function of x given by

$$F(x) = \int_{g_1(x)}^{g_2(x)} f(x, y)\, dy$$

for x in the interval $[a, b]$. It can be shown that F is continuous, so when $\max\limits_i \Delta x_i \to 0$ (equivalently, $m \to \infty$), the sums in (4.3.4) converge to

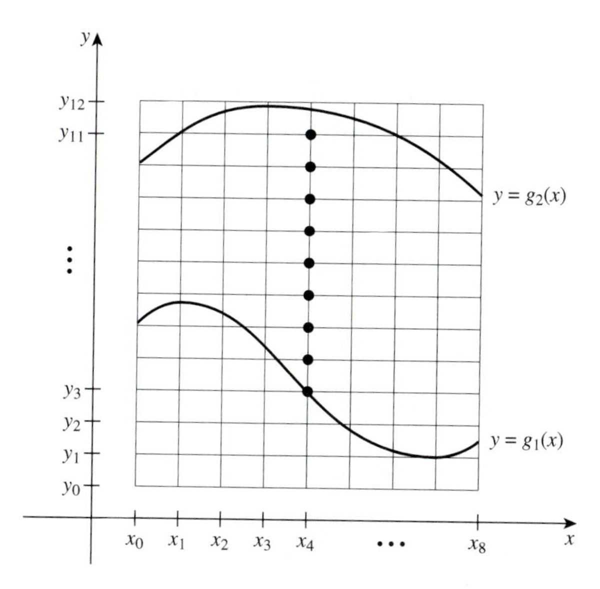

Figure 4.3.10 A schematic diagram for the iterated Riemann sum in (4.3.3)

$\int_a^b F(x)\,dx$. That is,

$$\lim_{m\to\infty}\lim_{n\to\infty} I(f,\,P) = \int_a^b \left(\int_{g_1(x)}^{g_2(x)} f(x,y)\,dy\right) dx. \tag{4.3.5}$$

We usually write the expression on the right without parentheses as

$$\int_a^b \int_{g_1(x)}^{g_2(x)} f(x,y)\,dy\,dx. \tag{4.3.6}$$

An expression such as (4.3.6) is called an **iterated integral**.

We have proven the following basic theorem.

Theorem 4.3.2 *If $f(x,\,y)$ is continuous on a y-simple region*

$$R = \{(x,\,y) | a \le x \le b,\ g_1(x) \le y \le g_2(x)\}$$

then

$$\iint\limits_R f\,dA = \int_a^b \int_{g_1(x)}^{g_2(x)} f(x,y)\,dy\,dx.$$

By turning our perspective sideways, we can see that this same argument gives an analogous result for x-simple regions.

Theorem 4.3.3 *If $f(x, y)$ is continuous on an x-simple region*

$$R = \{(x, y)|c \le y \le d,\ h_1(y) \le x \le h_2(y)\}$$

then

$$\iint\limits_R f\,dA = \int_c^d \int_{h_1(y)}^{h_2(y)} f(x, y)\,dx\,dy.$$

Example 4.3.4 Evaluate the iterated integral $\displaystyle\int_1^2 \int_{1-x^2}^{1+x^2} xy\ \ dy\,dx.$

Solution We simply have the chain of equalities

$$\begin{aligned}
\int_1^2 \int_{1-x^2}^{1+x^2} xy\ \ dy\,dx &= \int_1^2 \left(\frac{1}{2}xy^2 \Big|_{y=1-x^2}^{y=1+x^2} \right) dx \\
&= \int_1^2 \frac{x}{2}\left((1+x^2)^2 - (1-x^2)^2\right) dx \\
&= \int_1^2 2x^3\,dx = \frac{1}{2}x^4\Big|_1^2 = \frac{15}{2}. \quad \blacklozenge
\end{aligned}$$

Example 4.3.5 Let R be the region in the xy-plane bounded by the curve $y = \ln x$ and the lines $y = 0$ and $x = 2$. Find the volume of the solid T lying above R and beneath the surface $z = e^{y-x}$.

Solution The region is shown in Figure 4.3.11. It is both x- and y-simple. Treating it as y-simple, we describe it by the inequalities

$$1 \le x \le 2$$

$$0 \le y \le \ln x.$$

Since e^{y-x} is nonnegative, the double integral of e^{y-x} over R will represent the volume of T. By Theorem 4.3.2,

$$\begin{aligned}
\iint\limits_R e^{y-x}\,dA &= \int_1^2 \int_0^{\ln x} e^{y-x}\,dy\,dx = \int_1^2 e^{y-x}\big|_0^{\ln x}\,dx \\
&= \int_1^2 (e^{\ln x - x} - e^{-x})\,dx \\
&= \int_1^2 (x-1)e^{-x}\,dx.
\end{aligned}$$

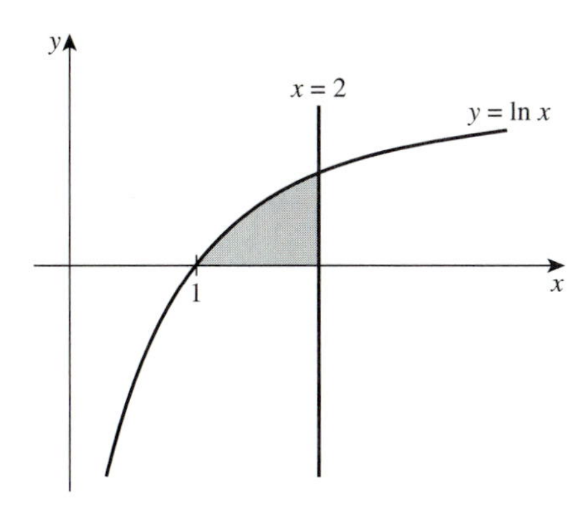

Figure 4.3.11 The region in the xy-plane bounded by the curve $y = \ln x$ and the lines $y = 0$ and $x = 2$

Integrating by parts, we obtain

$$\begin{aligned} -(x-1)e^{-x}\Big|_1^2 - \int_1^2 -e^{-x}\,dx. &= -e^{-2} - (e^{-2} - e^{-1}) \\ &= e^{-1} - 2e^{-2} \approx 0.097. \quad \blacklozenge \end{aligned}$$

If f and g are two functions that satisfy $f(x,\, y) \leq g(x,\, y)$ for all $(x,\, y)$ in a region R of the xy-plane, then it makes sense to ask for the volume of the solid T lying over R between the surfaces $z = f(x,\, y)$ and $z = g(x,\, y)$. In Figure 4.3.12 we see that if we partition R into small rectangles with areas $\Delta x_i\, \Delta y_j$, choose a point $(x_{ij},\, y_{ij})$ in each of these rectangles, and construct a box with height $g(x_{ij},\, y_{ij}) - f(x_{ij},\, y_{ij})$, then the sum of the volumes $(g(x_{ij},\, y_{ij}) - f(x_{ij},\, y_{ij}))\, \Delta x_i\, \Delta y_j$ of these boxes approximates the volume of the solid T. This leads us to define the **volume** of T as

$$\begin{aligned} V(T) &= \lim_{\delta \to 0+} \sum_{\substack{(x_{ij},\, y_{ij}) \in R \\ N(P) < \delta}} (g(x_{ij},\, y_{ij}) - f(x_{ij},\, y_{ij}))\, \Delta x_i\, \Delta y_j \\ &= \iint\limits_R (g(x,\, y) - f(x,\, y))\, dA. \end{aligned}$$

Example 4.3.6 Find the volume of the solid in the first octant bounded by the planes $y = 0$, $x = 0$, $x + y = 1$, $z = -x - y$, and the paraboloid $z = x^2 + y^2$.

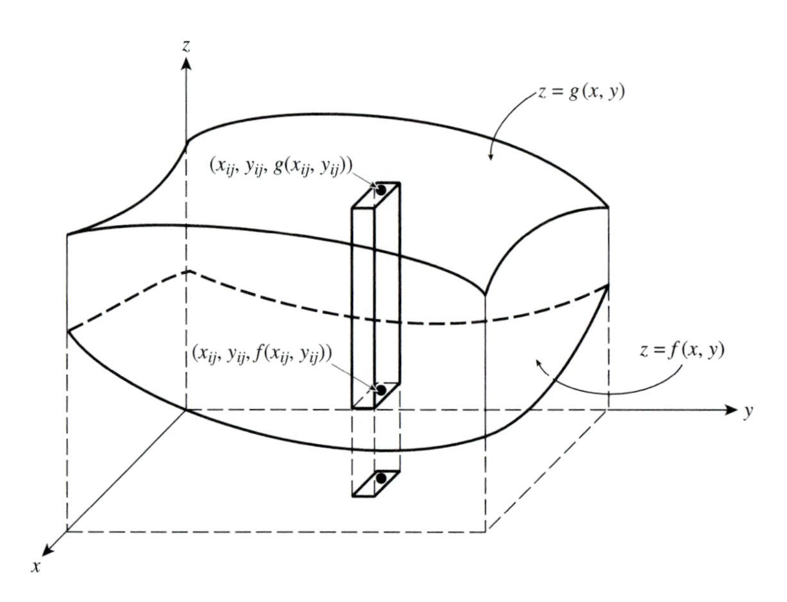

Figure 4.3.12 A solid between $z = f(x, y)$ and $z = g(x, y)$ corresponding to a region R

Solution The three planes $x = 0$, $y = 0$, and $x+y = 1$ form a wedge-shaped cylinder which cuts off the region $R = \{(x, y) \,|\, 0 \leq x \leq 1, 0 \leq y \leq 1 - x\}$ in the xy-plane. See Figure 4.3.13. The top of the solid is the paraboloid $z = x^2 + y^2$, and the bottom is the plane $z = -x - y$. So the volume is

$$\begin{aligned}
&\iint_R \left(x^2 + y^2 - (-x - y)\right) dA \\
&= \int_0^1 \int_0^{1-x} \left((x^2 + x) + y^2 + y\right) dy\, dx \\
&= \int_0^1 \left[(x^2 + x)(1 - x) + \frac{1}{3}(1 - x)^3 + \frac{1}{2}(1 - x)^2\right] dx \\
&= \int_0^1 x(1 - x^2)\, dx + \left(-\frac{1}{3 \cdot 4}(1 - x)^4 - \frac{1}{2 \cdot 3}(1 - x)^3\right)\Big|_0^1 \\
&= -\frac{1}{4}(1 - x^2)^2\Big|_0^1 + \left(\frac{1}{12} + \frac{1}{6}\right) \\
&= \frac{1}{4} + \frac{1}{4} = \frac{1}{2}. \quad \blacklozenge
\end{aligned}$$

Double integrals can also represent areas. For instance, let R be a y-

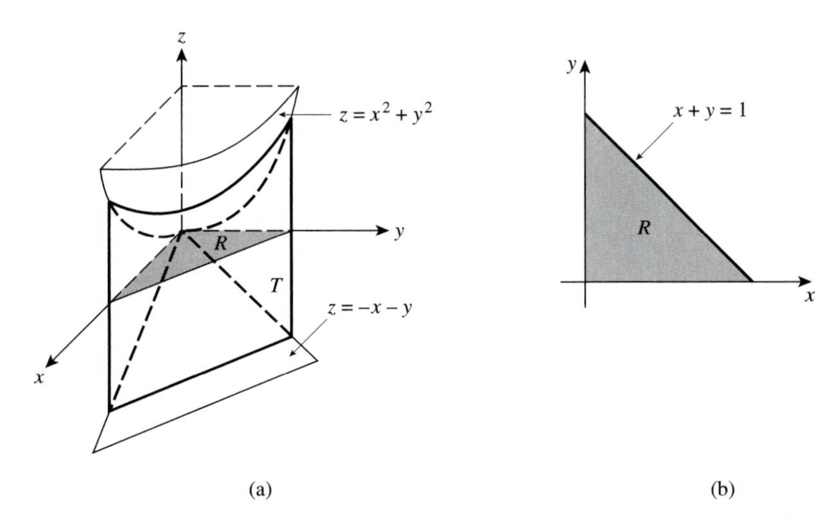

Figure 4.3.13 The solid described in Example 4.3.6

simple region in the xy-plane, and consider the double integral $\iint_R 1\,dA$. Since R is y-simple, the integral can be written as the iterated integral $\int_a^b \int_{g_1(x)}^{g_2(x)} 1\,dy\,dx$ for some choice of functions g_1 and g_2. Evaluating the inner integral, this becomes $\int_a^b (g_2(x) - g_1(x))\,dx$, which we recognize as the area of the region R. If R is x-simple, then we see in the same way that $\iint_R 1\,dA$ is the area of R. For this reason we define the area of any region R in the xy-plane to be

$$A(R) = \iint_R dA. \tag{4.3.7}$$

Example 4.3.7 Find the area of the region bounded by the parabolas $y = x^2/4$ and $y = 5 - x^2$.

Solution The curves intersect when $5 - x^2 = x^2/4$. See Figure 4.3.14. This gives $x = \pm 2$, so the region is y-simple and described by the inequalities

$$-2 \leq x \leq 2$$

$$\frac{x^2}{4} \leq y \leq 5 - x^2.$$

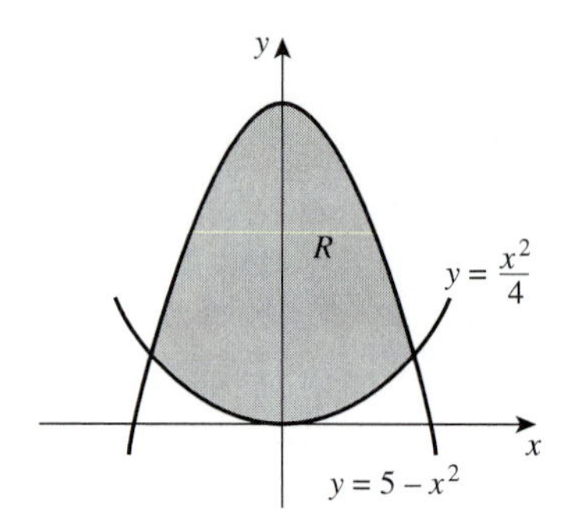

Figure 4.3.14 The region bounded by the parabolas $y = x^2/4$ and $y = 5 - x^2$

Thus the area is

$$\int_{-2}^{2}\int_{x^2/4}^{5-x^2} 1\,dy\,dx = \int_{-2}^{2}\left(5 - \frac{5}{4}x^2\right)dx = \frac{40}{3}. \quad \blacklozenge$$

Occasionally, treating a region as x-simple may lead to an integral that cannot be integrated in elementary terms. However, if the region is y-simple, we may be able to represent the double integral in a way that does give a tractable integral to evaluate. Rewriting an iterated integral in this way is called **reversing the order of integration**.

Example 4.3.8 Evaluate $\displaystyle\int_0^1\int_x^1 e^{y^2}\,dy\,dx$.

Solution The function e^{y^2} does not have an antiderivative that can be represented in terms of finitely many elementary functions, so we must look for other means to evaluate the integral. The region of integration is given by

$$\begin{aligned} 0 &\le x \le 1 \\ x &\le y \le 1. \end{aligned}$$

See Figure 4.3.15. This region is also x-simple and described by

$$\begin{aligned} 0 &\le y \le 1 \\ 0 &\le x \le y. \end{aligned}$$

Thus

$$\int_0^1\int_x^1 e^{y^2}\,dy\,dx = \int_0^1\int_0^y e^{y^2}\,dx\,dy = \int_0^1 xe^{y^2}\Big|_{x=0}^{x=y}\,dx$$

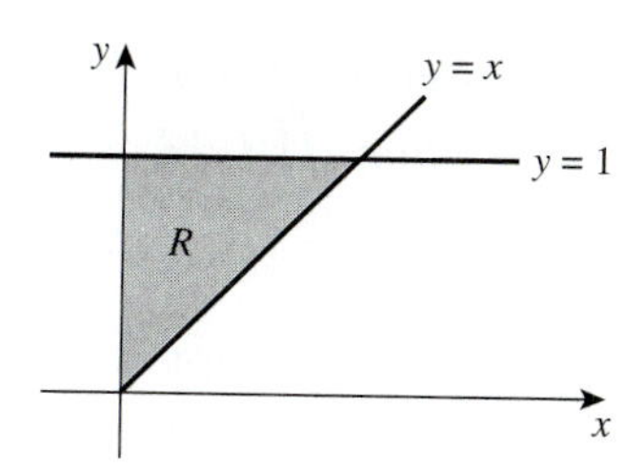

Figure 4.3.15 Regarding this region as x-simple, we can easily evaluate $\iint_R e^{y^2}\,dA$.

$$= \int_0^1 ye^{y^2}\,dy = \frac{1}{2}e^{y^2}\Big|_0^1 = \frac{1}{2}(e-1). \quad \blacklozenge$$

There are several properties of double integrals that are used frequently. We summarize them in the following theorem.

Theorem 4.3.4 *1. If f is integrable on R and c is a constant, then*

$$\iint_R cf\,dA = c\iint_R f\,dA.$$

2. If f and g are integrable on R, then

$$\iint_R (f+g)\,dA = \iint_R f\,dA + \iint_R g\,dA.$$

3. If f is integrable on R and $R = R_1 \cup R_2$, then

$$\iint_R f\,dA = \iint_{R_1} f\,dA + \iint_{R_2} f\,dA.$$

4. If f and g are integrable on R and $g(x, y) \le f(x, y)$ for all $(x, y) \in R$, then

$$\iint_R g\,dA \le \iint_R f\,dA.$$

5. (Mean Value Theorem for Double Integrals) If f is continuous on a region R with area $A(R)$, then there exists a point $(a_1, a_2) \in R$ such that

$$\iint_R f\,dA = A(R)f(a_1, a_2).$$

6. *If f is continuous and satisfies $f(x, y) > 0$ for all $(x, y) \in R$, where R has nonempty interior, then* $\iint\limits_R f\, dA > 0$.

The proofs of all of these properties, except 5 and 6, involve verifying that the claimed equality or inequality holds for all Riemann sums and hence for the integrals. Parts 5 and 6 are consequences of the Intermediate Value Theorem.

Example 4.3.9 Use only the properties of double integrals to evaluate $\iint\limits_R (x+y)\, dA$ where R is the square with vertices $(2, 0)$, $(0, 2)$, $(-2, 0)$, and $(0, -2)$.

Solution The graph of the integrand $f(x, y) = x + y$ is a plane that passes through the line $x + y = 0$ in the xy-plane. The value of this function at a point (x, y) above this line is positive, and the value is negative if the point is below the line. Indeed, $f(-x, -y) = -f(x, y)$ for all choices of (x, y).

Now the region of integration can be subdivided into two rectangles: R_1 with vertices $(2, 0)$, $(0, 2)$, $(-1, 1)$, and $(1, -1)$, and R_2 with vertices $(1, -1)$, $(-1, 1)$, $(-2, 0)$, and $(0, -2)$. See Figure 4.3.16. Each point

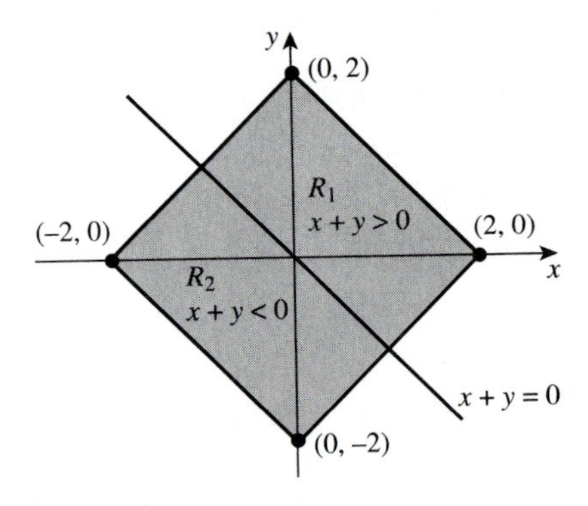

Figure 4.3.16 $\iint_R (x + y)\, dA$ is zero because the integrand is an "odd" function with respect to the line $y = -x$ and R is symmetric about this line.

$(x, y) \in R_2$ corresponds to exactly one point $(-x, -y)$ in R_1, so

$$\iint\limits_{R_2} f(x, y)\, dA \quad = \quad \iint\limits_{R_1} f(-x, -y)\, dA$$

$$= \iint\limits_{R_1} -f(x, y)\, dA$$

$$= -\iint\limits_{R_1} f(x, y)\, dA.$$

In this last step we have used property (a) in Theorem 4.3.4. By property (c) in that theorem,

$$\begin{aligned}\iint\limits_{R} f(x, y)\, dA &= \iint\limits_{R_1} f(x, y)\, dA + \iint\limits_{R_2} f(x, y)\, dA \\ &= \iint\limits_{R_1} f(x, y)\, dA - \iint\limits_{R_1} f(x, y)\, dA = 0.\end{aligned}$$

In fact, our reasoning here would apply to any region R that is symmetric about the line $x + y = 0$. ♦

Exercises 4.3

In Exercises 1 through 3, find an approximation for the double integral by using the indicated partition and choosing (x_{ij}, y_{ij}) to be the midpoint of each partition rectangle.

1. $\iint\limits_{R_1} x^2 y\, dA$; R_1 is the rectangle $[0, 2] \times [0, 4]$; P_1 is the collection of rectangles $\left[\frac{i-1}{2}, \frac{i}{2}\right] \times [j - 1, j]$, where $1 \le i \le 4$ and $1 \le j \le 4$.

2. $\iint\limits_{R_2} x^2 y\, dA$; R_2 is the triangular region bounded by $y = 2x$, $x = 2$, and the x-axis; P_2 is the collection of rectangles in P_1 of Exercise 1 whose midpoints lie in R_2.

3. $\iint\limits_{R_3} x^2 y\, dA$; R_3 is the region bounded by $y = x^2$, $y = 4$, and the y-axis; P_3 is the collection of rectangles in P_1 of Exercise 1 whose midpoints lie in R_3.

In Exercises 4 and 5, use the algorithm in Figure 4.3.6, appropriately modified, to approximate the double integral.

4. $\displaystyle\iint_R \sqrt{1+x^2+y^2}\, dA$; R is the disk of radius 2 centered at $(2, 0)$.

5. $\displaystyle\iint_R e^{x^2+y^2}\, dA$; R is the triangle bounded by the lines $y = x$, $y = 2x$, and $x = 3$.

6. An ecologist wants to approximate the volume of water in the pond shown in Figure 4.3.17. From her boat she has dropped a plumb line to measure the depth of the water at a grid of points also shown in the figure. These are regularly spaced at 5-meter intervals in both the north-south and east-west directions, and the depth in meters is indicated at each point. Find an estimate for the volume of water that uses all of the data she has gathered.

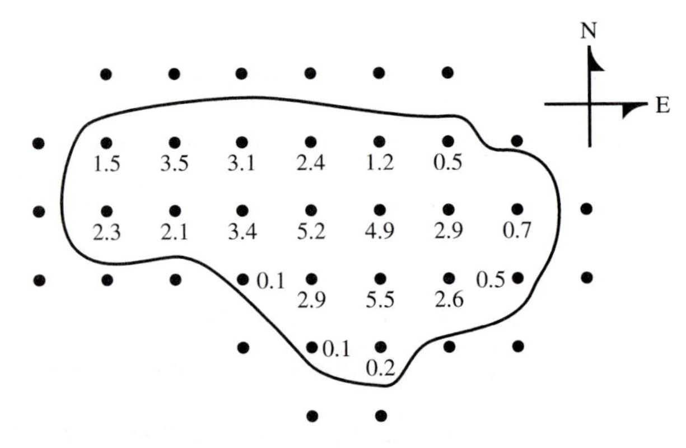

Figure 4.3.17 Depth measurements in a pond

In Exercises 7 through 11, evaluate the iterated integral and sketch the region of integration.

7. $\displaystyle\int_{-2}^{0}\int_{2}^{4} (3x^2+2y^3)\, dy\, dx$

8. $\displaystyle\int_{0}^{\pi}\int_{x}^{\pi} y \sin x\, dy\, dx$

9. $\displaystyle\int_{-1/4}^{1}\int_{-y}^{1-y^2} (5x-y)\, dx\, dy$

10. $\displaystyle\int_{0}^{1}\int_{e^{-x}}^{e^{x}} \frac{\ln y}{y}\, dy\, dx$

11. $\int_0^{\pi/8} \int_0^y \sec^2(x+y)\,dx\,dy$

In Exercises 12 through 15, find the double integral of the function over the indicated region.

12. $f(x, y) = (x - 2y)^2$, R is the rectangle with sides parallel to the axes and opposite corners $(-1, -1)$ and $(5, 2)$.

13. $f(x, y) = x/(y+1)^2$, R is the region bounded by the parabola $y = x^2$ and the line $y = 4x$.

14. $f(x, y) = ye^x$, R is the region bounded by the parabola $x = y^2$ and the line $x = 5y$.

15. $f(x, y) = x \cos y \sin y$, R is the region bounded by the lines $x = 1$, $x = -1$, the x-axis, and $y = \arctan x$.

In Exercises 16 through 18, find the volume of the indicated solid region.

16. The solid bounded above by the paraboloid $z = x^2/4 + y^2/9 + 1$, below by the xy-plane, and lying above the square $S = \{(x, y) \mid 0 \le x \le 1,\ 0 \le y \le 1\}$.

17. The solid bounded by the plane $z = 1 - x - y$ and the coordinate planes.

18. The solid bounded above by the sphere $x^2 + y^2 + z^2 = 16$ and below by the paraboloid $z = \frac{1}{6}(x^2 + y^2)$. [Set up the integral but do not evaluate.]

In Exercises 19 through 22, use *double integrals* to find the area of each of the indicated region.

19. The quadrilateral with vertices $(0, 0)$, $(4, 0)$, $(3, 1)$, $(0, 2)$.

20. A disk of radius 10.

21. The region bounded by the curves $y = x^3$ and $x = y^3$.

22. The region bounded by $x = \sin y$ and $y^2 - \pi^2 = x$.

In Exercises 23 through 25, find the mass of the region with the indicated density function.

23. R is the unit square and the density is proportional to the square of the distance from the origin.

24. R is the region in the xy-plane bounded by the curves $y = x^2 + 1$ and $y = x + 3$ and the density at a point is proportional to the distance of that point from the x-axis.

25. R is the portion of the annulus in the first quadrant bounded by the circles of radius 1 and 3 centered at the origin, and the density is inversely proportional to the square of the distance from the origin. (Set up the integral(s) but do not evaluate.)

In Exercises 26 through 31, reverse the order of integration.

26. $\int_0^2 \int_x^2 f(x, y)\, dy\, dx$

27. $\int_0^1 \int_1^{e^y} f(x, y)\, dx\, dy$

28. $\int_{\pi/2}^{\pi} \int_0^{\sin x} f(x, y)\, dy\, dx$

29. $\int_{-4}^{4} \int_{-\sqrt{16-y^2}}^{\sqrt{16-y^2}} f(x, y)\, dx\, dy$

30. $\int_{-1}^{0} \int_{-x}^{1} f(x, y)\, dy\, dx + \int_0^1 \int_{\sqrt{x}}^{1} f(x, y)\, dy\, dx$

31. $\int_{-1}^{1} \int_{y-3}^{y^2} f(x, y)\, dx\, dy$

In Exercises 32 through 36, sketch the region of integration and, without evaluating, determine whether the integral is positive, negative, or zero.

32. $\int_{-1}^{1} \int_{-1}^{1} x^2 y\, dy\, dx$

33. $\int_{-1}^{1} \int_{-1}^{1} x^2 y^2\, dx\, dy$

34. $\int_0^{\pi/2} \int_0^{\pi/2 - x} \sin(x + y)\, dy\, dx$

35. $\int_{-1}^{1} \int_0^{\sqrt{(1-y^2)/2}} (2x^2 + y^2)\, dx\, dy$

36. $\int_1^3 \int_1^2 (1 - x^2 - y^2)\, dy\, dx$

Using only (4.3.7) and the properties of integrals listed in Theorem 4.3.4 (i.e., without evaluating an iterated integral), evaluate the double integrals in Exercises 37 through 41.

37. $\displaystyle\iint_R (2x - y)\, dA$, where R is the disk of radius 1 centered at the origin

38. $\displaystyle\iint_R x^3y^2\, dA$, where R is the triangle with vertices (1, 0), (0, 1), and (−1, 0)

39. $\displaystyle\iint_R 25\, dA$ where R is the square with vertices (0, 0), (3, 1), (2, 4), and (−1, 3)

40. $\displaystyle\iint_R (-8)\, dA$, where R is the disk of radius 4 centered at (−5, 15)

41. $\displaystyle\iint_R (1 + (x - y)^3)\, dA$, where R is the triangle with vertices (0, 0), (3, 0), and (0, 3)

In Exercises 42 and 43, write the iterated integral corresponding to the Riemann sum and describe the region of integration.

42. $\displaystyle\sum_{j=1}^{n}\sum_{i=1}^{m} x_i^2 y_j^3\, \Delta x_i\, \Delta y_j$, where $x_i = 1 + i/m$, $i = 0, 1, \ldots, m$ and $y_j = j/n$, $j = 0, 1, \ldots, n$.

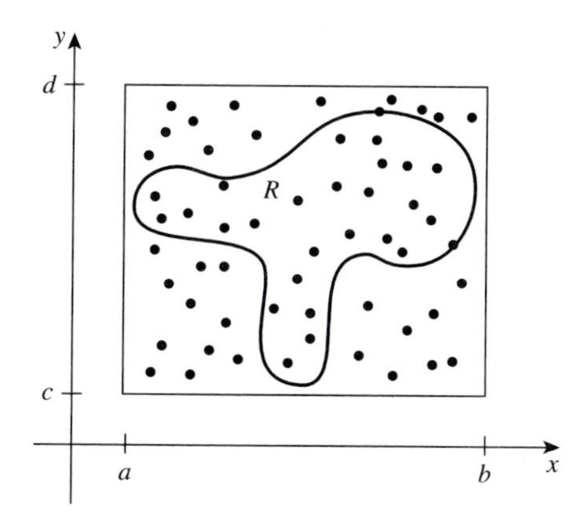

Figure 4.3.18 The ratio of the number of randomly selected points inside R to the total number of such points in the rectangle multiplied by the area of the rectangle is approximately the area of R.

43. $\sum_{j=1}^{n}\sum_{i=1}^{m}\Big[\Big(\frac{2x_i-1}{2}\Big)^2+\frac{2y_j-1}{2}\Big]\,\Delta x_i\,\Delta y_j$ where $x_i = -1+2i/m$, $i = 0, 1, \ldots, m$, and $y_j = 2+3j/n$, $j = 0, 1, \ldots, n$

44. It is possible to calculate area by throwing darts! Consider a region R in the xy-plane enclosed by a rectangle $[a, b] \times [c, d]$ as shown in Figure 4.3.18. Now choose, by throwing darts or using a random number generator on a computer, several points in the rectangle at random. Some will lie inside R and others will not. The ratio of the number of points in R to the total number of points in $[a, b]\times[c, d]$ is approximately the ratio of the area of R to the area of the rectangle. That is, if $A(R)$ is the area of R, $N(R)$ is the number of points landing in R, and n is the total number of points chosen, then

$$\frac{A(R)}{(b-a)(d-c)} \approx \frac{N(R)}{n}.$$

So $A(R) \approx (b-a)(d-c)N(R)/n$. Use this method to approximate the areas in Exercises 19 through 22.

4.4 Triple Integrals

We calculate the mass of an object whose density is constant throughout simply by multiplying the density by the object's volume. However, if the density of an object varies from point to point then some other procedure for finding mass must be employed. A reasonable approach is first to assume that the densities at points nearby to one another are nearly the same. Then, by subdividing the object into many small portions on each of which the density is practically constant, we may approximate the mass by adding up the products of these densities and volumes. This idea is a motivation for the **triple integral** of a function f of three variables.

Consider a solid region S in $\mathbb{R}^3$ that is bounded by a piecewise smooth boundary surface (i.e., a surface that can be decomposed into finitely many pieces, each of which is the graph of a function of two variables with continuous partials over a simple region in one of the coordinate planes.) See Figure 4.4.1. Let f be a real-valued function defined on S. If f is nonnegative, we can think of $f(x, y, z)$ as the density of the solid at the point (x, y, z). Surround S with a box $B = [a_1, b_1] \times [a_2, b_2] \times [a_3, b_3]$, partition each interval as

$$a_1 = x_0 < x_1 < \cdots < x_{n_1} = b_1$$

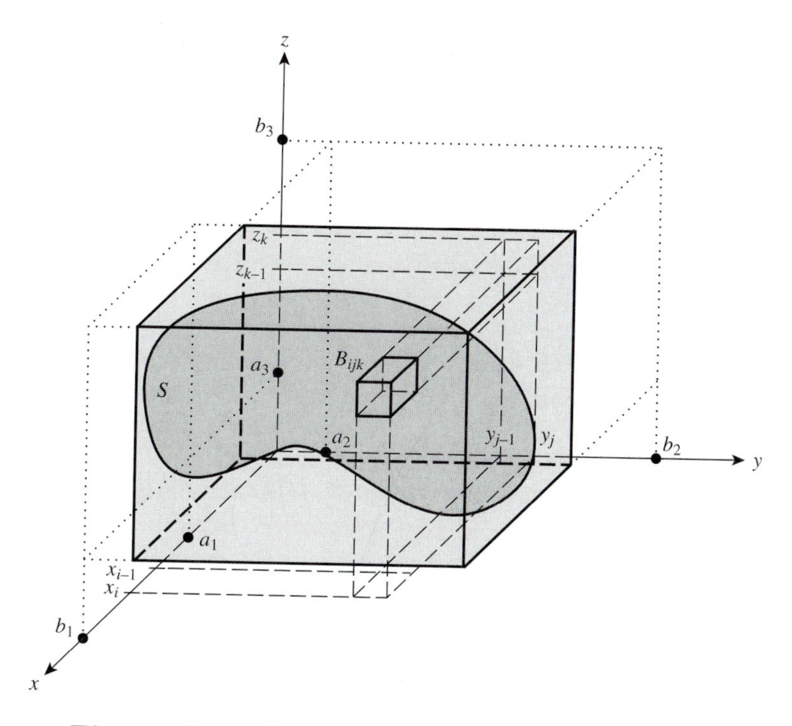

Figure 4.4.1 Partitioning a solid region in space

$$a_2 = y_0 < y_1 < \cdots < y_{n_2} = b_2$$
$$a_3 = z_0 < z_1 < \cdots < z_{n_3} = b_3,$$

and let $B_{ijk} = [x_{i-1}, x_i] \times [y_{j-1}, y_j] \times [z_{k-1}, z_k]$ for $1 \le i \le n_1$, $1 \le j \le n_2$, $1 \le k \le n_3$. Again, see Figure 4.4.1. For each choice of the indices, choose a point $(x_{ijk}, y_{ijk}, z_{ijk}) \in B_{ijk}$ and form the sum

$$I(f, P) = \sum_{(x_{ijk}, y_{ijk}, z_{ijk}) \in S} f(x_{ijk}, y_{ijk}, z_{ijk})\, \Delta x_i\, \Delta y_j\, \Delta z_k \tag{4.4.1}$$

where, as usual, $\Delta x_i = x_i - x_{i-1}$, $\Delta y_j = y_j - y_{j-1}$, $\Delta z_k = z_k - z_{k-1}$. Imagine carrying out this process for all partitions P of norm less than some small positive number δ. [The norm $N(P)$ of a partition P here is the maximum of the lengths of the diagonals of the rectangular boxes of P.] Then the values of the sums in (4.4.1) form a "cloud" of numbers which is smaller and more rarefied for smaller values of δ. If this cloud "condenses" to a single number as $\delta \to 0$, then we say that f is **integrable** over S and we call this number the **triple integral** of f over S. In symbols,

$$\iiint\limits_S f\,dV = \lim_{\delta\to 0} \sum_{\substack{(x_{ijk},\, y_{ijk},\, z_{ijk})\in S \\ N(P)<0}} f(x_{ijk}, y_{ijk}, z_{ijk})\Delta x_i \Delta y_j \Delta z_k. \qquad (4.4.2)$$

Just as for double integrals we have a simple criterion for integrability of a function of three variables. We state it here without proof.

Theorem 4.4.1 *If $f : D \subset \mathbb{R}^3 \to \mathbb{R}$ and S is a solid region contained in D that has a piecewise smooth boundary, then f is integrable on S.*

Example 4.4.1 Approximate the integral $\iiint\limits_S (xy - z^2)\,dV$, where S is the solid between the spheres $x^2 + y^2 + z^2 = 1$ and $x^2 + y^2 + z^2 = 4$.

Solution The integrand is continuous, and the solid has a boundary consisting of two smooth pieces, the inner and outer spheres. By Theorem 4.4.1, the integral exists.

The solid region S is contained in the cube $B = [-2, 2] \times [-2, 2] \times [-2, 2]$, so we form a partition by subdividing the x-, y-, and z-intervals into, respectively, n_1, n_2, and n_3 subintervals of equal length. This gives

$$\begin{aligned}
\Delta x &= \Delta x_i = \frac{2-(-2)}{n_1} = \frac{4}{n_1}, \qquad i = 1, \ldots, n_1 \\
\Delta y &= \Delta y_j = \frac{4}{n_2}, \qquad j = 1, \ldots, n_2 \\
\Delta z &= \Delta z_k = \frac{4}{n_3}, \qquad k = 1, \ldots, n_3
\end{aligned}$$

and

$$\begin{aligned}
x_i &= -2 + i\Delta x, \quad i = 1, \ldots, n_1 \\
y_j &= -2 + j\Delta y, \quad j = 1, \ldots, n_2 \\
z_k &= -2 + k\Delta z, \quad k = 1, \ldots, n_3.
\end{aligned}$$

We choose x_{ijk} to be the midpoint of $[x_{i-1}, x_i]$, and similarly, y_{ijk} and z_{ijk} are the midpoints of $[y_{j-1}, y_j]$ and $[z_{k-1}, z_k]$. We can modify the algorithm for double integrals shown in Figure 4.3.6 by adding another loop to take care of the z-coordinate. The result is the triple integral approximation algorithm shown in Figure 4.4.2. In Table 4.4.1 we see the approximations

```
a1 = -2
b1 = 2
a2 = -2
b2 = 2
a3 = -2
b3 = 2
f(x,y) = x*y - z^2
sum = 0
input n1, n2, n3
dx = (b1 - a1) / n1
dy = (b2 - a2) / n2
dz = (b3 - a3) / n3
x = a1 - dx / 2
for i = 1 to n1 do
     x = x + dx
     y = a2 - dy / 2
     for j = 1 to n2 do
          y = y + dy
          z = a3 - dz / 2
          for k = 1 to n3 do
               z = z + dz
               if (1 <= x^2 + y^2 + z^2)
                         and (x^2 + y^2 + z^2 <= 4)
                    then sum = sum + f(x,y)
output sum * dx * dy * dz
```

Figure 4.4.2 Algorithm to approximate $\iiint_S (xy - z^2)\,dV$, where S is the solid bounded by the spheres $x^2 + y^2 + z^2 = 1$ and $x^2 + y^2 + z^2 = 4$

obtained by using successively larger values of n_1, n_2, and n_3. The numbers tend toward -25.97. ♦

As for double integrals, triple integrals can often be evaluated as iterated integrals if the solid is of a simple form.

A solid $S \subset \mathbb{R}^3$ is z-**simple** if it is of the form $S = \{(x, y, z) \mid (x, y) \in R, g_1(x, y) \le z \le g_2(x, y)\}$ for some functions g_1 and g_2 defined on a region R in the xy-plane. Another way to describe such a solid is that its "sides" are part of a cylinder that is parallel to the z-axis. See Figure 4.4.3.

n1	n2	n3	Riemann sum
10	10	10	−28.631040
20	20	20	−26.267520
40	40	40	−26.015080
60	60	60	−25.954236
80	80	80	−25.970864
100	100	100	−25.996902
120	120	120	−25.979332
140	140	140	−25.960719
160	160	160	−25.965684
180	180	180	−25.971286
200	200	200	−25.970604

Table 4.4.1 Values generated by the approximation algorithm

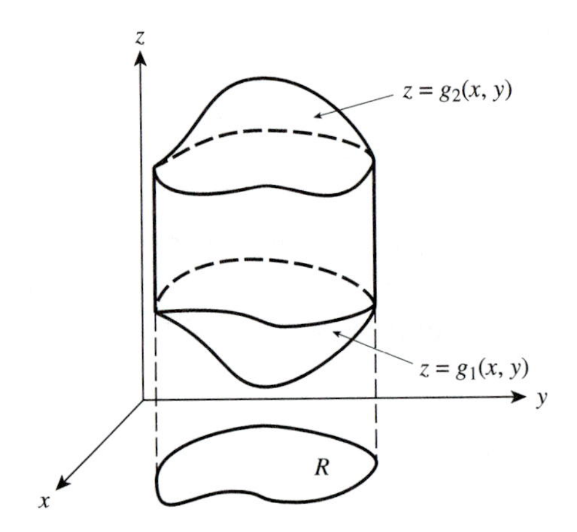

Figure 4.4.3 A typical z-simple solid

Similarly, we say that a solid is y-**simple** if it is of the form $S = \{(x,\, y,\, z)|(x,\, z) \in R,\, h_1(x,\, z) \le y \le h_2(x,\, z)\}$ for some functions h_1 and h_2 defined on a region R in the xz-plane. An x-simple solid is defined similarly.

Example 4.4.2 The intersection of the solid bounded by the sphere $x^2 + y^2 + z^2 = 9$ and the cylinder $x^2 + z^2 = 4$ is y-simple. See Figure 4.4.4. For any point $(x,\, y,\, z)$ in this solid, the y-coordinate is no less than that of the corresponding point $(x,\, -\sqrt{9 - x^2 - z^2},\, z)$ on the left side of the sphere and no more than that of the point $(x,\ \sqrt{9-x^2-z^2},\ z)$ on the right side. So

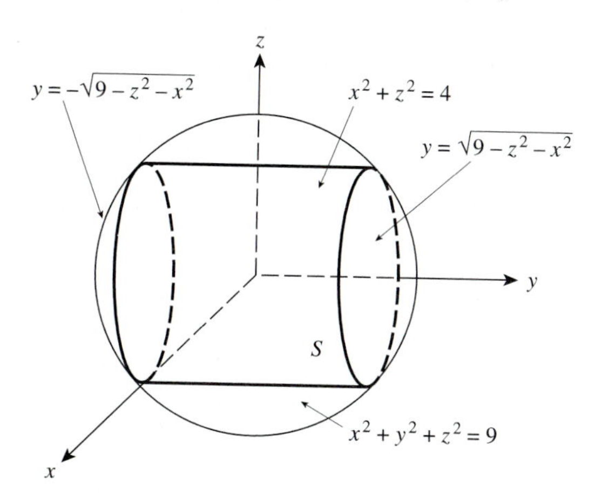

Figure 4.4.4 The solid inside both the sphere $x^2+y^2+z^2 = 9$ and the cylinder $x^2 + z^2 = 4$ is y-simple.

the solid is the set of points that satisfy the inequalities

$$-\sqrt{9-x^2-z^2} \le y \le \sqrt{9-x^2-z^2}, \quad (x,\ z) \in R,$$

where R is the disk $x^2 + z^2 \le 4$ in the xz-plane. ♦

When the solid is x-, y-, or z-simple, the triple integral can be written as an iterated integral. The reasoning that leads to this conclusion is directly analogous to the same result for double integrals. We summarize the facts in the following theorem.

Theorem 4.4.2 *If f is continuous on a solid $S \subset \mathbb{R}^3$ that is z-simple then*

$$\iiint\limits_S f\,dV = \iint\limits_R \left(\int_{g_1(x,y)}^{g_2(x,y)} f(x,y,z)\,dz \right) dA. \tag{4.4.3}$$

If S is y-simple, then

$$\iiint\limits_S f\,dV = \iint\limits_R \left(\int_{h_1(x,z)}^{h_2(x,z)} f(x,y,z)\,dy \right) dA, \tag{4.4.4}$$

and if S is x-simple, then

$$\iiint\limits_S f\,dV = \iint\limits_R \left(\int_{k_1(y,z)}^{k_2(y,z)} f(x,y,z)\,dx \right) dA. \tag{4.4.5}$$

Example 4.4.3 Evaluate the integral $\iiint\limits_S x\sin(y+z)\,dV$, where S is the solid bounded by the planes $x=0$, $x=2$, $y=0$, $y=\pi/2$, $z=0$, and $z=\pi/2$.

Solution The solid is just the box shown in Figure 4.4.5. It consists of the

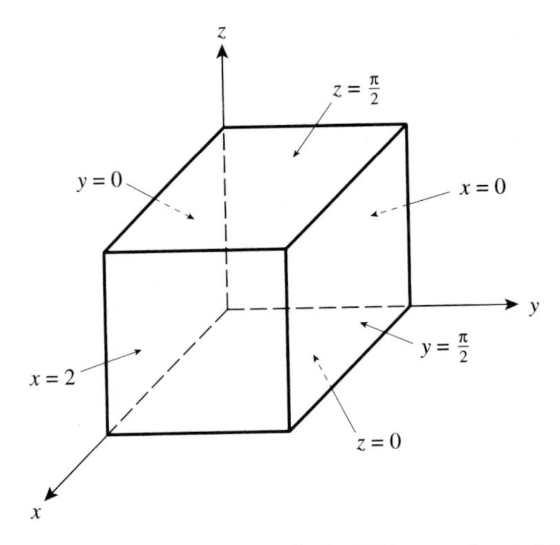

Figure 4.4.5 The solid of Example 4.4.3

points (x, y, z) that satisfy

$$\begin{aligned} 0 \le x &\le 2, \\ 0 \le y &\le \frac{\pi}{2}, \\ 0 \le z &\le \frac{\pi}{2}. \end{aligned}$$

Thus

$$\begin{aligned} \iiint\limits_S x\sin(y+z)\,dV &= \int_0^2 \int_0^{\pi/2} \left(\int_0^{\pi/2} x\sin(y+z)\,dz \right) dy\,dx \\ &= \int_0^2 \int_0^{\pi/2} x\,[\sin y + \cos y]\;dy\,dx \\ &= \int_0^2 \int_0^{\pi/2} x[\sin y + \cos y]\,dy\,dx \\ &= \int_0^2 x(-\cos y + \sin y)\Big|_0^{\pi/2}\;dx \end{aligned}$$

$$= \int_0^2 2x\,dx = 4. \quad \blacklozenge$$

Example 4.4.4 Evaluate $\iiint\limits_S f\,dV$, where $f(x, y, z) = xy + yz$ and S is the "unit tetrahedron" bounded by the coordinate planes and the plane $x + y + z = 1$.

Solution The solid is evidently x-, y-, and z-simple. See Figure 4.4.6. We

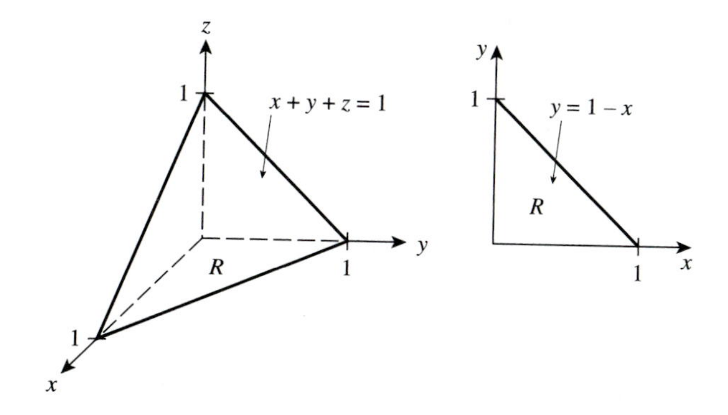

Figure 4.4.6 The unit tetrahedron as a z-simple solid

choose to treat it as z-simple; it consists of the points (x, y, z) satisfying

$$0 \leq z \leq 1 - x - y, \quad (x, y) \in R,$$

where R is the triangle in the xy-plane bounded by the coordinate axes and the line $x + y = 1$. Treating R as y-simple, we can describe it by the inequalities

$$\begin{aligned} 0 \leq y &\leq 1 - x \\ 0 \leq x &\leq 1. \end{aligned}$$

Thus

$$\iiint\limits_S (xy + yz)\,dV$$

$$
\begin{aligned}
&= \int_0^1 \int_0^{1-x} \int_0^{1-x-y} (xy + yz)\, dz\, dy\, dx \\
&= \int_0^1 \int_0^{1-x} \left[xy(1-x-y) + \frac{1}{2} y(1-x-y)^2 \right] dy\, dx \\
&= \int_0^1 \int_0^{1-x} \left[\frac{1}{2} y^3 - y^2 + \frac{1}{2} y - \frac{1}{2} x^2 y \right] dy\, dx \\
&= \int_0^1 \left[\frac{1}{8}(1-x)^4 - \frac{1}{3}(1-x)^3 + \frac{1}{4}(1-x)^2 - \frac{1}{4} x^2 (1-x)^2 \right] dx \\
&= \left[-\frac{1}{40}(1-x)^5 + \frac{1}{12}(1-x)^4 - \frac{1}{12}(1-x)^3 - \frac{1}{4}\left(\frac{1}{3}x^3 - \frac{1}{2}x^4 + \frac{1}{5}x^5 \right) \right] \Bigg|_0^1 \\
&= 0 - \frac{1}{4}\left(\frac{1}{3} - \frac{1}{2} + \frac{1}{5} \right) - \left(-\frac{1}{40} + \frac{1}{12} - \frac{1}{12} \right) \\
&= \frac{1}{40} + \frac{1}{120} = \frac{1}{30}. \quad \blacklozenge
\end{aligned}
$$

One motivation for the definition of the triple integral was that if $f(x, y, z)$ represents the density of the solid S at a point (x, y, z), then the mass of the solid is approximated by the Riemann sums in (4.4.1). Thus the **mass** of a solid S with density function f is defined to be $\iiint_S f\, dV$.

Another interpretation for triple integrals is volume. Since the Riemann sums for the constant function 1 on a solid S represent the total volume of partitions that approximate S, the **volume** of S is the limit of these partition volumes, that is,

$$
V(S) = \iiint_S 1\, dV = \iiint_S dV. \tag{4.4.6}
$$

Example 4.4.5 Find an expression for the mass of the solid bounded by the sphere $x^2 + y^2 + z^2 = 1$ whose density at each point is proportional to the distance from the origin.

Solution Since the density is proportional to the distance from the origin, it has the form $f(x, y, z) = k\sqrt{x^2 + y^2 + z^2}$ for some constant k. Now the solid is x-simple, and we can describe it by

$$
-\sqrt{1 - y^2 - z^2} \le x \le \sqrt{1 - y^2 - z^2}, \quad (y, z) \in R,
$$

where R is the unit disk in the yz-plane. But R is given by the inequalities

$$
-\sqrt{1 - z^2} \le y \le \sqrt{1 - z^2}, \quad -1 \le z \le 1,
$$

so the mass is given by the triple integral

$$\iiint_S f(x, y, z)\, dV = \int_{-1}^{1} \int_{-\sqrt{1-z^2}}^{\sqrt{1-z^2}} \int_{-\sqrt{1-y^2-z^2}}^{\sqrt{1-y^2-z^2}} k\sqrt{x^2 + y^2 + z^2}\, dx\, dy\, dz.$$

We could evaluate this integral by brute force using integral tables and a calculator, but we will instead wait until Section 4.8 where we transform coordinates. ♦

Triple integrals possess all the properties that we listed for double integrals. These are summarized in the following theorem.

Theorem 4.4.3 *Let f and g be integrable functions, and let c be a constant.*

1. $\displaystyle\iiint_S (f + g)\, dV = \iiint_S f\, dV + \iiint_S g\, dV.$

2. $\displaystyle\iiint_S cf\, dV = c \iiint_S f\, dV$

3. $\displaystyle\iiint_{S_1 \cup S_2} f\, dV = \iiint_{S_1} f\, dV + \iiint_{S_2} f\, dV,$ *where* $S_1 \cap S_2 = \emptyset$.

4. *(Mean Value Theorem for Triple Integrals) If f is continuous on S, then there exists a point $(x_0, y_0, z_0) \in S$ such that*
$$\iiint_S f\, dV = f(x_0, y_0, z_0)V(S).$$

5. *If $f(x, y, z) \le g(x, y, z)$ for $(x, y, z) \in S$, then* $\displaystyle\iiint_S f\, dV \le \iiint_S g\, dV.$

6. *If f is continuous and satisfies $f(x, y, z) > 0$ for all $(x, y, z) \in S$, where S has a nonempty interior, then* $\displaystyle\iiint_S f\, dV > 0.$

Example 4.4.6 Without calculating the integral, explain why

$$\iiint_S (x + y + z)\, dV = 0,$$

where S is the solid between the spheres $x^2+y^2+z^2=1$ and $x^2+y^2+z^2=4$.

Solution The integrand $f(x, y, z) = x + y + z$ is an "odd" function in that $f(-x, -y, -z) = -x - y - z = -(x + y + z) = -f(x, y, z)$. If we slice S into halves S_1 and S_2 with a plane through the origin, then for each point (x, y, z) in S_2 there corresponds exactly one point $(-x, -y, -z)$ in S_1. From this it follows, by the oddness of the integrand, f, and property 2 in Theorem 4.4.3 that

$$\begin{aligned}\iiint\limits_{S_2} f(x, y, z)\, dV &= \iiint\limits_{S_1} f(-x, -y, -z)\, dV = \iiint\limits_{S_1} -f(x, y, z)\, dV \\ &= -\iiint\limits_{S_1} f(x, y, z)\, dV.\end{aligned}$$

By property 3 in Theorem 4.4.3,

$$\begin{aligned}\iiint\limits_{S}(x+y+z)\, dV &= \iiint\limits_{S_1}(x+y+z)\, dV + \iiint\limits_{S_2}(x+y+z)\, dV \\ &= \iiint\limits_{S_1}(x+y+z)\, dV - \iiint\limits_{S_1}(x+y+z)\, dV = 0. \quad \blacklozenge\end{aligned}$$

Exercises 4.4

In Exercises 1 through 3, use the algorithm in Figure 4.4.2 to find an approximation for the integrals that you feel is accurate to two significant digits.

1. $\displaystyle\iiint\limits_{S} \sin(x^2 + y^2 + z^2)\, dV$; S is the box bounded by the planes $x = 0$, $x = 2$, $y = -\pi/2$, $y = \pi$, $z = 0$, $z = 1$.

2. $\displaystyle\iiint\limits_{S} e^{-(x^2+y^2+z^2)}\, dV$; S is the ball of radius 1 centered at the origin.

3. $\displaystyle\iiint\limits_{S} xyz\, dV$; S is the solid bounded by the cone $z = \sqrt{x^2 + y^2}$ and the paraboloid $z = 1 - x^2 - y^2$.

In Exercises 4 through 10, evaluate the triple iterated integral and describe the solid region over which you are integrating. In some instances, you may find it helpful to consult an integral table or use software that does symbolic integration.

4. $\displaystyle\int_0^1\int_1^2\int_2^3 \frac{ze^x}{y}\,dz\,dy\,dx$

5. $\displaystyle\int_1^3\int_0^5\int_{-1}^0 \left(\frac{1}{x}+y+z^2\right)dz\,dy\,dx$

6. $\displaystyle\int_{-1}^1\int_{-1}^1\int_x^{y+1} xy\,dz\,dx\,dy$

7. $\displaystyle\int_0^3\int_0^{6-2y}\int_0^{(6-2y-x)/3} y\,dz\,dx\,dy$

8. $\displaystyle\int_0^1\int_0^1\int_0^{1+x^2} xyz\,dz\,dx\,dy$

9. $\displaystyle\int_0^1\int_0^x\int_0^{1+x+y} dz\,dy\,dx$

10. $\displaystyle\int_0^2\int_0^{\sqrt{4-x^2}}\int_0^{\sqrt{4-x^2-y^2}} xy\,dz\,dy\,dx$

In Exercises 11 through 14, evaluate the triple integral of the given function over the indicated solid region. In some instances, to evaluate the integral you set up, you may find it helpful to consult an integral table or use software that does symbolic integration.

11. $f(x, y, z) = \sqrt{x+y+z}$; S is the solid bounded by the xy- and xz-planes, and the planes $y = 2$, $x = 1$, $x = 4$, and $z = 5$.

12. $f(x, y, z) = x$; S is the solid bounded by the paraboloids $z = x^2 + y^2$ and $z = 1 - (x^2 + y^2)$.

13. $f(x, y, z) = xy$; S is the solid bounded by the coordinate planes and the plane $x + 2y + 4z = 8$.

14. $f(x, y, z) = x^2$; S is the solid bounded by the paraboloids $z = x^2 + y^2$ and $z = 8 - x^2 - y^2$. (*Hint*: an integration by parts may be necessary somewhere along the way.)

In Exercises 15 through 17, find the volume of the indicated solid region by evaluating a triple integral.

15. The tetrahedron bounded by the coordinate planes and the plane $x + y + z = 1$.

16. A sphere of radius r.

17. The solid inside the cylinders $x^2+y^2=a^2$ and $x^2+z^2=a^2$, where a is any positive number.

In Exercises 18 through 20, find the mass of the solid with the given density.

18. S is the cube bounded by the planes $x=\pm 1$, $y=\pm 1$, $z=\pm 1$ and the density at a point is proportional to the distance from the point to the xy - plane.

19. S is the tetrahedron in Exercise 15 and the density is xyz.

20. S is the solid bounded by the planes $x=1$, $x=2$, $y=1$, $y=2$, $z=x+y$ and $z=-2x-3y$; the density is proportional to the distance from the xy-plane.

In each of the triple integrals in Exercises 21 through 24, the solid of integration is x-, y-, and z-simple. Change the order of integration to represent the integral in terms of the other two types of solids.

21. $\int_{-1}^{0}\int_{0}^{1}\int_{1}^{2} f(x, y, z)\, dy\, dz\, dx.$

22. $\int_{0}^{2}\int_{0}^{4-2x}\int_{0}^{8-4x-2y} f(x, y, z)\, dz\, dy\, dx.$

23. $\int_{0}^{2}\int_{0}^{\sqrt{1-z^2/4}}\int_{0}^{3\sqrt{1-x^2-z^2/4}} f(x, y, z)\, dy\, dx\, dz.$

24. $\int_{0}^{2}\int_{x}^{2}\int_{0}^{5} f(x, y, z)\, dz\, dy\, dx.$

Using Theorem 4.4.3, determine whether the following integrals are positive, negative, or zero.

25. $\iiint\limits_S x^3\, dV$; S is the cube $[-4, 4]\times[-4, 4]\times[-4, 4]$.

26. $\iiint\limits_S z\, dV$; S is the solid bounded by the paraboloid $z=-(x^2+y^2)$ and the plane $z=-4$.

27. $\iiint\limits_S (2x-3y+8z)\, dV$; S is the ball of radius 5 centered at the origin.

28. $\iiint\limits_S \sqrt{x^2+y^2+z^2}\, dV$; S is the ball of radius 12 centered at the point $(5, -3, -7)$.

29. Prove that if f is continuous, $f(x, y, z) \geq 0$ for all (x, y, z) in a solid region S, and $\iiint\limits_S f\, dV = 0$, then $f(x, y, z) = 0$ for all $(x, y, z) \in S$.

4.5 Parametrized Surfaces and Surface Area

To this point we have generally regarded a "surface" as the graph of a function of two variables or the graph of an equation in three variables. However, our ordinary use of the term *surface* compels us to call the object in Figure 4.5.1 by this name, even though it is not the graph of a function of two variables. The helicoid in Figure 4.5.1 was generated on a computer by using a **parametrization**. Analogous to the way in which we parametrize curves, we specify the points (x, y, z) on the surface by defining each coordinate as a function of two variables, s and t. As these parameters range over their prescribed values, the point (x, y, z) "traces out" the surface in space. The computer automatically "hides" the portions of the surface that are obscured by other parts.

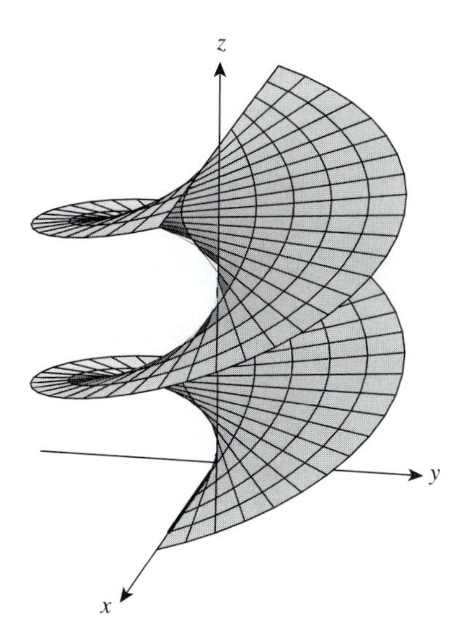

Figure 4.5.1 A parametrized surface

Our objective in this section is to build some intuition for parametrizations of surfaces and to come up with a reasonable formula for the area of such surfaces. This will prepare the way for our discussion of surface integrals in the next section.

Given a function $\mathbf{f}: R \subset \mathbb{R}^2 \to \mathbb{R}^3$, we define the **surface parametrized by f** to be the set of points $M = \{\mathbf{f}(s, t) | (s, t) \in R\}$. That is, M is the image of R under $\mathbf{f}$. The equation $\mathbf{x} = \mathbf{f}(s, t)$ is a **parametrization** of M. We say that the parametrization is **smooth** if the Jacobian matrix

$$J\mathbf{f}(s, t) = \begin{bmatrix} \partial f_1/\partial s & \partial f_1/\partial t \\ \partial f_2/\partial s & \partial f_2/\partial t \\ \partial f_3/\partial s & \partial f_3/\partial t \end{bmatrix}$$

has continuous entries and the vector $\mathbf{n} = \partial\mathbf{f}/\partial s \times \partial\mathbf{f}/\partial t$ is never zero. The requirement that $J\mathbf{f}(s, t)$ be continuous is to ensure a continuously varying normal $\mathbf{n}$ to the surface, and the nonvanishing cross product is to make certain that the normal never becomes zero.

In Figure 4.5.2, for instance, is the surface parametrized by

$$\mathbf{x} = (\sin^3 s \cos^3 t,\ \sin^3 s \sin^3 t,\ \cos^3 s) \tag{4.5.1}$$

where $0 \le s \le \pi$ and $0 \le t \le 2\pi$. It is not smooth because, for example,

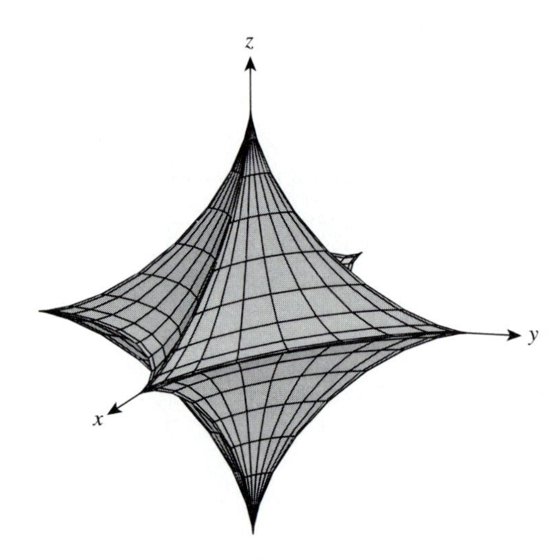

Figure 4.5.2 The surface parametrized by $\mathbf{x} = (\sin^3 s \cos^3 t,\ \sin^3 s \sin^3 t,\ \cos^3 s)$ is not smooth.

$\partial\mathbf{f}/\partial s \times \partial\mathbf{f}/\partial t$ vanishes when $s = \pi/2$ and $t = \pi/2$; this is the sharp point on the surface at the point $(0, 1, 0)$.

Example 4.5.1 Determine a parametrization for the **helicoid** shown in Figure 4.5.1.

Solution The surface appears to be generated, or swept out, by a "stick" that rotates counterclockwise about its center at a constant rate as that center slides upward at a constant rate along a "string" coinciding with the z-axis. The two ends of the stick trace out helical curves that are 180° out of phase with one another. Our idea is first to find parametrizations for these curves, and then with a second variable, to parametrize the line segment between them.

One end of the stick starts out at the point (1, 0, 0) and winds twice around the z-axis as it rises. Recalling similar helixes in Section 1.8, we see that it is parametrized by

$$\mathbf{x} = (\cos t,\ \sin t,\ kt) \quad 0 \le t \le 4\pi,$$

for some choice of the constant k. The other end of the stick starts at $(-1, 0, 0)$ and winds twice about the z-axis also; its z-coordinate is the same as that of the first end, but its rotation is "delayed" by π units. So the second end is parametrized by

$$\begin{aligned} \mathbf{x} &= (\cos(t-\pi),\ \sin(t-\pi),\ kt) \\ &= (-\cos t,\ -\sin t,\ kt), \quad 0 \le t \le 4\pi. \end{aligned}$$

For each value of t in $[0, 4\pi]$, the "stick" itself is the line segment from $(-\cos t,\ -\sin t,\ kt)$ to $(\cos t,\ \sin t,\ kt)$. A direction vector for this line is $(\cos t,\ \sin t,\ kt) - (-\cos t, -\sin t,\ kt) = (2\cos t,\ 2\sin t,\ 0)$, and the segment is then given by

$$\begin{aligned} \mathbf{x} &= (-\cos t,\ -\sin t,\ kt) + s(2\cos t,\ 2\sin t,\ 0) \\ &= ((2s-1)\cos t,\ (2s-1)\sin t,\ kt)\,, \quad 0 \le s \le 1. \end{aligned}$$

Now we have a parametrization for the helix, and by adjusting the constant k to be larger or smaller we can stretch out or compress the surface. Try this out on your computer! ♦

Example 4.5.2 Find two different parametrizations of the upper half of the sphere $x^2 + y^2 + z^2 = 1$.

Solution The upper half of the sphere is just the graph of $z = \sqrt{1 - x^2 - y^2}$, where x and y satisfy the inequalities

$$\begin{aligned} -\sqrt{1-x^2} \ \le\ &y \ \le\ \sqrt{1-x^2} \\ -1 \ \le\ &x \ \le\ 1. \end{aligned}$$

By setting $x = s$ and $y = t$, we have $z = \sqrt{1 - s^2 - t^2}$, so the surface is parametrized (in a sort of trivial way) by

$$\mathbf{x} = \left(s,\, t,\, \sqrt{1 - s^2 - t^2}\right),$$

where t ranges over $\left[-\sqrt{1 - s^2},\, \sqrt{1 - s^2}\right]$ for each s in the interval $[-1,\, 1]$.

Another way to parametrize the hemisphere is to use the transformations from spherical to rectangular coordinates (see Section 1.3). Each point on this hemisphere has ρ-coordinate 1, θ-coordinate in the range $[0,\, 2\pi]$, and ϕ-coordinate in the range $[0,\, \pi/2]$. Putting this value of ρ into (1.3.7), (1.3.8), and (1.3.9), we have

$$x = 1 \cdot \sin\phi\cos\theta \tag{4.5.2}$$
$$y = 1 \cdot \sin\phi\sin\theta \tag{4.5.3}$$
$$z = 1\cos\phi. \tag{4.5.4}$$

That is, the hemisphere is parametrized by

$$\mathbf{x} = (\sin\phi\cos\theta,\, \sin\phi\sin\theta,\, \cos\phi),$$

$$\begin{array}{ccccc} 0 & \leq & \theta & < & 2\pi \\ 0 & \leq & \phi & \leq & \pi/2. \end{array} \quad \blacklozenge$$

To obtain a reasonable definition of the area of a surface M that is parametrized by $\mathbf{x} = \mathbf{f}(s,\, t)$, where $(s,\, t) \in R \subset \mathbb{R}^2$, we reason as follows. If we partition R into many small rectangles, then M is partitioned into many pieces, each of which is the image under $\mathbf{f}$ of one of these small rectangles. See Figure 4.5.3.

If $\mathbf{f}$ is differentiable on R, then on each of these small rectangles, $\mathbf{f}$ has a good linear approximation, so the image of a small rectangle under $\mathbf{f}$ closely resembles the image of the same rectangle under the linear approximation. See Figure 4.5.4. Since the images under the linear approximation are parallelograms, we need simply to sum their areas to approximate the area of M. If we do it right, we will obtain a Reimann sum, which will lead us to an integral for the surface area of M.

Let $R_{ij} = [s_{i-1},\, s_i] \times [t_{j-1},\, t_j]$ be a typical partition rectangle for R. Now the linear approximation of $\mathbf{f}$ at $(s_{i-1},\, t_{j-1})$ is

$$\mathbf{L}(s,\, t) = \mathbf{f}(s_{i-1},\, t_{j-1}) + \mathbf{Df}(s_{i-1},\, t_{j-1})\,(s - s_{i-1},\, t - t_{j-1})$$

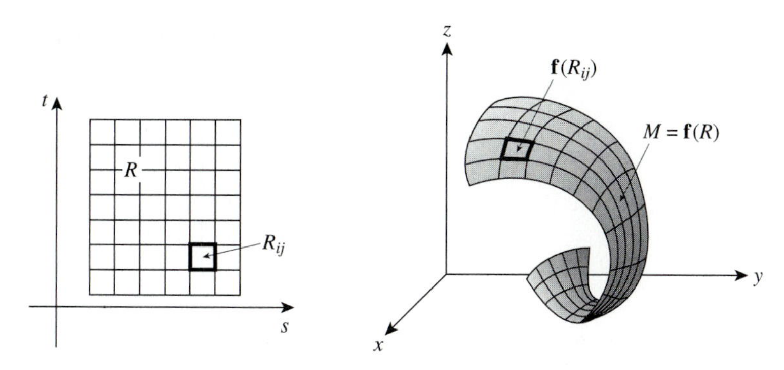

Figure 4.5.3 A region R in the st-plane and its image surface under $\mathbf{f}$ in $\mathbb{R}^3$

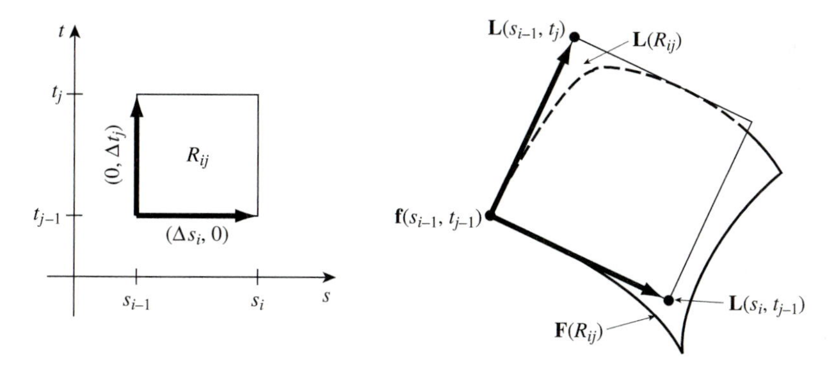

Figure 4.5.4 The image of a small rectangle under $\mathbf{f}$ closely resembles the image of the rectangle under the linear approximation to $\mathbf{f}$.

$$= \mathbf{f}(s_{i-1}, t_{j-1}) + \begin{bmatrix} \partial f_1/\partial s & \partial f_1/\partial t \\ \partial f_2/\partial s & \partial f_2/\partial t \\ \partial f_3/\partial s & \partial f_3/\partial t \end{bmatrix} \begin{bmatrix} s - s_{i-1} \\ t - t_{j-1} \end{bmatrix},$$

where we understand that the partials in the matrix are all evaluated at (s_{i-1}, t_{j-1}). The image of R_{ij} under $\mathbf{L}$ is the parallelogram spanned by the image of the vectors

$$(s_i, t_{j-1}) - (s_{i-1}, t_{j-1}) = (s_i - s_{i-1}, 0) = (\Delta s_i, 0)$$

and

$$(s_{i-1}, t_j) - (s_{i-1}, t_{j-1}) = (0, t_j - t_{j-1}) = (0, \Delta t_i)$$

under $\mathbf{L}-\mathbf{f}$. Again, see Figure 4.5.4. That is, $\mathbf{L}(R_{ij})$ is the parallelogram spanned by

$$\begin{aligned}
\mathbf{L}(s_i, t_{j-1}) - \mathbf{f}(s_{i-1}, t_{j-1}) &= \begin{bmatrix} \partial f_1/\partial s & \partial f_1/\partial t \\ \partial f_2/\partial s & \partial f_2/\partial t \\ \partial f_3/\partial s & \partial f_3/\partial t \end{bmatrix} \begin{bmatrix} s_i - s_{i-1} \\ 0 \end{bmatrix} \\
&= \begin{bmatrix} (s_i - s_{i-1})\partial f_1/\partial s \\ (s_i - s_{i-1})\partial f_2/\partial s \\ (s_i - s_{i-1})\partial f_3/\partial s \end{bmatrix} = \Delta s_i \begin{bmatrix} \partial f_1/\partial s \\ \partial f_2/\partial s \\ \partial f_3/\partial s \end{bmatrix} \\
&\equiv \Delta s_i \frac{\partial \mathbf{f}}{\partial s}
\end{aligned}$$

and

$$\begin{aligned}
\mathbf{L}(s_{i-1}, t_j) - \mathbf{f}(s_{i-1}, t_{j-1}) &= \begin{bmatrix} (t_j - t_{j-1})\partial f_1/\partial t \\ (t_i - t_{j-1})\partial f_2/\partial t \\ (t_i - t_{j-1})\partial f_3/\partial t \end{bmatrix} = \Delta t_j \begin{bmatrix} \partial f_1/\partial t \\ \partial f_2/\partial t \\ \partial f_3/\partial t \end{bmatrix} \\
&\equiv \Delta t_j \frac{\partial \mathbf{f}}{\partial t}.
\end{aligned}$$

The area of this parallelogram is, by our work in Section 1.6,

$$\left\| \Delta s_i \frac{\partial \mathbf{f}}{\partial s} \times \Delta t_i \frac{\partial \mathbf{f}}{\partial t} \right\| = \left\| \frac{\partial \mathbf{f}}{\partial s} \times \frac{\partial \mathbf{f}}{\partial t} \right\| \Delta s_i \, \Delta t_j. \tag{4.5.5}$$

[Remember that the partials here are all evaluated at (s_{i-1}, t_{j-1}).] Adding up the expressions in (4.5.5) for each partition rectangle, we obtain

$$\sum_{R_{ij}} \left\| \frac{\partial \mathbf{f}}{\partial s} \times \frac{\partial \mathbf{f}}{\partial t} \right\| \Delta s_i \, \Delta t_j,$$

which is a Riemann sum for the double integral $\iint_R \left\| \frac{\partial \mathbf{f}}{\partial s} \times \frac{\partial \mathbf{f}}{\partial t} \right\| ds\, dt$. A different parametrization of M could conceivably yield a different value for this integral, but in fact under these condition it does not: the value of this integral is independent of the particular parametrization of M we use.

Definition 4.5.1 Let M be a surface in $\mathbb{R}^3$ that is parametrized by $\mathbf{x} = \mathbf{f}(s, t)$, where $(s, t) \in R \subset \mathbb{R}^2$, and $\mathbf{f}$ has continuous partials. The **surface area of** M, denoted by $\sigma(M)$, is

$$\sigma(M) = \iint_R \left\| \frac{\partial \mathbf{f}}{\partial s} \times \frac{\partial \mathbf{f}}{\partial t} \right\| ds\, dt. \tag{4.5.6}$$

Example 4.5.3 Find the surface area of the helicoid given in Example 4.5.1 when $k = \sqrt{3}$.

Solution We first calculate

$$\begin{aligned} \frac{\partial \mathbf{f}}{\partial s} &= (2\cos t,\ 2\sin t,\ 0) \\ \frac{\partial \mathbf{f}}{\partial t} &= (-(2s-1)\sin t,\ (2s-1)\cos t,\ k). \end{aligned}$$

Then (check this)

$$\frac{\partial \mathbf{f}}{\partial s} \times \frac{\partial \mathbf{f}}{\partial t} = (2k\sin t,\ -2k\cos t,\ 2(2s-1))$$

and

$$\begin{aligned} \left\| \frac{\partial \mathbf{f}}{\partial s} \times \frac{\partial \mathbf{f}}{\partial t} \right\| &= \sqrt{4k^2\sin^2 t + 4k^2\cos^2 t + 4(2s-1)^2} \\ &= 2\sqrt{k^2 + (2s-1)^2}. \end{aligned}$$

By Definition 4.5.1, the area is

$$\begin{aligned} \sigma(M) &= \int_0^1 \int_0^{4\pi} 2\sqrt{k^2 + (2s-1)^2}\, dt\, ds \\ &= 8\pi \int_0^1 \sqrt{k^2 + (2s-1)^2}\, ds \\ &= 8\pi \cdot \frac{1}{4}\left(2\sqrt{1+k^2} + k^2 \ln\left(\frac{\sqrt{1+k^2}+1}{\sqrt{1+k^2}-1}\right)\right). \end{aligned}$$

In the last step, we have used a computer algebra system to do the integration; a table would work as well. When $k = \sqrt{3}$, this becomes

$$\sigma(M) = 2\pi(4 + 3\ln 3). \quad \blacklozenge$$

Example 4.5.4 Find the surface area of a sphere S of radius R.

Solution You undoubtedly have been told that the surface area of a sphere is $4\pi R^2$, but you may not have known *why*.

The sphere is parametrized by

$$\mathbf{x} = (R\sin\phi\cos\theta,\ R\sin\phi\sin\theta,\ R\cos\phi),$$

where $0 \le \theta \le 2\pi$ and $0 \le \phi \le \pi$. We have

$$\begin{aligned}\frac{\partial \mathbf{x}}{\partial \theta} &= (-R\sin\phi\sin\theta,\ R\sin\phi\cos\theta, 0)\\ \frac{\partial \mathbf{x}}{\partial \phi} &= (R\cos\phi\cos\theta,\ R\cos\phi\sin\theta, -R\sin\phi),\end{aligned}$$

so

$$\frac{\partial \mathbf{x}}{\partial \theta} \times \frac{\partial \mathbf{x}}{\partial \phi} = \left(-R^2\sin^2\phi\cos\theta,\ -R^2\sin^2\phi\sin\theta,\ -R^2\sin\phi\cos\phi\right)$$

and

$$\begin{aligned}\left\|\frac{\partial \mathbf{x}}{\partial \theta} \times \frac{\partial \mathbf{x}}{\partial \phi}\right\| &= R^2\sqrt{\sin^4\phi\cos^2\theta + \sin^4\phi\sin^2\theta + \sin^2\phi\cos^2\phi}\\ &= R^2\sqrt{\sin^4\phi + \sin^2\phi\cos^2\phi}\\ &= R^2\sin\phi\sqrt{\sin^2\phi + \cos^2\phi}\\ &= R^2\sin\phi.\end{aligned}$$

Thus the surface area is

$$\sigma(S) = \int_0^{2\pi}\int_0^{\pi} R^2\sin\phi\, d\phi\, d\theta = R^2\int_0^{2\pi} 2\, d\theta = 2R^2\cdot 2\pi = 4\pi R^2. \quad \blacklozenge$$

Whenever a surface is the graph of a function of two variables, then (4.5.6) has a simpler form. For instance, if the surface is given by $z = h(x, y)$ for $(x, y) \in R$, then the natural parametrization of the surface is $x = s$, $y = t$, $z = h(s, t)$; that is,

$$\mathbf{x} = \mathbf{f}(s, t) = (s, t, h(s, t)).$$

Then

$$\begin{aligned}\left\|\frac{\partial \mathbf{f}}{\partial s}\times\frac{\partial \mathbf{f}}{\partial t}\right\| &= \left\|\left(1,\,0,\,\frac{\partial h}{\partial s}\right)\times\left(0,\,1,\,\frac{\partial h}{\partial t}\right)\right\| \\ &= \left\|\left(\frac{-\partial h}{\partial s},\,\frac{-\partial h}{\partial t},\,1\right)\right\| \\ &= \sqrt{\left(\frac{\partial h}{\partial s}\right)^2+\left(\frac{\partial h}{\partial t}\right)^2+1} \\ &= \sqrt{\left(\frac{\partial h}{\partial x}\right)^2+\left(\frac{\partial h}{\partial y}\right)^2+1}.\end{aligned}$$

So the surface area in this case is

$$\sigma(M)=\iint\limits_R \sqrt{\left(\frac{\partial h}{\partial x}\right)^2+\left(\frac{\partial h}{\partial y}\right)^2+1}\,dy\,dx. \qquad (4.5.7)$$

Example 4.5.5 Find the area of the portion of the surface $z = 1-\frac{2}{3}y^{3/2}$ that lies above the triangle in the xy-plane with vertices $(0,\,0)$, $(1,\,0)$, and $(0,\,1)$.

Solution The surface is the graph of $h(x,\,y) = 1-\frac{2}{3}y^{3/2}$, so we use (4.5.7). First, we have $\partial z/\partial x = 0$ and $\partial z/\partial y = -y^{1/2}$, so the integrand will just be

$$\sqrt{0^2+(-y^{1/2})^2+1}=\sqrt{1+y}.$$

The triangle is bounded by the coordinate axes and the line $y = 1-x$, so it is described by the inequalities

$$\begin{aligned}0 \;\leq\; y \;&\leq\; 1-x\\ 0 \;\leq\; x \;&\leq\; 1.\end{aligned}$$

By (4.5.7),

$$\begin{aligned}\sigma(M) &= \int_0^1\int_0^{1-x}\sqrt{1+y}\,dy\,dx = \int_0^1 \frac{2}{3}(1+y)^{3/2}\Big|_0^{1-x}\,dx \\ &= \frac{2}{3}\int_0^1\left((2-x)^{3/2}-1\right)dx = \frac{2}{3}\left[-\frac{2}{5}(2-x)^{5/2}-x\right]\Big|_0^1 \\ &= \frac{2}{3}\left[-\frac{2}{5}-1-\left(-\frac{2}{5}2^{5/2}\right)\right] = \frac{2}{15}\left[2^{7/2}-7\right].\end{aligned}$$ ♦

A surface M in $\mathbb{R}^3$ is said to be **orientable** if a unit normal vector $\mathbf{n}$ can be constructed at each point (x, y, z) on M so that $\mathbf{n}(x, y, z)$ is a continuous function. This definition is an attempt to capture the distinction between a surface with "two sides"—such as a parallelogram or a sphere—and a surface with only one side. The famed **Möbius strip** is an example of such a "one-sided" surface. A physical model for it can be constructed by putting a half-twist in a strip of paper and then taping the two ends of the strip together. See Figure 4.5.5. If we select a point on the strip and a

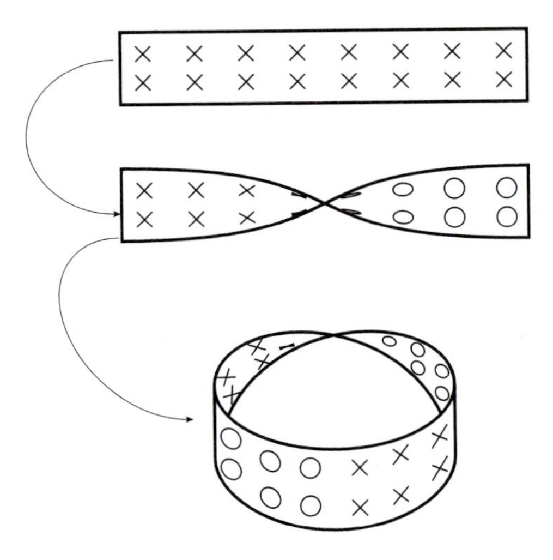

Figure 4.5.5 Steps in constructing a model for the Möbius strip

unit vector perpendicular to the surface at that point, we should be able to follow any path along the surface emanating from that point and have the corresponding unit normal vary continuously. Take a look at Figure 4.5.6. If we start at the point P_1 and call the unit normal at that point $\mathbf{n}$, then as we move along the curve shown through P_2, P_3, and P_4, we see the only choices for a continuously varying unit normal at these points are $\mathbf{n}_2$, $\mathbf{n}_3$, and $\mathbf{n}_4$. This leads us to $\mathbf{n}_5$ for the normal at P_1, contrary to our original choice of $\mathbf{n}_1$. Thus the surface is not orientable.

If M is a smooth orientable surface, then any smooth parametrization of it, $\mathbf{x} = \mathbf{f}(s, t)$, provides an orientation: The unit normal given by

$$\mathbf{n} = \frac{\dfrac{\partial \mathbf{f}}{\partial s} \times \dfrac{\partial \mathbf{f}}{\partial t}}{\left\| \dfrac{\partial \mathbf{f}}{\partial s} \times \dfrac{\partial \mathbf{f}}{\partial t} \right\|}$$

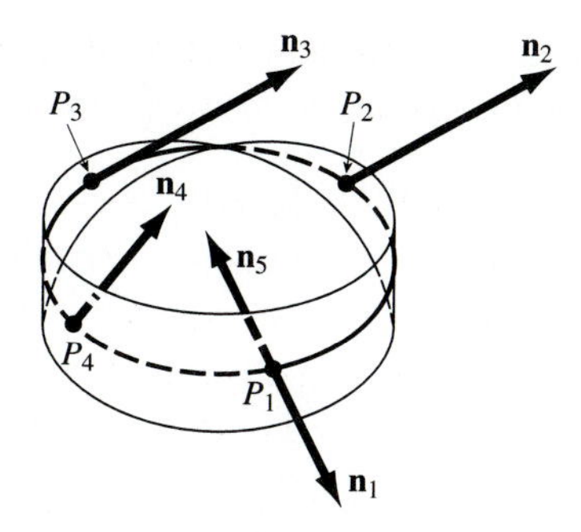

Figure 4.5.6 The Möbius strip is not orientable because any attempt to define a continuous unit normal to the surface forces us to conclude that $\mathbf{n}_1 = \mathbf{n}_5$.

is a continuous function of the points on M. Any parametrization then induces one of two possible orientations of the surface.

A special case of Definition 4.5.1 gives us a formula for the area of a region in a plane. Suppose, for example, that $\mathbf{f}(s, t) = (f_1(s, t), f_2(s, t), 0)$. Then

$$\begin{aligned} \left\| \frac{\partial \mathbf{f}}{\partial s} \times \frac{\partial \mathbf{f}}{\partial t} \right\| &= \|(0, 0, (f_1)_s(f_2)_t - (f_2)_s(f_1)_t)\| \\ &= \left| \det \begin{bmatrix} (f_1)_s & (f_1)_t \\ (f_2)_s & (f_2)_t \end{bmatrix} \right| \\ &= \left| \frac{\partial(f_1, f_2)}{\partial(s, t)} \right|. \end{aligned}$$

Then Definition 4.5.1 says that the area of the image M of R under a mapping (f_1, f_2) from R to $\mathbb{R}^2$ is

$$A(M) = \iint\limits_R \left| \frac{\partial(f_1, f_2)}{\partial(s, t)} \right| ds\, dt. \tag{4.5.8}$$

Does this formula for the area of a region agree with the definition we gave in Section 4.3? Yes! A region M in $\mathbb{R}^2$ is the image of itself under the identity mapping $(f_1(s, t), f_2(s, t)) = (s, t)$. In this case, $\left| \dfrac{\partial(f_1, f_2)}{\partial(s, t)} \right| = 1$, so $A(M) = \displaystyle\iint\limits_M 1\, ds\, dt$.

Example 4.5.6 Find the area of the polar sector $1 \leq r \leq 4$, $\pi/6 \leq \theta \leq \pi/2$ in the xy-plane.

Solution The sector M is the image of the rectangle $R = [1,\, 4] \times [\pi/6,\, \pi/2]$ in the $r\theta$-plane under the mapping $(x,\, y) = (r\cos\theta,\, r\sin\theta)$. Here

$$\frac{\partial(r\cos\theta,\, r\sin\theta)}{\partial(r,\, \theta)} = \begin{vmatrix} \cos\theta & -r\sin\theta \\ \sin\theta & r\cos\theta \end{vmatrix} = r,$$

so

$$A(M) = \int_{\pi/6}^{\pi/2} \int_1^4 |r|\, dr\, d\theta = \int_{\pi/6}^{\pi/2} \frac{1}{2} r^2 \bigg|_1^4 d\theta = \frac{15}{2} \cdot \left(\frac{\pi}{2} - \frac{\pi}{6}\right) = \frac{5\pi}{2}. \quad \blacklozenge$$

We can also generalize the notion of a surface and its area to $\mathbb{R}^n$. A two-dimensional surface in $\mathbb{R}^n$ is the image of a region $R \subset \mathbb{R}^2$ under a continuous function $\mathbf{f} : \mathbb{R}^2 \to \mathbb{R}^n$. The terms **simple** and **smooth** carry over exactly from the $\mathbb{R}^3$ environment. For example,

$$\mathbf{x} = (2s - t,\, s + 5t - 1,\, 3s + t,\, t - s,\, t)$$

for $(s,\, t) \in [-2, 2] \times [3,\, 5]$ is a parametrization of a "parallelogram" that lies in $\mathbb{R}^5$.

The key to making a reasonable generalization of (4.5.6) to $\mathbb{R}^n$ is to note that if $\mathbf{x} = \mathbf{f}(s,\, t) = (f_1(s,\, t),\, f_2(s,\, t),\, f_3(s,\, t))$ is a parametrization of a surface M in $\mathbb{R}^3$ then the integrand in (4.5.6) can be written as

$$\left\| \frac{\partial \mathbf{f}}{\partial s} \times \frac{\partial \mathbf{f}}{\partial t} \right\| = \sqrt{\|\mathbf{f}_s\|^2\, \|\mathbf{f}_t\|^2 - (\mathbf{f}_s \cdot \mathbf{f}_t)^2}. \tag{4.5.9}$$

[Compare this with (1.6.4) in Chapter 1.] The quantity under the radical in (4.5.9) can be calculated no matter which Euclidean space $\mathbf{f}$ maps into. So if $\mathbf{f} : R \subset \mathbb{R}^2 \to \mathbb{R}^n$, we define the area of the two-dimensional surface $M = \mathbf{f}(R)$ to be

$$\sigma(M) = \iint\limits_R \sqrt{\|\mathbf{f}_s\|^2\, \|\mathbf{f}_t\|^2 - (\mathbf{f}_s \cdot \mathbf{f}_t)^2}\, ds\, dt. \tag{4.5.10}$$

Example 4.5.7 Consider the surface lying in $\mathbb{R}^4$ that is parametrized by

$$\mathbf{x} = \left(\frac{1}{\sqrt{2}} \cos t,\, \sin t \cos s,\, \sin t \sin s,\, \frac{1}{\sqrt{2}} \cos t\right), \tag{4.5.11}$$

$$0 \le s \le 2\pi, \quad 0 \le t \le \pi.$$

Calculate its area.

Solution We first calculate

$$\frac{\partial \mathbf{f}}{\partial s} = (0,\ -\sin t \sin s,\ \sin t \cos s,\ 0)$$

and

$$\frac{\partial \mathbf{f}}{\partial t} = \left(-\frac{1}{\sqrt{2}} \sin t,\ \cos t \cos s,\ \cos t \sin s,\ -\frac{1}{\sqrt{2}} \sin t\right).$$

So

$$\left\|\frac{\partial \mathbf{f}}{\partial s}\right\| = \sqrt{\sin^2 t} = |\sin t|, \quad \left\|\frac{\partial \mathbf{f}}{\partial t}\right\| = 1, \quad \text{and} \quad \frac{\partial \mathbf{f}}{\partial s} \cdot \frac{\partial \mathbf{f}}{\partial t} = 0.$$

Thus the area is

$$\begin{aligned}
\sigma(M) &= \int_0^{2\pi}\int_0^{\pi} \sqrt{\sin^2 t \cdot 1^2 - 0^2}\, dt\, ds \\
&= \int_0^{2\pi}\int_0^{\pi} |\sin t|\, dt\, ds = \int_0^{2\pi}\int_0^{\pi} \sin t\, dt\, ds \\
&= \int_0^{2\pi} (-\cos t)\big|_0^{\pi}\, ds = 4\pi. \quad \blacklozenge
\end{aligned}$$

Exercises 4.5

In Exercises 1 to 8, find a parametrization for the surface.

1. The lower half of the sphere $\rho = 1$.
2. The portion of the cylinder $x^2 + y^2 = 9$ between $z = -1$ and $z = 2$.
3. The portion of the cylinder $y^2 + z^2 = 1$ between $x = 0$ and $x = 5$.
4. The hyperboloid of one sheet $x^2 + y^2 - z^2 = 1$ between $z = 0$ and $z = 1$. (See Exercise 22 in Section 1.2.)
5. The portion of the plane $x - 2y + 4z = 8$ lying above the rectangle $[-1, 0] \times [3, 7]$ in the xy-plane.
6. The portion of the plane $x + 6y + z = 12$ in the first octant.
7. The portion of the surface $r = \theta$, $0 \le \theta \le 6\pi$, in cylindrical coordinates between the planes $w = -2$ and $w = 2$.
8. The portion of the surface $\rho = \phi$, $0 \le \phi \le \pi/2$, in spherical coordinates where the θ-coordinate is between 0 and π.

9. Find a parametrization of the portion of the plane $x+y+z=1$ that lies above the parallelogram in the xy-plane bounded by the lines $y-x=0$, $y-x=1$, $y+2x=0$ and $y+2x=1$. (*Hint*: Let $s=y-x$, $t=y+2x$.)

10. Find a parametrization for the portion of the surface $z=(x^2+y^2)/x^2$ that lies above the region in the xy-plane bounded by the curves $y=\frac{1}{3}x$, $y=3x$, $y=1/x$ and $y=4/x$. (*Hint*: let $s=y/x$, $t=xy$.)

In Exercises 11 through 14, find a single equation in x, y, and z for the surfaces parametrized.

11. $\mathbf{x}=(s+t,\ 3t,\ s-t)$

12. $\mathbf{x}=(s^2+t^2,\ s,\ t)$

13. $\mathbf{x}=(2(s-1),\ (s-1)^2-(t+2)^2,\ t+2)$

14. $\mathbf{x}=(4\cos\theta,\ 4\sin\theta,\ w)$

Use a graphics program to help you visualize the parametrized surfaces in Exercises 15 through 18.

15. $\mathbf{x}=(t^2-1,\ t^3-t,\ s),\quad 0\le s\le 1,\ -2\le t\le 2$

16. $\mathbf{x}=(\theta\cos\theta,\ \theta\sin\theta,\ w),\quad 0\le\theta\le 2\pi,\ 0\le w\le 1$

17. $\mathbf{x}=(\theta\cos\theta\sin\phi,\ \theta\sin\theta\sin\phi,\ \theta\cos\phi),\quad 0\le\theta\le 2\pi,\ 0\le\phi\le\pi$. This is the surface with equation $\rho=\theta$.

18. $\mathbf{x}=(2\cos\theta+r\cos(\theta/2),\ 2\sin\theta+r\cos(\theta/2),\ r\sin(\theta/2)),\quad -\frac{1}{2}\le r\le\frac{1}{2},\ 0\le\theta\le 2\pi$. This is a parametrization for the **Möbius strip**.

In Exercises 19 through 23, set up the double integral that represent the area of each of the parametrized surface.

19. $\mathbf{x}=(2s-3t+1,\ t,\ s+t),\quad -1\le s\le 1,\ 0\le t\le 1$

20. $\mathbf{x}=(e^{s+t},\ 2s-t,\ s+t),\quad 0\le s\le 1,\ 0\le t\le \ln 2$

21. $\mathbf{x}=(\tan^{-1}s+t,\ \dfrac{1}{3}s^3-s-t,\ s+t),\quad 0\le s\le 1,\ 0\le t\le 1$

22. $\mathbf{x}=\left(\ln st,\ 0,\ \dfrac{1}{s+t}\right),\quad 1\le s\le e,\ 1\le t\le 2e$

23. $\mathbf{x}=(r\cos\theta,\ r\sin\theta,\ r^2),\quad 0\le r\le 1,\ 0\le\theta\le 2\pi$

Find the area of the surface M in Exercises 24 through 27.

24. M is the portion of the plane $z=8x+5y+2$ that lies above the region in the xy-plane bounded by the parabola $y^2=x$ and the line $x=4$.

25. M is the portion of the surface $z = \sqrt{2xy}$ bounded by the planes $x = 1$, $x = 3$, $y = 0$ and $y = 1$.

26. M is the portion of the surface $y = \sin x + \cos z$ cut off by the cylinder $x^2 + z^2 = 6$. (Set up but do not evaluate the integral.)

27. M is the portion of the cone $x = \sqrt{y^2 + z^2}$ cut off by the cylinder $(y-1)^2 + z^2 = 1$.

28. A **surface of revolution** is obtained by revolving a curve in a plane about a line that lies in the same plane. For instance, if C is a curve in the xy-plane parametrized by $\mathbf{x} = (f_1(s), f_2(s), 0)$, for $a \le s \le b$, then the surface of revolution obtained by revolving C about the x-axis is parametrized by $\mathbf{x} = (f_1(s), f_2(s)\cos t, f_2(s)\sin t)$, $a \le s \le b$, $0 \le t \le 2\pi$. (See Figure 4.5.7.)

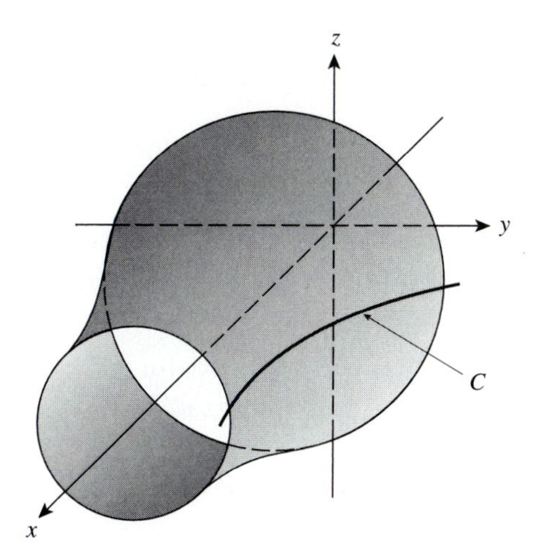

Figure 4.5.7 A surface obtained by revolving a curve C in the xy-plane about the x-axis

(a) Find a parametrization for the surface of revolution obtained by revolving a curve C in the xy-plane about the y-axis.

(b) Find parametrizations for the analogous surfaces of revolution obtained by revolving curves in the other coordinate planes about the corresponding axes.

29. Use Exercise 28 to show that the area of a surface of revolution parametrized by

$$\mathbf{x} = (f_1(s),\ f_2(s)\cos t,\ f_2(s)\sin t),$$

$$a \le s \le b,\ 0 \le t \le 2\pi$$

is

$$\sigma(M) = 2\pi \int_a^b |f_2(s)| \sqrt{f_1'(s)^2 + f_2'(s)^2}\, ds.$$

30. A **torus** is obtained by revolving a circle of radius r about a line in the same plane that lies a distance R (where $R > r$) from the center of the circle.

(a) Find a parametrization for the torus. (Check yourself with a graphics program.)

(b) Find the surface area of the torus.

31. Show that if the curve $y = f(x)$, $a \le x \le b$ is revolved about the x-axis to obtain a surface of revolution, then the area of the surface is

$$A = 2\pi \int_a^b |f(x)| \sqrt{1 + (f'(x))^2} dx.$$

What is the area of the surface of revolution obtained by revolving $z = f(y)$, $c \le y \le d$ about the y-axis?

32. Find the parametrization for the surface obtained by revolving the astroid $x^{2/3} + y^{2/3} = 1$ about the x-axis. (*Hint*: See the exercises in section 1.8.) Find its area.

33. Let $\mathbf{g}_1$ and $\mathbf{g}_2$ be continuous vector-valued functions. Find a formula for the area of the surface parametrized by

$$\mathbf{x} = \mathbf{g}_1(t) + s(\mathbf{g}_2(t) - \mathbf{g}_1(t)),$$

$$0 \le s \le 1, \quad a \le t \le b.$$

34. Let R be a region in the xy-plane that is described in terms of polar coordinates by $f_1(\theta) \le r \le f_2(\theta)$ and $\theta_1 \le \theta \le \theta_2$, where $f_1 \ge 0$. See Figure 4.5.8. By regarding R as the image of the region $Q = \{(r, \theta) \mid f_1(\theta) \le r \le f_2(\theta),\ \theta_1 \le \theta \le \theta_2\}$, show that the area of R is

$$A = \frac{1}{2} \int_{\theta_1}^{\theta_2} \left(f_2(\theta)^2 - f_1(\theta)^2 \right) d\theta. \qquad (4.5.12)$$

In Exercises 35 through 38, use Exercise 34 to find the area of the given region. In each case, sketch both of the regions R and Q.

35. The region in polar coordinates bounded by $r = 1$, $r = 5$, $\theta = \pi/8$, and $\theta = \pi/2$.

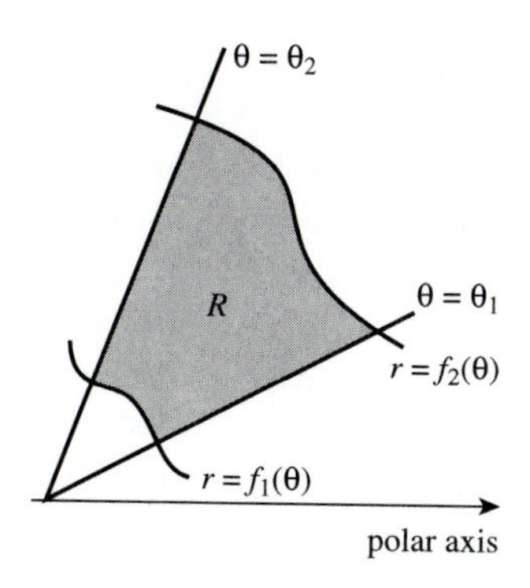

Figure 4.5.8 A region in the xy-plane described in terms of polar coordinates

36. The region in polar coordinates bounded by $\theta = \pi/4$, and $\theta = \pi/2$, $r = 0$, and $r = \theta$.

37. The region between $\theta = 0$ and $\theta = \pi/2$ bounded by the circles $r = 1$ and $r = \cos\theta$.

38. The region between the cardioid $r = 1 + \sin\theta$ and the circle $r = 3$.

39. Find the area of the parallelogram bounded by the lines $x - 2y = 0$, $x - 2y = 3$, $x + y = 1$, and $x + y = 2$. (*Hint*: Let $s = x - 2y$ and $t = x + y$ and regard the parallelogram as the image of the rectangle $[0,\, 3] \times [1,\, 2]$ in the st-plane under an appropriate transformation.)

40. Find the area of the region bounded by the hyperbolas $xy = 1$ and $xy = 6$, and the lines $y/x = 1/2$ and $y/x = 2$.

41. Verify that the function in (4.5.1) satisfies $\partial\mathbf{f}/\partial s \times \partial\mathbf{f}/\partial t = \mathbf{0}$ when $s = \pi/2$, $t = \pi/2$.

42. Find the area of the surface in $\mathbb{R}^4$ parametrized by

$$\mathbf{x} = (3s + t,\ s - t - 1,\ -2s + 8t + 5,\ 4s - 4t),$$

$$0 \le s \le 1, \quad -2 \le t \le 1.$$

43. Find the area of the surface in $\mathbb{R}^5$ parametrized by

$$\mathbf{x} = (\cos s,\ t,\ -\sin s,\ -t,\ s),$$

$$0 \le s \le \pi/2, \quad 0 \le t \le 3.$$

4.6 Surface Integrals

In this section we are concerned with two different notions of integration over surfaces. First we obtain a formula for integrating a scalar-valued function of three variables over a (two-dimensional) surface in $\mathbb{R}^3$. This formula is motivated by our desire for a reasonable way to measure the mass of a surface whose per-unit-area density varies from point to point. The second objective is to obtain a formula for the integral of a vector field on $\mathbb{R}^3$ over a surface. When the value of the vector field at a point (x, y, z) represents the velocity of particles in a fluid as they pass through the point (x, y, z), our vector field integral will give a reasonable measure of the volume-per-unit-time rate of flow of the fluid through the surface. This notion of **flux integral** is an important tool for formulating questions in fluid mechanics and also in electromagnetism, where electric and magnetic fields are regarded as types of flows.

Surface Integrals of Scalar-Valued Functions

Let $g : D \subset \mathbb{R}^3 \to \mathbb{R}$ be a continuous function of three variables and let M be a smooth surface that lies in D. If g is nonnegative, then for any point (x, y, z) on M we think of $g(x, y, z)$ as the per-unit-area density of the surface at (x, y, z). Let $\mathbf{x} = \mathbf{f}(s, t)$, where (s, t) ranges over some region R in $\mathbb{R}^2$, be a smooth parametrization of M. Further, let P be a partition of R consisting of rectangles R_{ij} with areas $\Delta s_i\, \Delta t_j$, and selected points $\mathbf{s}_{ij} \in R_{ij} \cap R$. Then, with exception of a little raggedness around the edges, M consists of the union of the images of R_{ij} under $\mathbf{f}$. See Figure 4.6.1. Since $\mathbf{f}$ is, in particular, continuous, the value of $\mathbf{f}$ at any point in R_{ij} is approximately $\mathbf{f}(\mathbf{s}_{ij})$ whenever the rectangles R_{ij} have uniformly small diagonals. Because g is continuous, its value at any point $\mathbf{f}(\mathbf{s}_{ij})$ on $\mathbf{f}(R_{ij})$ is approximately $g(\mathbf{f}(\mathbf{s}_{ij}))$. So if g represents the per-unit-area density on M, then $g(\mathbf{f}(\mathbf{s}_{ij}))\sigma(\mathbf{f}(R_{ij}))$ approximates the mass of the portion $\mathbf{f}(R_{ij})$ of the surface. The sum of these values

$$\sum_{\mathbf{s}_{ij} \in R} g\left(\mathbf{f}\left(\mathbf{s}_{ij}\right)\right) \sigma\left(\mathbf{f}\left(R_{ij}\right)\right) \tag{4.6.1}$$

is approximately what we would like to call the mass of M. We want to rewrite (4.6.1) as a Riemann sum and then define the integral of $\mathbf{g}$ over M to be the corresponding integral.

By (4.5.6),

$$\sigma\left(\mathbf{f}\left(R_{ij}\right)\right) = \iint_{R_{ij}} \|\mathbf{f}_s \times \mathbf{f}_t\| \, ds\, dt, \tag{4.6.2}$$

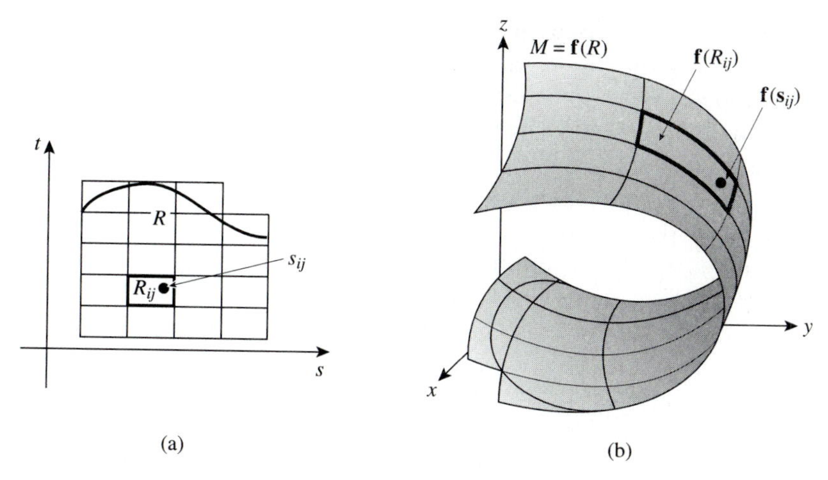

Figure 4.6.1 The image of a region R under $\mathbf{f}$ and the image of a partition of R under $\mathbf{f}$

and by the Mean Value Theorem for Double Integrals (Theorem 4.3.4, part 5), there exists a point $\mathbf{s}_{ij} \in R_{ij}$ such that

$$\iint_{R_{ij}} \|\mathbf{f}_s \times \mathbf{f}_t\|\, dA = \|\mathbf{f}_s(\mathbf{s}_{ij}) \times \mathbf{f}_t(\mathbf{s}_{ij})\|\ \Delta s_i\, \Delta t_j. \tag{4.6.3}$$

Using (4.6.3) and (4.6.2), we can now write (4.6.1) as

$$\sum_{x_{ij} \in R} g(\mathbf{f}(\mathbf{s}_{ij}))\, \|\mathbf{f}_s(\mathbf{s}_{ij}) \times \mathbf{f}_t(\mathbf{s}_{ij})\|\ \Delta s_i\, \Delta t_j, \tag{4.6.4}$$

where we agree to choose $\mathbf{s}_{ij}$ as the point in R_{ij} whose existence is guaranteed by the Mean Value Theorem. This is a Riemann sum for the double integral $\iint_R g(\mathbf{f}(\mathbf{s}))\|\mathbf{f}_s(\mathbf{s}) \times \mathbf{f}_t(\mathbf{s})\|\, ds\, dt$, where $\mathbf{s} = (s,\, t)$. Thus we are led to make the following definition.

Definition 4.6.1 Let $g : D \subset \mathbb{R}^3 \to \mathbb{R}$ be a continuous function and let M be a smooth surface lying in D that is parametrized by $\mathbf{x} = \mathbf{f}(s,\, t)$, where $(s,\, t) \in R \subset \mathbb{R}^2$. The **surface integral of** g **over** M is

$$\iint_M g\, d\sigma = \iint_R g(\mathbf{f}(s,\, t))\|\mathbf{f}_s(s,\, t) \times \mathbf{f}_t(s,\, t)\|\, ds\, dt. \tag{4.6.5}$$

It is essential to point out that the surface integral defined in (4.6.5) might have different values for different parametrizations $\mathbf{f}$. In fact, the value of the integral is the same no matter which smooth parametrization $\mathbf{f}$ is used. To show this involves the same sort of reasoning we used to show that the line integral of a scalar-valued function is independent of the parametrization.

Whenever the surface M is the graph of a function of two variables [e.g., $z = h(x, y)$ for $(x, y) \in R \subset \mathbb{R}^2$] then, analogous to (4.5.7), the surface integral can be calculated by the formula

$$\iint_M g\, d\sigma = \iint_R g(x, y, h(x, y))\sqrt{\left(\frac{\partial h}{\partial x}\right)^2 + \left(\frac{\partial h}{\partial y}\right)^2 + 1}\, dx\, dy. \tag{4.6.6}$$

Example 4.6.1 Evaluate the surface integral $\iint_M g\, d\sigma$, where $g(x, y, z) = x/\sqrt{4y+5}+z$ and M is the surface parametrized by $\mathbf{x} = (s, t^2-1, t)$, $(s, t) \in [0, 1] \times [-1, 1]$.

Solution We simply use (4.6.5). First we calculate the integrand:

$$\begin{aligned}\frac{\partial \mathbf{f}}{\partial s} \times \frac{\partial \mathbf{f}}{\partial t} &= \frac{\partial}{\partial s}\left(s, t^2 - 1, t\right) \times \frac{\partial}{\partial t}\left(s, t^2 - 1, t\right)\\ &= (1, 0, 0) \times (0, 2t, 1)\\ &= (0, -1, 2t),\end{aligned}$$

so $\left\|\frac{\partial \mathbf{f}}{\partial s} \times \frac{\partial \mathbf{f}}{\partial t}\right\| = \sqrt{1+4t^2}$. Also,

$$\begin{aligned}g(\mathbf{f}(s, t)) &= g(s, t^2 - 1, t)\\ &= \frac{s}{\sqrt{4(t^2-1)+5}} + t\\ &= \frac{s}{\sqrt{4t^2+1}} + t.\end{aligned}$$

Putting these expressions into (4.6.5), we obtain

$$\begin{aligned}\iint_M g\, d\sigma &= \int_0^1 \int_{-1}^1 \left(\frac{s}{\sqrt{4t^2+1}} + t\right)\sqrt{4t^2+1}\, dt\, ds\\ &= \int_0^1 \int_{-1}^1 \left(s + t\sqrt{t^2+1}\right) dt\, ds\\ &= \int_0^1 (st + \frac{1}{3}(t^2+1)^{3/2})\Big|_{-1}^{1} ds\\ &= \int_0^1 2s\, ds = s^2\Big|_0^1 = 1. \quad \blacklozenge\end{aligned}$$

Example 4.6.2 Evaluate the surface integral of $g(x, y, z) = (x^2 + y^2)z$ over the upper half of the sphere of radius 1 centered at the origin. See Figure 4.6.2.

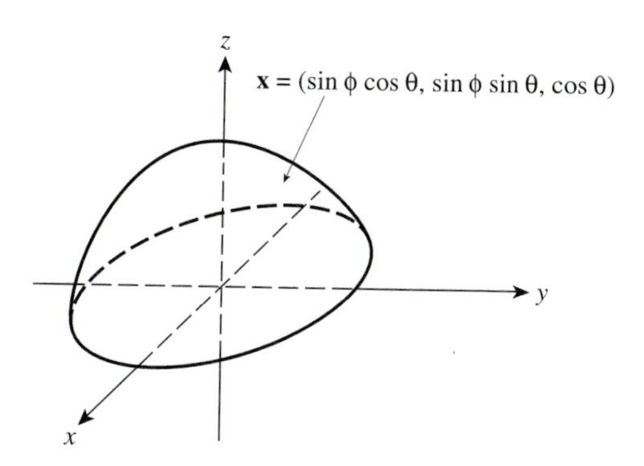

Figure 4.6.2 The upper half of the unit sphere

Solution There are several ways to approach this problem, but the term "sphere" suggests spherical coordinates. Indeed, the upper half of the unit sphere is parametrized by

$$\mathbf{x} = (\sin\phi\cos\theta,\ \sin\phi\sin\theta,\ \cos\phi),$$

where $0 \le \phi \le \pi/2$ and $0 \le \theta < 2\pi$. Thus

$$\frac{\partial \mathbf{x}}{\partial \theta} = (-\sin\phi\sin\theta,\ \sin\phi\cos\theta,\ 0)$$

$$\frac{\partial \mathbf{x}}{\partial \phi} = (\cos\phi\cos\theta,\ \cos\phi\sin\theta,\ -\sin\phi),$$

and consequently,

$$\begin{aligned}\left\|\frac{\partial \mathbf{x}}{\partial \theta} \times \frac{\partial \mathbf{x}}{\partial \phi}\right\| &= \left\|(-\sin^2\phi\cos\theta, -\sin^2\phi\sin\theta, -\sin\phi\cos\phi)\right\| \\ &= |\sin\phi|(\sin^2\phi\cos^2\theta + \sin^2\phi\sin^2\theta + \cos^2\phi)^{1/2} \\ &= \sin\phi. \end{aligned} \qquad (4.6.7)$$

The last step follows since $\sin\phi$ is nonnegative in the range of ϕ values we are considering. By (4.6.5),

$$\iint_M g\,d\sigma = \int_0^{2\pi}\int_0^{\pi/2} g(\sin\phi\cos\theta,\ \sin\phi\sin\theta,\ \cos\phi)\sin\phi\,d\phi\,d\theta$$

$$= \int_0^{2\pi}\int_0^{\pi/2} (\sin^2\phi\cos^2\theta + \sin^2\phi\sin^2\theta)\cos\phi\sin\phi\,d\phi\,d\theta$$

$$= \int_0^{2\pi}\int_0^{\pi/2} \sin^3\phi\cos\phi\,d\phi\,d\theta = 2\pi\cdot\frac{1}{4}\sin^4\phi\bigg|_0^{\pi/2} = \frac{\pi}{2}. \quad \blacklozenge$$

Example 4.6.3 Evaluate $\iint\limits_M xy\sqrt{z+1/4}\,d\sigma$, where M is the portion of the paraboloid $z = x^2+y^2$ that lies above the square region $R = [0,1]\times[1,2]$ in the xy-plane.

Solution The surface is the graph of $h(x,\,y) = x^2+y^2$, so we can use the formula (4.6.6). Since

$$\frac{\partial h}{\partial x} = 2x$$

and

$$\partial h/\partial y = 2y,$$

we have

$$\begin{aligned}
\iint\limits_M xy\sqrt{z+1/4}\;d\sigma &= \int_0^1\int_1^2 xy\sqrt{x^2+y^2+1/4}\;\sqrt{4x^2+4y^2+1}\;dy\,dx \\
&= \int_0^1\int_1^2 2xy(x^2+y^2+1/4)\,dy\,dx \\
&= \int_1^2 \frac{x}{2}(x^2+y^2+1/4)^2\,\bigg|_1^2\;dx \\
&= \frac{1}{2}\int_0^1 \left[x(x^2+17/4)^2 - x(x^2+\frac{5}{4})^2\right]\,dx \\
&= \frac{1}{2}\left(\frac{1}{2}\cdot\frac{1}{3}\left(x^2+\frac{17}{4}\right)^3 - \frac{1}{2}\cdot\frac{1}{3}\left(x^2+\frac{5}{4}\right)^3\right)\bigg|_0^1 \\
&= \frac{1}{12}\left[\left(\frac{21}{4}\right)^3 - \left(\frac{9}{4}\right)^3 - \left(\frac{17}{4}\right)^3 + \left(\frac{5}{4}\right)^3\right] \\
&= \frac{3744}{12\cdot 64} = \frac{39}{8}. \quad \blacklozenge
\end{aligned}$$

Surface Integrals of Vector Fields

The concept of the surface integral of a *vector field* is analogous to that of a line integral of a vector field. Its definition is motivated by physical considerations. Imagine that a liquid or gas consisting of infinitely small particles is flowing through a region of space. Suppose that the flow is steady, in the sense that the velocity of each particle that passes through the point (x, y, z) has the same velocity. Then associated with each point (x, y, z) in this region of space we have a velocity vector $\mathbf{F}(x, y, z)$, and if we assume that this velocity field is continuous, then velocity vectors at nearby points are approximately equal. Now imagine placing a surface in this flow which is completely permeable by the fluid particles. We distinguish between the two sides of the surface by agreeing that one is positive and the other negative. Over a short time interval, fluid particles will have moved from one side of the surface to the other and it makes sense to ask, "What is the net amount of fluid to pass from the negative to the positive side of the surface during this time?" and "What is the instantaneous time rate of flow through this surface in the negative-to-positive direction?" We define the integral of $\mathbf{F}$ over the surface so that it gives a reasonable measure of this flow rate.

Let $\mathbf{F} : D \subset \mathbb{R}^3 \to \mathbb{R}^3$ be continuous, and let M be a smooth oriented surface lying in D that is parametrized by $\mathbf{x} = \mathbf{f}(s, t)$, where $(s, t) \in R \subset \mathbb{R}^2$. Let $R_{ij} = [s_{i-1}, s_i] \times [t_{j-1}, t_j]$ be a partition rectangle of R. If $\mathbf{F}$ represents the velocity field for a fluid in steady motion, we want to obtain a reasonable expression for the volume of fluid passing through $\mathbf{f}(R_{ij})$ per unit time. Since $\mathbf{f}$ is smooth and R_{ij} is small, $\mathbf{f}(R_{ij})$ is approximated by the image of R_{ij} under the linear approximation of $\mathbf{f}$ at (s_{i-1}, t_{j-1}). This image is—by the same reasoning that led to our surface area formula—the parallelogram spanned by the vectors $\dfrac{\partial \mathbf{f}}{\partial s}(s_i, t_j)\,\Delta s_i$ and $\dfrac{\partial \mathbf{f}}{\partial s}(s_i, t_j)\,\Delta t_j$ Since $\mathbf{F}$ is continuous, the velocity of each fluid particle passing through $\mathbf{f}(R_{ij})$ is approximately $\mathbf{F}(\mathbf{f}(s_i, t_j))$, and for a short time of duration h, each of these particles undergoes a displacement of approximately $h\mathbf{F}(\mathbf{f}(s_i, t_j))$. Thus the total volume of fluid that passes through $\mathbf{f}(R_{ij})$ in time h is approximated by the volume of the parallelepiped spanned by the vectors $\dfrac{\partial \mathbf{f}}{\partial s}(s_i, t_j)\,\Delta s_i$, $\dfrac{\partial \mathbf{f}}{\partial t}(s_i, t_j)\,\Delta t_i$, and $h\mathbf{F}(\mathbf{f}(s_i, t_j))$. We multiply this volume by $+1$ if $\mathbf{F}$ points toward the positive side of the surface (i.e., toward $\partial \mathbf{f}/\partial s \times \partial \mathbf{f}/\partial t$), and by -1 if $\mathbf{F}$ points toward the opposite side. (Since our surface is orientable, this can be done.) The quantity of interest then is the scalar triple product $h\mathbf{F}(\mathbf{f}) \cdot \dfrac{\partial \mathbf{f}}{\partial s} \times \dfrac{\partial \mathbf{f}}{\partial t}\,\Delta s_i\,\Delta t_j$. Adding these signed volumes for all the rectangles in our partition, we obtain

$$\sum_{(s_i, t_j) \in R} h\mathbf{F}\left(f(s_i, t_j)\right) \cdot \left(\frac{\partial \mathbf{f}}{\partial s}(s_i, t_j) \times \frac{\partial \mathbf{f}}{\partial t}(s_i, t_j) \right) \Delta s_i \, \Delta t_j$$

as an approximation for the *net* volume of fluid passing through the surface from the negative to the positive side in time h. This is a Riemann sum for the double integral

$$h \iint\limits_R \mathbf{F}\left(\mathbf{f}(s, t)\right) \cdot \left(\frac{\partial \mathbf{f}}{\partial s}(s, t) \times \frac{\partial \mathbf{f}}{\partial t}(s, t) \right) ds \, dt.$$

The instantaneous rate of flow through the surface is the limit as $h \to 0$ of this expression divided by h. This discussion motivates the following definition.

Definition 4.6.2 Let $\mathbf{F} : D \subset \mathbb{R}^3 \to \mathbb{R}^3$ be a continuous vector field and let $M \subset D$ be a smooth surface parametrized and oriented by $\mathbf{x} = \mathbf{f}(s, t)$ where $(s, t) \in R$ and R is a region in $\mathbb{R}^2$. The **flux** (or **flux integral**) of $\mathbf{F}$ over M is

$$\iint\limits_M \mathbf{F} \cdot \mathbf{n} \, d\sigma = \iint\limits_R \mathbf{F}\left(\mathbf{f}(s, t)\right) \cdot \left(\frac{\partial \mathbf{f}}{\partial s}(s, t) \times \frac{\partial \mathbf{f}}{\partial t}(s, t) \right) ds \, dt. \tag{4.6.8}$$

The reason for the notation $\mathbf{F} \cdot \mathbf{n}$ on the left is that

$$\begin{aligned} \mathbf{F} \cdot \frac{\partial \mathbf{f}}{\partial s} \times \frac{\partial \mathbf{f}}{\partial t} &= \mathbf{F} \cdot \left(\frac{\partial \mathbf{f}}{\partial s} \times \frac{\partial \mathbf{f}}{\partial t} \Big/ \left\| \frac{\partial f}{\partial s} \times \frac{\partial f}{\partial t} \right\| \right) \left\| \frac{\partial \mathbf{f}}{\partial s} \times \frac{\partial \mathbf{f}}{\partial t} \right\| \\ &= \mathbf{F} \cdot \mathbf{n} \left\| \frac{\partial \mathbf{f}}{\partial s} \times \frac{\partial \mathbf{f}}{\partial t} \right\|, \end{aligned}$$

where $\mathbf{n} = \dfrac{\partial \mathbf{f}}{\partial s} \times \dfrac{\partial \mathbf{f}}{\partial t} \Big/ \left\| \dfrac{\partial \mathbf{f}}{\partial s} \times \dfrac{\partial \mathbf{f}}{\partial t} \right\|$ is the unit normal to the surface. Writing it this way, we see that the flux integral is just the surface integral of the scalar-valued function $\mathbf{F} \cdot \mathbf{n}$.

If a surface M is not smooth but consists of finitely many pieces that are, we define the flux integral of a vector field $\mathbf{F}$ over M to be the sum of the integrals over the pieces.

One other remark is in order at this point. It is a fact, which we do not prove here, that the integral defined in (4.6.8) is, except for a factor of ± 1, independent of the parametrization. Another way to say this is that if

$\mathbf{x} = \mathbf{g}(s, t)$ is some other parametrization for M with the same orientation as that induced by $\mathbf{f}$, then replacing $\mathbf{f}$ by $\mathbf{g}$ in the formula does not change the value of the integral.

Example 4.6.4 Evaluate $\displaystyle\iint_M \mathbf{F}\cdot\mathbf{n}\,d\sigma$, where $\mathbf{F}(x, y, z) = (3y, -z, x^2)$ and M is the surface parametrized and oriented by

$$\mathbf{x} = \left(st,\ s+t,\ \frac{1}{2}\left(s^2 - t^2\right)\right) \quad 0 \le s \le 1, \quad 0 \le t \le 3.$$

Solution The normal vector is

$$\frac{\partial \mathbf{f}}{\partial s} \times \frac{\partial \mathbf{f}}{\partial t} = (t,\ 1,\ s) \times (s,\ 1,\ -t) = (-t-s,\ s^2+t^2,\ t-s),$$

so by (4.6.8),

$$\begin{aligned}
&\iint_M \mathbf{F}\cdot\mathbf{n}\,d\sigma \\
&= \int_0^1\int_0^3 \mathbf{F}\left(st,\ s+t,\ \frac{1}{2}\left(s^2-t^2\right)\right)\cdot\left(-t-s,\ s^2+t^2,\ t-s\right)\,dt\,ds \\
&= \int_0^1\int_0^3 \left(3(s+t),\ \frac{1}{2}\left(t^2-s^2\right),\ s^2t^2\right)\cdot\left(-(t+s),\ s^2+t^2,\ t-s\right)\,dt\,ds \\
&= \int_0^1\int_0^3 \left(-3(s+t)^2 + \frac{1}{2}\left(t^4-s^4\right) + s^2t^3 - s^3t^2\right)\,dt\,ds \\
&= \int_0^1 \left(-\frac{27}{10} - 27s + \frac{45}{4}s^2 - 9s^3 - \frac{3}{2}s^4\right)\,ds \\
&= -15. \quad \blacklozenge
\end{aligned}$$

If a surface M is given by $z = h(x, y)$ for $(x, y) \in R \subset \mathbb{R}^2$, then (4.6.8) becomes

$$\iint_M \mathbf{F}\cdot\mathbf{n}\,d\sigma = \iint_R \mathbf{F}(x,\ y,\ h(x,\ y))\cdot\left(-\frac{\partial h}{\partial x},\ -\frac{\partial h}{\partial y},\ 1\right)\,dx\,dy \qquad (4.6.9)$$

for an *upward*-pointing normal $\mathbf{n}$; the sign is reversed if the normal is pointing downward.

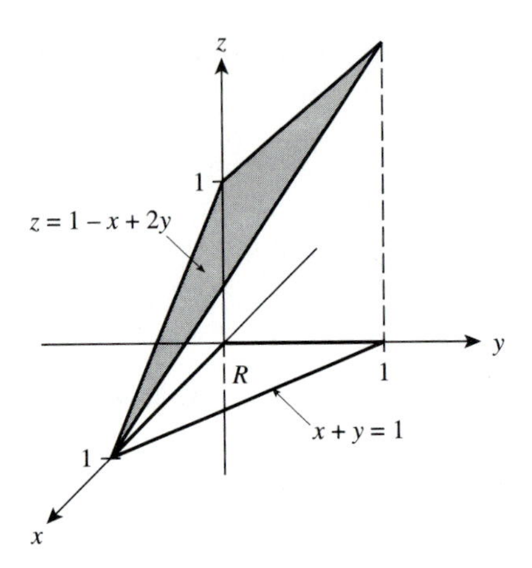

Figure 4.6.3 The surface for Example 4.6.5

Example 4.6.5 Find the surface integral of $\mathbf{F} = (y,\ z,\ x^2)$ over the portion of the plane $x - 2y + z = 1$ that is cut off by the coordinate planes and the plane $x + y = 1$. See Figure 4.6.3.

Solution The surface is the portion of the graph of $z = 1 - x + 2y$ corresponding to $(x,\ y)$ in the region R in the xy-plane bounded by the coordinate axes and the line $x + y = 1$. Since $\partial z/\partial x = -1$ and $\partial z/\partial y = 2$, by (4.6.9) we have

$$\begin{aligned}
\iint_M \mathbf{F}\cdot\mathbf{n}\,d\sigma &= \iint_R (y,\ 1 - x + 2y,\ x^2)\cdot(-(-1),\ -2,\ 1)\,dA \\
&= \int_0^1\int_0^{1-x}\left(-2 + 2x - 3y + x^2\right)\,dy\,dx \\
&= \int_0^1\left[\left(-2 + 2x + x^2\right)(1 - x) - \frac{3}{2}(1 - x)^2\right]\,dx \\
&= \int_0^1\left[-\frac{7}{2} + 7x - \frac{5}{2}x^2 - x^3\right]\,dx \\
&= -\frac{7}{2} + \frac{7}{2} - \frac{5}{6} - \frac{1}{4} = \frac{-13}{12}.
\end{aligned}$$

Is a negative value reasonable for this integral? Yes! If we regard $\mathbf{F}$ as the velocity field for some fluid flow, this result just says that the net flow per unit time is actually from above the surface into the region below it. ♦

Example 4.6.6 Find the flux out of the unit sphere for the vector field

$$\mathbf{F}(x,\, y,\, z) = \frac{1}{x^2 + y^2 + z^2}(\mathbf{i} + \mathbf{j} + \mathbf{k}).$$

Solution Let us parametrize the sphere by

$$(x,\, y,\, z) = \mathbf{f}(\theta,\, \phi) = (\sin\phi\cos\theta,\, \sin\phi\sin\theta,\, \cos\phi),$$
$$0 \le \theta \le 2\pi$$
$$0 \le \phi \le \pi.$$

Then, from (4.6.7), we have

$$\frac{\partial \mathbf{f}}{\partial \theta} \times \frac{\partial \mathbf{f}}{\partial \phi} = (-\sin^2\phi\cos\theta,\, -\sin^2\phi\sin\theta,\, -\sin\phi\cos\phi).$$

This normal points inward, so we take its negative for the outward normal. Now

$$\begin{aligned} \mathbf{F}(\sin\phi\cos\theta,\, \sin\phi\sin\theta,\, \cos\phi) &= \frac{(1,\, 1,\, 1)}{\sin^2\phi\cos^2\theta + \sin^2\phi\sin^2\theta + \cos^2\phi} \\ &= (1,\, 1,\, 1), \end{aligned}$$

so the flux of $\mathbf{F}$ is (see Figure 4.6.4)

$$\begin{aligned} &\iint_M \mathbf{F}\cdot\mathbf{n}\,d\sigma \\ &= \iint_M \mathbf{F}\cdot\frac{\partial \mathbf{f}}{\partial \theta} \times \frac{\partial \mathbf{f}}{\partial \phi}\,d\theta\,d\phi \\ &= \int_0^{\pi}\int_0^{2\pi} (1,\, 1,\, 1)\cdot(\sin^2\phi\cos\theta,\, \sin^2\phi\sin\theta,\, \sin\phi\cos\phi)\,d\theta\,d\phi \\ &= \int_0^{\pi}\int_0^{2\pi} (\sin^2\phi(\cos\theta + \sin\theta) + \sin\phi\cos\phi)\,d\theta\,d\phi \\ &= \int_0^{\pi} 2\pi\sin\phi\cos\phi\,d\phi = \pi\sin^2\phi\Big|_0^{\pi} = 0. \end{aligned}$$

Is this reasonable? We can verify geometrically that it is. At each point on the unit sphere the vector field has the value $\mathbf{i}+\mathbf{j}+\mathbf{k}$. On half the sphere these arrows point outward, and on the other half they point inward so that the flux integrals over the respective halves exactly cancel. ♦

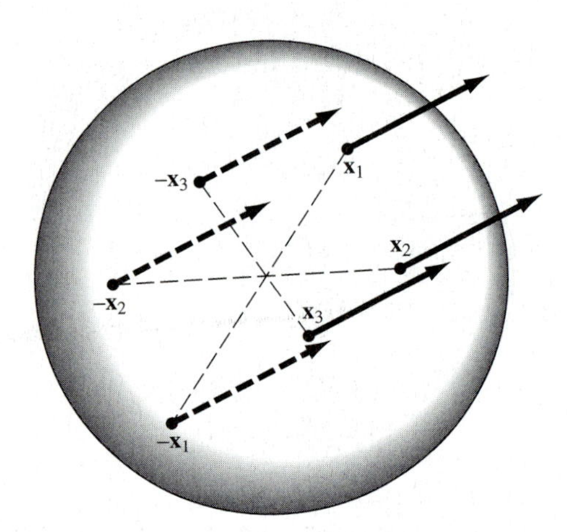

Figure 4.6.4 The flux of the vector field $\mathbf{F}(x, y, z) = (\mathbf{i}+\mathbf{j}+\mathbf{k})/(x^2+y^2+z^2)$ through the unit sphere is zero because the flow inward over the back side of the sphere exactly balances the flow outward through the front side.

Equation (4.6.8) is often written in **differential form**. You should verify that the cross product in (4.6.8) is the same as the vector

$$\left(\begin{vmatrix} f_{2s} & f_{2t} \\ f_{3s} & f_{3t} \end{vmatrix}, \begin{vmatrix} f_{3s} & f_{3t} \\ f_{1s} & f_{1t} \end{vmatrix}, \begin{vmatrix} f_{1s} & f_{1t} \\ f_{2s} & f_{2t} \end{vmatrix} \right). \tag{4.6.10}$$

If the vector field's components are $\mathbf{F} = (F_1, F_2, F_3)$, then the integrand of the double integral on the right-hand side of (4.6.8) is the dot product of $\mathbf{F}$ with the vector in (4.6.10) and can be written

$$F_1 \begin{vmatrix} f_{2s} & f_{2t} \\ f_{3s} & f_{3t} \end{vmatrix} + F_2 \begin{vmatrix} f_{3s} & f_{3t} \\ f_{1s} & f_{1t} \end{vmatrix} + F_3 \begin{vmatrix} f_{1s} & f_{1t} \\ f_{2s} & f_{2t} \end{vmatrix}.$$

Here it is understood that the component functions F_1, F_2, and F_3 are evaluated at $\mathbf{f}(s, t)$ and that the partials are all evaluated at (s, t). We think of the components of $\mathbf{f}$ as $x = f_1(s, t)$, $y = f_2(s, t)$, and $z = f_3(s, t)$, so their total differentials are

$$\begin{aligned} dx &= \frac{\partial f_1}{\partial s}\, ds + \frac{\partial f_1}{\partial t}\, dt, \\ dy &= \frac{\partial f_2}{\partial s}\, ds + \frac{\partial f_2}{\partial t}\, dt, \end{aligned}$$

and

$$dz = \frac{\partial f_3}{\partial s}\, ds + \frac{\partial f_3}{\partial t}\, dt.$$

We define the **wedge product** of the differentials by

$$dy \wedge dz = \begin{vmatrix} f_{2s} & f_{2t} \\ f_{3s} & f_{3t} \end{vmatrix} ds\, dt \tag{4.6.11}$$

$$dz \wedge dx = \begin{vmatrix} f_{3s} & f_{3t} \\ f_{1s} & f_{1t} \end{vmatrix} ds\, dt \tag{4.6.12}$$

$$dx \wedge dy = \begin{vmatrix} f_{1s} & f_{1t} \\ f_{2s} & f_{2t} \end{vmatrix} ds\, dt. \tag{4.6.13}$$

Also, we define

$$\begin{aligned} dz \wedge dy &= -dy \wedge dz \\ dx \wedge dz &= -dz \wedge dx \\ dy \wedge dx &= -dx \wedge dy. \end{aligned}$$

(See if you can draw and analogy between these formulas and those for $\mathbf{i} \times \mathbf{j}$, $\mathbf{j} \times \mathbf{k}$, and $\mathbf{k} \times \mathbf{i}$.) With this notation, the integral defined in (4.6.8) is symbolized by

$$\iint_M \mathbf{F} \cdot \mathbf{n}\, d\sigma = \iint_M F_1\, dy \wedge dz + F_2\, dz \wedge dx + F_3\, dx \wedge dy. \tag{4.6.14}$$

You can think of (4.6.14) as (4.6.8) "multiplied out." Compare this formula with the differential form of a line integral [see (4.2.5)]. This notation is compact and can facilitate computations, as we see in the next example.

Example 4.6.7 Evaluate $\iint_M z^2\, dy \wedge dz + y\, dz \wedge dx + (z + x)\, dx \wedge dy$ where M is the surface consisting of the unit square in the xy-plane along with the unit square in the yz-plane. See Figure 4.6.5. Assume that M is oriented so that the normal points into the first octant.

Solution The surface consists of two pieces, M_1 and M_2, so we calculate the integrals over each piece separately and then add. First, M_1 is a portion of $z = 0$; consequently, $dz = 0$, so $dy \wedge dz = 0$ and $dz \wedge dx = 0$. So the flux integral over M_1 is just $\iint_M (z + x)\, dx \wedge dy$. Because M_1 is the graph of $h(x, y) = 0$, we let the parameters be $s = x$ and $t = y$. By (4.6.13),

$$dx \wedge dy = \begin{vmatrix} \partial x/\partial x & \partial x/\partial y \\ \partial y/\partial x & \partial y/\partial y \end{vmatrix} dx\, dy = \begin{vmatrix} 1 & 0 \\ 0 & 1 \end{vmatrix} dx\, dy = dx\, dy.$$

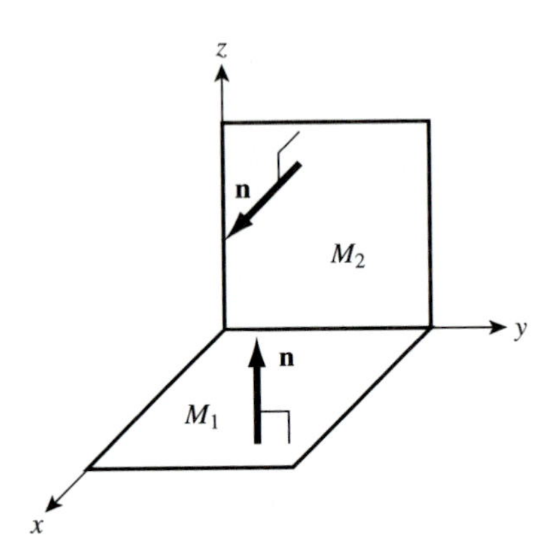

Figure 4.6.5 The surface for Example 4.6.7

Since this determinant is the z-component of the normal and it is positive, the normal points into the first octant as we required. So

$$\iint_{M_1} (z+x)\, dx \wedge dy = \int_0^1 \int_0^1 (0+x)\, dx\, dy = \int_0^1 \frac{1}{2}\, dy = \frac{1}{2}. \tag{4.6.15}$$

Now M_2 is a portion of $x = 0$, so $dx = 0$ and thus $dz \wedge dx = 0$ and $dx \wedge dy = 0$. This means that the integral over M_2 reduces to $\iint_{M_2} z^2\, dy \wedge dz$.

As for the first integral, we have

$$dy \wedge dz = \begin{vmatrix} \partial y/\partial y & \partial y/\partial z \\ \partial z/\partial y & \partial z/\partial z \end{vmatrix} dy\, dz = \begin{vmatrix} 1 & 0 \\ 0 & 1 \end{vmatrix} dy\, dz = dy\, dz.$$

This determinant is the x-component of the normal, and since it is positive, the normal points into the first octant. Now

$$\iint_{M_2} z^2\, dy \wedge dz = \int_0^1 \int_0^1 z^2\, dy\, dz = \int_0^1 \frac{1}{3}\, dz = \frac{1}{3}. \tag{4.6.16}$$

Adding (4.6.15) and (4.6.16), we have

$$\iint_{M_2} z^2\, dy \wedge dz + y\, dz \wedge dx + (z+x)\, dx \wedge dy = \frac{1}{2} + \frac{1}{3} = \frac{5}{6}. \quad \blacklozenge$$

Example 4.6.8 Evaluate $\displaystyle\iint_M xy\,dy \wedge dz - z^2\,dx \wedge dy$, where M is the surface of the unit box $[0,\,1] \times [0,\,1] \times [0,\,1]$ in $\mathbb{R}^3$ with outward pointing normal.

Solution M consists of six pieces (see Figure 4.6.6), the back B in the plane $x = 0$, front F in $x = 1$, left side L in $y = 0$, right side R in $y = 1$, floor P in $z = 0$, and ceiling C in $z = 1$. We just calculate the corresponding six integrals and add.

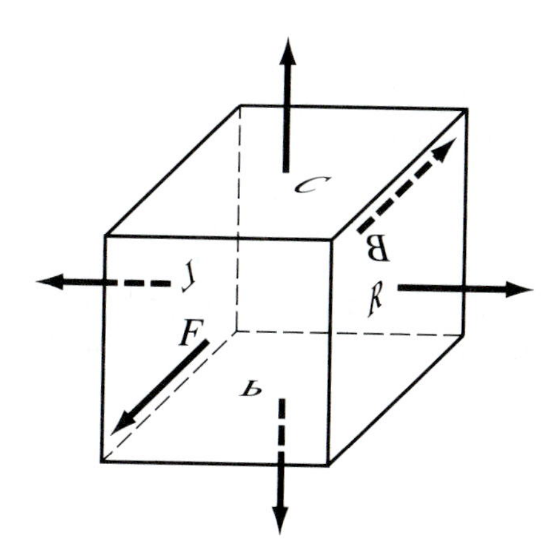

Figure 4.6.6 The surface for Example 4.6.8

On B, we have $x = 0$, so $dx = 0$ and thus $dx \wedge dy = 0$. Since the normal on B points in the negative x-direction, $dy \wedge dz = -dy\,dz$. Therefore,

$$\iint_B xy\,dy \wedge dz - z^2\,dx \wedge dy = \int_0^1 \int_0^1 0 \cdot y(-dy\,dz) = 0$$

On F, we have $x = 1$, so $dx = 0$ and $dx \wedge dy = 0$. Since the normal points in the positive direction, $dy \wedge dz = dy\,dz$. So, shortening our notation by writing ω for the differential form $xy\,dy \wedge dz - z^2\,dx \wedge dy$,

$$\iint_F \omega = \int_0^1 \int_0^1 1y\,dy\,dz = \frac{1}{2}\int_0^1 dz = \frac{1}{2}.$$

You can check that the four remaining integrals are

$$\iint_L \omega \quad = \quad \iint_L 0 = 0,$$

$$\begin{aligned}\iint_R \omega &= \iint_R 0 = 0,\\ \iint_P \omega &= \iint_P -z^2\, dx \wedge dy = \int_0^1 \int_0^1 (-0^2)(-dx\, dy) = 0,\end{aligned}$$

and

$$\iint_C \omega = \iint_C -1^2(dx\, dy) = -\int_0^1 \int_0^1 dx\, dy = -1.$$

Thus $\iint_M \omega = 0 + \frac{1}{2} + 0 + 0 + 0 - 1 = -\frac{1}{2}$. If, for example, $\mathbf{F} = (xy,\, 0,\, -z^2)$ is the velocity of a fluid measured in meters per second and the box over which we are integrating is measured in meters, then the integral says that the *net* flow per unit time is $\frac{1}{2}$m^3/sec *inward* through the box. In other words, the fluid is being compressed at some points inside the box. ♦

Exercises 4.6

In Exercise 1 through 9, evaluate the surface integral over the indicated surface. In each case describe or sketch the surface and interpret the values of the integral.

1. $\iint_M xy\, d\sigma$, where M is the planar surface parametrized by $\mathbf{x} = (s,\, t,\, 3 - s - t)$ $0 \le s \le 1$, $0 \le t \le 1$.

2. $\iint_M \sqrt{x^2 + y^2}\, d\sigma$, where M is the cylindrical surface parametrized by $\mathbf{x} = (2\cos s,\, 2\sin s,\, t)$, $0 \le s \le 2\pi$, $-1 \le t \le 1$.

3. $\iint_M \tan^{-1}\left(\frac{y}{x}\right) d\sigma$, where M is the surface parametrized by $\mathbf{x} = (r\cos\theta,\, r\sin\theta,\, \sqrt{1 - r^2})$, $\frac{1}{2} \le r \le 1$, $0 \le \theta \le \pi/2$.

4. $\iint_M z^2\sqrt{x^2 + y^2}\, d\sigma$, where M is parametrized by $\mathbf{x} = (\theta\cos\theta,\, \theta\sin\theta,\, w)$, $0 \le \theta \le 6\pi$, $-2 \le w \le 2$.

5. $\iint\limits_M x\,z\,d\sigma$, where M is the portion of the hyperbolic paraboloid $z = y^2 - x^2$ bounded by the planes $x = 0$, $y = 1$, and $y = x$.

6. $\iint\limits_M (y + z)\,d\sigma$, where M is the part of the plane $2x + y + z = 3$ that lies above the region $R = \{(x, y) | 0 \le x \le \frac{1}{3},\quad 0 \le y \le 1 - 3x\}$ in the xy-plane.

7. $\iint\limits_M y\,z\,d\sigma$, where M is the portion of $x = y^2 + 2z^2$ cut off by the planes $y = 1$, $y = 5$, $z = -2$, and $z = 4$.

8. $\iint\limits_M xyz\,d\sigma$, where M is the portion of the surface $y = z + 8x$ cut off by the planes $z = 0$, $x = 0$, and $2x + 3z = 6$.

9. $\iint\limits_M \sqrt{1 - x^2 - y^2}\,d\sigma$, where M is the upper half of the sphere given by $x^2 + y^2 + z^2 = 1$.

10. Find the mass of a sheet in the shape of the surface $z = \sqrt{x^2 + y^2}$, $0 \le x \le 1$, $0 \le y \le 1$ if the density at each point is $g(x, y, z) = 1 + z^2$.

Evaluate the flux integrals in Exercises 11 through 16 and interpret the value of the integral.

11. $\iint\limits_M \mathbf{F} \cdot \mathbf{n}\,d\sigma$, where $\mathbf{F} = (y - z,\ x - z,\ x - y)$ and M is the planar surface parametrized and oriented by $\mathbf{x} = (s - t,\ s + t,\ 3s - t)$ such that s and t satisfy $0 \le s \le 1$ and $1 - s \le t \le 1$.

12. $\iint\limits_M \mathbf{F} \cdot \mathbf{n}\,d\sigma$, where $\mathbf{F} = (x,\ -z,\ -y)$ and M is the surface of revolution parametrized and oriented by $\mathbf{x} = (s,\ (s^3 - s)\cos t,\ (s^3 - s)\sin t)$ $-1 \le s \le 1$, $0 \le t \le 2\pi$.

13. $\iint\limits_M \mathbf{F} \cdot \mathbf{n}\,d\sigma$, where $\mathbf{F} = (1,\ -1,\ x^2 + y^2)$ and M is the ruled surface parametrized and oriented by $\mathbf{x} = (\cos t,\ \sin t,\ 0) + s(-\cos t,\ -\sin t,\ t)$, $0 \le s \le 1$, $0 \le t \le \pi$.

14. $\iint\limits_M \mathbf{F} \cdot \mathbf{n}\,d\sigma$, where $\mathbf{F} = (-x,\ -y,\ -z)$ and M is the sphere parametrized and oriented by $(x, y, z) = (\sin\phi\cos\theta, \sin\phi\sin\theta, \cos\phi)$, where $0 \le \theta \le 2\pi$, $0 \le \phi \le \pi$.

15. $\iint_M \mathbf{F}\cdot\mathbf{n}\,d\sigma$, where $\mathbf{F} = (3x+z^2, y^2-x^2, xz+5)$ and M is the surface of the box bounded by the coordinate planes and the planes $x = 1$, $y = 3$, $z = 2$. Take $\mathbf{n}$ to be pointing outward.

16. $\iint_M \mathbf{F}\cdot\mathbf{n}\,d\sigma$, where $\mathbf{F} = (y^2, x^3, z)$ and M is the boundary of the cylindrical solid bounded by $x^2+y^2 = 4$, $z = 2$, and $z = 0$. Use the outward normal.

In Exercises 17 through 19, write the surface integrals using differential notation.

17. $\iint_M \mathbf{F}\cdot\mathbf{n}\,d\sigma$, where $\mathbf{F}(x, y, z) = (x\sin y, z\sin y, x\cos z)$.

18. $\iint_M \mathbf{F}\cdot\mathbf{n}\,d\sigma$, where $\mathbf{F}(x, y, z) = (0, 0, x+e^{yz})$.

19. $\iint_M \mathbf{F}\cdot\mathbf{n}\,d\sigma$, where $\mathbf{F} = \mathbf{n}$ and $\mathbf{n}$ is the outward-pointing normal to the sphere $x^2+y^2+z^2 = r^2$.

Evaluate the surface integrals in Exercises 20 through 26.

20. $\iint_M x\,dy\wedge dz + y\,dz\wedge dx + z\,dx\wedge dy$, where M is parametrized and oriented by $(x, y, z) = (s^3-t, s^2-t^2, s-t^3)$, $-1 \le s \le 1$, $0 \le t \le 1$.

21. $\iint_M (x-z)\,dy\wedge dz$, where M is parametrized and oriented by $(x, y, z) = (te^s, t-s, te^{-s})$, $0 \le s \le \ln 2$, $0 \le t \le 4$.

22. $\iint_M y\cos z\,dy\wedge dz - dz\wedge dx$, where M is parametrized and oriented by $(x, y, z) = (st, s\sin t, t)$, $0 \le s \le 2$, $0 \le t \le \pi$.

23. $\iint_M y\,dy\wedge dz + z\,dz\wedge dx + x\,dx\wedge dy$, where M is a portion of the surface $z = xy^2$ cut off by the planes $x = 0$, $y = 0$ $x + 2y = 4$. (Assume that the normal to M points in the positive z-direction.)

24. $\iint\limits_M dy \wedge dz + dx \wedge dy$, where M is the sphere $x^2 + y^2 + z^2 = 25$, positively oriented outward.

25. $\iint\limits_M dz \wedge dx - dx \wedge dy$, where M is the unit box with outward normal.

26. $\iint\limits_M x\,dy \wedge dz + y\,dz \wedge dx + z\,dx \wedge dy$, where M is the surface of the cylindrical region $x^2 + y^2 \leq 25$, $-1 \leq z \leq 2$.

27. Our intuitive idea of **heat** is that it measures energy stored in a solid, liquid, or gas. For instance, there is far more heat in a 4-liter pot of water at 100 C than in a 10-ml vial of water at the same temperature. We can imagine that the amount of heat in one small part of a solid will be different from that in another part, and that if we measure the amount of heat in one small region near a point (x, y, z) as time t progresses, we'll observe the amount of heat to change. In thermal equilibrium, however, the amount of heat near (x, y, z) does not change in time [though if we look at another point (x', y', z'), the amount of heat near that point may be different from that near (x, y, z)].

For a solid region S in a three-dimensional body with an equilibrium temperature distribution $T(x, y, z)$, the amount of heat in S is defined to be

$$H(S) = \iiint\limits_S kT(x, y, z)\,dV,$$

where k is a constant (the product of the **specific heat** and density of the material.). In effect, we can think of heat as "flowing" through the body; it is a virtual fluid. By physical experiments we can determine that the velocity of this heat fluid at each point (x, y, z) is given by $-K\,\vec{\nabla}T(x, y, z)$, where K is a constant known as **thermal conductivity**. This velocity field says that heat flows in the direction of greatest decrease in temperature. The amount of heat flowing through a surface M, per unit time, is just the flux integral of the heat fluid's velocity field:

$$\{\text{rate of heat flow through } M\} = \iint\limits_M -K\vec{\nabla}T \cdot \mathbf{n}\,d\sigma.$$

Let $T(x, y, z) = x^2 + y^2 - 2z^2$ be a temperature distribution. Find the rate of heat flow through

(a) the portion of the paraboloid $z = x^2 + y^2$, $-1 \leq x \leq 1$, $-1 \leq y \leq 1$.

(b) the sphere $x^2 + y^2 + z^2 = r^2$.

(c) the box bounding $[a, b] \times [c, d] \times [e, f]$.

28. We can extend the formula in (4.6.5) directly to n dimensions. Let $g : \mathbb{R}^n \to \mathbb{R}$ and let M be a two-dimensional "surface" in $\mathbb{R}^n$ parametrized by $\mathbf{x} = \mathbf{f}(s, t)$, where $(s, t) \in R \subset \mathbb{R}^2$. Then the surface integral of g over M is

$$\iint_M g\, d\sigma = \iint_R g(\mathbf{f}(s, t))\sqrt{\|\mathbf{f}_s\|^2 \|\mathbf{f}_t\|^2 - (\mathbf{f}_s \cdot \mathbf{f}_t)^2}\, ds\, dt.$$

Evaluate $\iint_M g\, d\sigma$, where $g(x_1, x_2, x_3, x_4) = x_1 + 3x_2 - x_3^2 + x_4$ and M is parametrized by $\mathbf{x} = \left(s - t,\ 1 + 2s + 4t,\ 5 + \frac{1}{2}s + \frac{1}{3}t,\ s + t\right)$, $0 \le s \le 1$, $0 \le t \le 3$.

4.7 Change of Variables in Double Integrals

In applications, double integrals often arise in which the region of integration is more readily described in coordinates other than rectangular. For instance, let R be the region in the first quadrant of the xy-plane between the lines $y = x/\sqrt{3}$ and $y = \sqrt{3}x$, outside the circle $x^2 + y^2 = 1$ and inside the circle $x^2 + y^2 = 4$. If we are to evaluate $\iint_R \sqrt{x^2 + y^2}\, dA$, then the direct

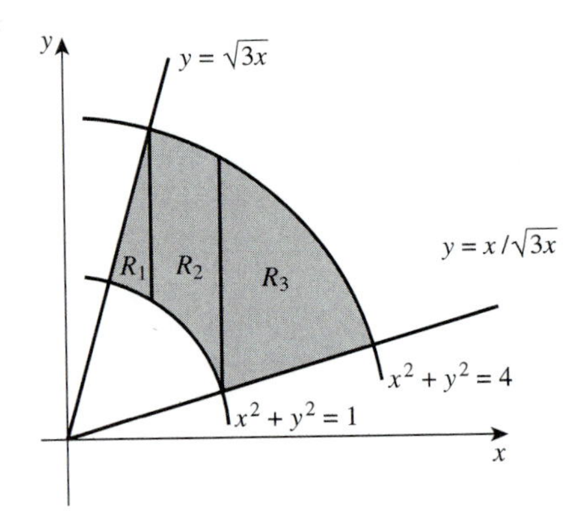

Figure 4.7.1 The integral $\iint_R (x^2 + y^2)^{1/2}\, dA$ can be evaluated by adding the integrals over R_1, R_2, and R_3.

approach is to write the double integral as the sum of three integrals over the y-simple regions shown in Figure 4.7.1. But since R is simply described by the inequalities

$$1 \leq r \leq 2$$
$$\frac{\pi}{6} \leq \theta \leq \frac{\pi}{3}$$

in polar coordinates, it is likely that representing the integral in polar coordinates might remove some of these complications.

Our goal in this section, then, is to derive and illustrate a "substitution" formula for double integrals that will work in this situation and in others.

Consider a double integral $\iint\limits_R g(x, y)\, dx\, dy$, where g is a continuous function and R is the image of a region R^* in the st-plane under a smooth one-to-one function $\mathbf{f} : R^* \to R$. In other words, let R be parametrized by $\mathbf{x} = \mathbf{f}(s, t)$ for $(s, t) \in R^*$. See Figure 4.7.2. Let P^* be a partition of R^*

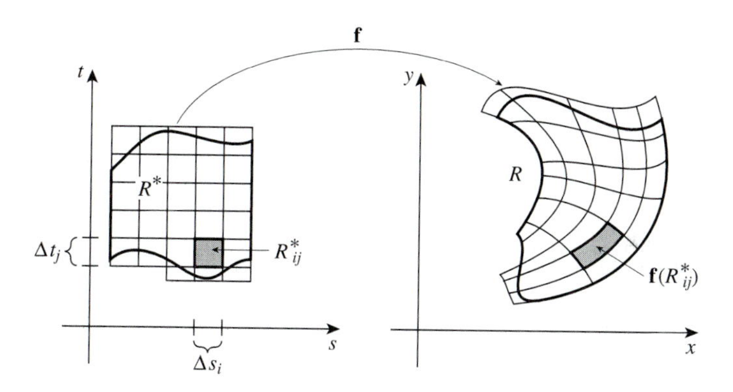

Figure 4.7.2 A region R in the xy-plane as the image of a region R^* in the st-plane under $\mathbf{f}$

into rectangles R_{ij} each with area $\Delta s_i\, \Delta t_j$. By the additive property, part 3 of Theorem 4.3.4, the integral is the sum of the integrals of g over all the pieces $\mathbf{f}(R^*_{ij})$:

$$\iint\limits_R g(x, y)\, dx\, dy = \sum_{R^*_{ij} \in P^*} \iint\limits_{\mathbf{f}(R^*_{ij})} g(x, y)\, dx\, dy. \tag{4.7.1}$$

By the Integral Mean Value Theorem, part 5 of Theorem 4.3.4, there is a point $\mathbf{f}(s_{ij}, t_{ij}) \in R_{ij}$ such that (see Figure 4.7.3)

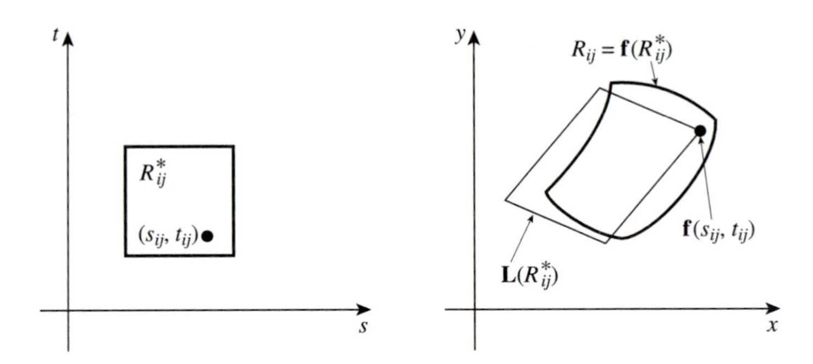

Figure 4.7.3 The area of $\mathbf{L}(R^*_{ij})$ approximates the area of $\mathbf{f}(R^*_{ij})$ when R^*_{ij} is small.

$$\begin{aligned}\iint_{\mathbf{f}(R^*_{ij})} g(x, y)\,dx\,dy &= g(\mathbf{f}(s_{ij}, t_{ij}))A(\mathbf{f}(R^*_{ij}))\\ &= g(\mathbf{f}(s_{ij}, t_{ij}))\iint_{R^*_{ij}}\left|\frac{\partial\mathbf{f}(s, t)}{\partial(s, t)}\right| ds\,dt. \qquad (4.7.2)\end{aligned}$$

Here we have used (4.5.8) to represent the area of $\mathbf{f}(R^*_{ij})$. Now since $\mathbf{f}$ is smooth, and the partition rectangles are small, the area of $\mathbf{f}(R^*_{ij})$ is approximated by the area of $\mathbf{L}(R^*_{ij})$, where

$$\mathbf{L}(s, t) = \mathbf{f}(s_{ij}, t_{ij}) + J\mathbf{f}(s_{ij}, t_{ij})\,((s, t) - (s_{ij}, t_{ij}))$$

is the linear approximation of $\mathbf{f}$ at the point (s_{ij}, t_{ij}):

$$\iint_{R^*_{ij}}\left|\frac{\partial\mathbf{f}(s, t)}{\partial(s, t)}\right| ds\,dt \approx \left|\frac{\partial\mathbf{f}(s_{ij}, t_{ij})}{\partial(s, t)}\right| \Delta s_i\,\Delta t_j. \qquad (4.7.3)$$

Combining (4.7.3), (4.7.2), and (4.7.1), we have

$$\iint_R g(x, y)\,dx\,dy \approx \sum_{R^*_{ij}\in P^*} g(\mathbf{f}(s_{ij}, t_{ij}))\left|\frac{\partial\mathbf{f}(s_{ij}, t_{ij})}{\partial(s, t)}\right| \Delta s_i\,\Delta t_j. \qquad (4.7.4)$$

The right-hand side of (4.7.4) is a Riemann sum for $\iint_{R^*} g(\mathbf{f}(s, t))\left|\frac{\partial\mathbf{f}(s, t)}{\partial(s, t)}\right| ds\,dt$, so as we consider finer partitions P^*, the right-hand side of (4.7.4) approaches this integral. Simultaneously, the approximation in (4.7.4) gets better and better, so

$$\iint_R g(x, y)\,dx\,dy = \lim_{\delta\to 0+} \sum_{\substack{R^*_{ij}\in P^*\\ N(P^*)<\delta}} g(\mathbf{f}(s_{ij}, t_{ij}))\left|\frac{\partial\mathbf{f}(s_{ij}, t_{ij})}{\partial(s, t)}\right| \Delta s_i\,\Delta t_j$$

$$= \iint_{R^*} g(\mathbf{f}(s, t))\left|\frac{\partial\mathbf{f}(s, t)}{\partial(s, t)}\right| ds\,dt.$$

This proves the following theorem.

Theorem 4.7.1 *(Change of variables formula for double integrals) Let R be a region in the xy-plane that is parametrized by $(x, y) = \mathbf{f}(s, t)$ for $(s, t) \in R^*$, where $\mathbf{f}$ is a smooth one-to-one function and R^* is a region in the st-plane. For any function g that is continuous on R,*

$$\iint_R g(x, y)\,dx\,dy = \iint_{R^*} g(\mathbf{f}(s, t))\left|\frac{\partial\mathbf{f}(s, t)}{\partial(s, t)}\right| ds\,dt. \tag{4.7.5}$$

In the following examples, we illustrate how Theorem 4.7.1 can be used.

Example 4.7.1 Let us return to the original question posed in this section, to evaluate $\iint_R \sqrt{x^2+y^2}\,dx\,dy$, where R is the region in the first quadrant described using polar coordinates by

$$1 \le r \le 2$$
$$\frac{\pi}{6} \le \theta \le \frac{\pi}{3}.$$

Since R is the image of the region $R^* = \{(r, \theta)\,|\,1 \le r \le 2,\ \pi/6 \le \theta \le \pi/3\}$ under the transformation $\mathbf{f}(r, \theta) = (r\cos\theta, r\sin\theta)$, we can use Theorem 4.7.1. The Jacobian determinant of $\mathbf{f}$ is

$$\frac{\partial\mathbf{f}(r, \theta)}{\partial(r, \theta)} = \begin{vmatrix} \cos\theta & -r\sin\theta \\ \sin\theta & r\cos\theta \end{vmatrix} = r\cos^2\theta + r\sin^2\theta = r,$$

so, by (4.7.5),

$$\iint_R \sqrt{x^2+y^2}\,dx\,dy = \iint_{R^*} \sqrt{(r\cos\theta)^2 + (r\sin\theta)^2}\,r\,dr\,d\theta$$

$$= \int_{\pi/6}^{\pi/3}\int_1^2 r^2\,dr\,d\theta$$

$$= \int_{\pi/6}^{\pi/3} \frac{1}{3}r^3\Big|_1^2\,d\theta = \frac{7}{3}\int_{\pi/6}^{\pi/3} d\theta = \frac{7\pi}{18}. \quad \blacklozenge$$

Changing variables to polar coordinates is common enough that we should write down a formula for this special case. In the example we already saw that

$$\frac{\partial(r\cos\theta,\, r\sin\theta)}{\partial(r,\,\theta)} = r,$$

so if R is a region in the xy-plane which is the image of $R^* = \{(r,\,\theta)\,|\, f_1(\theta) \le r \le f_2(\theta),\ \theta_1 \le \theta \le \theta_2\}$, then

$$\iint_R g(x,\,y)\,dx\,dy = \int_{\theta_1}^{\theta_2}\int_{f_1(\theta)}^{f_2(\theta)} g(r\cos\theta,\, r\sin\theta) r\,dr\,d\theta. \tag{4.7.6}$$

Example 4.7.2 Evaluate $\displaystyle\iint_R \frac{1}{\sqrt{1-x^2-y^2}}dx\,dy$, where R is the disk of diameter 1 centered at $(1/2,\,0)$.

Solution The boundary of R is the circle $(x-1/2)^2+y^2 = 1/4$. You should verify that in polar coordinates this circle is parametrized by $r = \cos\theta$, for $-\pi/2 \le \theta \le \pi/2$. Thus the disk is the image of $R^* = \{(r,\,\theta)\,|\, -\pi/2 \le r \le \cos\theta,\ 0 \le \theta \le \pi/2\}$ under the polar-to-rectangular coordinate transformation. See Figure 4.7.4. Thus, converting the integral to polar

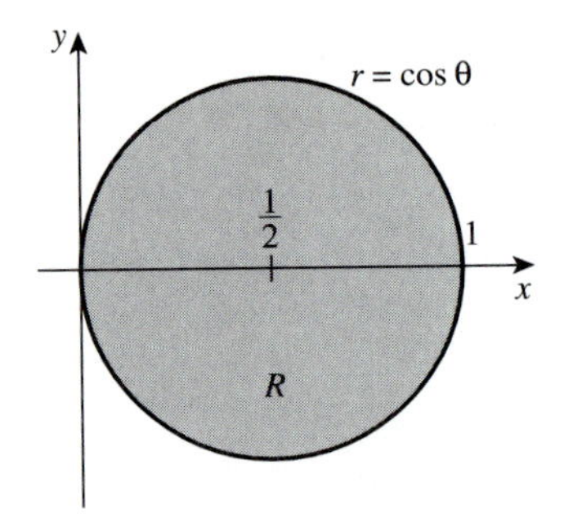

Figure 4.7.4 The disk $R = \{(x,\,y)\,|\,(x-1/2)^2+y^2 = 1/4\}$ is the image of $R^* = \{(r,\,\theta)\,|\,0 \le r \le \cos\theta,\ -\pi/2 \le \theta \le \pi/2\}$ under the polar-to-rectangular coordinate transformation.

coordinates and using (4.7.6), we have

$$\begin{aligned}\iint_R \frac{1}{\sqrt{1-x^2-y^2}}\,dx\,dy &= \int_{-\pi/2}^{\pi/2}\int_0^{\cos\theta} \frac{1}{\sqrt{1-r^2}} r\,dr\,d\theta \\ &= \int_{-\pi/2}^{\pi/2} -\frac{1}{2}\cdot 2\sqrt{1-r^2}\,\Big|_0^{\cos\theta}\, d\theta\end{aligned}$$

$$= \int_{-\pi/2}^{\pi/2} (1 - \sin\theta)\, d\theta$$

$$= (\theta + \cos\theta)\Big|_{-\pi/2}^{\pi/2} = \pi. \quad \blacklozenge$$

Sometimes we must determine the right change of variables from the context, as we see in the next example.

Example 4.7.3 Evaluate

$$\iint\limits_R \frac{x - y}{(x + 2y)^2}\, dy\, dx, \tag{4.7.7}$$

where R is the region bounded by the lines $2y+x = 2$, $2y+x = 4$, $y-x = -3$, and $y - x = 0$.

Solution The region is the parallelogram shown in Figure 4.7.5. If we let

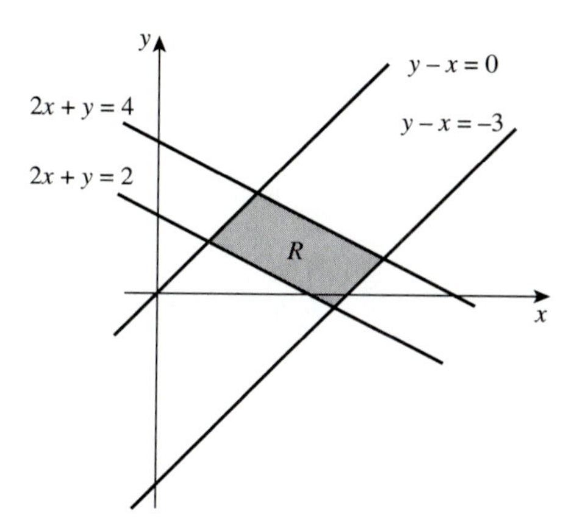

Figure 4.7.5 The parallelogram bounded by the lines $2y + x = 2$, $2y + x = 4$, $y - x = -3$, and $y - x = 0$

$s = 2y + x$ and $t = y - x$, then the region is neatly described by

$$\begin{aligned} 2 \leq\ & s \leq 4 \\ -3 \leq\ & t \leq 0. \end{aligned}$$

In other words, R is the image of $R^* = \{(s, t) \mid 2 \leq s \leq 4,\ -3 \leq t \leq 0\}$ under the transformation implicitly defined by these equations. Since our change of variables formula (4.7.5) requires the Jacobian of (x, y) with respect to (s, t), it appears that we must first solve for x and y in terms of s and t. This is not much trouble for the two linear equations

$$2y + x = s \tag{4.7.8}$$

$$y - x = t. \tag{4.7.9}$$

Adding (4.7.8) and (4.7.9), we obtain $3y = s + t$, so

$$y = \frac{1}{3}s + \frac{1}{3}t. \tag{4.7.10}$$

Substituting (4.7.10) into (4.7.8), we have $x + \frac{2}{3}s + \frac{2}{3}t = s$;

$$x = \frac{1}{3}s - \frac{2}{3}t. \tag{4.7.11}$$

Then we use (4.7.11) and (4.7.10) to calculate

$$\frac{\partial(x, y)}{\partial(s, t)} = \begin{vmatrix} 1/3 & -2/3 \\ 1/3 & 1/3 \end{vmatrix} = \frac{1}{9} + \frac{2}{9} = \frac{1}{3}.$$

By (4.7.5), our integral is

$$\begin{aligned}
\iint\limits_R \frac{x - y}{(x + 2y)^2}\, dy\, dx &= \int_{-3}^{0}\int_{2}^{4} -\frac{t}{s^2}\cdot\frac{1}{3}\, ds\, dt \\
&= \frac{1}{3}\int_{-3}^{0} t \cdot \frac{1}{s}\Big|_2^4\, dt = -\frac{1}{12}\int_{-3}^{0} t\, dt \\
&= -\frac{1}{24}t^2\Big|_{-3}^{0} = \frac{3}{8}. \quad \blacklozenge
\end{aligned}$$

In Example 4.7.3, we first inverted the transformation relating x and y to s and t, and then we calculated $\partial(x, y)/\partial(s, t)$ directly. There is an alternative approach.

Suppose that x and y are related implicitly to s and t by the vector equation

$$\begin{bmatrix} s \\ t \end{bmatrix} = \mathbf{f}(x, y), \tag{4.7.12}$$

where $\mathbf{f}$ is given. Now let $\mathbf{F}$ be the (unknown) function that gives x and y explicitly in terms of s and t. Then

$$\begin{bmatrix} x \\ y \end{bmatrix} = \mathbf{F}(s,\, t). \tag{4.7.13}$$

Substituting (4.7.13) into (4.7.12), we get

$$\begin{bmatrix} s \\ t \end{bmatrix} = \mathbf{f}(\mathbf{F}(s,\, t)). \tag{4.7.14}$$

When we calculate the Jacobian matrix on both sides of (4.7.14) and use the Chain Rule on the right, we have

$$\begin{aligned} J\begin{bmatrix} s \\ t \end{bmatrix} &= J(\mathbf{f}(\mathbf{F}(s,\, t))) \\ \begin{bmatrix} 1 & 0 \\ 0 & 1 \end{bmatrix} &= J\mathbf{f}(\mathbf{F}(s,\, t))J\mathbf{F}(s,\, t). \end{aligned} \tag{4.7.15}$$

Thus the Jacobian matrices are inverses of one another. By Exercise 22 in Section 2.2, their determinants are reciprocals:

$$\frac{\partial(x,\, y)}{\partial(s,\, t)} = \det(J\mathbf{F}(s,\, t)) = \frac{1}{\det(J\mathbf{f}(\mathbf{F}(s,\, t)))} = \frac{1}{\partial(s,\, t)/\partial(x,\, y)}. \tag{4.7.16}$$

Formula (4.7.16) can be put to good use in many instances.

Example 4.7.4 Evaluate $\displaystyle\iint_R (x^2 - y^2)\, dx\, dy$, where R is the region in the first quadrant of the xy-plane bounded by the lines $y - x = 0$, $y - x = 1$ and the curves $xy = 1$ and $xy = 2$.

Solution The region in question is shown in Figure 4.7.6. We change to st-coordinates, where $s = y - x$ and $t = xy$. Then R is the image of $R^* = \{(s,\, t)\,|\,0 \le s \le 1,\, 1 \le t \le 2\}$ under the transformation given implicitly by these equations. Proceeding directly to find x and y in terms of s and t becomes an arduous algebra problem, so we attempt to use (4.7.16):

$$\frac{\partial(x,\, y)}{\partial(s,\, t)} = \frac{1}{\partial(s,\, t)/\partial(x,\, y)} = \frac{1}{\begin{vmatrix} -1 & 1 \\ y & x \end{vmatrix}} = \frac{1}{-x - y}.$$

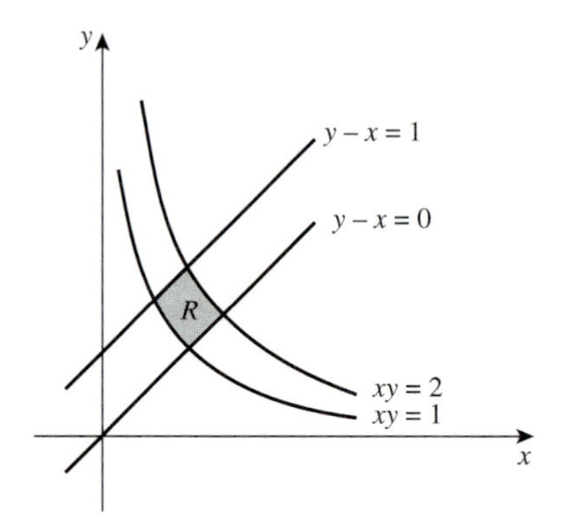

Figure 4.7.6 The region in the xy-plane bounded by the lines $y - x = 0$, $y - x = 1$ and the curves $xy = 1$ and $xy = 2$

Thus we have

$$\begin{aligned}
\iint\limits_{R}(x^2 - y^2)\,dx\,dy &= \iint\limits_{R^*}(x^2 - y^2)\cdot\left|\frac{1}{-x-y}\right|\,ds\,dt \\
&= \iint\limits_{R^*}\frac{(x^2 - y^2)}{x+y}\,ds\,dt \\
&= \iint\limits_{R^*}-(y-x)\,ds\,dt = \iint\limits_{R^*}-s\;ds\,dt \\
&= -\int_0^1\int_1^2 s\,dt\,ds = -\int_0^1 s\,ds = -\frac{1}{2}s^2\bigg|_0^1 = -\frac{1}{2}. \quad \blacklozenge
\end{aligned}$$

As a final example, we show how changing variables can be used to evaluate an **improper double integral**, that is a double integral in which the region of integration is unbounded. Let $f(x, y)$ be a nonnegative[2] continuous function on an unbounded region R in the xy-plane, and let (a, b) be any point in R. Then for any positive number c, f is integrable on the portion of R that lies in the disk D_c of radius c centered at (a, b). See Figure 4.7.7. That is, we can calculate $\iint_{R\cap D_c} f\,dA$ for each choice of c. As we take c to be larger and larger, these integrals either approach a number or they do not. If they do, we say that the improper integral $\iint_R f\,dA$ **converges**, and

[2]Somewhat surprisingly, we could run into trouble if we tried to define the improper double integral of continuous functions that take on both positive and negative values. See [5] for more details.

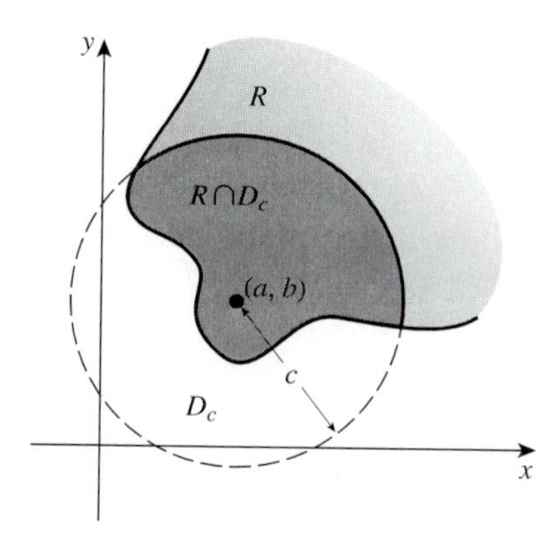

Figure 4.7.7 The improper integral of f over R is the limit as $c \to \infty$ of proper integrals over $R \cap D_c$.

we write

$$\iint_R f\,dA = \lim_{c\to\infty} \iint_{R\cap D_c} f\,dA; \tag{4.7.17}$$

otherwise, we say that it **diverges**.

Example 4.7.5 Evaluate $\displaystyle\iint_R e^{-x^2-y^2}\,dA$, where R is the first quadrant of the xy-plane.

Solution We consider first a proper double integral over $R \cap D_c$, where D_c is a disk of radius c centered at $(0,\,0)$. This is just the sector given in polar coordinates by $0 \le \theta \le \pi/2$ and $0 \le r \le c$. Converting the integral to polar coordinates, we have

$$\begin{aligned}\iint_{R\cap D_c} e^{-x^2-y^2}\,dA &= \int_0^{\pi/2}\int_0^c e^{-r^2} r\,dr\,d\theta \\ &= \int_0^{\pi/2} -\frac{1}{2}e^{-r^2}\Big|_0^c\,d\theta \\ &= \frac{1}{2}\left(1-e^{-c^2}\right)\int_0^{\pi/2} d\theta \\ &= \frac{1}{4}(1-e^{-c^2})\pi.\end{aligned}$$

Then

$$\iint_R e^{-x^2-y^2}\,dA = \lim_{c\to\infty}\iint_{R\cap D_c} e^{-x^2-y^2}\,dA$$

$$= \lim_{c\to\infty} \frac{1}{4}(1 - e^{-c^2})\pi = \frac{\pi}{4}. \quad \blacklozenge$$

If R is, for instance, the quarter-plane described by $a \le x$ and $b \le y$ and $\iint\limits_R f\,dA$ converges, then it follows that $\int_b^\infty f(x,\,y)\,dy$ converges for each choice of x in the interval $[a,\,\infty)$, and also that $\int_a^\infty \left(\int_b^\infty f(x,\,y)\,dy\right)dx$ converges. In other words, a double integral over a region of this special type can be evaluated as an iterated improper integral.

Example 4.7.6 Evaluate $\iint\limits_R \frac{1}{x^2y^3}\,dA$, where R is the region described by $1 \le x$ and $5 \le y$.

Solution If the integral converges, then it is equal to the iterated integral $\int_1^\infty \int_5^\infty \frac{1}{x^2y^3}\,dy\,dx$. Now

$$\int_5^\infty \frac{1}{x^2y^3}\,dy = \lim_{b\to\infty} \int_5^b \frac{1}{x^2y^3}\,dy = \lim_{b\to\infty} \frac{1}{x^2} \cdot \left.\frac{-1}{2y^2}\right|_5^b = \frac{1}{x^2}\frac{1}{50}$$

for each $x \ge 1$. Thus

$$\int_1^\infty \int_5^\infty \frac{1}{x^2y^3}\,dy\,dx = \int_1^\infty \frac{1}{50}\frac{1}{x^2}\,dx = \lim_{b\to\infty} \int_1^b \frac{1}{50}\frac{1}{x^2}\,dx = \frac{1}{50}. \quad \blacklozenge$$

Exercises 4.7

In Exercises 1 through 9, evaluate the double integral by converting to polar coordinates.

1. $\iint\limits_R \frac{y}{x}\,dA$; R is the region in the first quadrant bounded by the lines $y = 0$, $y = x$, and the circles $x^2 + y^2 = 2$ and $x^2 + y^2 = 25$.

2. $\iint\limits_R (x^2 + y^2)^{3/2}\,dA$; R is the half-disk $x^2 + y^2 \le 1$, $x \le 0$.

3. $\displaystyle\iint_R \frac{dx\,dy}{\sqrt{x^2+y^2}}$; R is the annular region given by $4 \le x^2 + y^2 \le 9$.

4. $\displaystyle\iint_R (x^2+y^2)\,dA$; R is the disk $x^2+y^2 \le 4x$.

5. $\displaystyle\iint_R \arctan y/x\,dA$; R is the sector in the first quadrant between the circles $\frac{1}{4} = x^2+y^2$ and $x^2+y^2 = 1$ and the lines $y = x/\sqrt{3}$ and $y = x$.

6. $\displaystyle\int_0^1 \int_0^{\sqrt{1-x^2}} x^2 y\,dy\,dx$

7. $\displaystyle\int_{-2}^2 \int_{-\sqrt{4-y^2}}^{\sqrt{4-y^2}} e^{-x^2-y^2}\,dx\,dy$

8. $\displaystyle\int_0^4 \int_x^4 (x^2+y^2)^{3/2}\,dy\,dx$

9. $\displaystyle\int_1^2 \int_{-y}^{y/\sqrt{3}} \frac{1+\sqrt{x^2+y^2}}{\sqrt{x^2+y^2}}\,dx\,dy$

10. Complete Exercise 18 in Section 4.3.

11. Complete Exercise 25 in Section 4.3.

12. Show that the volume of a right circular cone of base radius R and height h is $V = (\pi/3)R^2h$. (*Hint*: The cone is the solid bounded by $z = h$ and $z = k\sqrt{x^2+y^2}$ for an appropriate constant k. Use polar coordinates to evaluate the double integral.)

13. Show that the volume inside a sphere of radius R is $V = (4\pi/3)R^3$.

In Exercises 14 through 18, evaluate the double integral by making an appropriate change of variables.

14. $\displaystyle\iint_R (x+y)^2 dA$; R is the region bounded by the lines $x+y=0$, $x+y=1$, $2x-y=0$, and $2x-y=1$.

15. $\displaystyle\iint_R \frac{1}{y}\,dx\,dy$; R is the region bounded by $y^3 = x^2$, $y^3 = 6x^2$, $y = 2x$, and $y = 3x$.

16. $\displaystyle\iint_R \frac{x^2 \sin xy}{y}\,dx\,dy$; R is bounded by $x^2 = \pi y/2$, $x^2 = \pi y$, $y^2 = x/2$, and $y^2 = x$.

17. $\iint_R x^2\,dx\,dy$; R is bounded by $y=x$, $y=3x$, $y=-1-x$, and $y=-3-x$.

18. $\iint_R \dfrac{9x^2+8y^2}{xy}\,dx\,dy$; R is the region in the first quadrant bounded by the ellipses $x^2/4+y^2/9=1$ and $x^2/16+y^2/36=1$ and the parabolas $y=x^2/2$ and $y=2x^2$.

In Exercises 19 through 21, evaluate the improper double integrals.

19. $\iint_R e^{-x^2-y^2}\,dA$; R is the entire xy-plane.

20. $\iint_R \dfrac{1}{1+(x^2+y^2)^2}\,dA$; R is the portion of the first quadrant bounded by the x-axis and the line $y=x$.

21. $\iint_R \dfrac{1}{(x^2+y^2)^{3/2}}\,dA$; R is the half-plane to the right of the vertical line $x=1$.

22. Use the fact that

$$\left(\int_0^\infty e^{-x^2}\,dx\right)^2 = \left(\int_0^\infty e^{-x^2}\,dx\right)\cdot\left(\int_0^\infty e^{-y^2}\,dy\right) = \int_0^\infty\int_0^\infty e^{-x^2}e^{-y^2}\,dx\,dy$$

to show that $\int_0^\infty e^{-x^2}\,dx = \sqrt{\pi}/2$.

23. Show that

$$\int_{-\infty}^\infty e^{-x^2}\,dx = \sqrt{\pi}. \qquad (4.7.18)$$

24. The **Gaussian**[3] (or **normal**) probability distribution function is given by

$$N(x) = \frac{1}{\sqrt{2\pi\sigma}}e^{-(x-\mu)^2/2\sigma},$$

where μ is the **mean** of the distribution and σ is a positive number called the **standard deviation** of the distribution. (For a fixed μ you might want to graph this function for different values of σ.) By using

[3] Named after the same Karl Friedrich Gauss mentioned in Section 5.3.

the substitution $u^2 = (x-\mu)^2/2\sigma$ and formula (4.7.18), show that the Gaussian distribution satisfies $\int_{-\infty}^{\infty} N(x)\,dx = 1$.

25. The **Laplace[4] transform** of a function f defined on the interval $[0, \infty)$ is the function F defined by $F(u) = \int_0^{\infty} e^{-ux} f(x)\,dx$ for each value of u for which the integral converges.

(a) Show that if f has Laplace transform F and g has Laplace transform G, then

$$F(u)G(u) = \int_0^{\infty}\int_0^{\infty} e^{-u(x+y)} f(x)g(y)\,dx\,dy.$$

(b) By making the change of variables $s = x + y$ and $t = y$, show that

$$F(u)G(u) = \int_0^{\infty}\int_t^{\infty} e^{-us} f(s-t)g(t)\,ds\,dt.$$

(c) Change the order of integration to show that

$$F(u)G(u) = \int_0^{\infty}\int_0^{s} e^{-us} f(s-t)g(t)\,dt\,ds.$$

(d) Conclude that

$$F(u)G(u) = \int_0^{\infty} e^{-us} h(s)\,ds,$$

where $h(s) = \int_0^s f(s-t)g(t)\,dt$ is the **convolution** of f with g. This result is sometimes called the **convolution theorem**; it is often applied to solving differential equations that model the behavior of electrical circuits.

4.8 Change of Variables in Triple Integrals

It is often the case that a triple integral can be evaluated by making an appropriate substitution. For instance, the region of integration may be symmetric about the z-axis and the integrand may be a function of $x^2 + y^2$.

[4]The same Pierre-Simon Laplace for whom the partial differential equation $\partial^2 u/\partial x^2 + \partial^2 u/\partial y^2 = 0$ is named.

In this situation it would be natural to convert to cylindrical coordinates by the change of variables $x = r\cos\theta$, $y = r\sin\theta$, and $z = w$. What is the resulting integral? When such a substitution is made, the answer is basically the same as that for double integrals: The substitution introduces a factor of the Jacobian determinant into the integral, and it also requires us to "change the limits on the integral" (i.e., integrate over an appropriate region in $r\theta w$-space).

In general, if S is a solid region parametrized by $(x, y, z) = \mathbf{f}(s, t, u)$ for $(s, t, u) \in S^* \subset \mathbb{R}^3$, where $\mathbf{f}$ is smooth and one-to-one (see Figure 4.8.1) then, in exact analogy with the change of variables for double integrals,

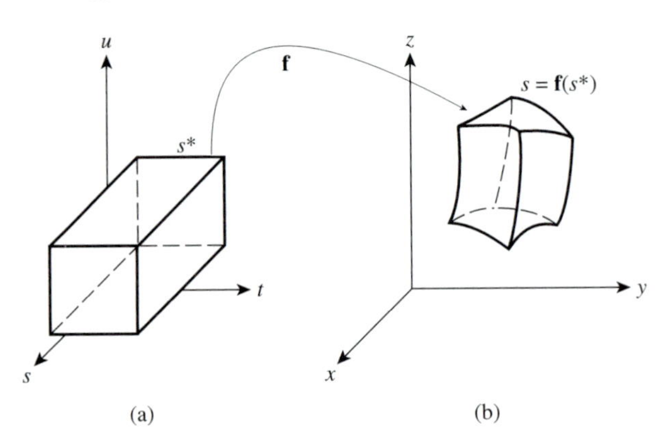

Figure 4.8.1 A solid region S as the image of S^* under a smooth function $\mathbf{f}$

$$\iiint_S g(x, y, z)\,dx\,dy\,dz = \iiint_{S^*} g(\mathbf{f}, (s, t, u)) \left| \frac{\partial \mathbf{f}(s, t, u)}{\partial(s, t, u)} \right| ds\,dt\,du \qquad (4.8.1)$$

for any continuous function g.

The proof of (4.8.1) follows the same line as that for double integrals. For a partition P^* of S^* consisting of small rectangular solids B_{ijk}, each with volume $\Delta s_i\,\Delta t_j\,\Delta u_k$ (see Figure 4.8.2), we have

$$\iiint_S g(x, y, z)\,dx\,dy\,dz = \sum_{B^*_{ijk} \in P^*} \iiint_{\mathbf{f}(B^*_{ijk})} g(x, y, z)\,dx\,dy\,dz. \qquad (4.8.2)$$

By the Integral Mean Value Theorem, part 4 of Theorem 4.4.3, there exists a point $\mathbf{s}_{ijk} = (s_{ijk}, t_{ijk}, u_{ijk}) \in B^*_{ijk}$ such that

$$\iiint_{\mathbf{f}(B^*_{ijk})} g(x, y, z)\,dx\,dy\,dz = g(\mathbf{f}(\mathbf{s}_{ijk}))\,\text{Volume}\,(\mathbf{f}(B^*_{ijk})). \qquad (4.8.3)$$

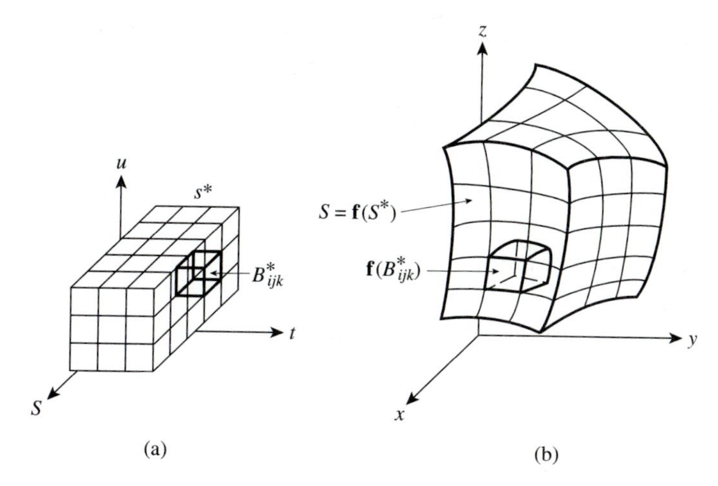

Figure 4.8.2 $\iiint_S g\,dV$ is equal to the sum $\sum_{B^*_{ijk}\in P^*}\iiint_{\mathbf{f}(B^*_{ijk})} g\,dV$ when S^* is a rectangular solid.

Now, because $\mathbf{f}$ is smooth,

$$\text{Volume}\,(\mathbf{f}(B^*_{ijk})) \approx \text{Volume}\,(\mathbf{L}(B^*_{ijk})), \tag{4.8.4}$$

where

$$\mathbf{L}(\mathbf{s}) = \mathbf{f}(\mathbf{s}_{ijk}) + \mathbf{Df}(\mathbf{s}_{ijk})(\mathbf{s} - \mathbf{s}_{ijk})$$

is the linear approximation of $\mathbf{f}$ at $\mathbf{s}_{ijk}$. See Figure 4.8.3. As we know from Section 2.4, a linear transformation maps a parallelepiped to another parallelepiped whose volume is that of the original scaled by the determinant of the matrix that represents the transformation. Thus

$$\begin{aligned}\text{Volume}\,(\mathbf{L}(B^*_{ijk})) &= \text{Volume}\,(\mathbf{Df}(\mathbf{s}_{ijk})(B^*_{ijk}))\\ &= |\det(J\mathbf{f}(\mathbf{s}_{ijk}))|\text{Volume}\,(B^*_{ijk})\\ &= \left|\frac{\partial \mathbf{f}(\mathbf{s}_{ijk})}{\partial(s,\,t,\,u)}\right| \Delta s_i\,\Delta t_j\,\Delta u_k. \end{aligned}\tag{4.8.5}$$

Now substituting (4.8.5) into (4.8.4), and then using these in (4.8.3) and (4.8.2), we obtain

$$\iiint\limits_S g(x,\,y,\,z)\,dx\,dy\,dz$$

$$\approx \sum_{B^*_{ijk}\in P^*} g(\mathbf{f}(\mathbf{s}_{ijk}))|\det(J\mathbf{f}(\mathbf{s}_{ijk}))|\text{Volume}\,(B^*_{ijk})$$

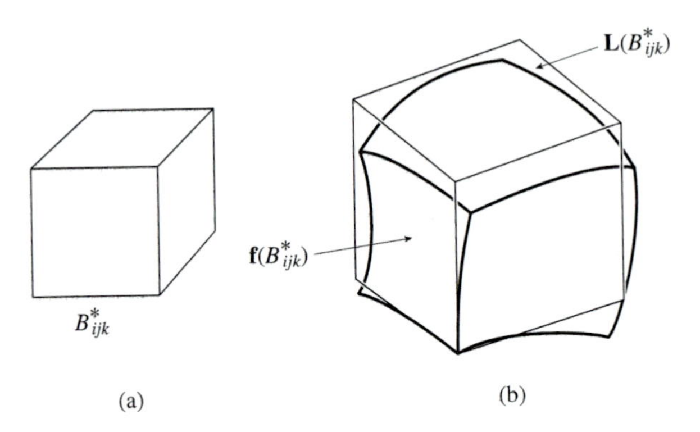

Figure 4.8.3 The image of a "small" rectangular solid B^*_{ijk} under $\mathbf{f}$ is approximated by the image of B^*_{ijk} under the linear approximation $\mathbf{L}$ of $\mathbf{f}$ at any point in B^*_{ijk}.

$$= \sum_{B^*_{ijk} \in P^*} g(\mathbf{f}(\mathbf{s}_{ijk})) \left| \frac{\partial \mathbf{f}(\mathbf{s}_{ijk})}{\partial(s,\, t,\, u)} \right| \Delta s_i \, \Delta t_j \, \Delta u_k. \tag{4.8.6}$$

The right-hand side of (4.8.6) is a Riemann sum for the triple integral

$$\iiint_{S^*} g(\mathbf{f},\, (s,\, t,\, u)) \left| \frac{\partial \mathbf{f}(s,\, t,\, u)}{\partial(s,\, t,\, u)} \right| ds\, dt\, du.$$

For partitions P^* whose norms are close to zero, the Riemann sums approach this triple integral, and simultaneously the approximation in (4.8.6) gets better and better:

$$\begin{aligned} \iiint_S g(x,\, y,\, z)\, dx\, dy\, dz &= \lim_{\delta \to 0+} \sum_{\substack{B^*_{ijk} \in P^* \\ N(P^*) < \delta}} g(\mathbf{f}(\mathbf{s}_{ijk}) \left| \frac{\partial \mathbf{f}(\mathbf{s}_{ijk})}{\partial(s,\, t,\, u)} \right| \Delta s_i \, \Delta t_j \, \Delta u_k \\ &= \iiint_{S^*} g(\mathbf{f},\, (s,\, t,\, u)) \left| \frac{\partial \mathbf{f}(s,\, t,\, u)}{\partial(s,\, t,\, u)} \right| ds\, dt\, du, \qquad (4.8.7) \end{aligned}$$

which is what we claimed was true in (4.8.1).

We illustrate the use of this formula with several examples. First we use it to obtain a change of variables formula for cylindrical coordinates.

Since the transformation from cylindrical to rectangular coordinates is given by

$$\begin{aligned} x &= r \cos\theta & (4.8.8) \\ y &= r \sin\theta & (4.8.9) \\ z &= w, & (4.8.10) \end{aligned}$$

we can calculate the Jacobian determinant directly:

$$\frac{\partial(x,\, y,\, z)}{\partial(r,\, \theta,\, w)} = \begin{vmatrix} \cos\theta & -r\sin\theta & 0 \\ \sin\theta & r\cos\theta & 0 \\ 0 & 0 & 1 \end{vmatrix} = r.$$

If S is a solid region in xyz-space that is parametrized by the equations (4.8.8), (4.8.9), and (4.8.10), for $(r,\, \theta,\, w)$ in a solid region S^* in $r\theta w$-space, then (4.8.1) gives us

$$\iiint_S f(x,\, y,\, z)\, dx\, dy\, dz = \iiint_{S^*} f(r\cos\theta,\, r\sin\theta,\, w)\, r dr\, d\theta\, dw. \qquad (4.8.11)$$

(We agree that the r-coordinate is nonnegative, so absolute value signs on r in this formula are unnecessary.)

Example 4.8.1 Evaluate $\displaystyle\int_{-4}^{4}\int_{-\sqrt{16-x^2}}^{\sqrt{16-x^2}}\int_0^{\sqrt{x^2+y^2}} (z+x)\, dz\, dy\, dx.$

Solution The region of integration is inside the cylinder $x^2 + y^2 \le 16$ between the plane $z = 0$ and the cone $z = \sqrt{x^2 + y^2}$. See Figure 4.8.4. In

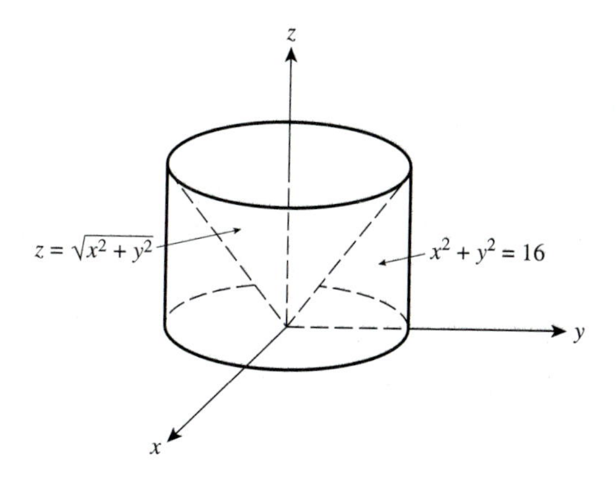

Figure 4.8.4 The solid inside the cylinder $x^2 + y^2 \le 16$ between the plane $z = 0$ and the cone $z = (x^2 + y^2)^{1/2}$

cylindrical coordinates it has the simple description

$$\begin{array}{ccccc} 0 & \le & r & \le & 4 \\ 0 & \le & \theta & \le & 2\pi \\ 0 & \le & w & \le & r \end{array}$$

Thus the integral is, by (4.8.11),

$$\begin{aligned}\int_0^4\int_0^{2\pi}\int_0^r (w + r\cos\theta)r\,dw\,d\theta\,dr &= \int_0^4\int_0^{2\pi}\left(\frac{1}{2}+\cos\theta\right)r^3\,d\theta\,dr \\ &= \int_0^4 \pi r^3\,dr = \frac{\pi}{4}r^4\Big|_0^4 = 64\pi. \quad \blacklozenge\end{aligned}$$

Next, we obtain a change of variables formula for spherical coordinates. The transformation from spherical coordinates to rectangular is given by the three equations

$$x = \rho\sin\phi\cos\theta, \tag{4.8.12}$$

$$y = \rho\sin\phi\sin\theta, \tag{4.8.13}$$

$$z = \rho\cos\phi, \tag{4.8.14}$$

so the Jacobian determinant is

$$\begin{aligned}\frac{\partial(x,\,y,\,z)}{\partial(\rho,\,\theta,\,\phi)} &= \begin{vmatrix}\sin\phi\cos\theta & -\rho\sin\phi\sin\theta & \rho\cos\phi\cos\theta \\ \sin\phi\sin\theta & \rho\sin\phi\cos\theta & \rho\cos\phi\sin\theta \\ \cos\phi & 0 & -\rho\sin\phi\end{vmatrix} \\ &= -\rho^2\sin\phi\cos^2\phi - \rho^2\sin^3\phi \\ &= -\rho^2\sin\phi.\end{aligned}$$

If S is a solid region in xyz-space that is parametrized by (4.8.12), (4.8.13), and (4.8.14) for $(\rho,\,\theta,\,\phi)$ in a solid region S^* of $\rho\theta\phi$-space, then (4.8.1) gives us (note that we take the *absolute value* of the Jacobian!)

$$\begin{aligned}&\iiint_S f(x,\,y,\,z)\,dx\,dy\,dz \qquad (4.8.15)\\ &= \iiint_{S^*} f(\rho\sin\phi\cos\theta,\ \rho\sin\phi\sin\theta,\ \rho\cos\phi)\rho^2\sin\phi\,d\rho\,d\theta\,d\phi.\end{aligned}$$

Example 4.8.2 Recall Example 4.4.5 in Section 4.4 where we had set up but not evaluated a triple integral to find the mass of a solid with a specified density. The challenge there was to evaluate the iterated integral

$$I = \int_{-1}^{1}\int_{-\sqrt{1-z^2}}^{\sqrt{1-z^2}}\int_{-\sqrt{1-y^2-z^2}}^{\sqrt{1-y^2-z^2}} k\sqrt{x^2+y^2+z^2}\,dx\,dy\,dz.$$

The solid region over which we integrate is the ball described by $\rho \leq 1$, $0 \leq \theta \leq 2\pi$, and $0 \leq \phi \leq \pi$. Thus, by (4.8.15),

$$\begin{aligned} I &= \int_0^1 \int_0^{\pi} \int_0^{2\pi} k\rho \cdot \rho^2 \sin\phi \, d\theta \, d\phi \, d\rho \\ &= 2\pi k \int_0^1 \int_0^{\pi} \rho^3 \sin\phi \, d\phi \, d\rho \\ &= 2\pi k \int_0^1 \rho^3(-\cos\phi)\Big|_0^{\pi} d\rho = 4\pi k \cdot \frac{1}{4}\rho^4\Big|_0^1 = \pi k. \end{aligned}$$

This integration is immensely simpler than carrying out the calculations using rectangular coordinates! ♦

In some cases, the change of variables may be given implicitly by a set of equations. For instance, S might be the solid bounded by the six planes $x+y+z=0$, $x+y+z=7$, $2x+y+3z=0$, $2x+y+3z=2$, $-x+y-z=1$, and $-x+y-z=8$. In this case it would be reasonable to let $s=x+y+z$, $t=2x+y+3z$, and $u=-x+y-z$ so that the S is the image of $S^* = [0,\, 7] \times [0,\, 2] \times [1,\, 8]$ under the function of s, t, and u defined implicitly by these equations. Here we can solve for x, y, and z in terms of s, t, and u to find this function, but this may be impractical or impossible in other situations. Nevertheless, we can sometimes still make a change of variables by using the three-variable analog to (4.7.16).

Suppose that $\mathbf{f}$ is a given differentiable function and that $\mathbf{F}$ is another differentiable function that satisfies

$$\mathbf{F}(\mathbf{f}(s,\, t,\, u)) = \begin{bmatrix} s \\ t \\ u \end{bmatrix}.$$

With

$$\begin{bmatrix} s \\ t \\ u \end{bmatrix} = \mathbf{f}(x,\, y,\, z)$$

and

$$\begin{bmatrix} x \\ y \\ z \end{bmatrix} = \mathbf{F}(s,\, t,\, u),$$

it follows (see the exercises) that

$$\frac{\partial(x,y,z)}{\partial(s,t,u)} = \frac{1}{\partial(s,t,u)/\partial(x,y,z)}. \tag{4.8.16}$$

We illustrate the use of (4.8.16) in the next example.

Example 4.8.3 Evaluate $\iiint_S x^2\,dx\,dy\,dz$, where S is the solid bounded by the six planes $x+y=1$, $x+y=2$, $2x+y+z=0$, $2x+y+z=1$, $x-2y=-1$, and $x-2y=1$.

Solution A reasonable change of coordinates is to let

$$\begin{aligned} s &= x+y, \\ t &= 2x+y+z, \\ u &= x-2y, \end{aligned}$$

so that S is the image of $S^* = \{(s,\,t,\,u) | 1\le s\le 2,\ \ 0\le t\le 1,\ \ -1\le u\le 1\}$ under the transformation whose inverse is given by these formulas. Since this is a linear transformation its inverse is also linear, and we could readily calculate it. Instead, we use (4.8.16). Because

$$\frac{\partial(s,t,u)}{\partial(x,y,z)} = \begin{vmatrix} 1 & 1 & 0 \\ 2 & 1 & 1 \\ 1 & -2 & 0 \end{vmatrix} = 3,$$

formula (4.8.16) gives us

$$\frac{\partial(x,y,z)}{\partial(s,t,u)} = \frac{1}{3}.$$

It is necessary to do a bit of algebra to find x in terms of s, t, and u; you should verify that $x = \frac{2}{3}s + \frac{1}{3}u$. Then, by (4.8.1),

$$\begin{aligned} \iiint_S x^2 dx\,dy\,dz &= \int_1^2\int_0^1\int_{-1}^1 \left(\frac{2}{3}s+\frac{1}{3}t\right)^2 \frac{1}{3}\,du\,dt\,ds \\ &= \frac{1}{3}\int_1^2\int_0^1 \left(\frac{2}{3}s+\frac{1}{3}t\right)^2 u\Big|_{-1}^{1}\,dt\,ds \\ &= \frac{2}{3}\int_1^2 \left(\frac{2}{3}s+\frac{1}{3}t\right)^3\Big|_0^1\,ds \\ &= \left[\frac{1}{4}\left(\frac{2}{3}s+\frac{1}{3}\right)^4 - \frac{1}{2}s^4\right]\Bigg|_1^2 \\ &= \frac{76}{81}. \quad \blacklozenge \end{aligned}$$

Exercises 4.8

In Exercises 1 through 5, evaluate the triple integrals by changing to cylindrical coordinates.

1. $\displaystyle\iiint_S \left(1 + \frac{x}{\sqrt{x^2+y^2}}\right) dx\,dy\,dz$; S is the solid bounded by the paraboloids $z = x^2 + y^2$, $z = 1 - x^2 - y^2$.

2. $\displaystyle\iiint_S z\,dx\,dy\,dz$; S is the solid above the cone $z = \sqrt{x^2+y^2}$ and beneath the paraboloid $z = 1 - x^2 - y^2$.

3. $\displaystyle\int_{-1}^{0}\int_{-\sqrt{1-x^2}}^{\sqrt{1-x^2}}\int_{-1}^{3-x^2-y^2} z\,dz\,dy\,dx$

4. $\displaystyle\int_{0}^{1}\int_{-\sqrt{1/4-(y-1/2)^2}}^{\sqrt{1/4-(y-1/2)^2}}\int_{x^2+y^2}^{4} (x+y)\,dz\,dx\,dy$

5. $\displaystyle\int_{0}^{1/\sqrt{2}}\int_{x}^{\sqrt{1-x^2}}\int_{x^2+y^2-1}^{\sqrt{x^2+y^2}} (x^2+y^2)\,dz\,dy\,dx$

In Exercises 6 through 12, evaluate the integrals by changing to spherical coordinates.

6. $\displaystyle\iiint_S \sqrt{x^2+y^2+z^2}\,dx\,dy\,dz$; S is the solid in the first octant bounded by the coordinate planes and the sphere $x^2 + y^2 + z^2 = 9$.

7. $\displaystyle\iiint_S (x+y)\,dx\,dy\,dz$; S is the solid bounded above by the sphere $x^2 + y^2 + z^2 = 16$ and below by the cone $z = \sqrt{3x^2 + 3y^2}$.

8. $\displaystyle\iiint_S \frac{x}{\sqrt{x^2+y^2+z^2}}\,dx\,dy\,dz$; S is the solid in the first octant bounded by the planes $y = x$ and $y = \sqrt{3}x$, the cone $z = \sqrt{x^2+y^2}$, the plane $z = 0$, and the spheres $x^2 + y^2 + z^2 = 2$ and $x^2 + y^2 + z^2 = 8$.

9. $\displaystyle\int_{-2}^{2}\int_{-\sqrt{4-y^2}}^{\sqrt{4-y^2}}\int_{-\sqrt{4-y^2-z^2}}^{\sqrt{4-y^2-z^2}} (x^2+y^2+z^2)^{3/2}\,dx\,dz\,dy$

10. $\displaystyle\int_0^{1/2}\int_y^{\sqrt{1/2-y^2}}\int_{\sqrt{x^2+y^2}}^{\sqrt{1-x^2-y^2}} \frac{1}{1+(x^2+y^2+z^2)^{3/2}}\,dz\,dx\,dy$

11. $\displaystyle\int_0^{3}\int_0^{\sqrt{9-x^2}}\int_0^{\sqrt{9-x^2-y^2}} xy\,dz\,dy\,dx$

12. $\displaystyle\int_0^{1/\sqrt{2}}\int_y^{\sqrt{1-y^2}}\int_0^{\sqrt{1-x^2-y^2}} (z+y)\,dz\,dx\,dy$

In Exercises 13 through 15, evaluate the triple integrals by making an appropriate change of coordinates.

13. $\displaystyle\iiint_S (x+z)^2\,dx\,dy\,dz$; S is the solid bounded by the planes $x+y+3z=0$, $x+y+3z=1$, $y+z=0$, $y+z=1$, $x+z=-1$, and $x+z=1$.

14. $\displaystyle\iiint_S \frac{yz+z^2}{x}\,dV$; S is the solid bounded by the planes $z=x/3$, $z=4x$, $y+z=2$, $y+z=5$, $x=6$, and $x=8$.

15. $\displaystyle\iiint_S \cos y\,dx\,dy\,dz$; S is the solid bounded by the surfaces $z=\sin y+1$, $z=\sin y-1$ and the planes $y=0$, $y=\pi/2$, $x+z=0$, and $x+z=1$.

16. Find the volume remaining when the solid bounded by the paraboloid $z=a(x^2+y^2)$ and the plane $z=h$ has a cylindrical hole of radius R bored through its center. (Here a, h, and R are constants.)

17. Find the volume of the solid remaining when a sphere of radius R has a cylindrical hole of radius r bored through its center.

18. Find the volume of the solid that remains when the portion of the ellipsoidal solid bounded by $x^2/9+y^2/9+z^2=1$ that lies inside the sphere $x^2+y^2+z^2=4$ is removed.

If a function g of three variables is continuous and nonnegative on an unbounded region S in $\mathbb{R}^3$, then the **improper integral** of g over S is defined by

$$\iiint_S g\,dV = \lim_{c\to\infty}\iiint_{S\cap B_c} g\,dV,$$

where B_c is a ball of radius c centered at any point $\mathbf{a}$ in S, provided that the limit exists. (It turns out that for a g such as this, it does not matter where the ball is centered. Also, it turns out that if we used cubes or other shapes instead of balls, the limit would be the same.) Use this definition to evaluate the improper triple integrals in Exercises 19 and 20.

19. $\displaystyle\iiint_S e^{-\sqrt{x^2+y^2+z^2}}\,dx\,dy\,dz$; S is all of $\mathbb{R}^3$.

20. $\displaystyle\iiint_S \frac{1}{(x^2+y^2+z^2+a^2)^{3/2}}\,dV$; S is the first octant (and a is a nonzero constant).

Chapter 5

Fundamental Theorems

In elementary calculus, the Fundamental Theorem is a statement of the connection between the apparently unrelated processes of differentiation and integration. Paraphrased, it says that "Integration is antidifferentiation." In this fact lies the true power of calculus as a tool for both reasoning and calculating.

The process of taking the line integral of certain vector fields, it turns out, amounts to a sort of antidifferentiation. This fact is the **Fundamental Theorem of Line Integrals**, the first in a list of fundamental theorems that generalize the "ordinary" Fundamental Theorem of elementary calculus.

When we restrict our attention to $\mathbb{R}^2$ and vector fields on it, we find that there is an important connection between the line integral of the field around a simple closed path and the double integral of a "derivative" (the two-dimensional curl) of the field over the region the curve bounds. This result is known as **Green's Theorem**. To illustrate its uses, we produce a formula for the area of a region in terms of a line integral over its boundary.

When the context is $\mathbb{R}^3$, the analog to Green's Theorem is the **Divergence Theorem** or Gauss's Theorem. This result provides a connection between the surface integral of a vector field over a surface that encloses a solid region of space and the triple integral of a "derivative" (the divergence) of the vector field over this solid. We illustrate the utility of this theorem by using it to deduce properties of solutions of partial differential equations.

Stokes' Theorem provides a connection between the line integral of a vector field around a closed path and the integral of a "derivative" (the three-dimensional curl) of the field over a surface for which the path is a "boundary." We apply Stokes' Theorem to answer some questions left unanswered in Section 5.1.

5.1 Path-Independent Vector Fields

In physical situations, the path integral of a force field represents the work done by the force on an object moving along that path as it is influenced by the vector field. The negative of this number is the work that must be done by forces that keep the object on the path. For instance, a spacecraft such as Voyager that is designed to follow a specified path among the planets under the influence of their combined gravitational fields must, at critical moments, fire small rockets to keep it on track. Under such simplifying assumptions as (1) that the gravitational field is time independent and (2) that the spacecraft's mass is negligible in comparison with those of the planets, the amount of energy necessary to move the Voyager along a given path would be the path integral of the planets' combined gravitational field.

It seems likely that to get from a point $\mathbf{a}$ to another $\mathbf{b}$ along one path might require more energy than going from $\mathbf{a}$ to $\mathbf{b}$ along a different path. However, it turns out that for some vector fields, the energy needed to get from one point to another is independent of the path followed.

Path-independence is equivalent to the vector field being the gradient of some scalar-valued function. In this section we obtain a way of detecting when a given vector field is the gradient of a scalar function. In addition, we will be able to integrate such gradient fields to obtain the corresponding scalar field.

To get started, we need to define a concept that is important throughout this section.

Definition 5.1.1 A set A in $\mathbb{R}^n$ is **connected** if, given any two points $\mathbf{a}$ and $\mathbf{b}$ in A, we can find a continuous path lying entirely in A that joins the two points.

This definition captures our intuitive notion that "connected" should mean "all in one piece." For example, $A = \{(x, y) \mid x^2 + y^2 \leq 1\}$ is a connected set: If $\mathbf{a}$ and $\mathbf{b}$ are points in this disk, then the path given by $\mathbf{x} = \mathbf{a} + t(\mathbf{b} - \mathbf{a})$, $0 \leq t \leq 1$ lies entirely inside the disk. On the other hand, the set $B = \{(x, y) \mid |y| > 1\}$ is not connected: Any continuous path joining, for example, the points $(0, -2)$ and $(0, 2)$ must, for some value of its parameter, have y-coordinate 0; but there are no points in B with y-coordinate 0.

Given a differentiable scalar-valued function $f : D \subset \mathbb{R}^n \to \mathbb{R}$, we can produce a vector field on $\mathbb{R}^n$ simply by taking the gradient of f. The following theorem states what happens when we take the path integral of a vector field obtained in this way.

Theorem 5.1.1 *(Fundamental Theorem of Line Integrals) Let $D \subset \mathbb{R}^n$ be an open connected set and let $f : D \to \mathbb{R}$ be a scalar-valued function whose gradient is continuous on D. If C is any smooth path lying in D that joins a point* $\mathbf{a}$ *to another point* $\mathbf{b}$*, then*

$$\int_C \vec{\nabla} f \cdot d\mathbf{x} = f(\mathbf{b}) - f(\mathbf{a}). \tag{5.1.1}$$

Proof The proof is a simple consequence of the Chain Rule and the ordinary Fundamental Theorem of Calculus. Let C be parametrized by $\mathbf{x} = \mathbf{g}(t)$, $\alpha \leq t \leq \beta$, so that $\mathbf{g}(\alpha) = \mathbf{a}$ and $\mathbf{g}(\beta) = \mathbf{b}$. Then

$$\begin{aligned}
\int_C \vec{\nabla} f \cdot d\mathbf{x} &= \int_\alpha^\beta \vec{\nabla} f\left(\mathbf{g}(t)\right) \cdot \mathbf{g}'(t)\, dt \\
&= \int_\alpha^\beta \frac{d}{dt} f\left(\mathbf{g}(t)\right)\, dt \\
&= f\left(\mathbf{g}(\beta)\right) - f\left(\mathbf{g}(\alpha)\right) \\
&= f(\mathbf{b}) - f(\mathbf{a}). \quad \blacklozenge
\end{aligned}$$

In differential notation, the formula in (5.1.1) is

$$\int_C \frac{\partial f}{\partial x} dx + \frac{\partial f}{\partial y} dy + \frac{\partial f}{\partial z} dz = f(\mathbf{b}) - f(\mathbf{a}). \tag{5.1.2}$$

Definition 5.1.2 A vector field $\mathbf{F} : D \subset \mathbb{R}^n \to \mathbb{R}^n$ is said to be **path independent** (or **conservative**) if the value of $\int_C \mathbf{F} \cdot d\mathbf{x}$ depends only on the endpoints of the path C.

Theorem 5.1.1 says that $\vec{\nabla} f$ is a path-independent vector field. In fact, gradient fields are the only path-independent fields. This is the content of the next theorem.

Theorem 5.1.2 *Let $D \subset \mathbb{R}^n$ be an open connected set and let $\mathbf{F} : D \to \mathbb{R}^n$ be a continuous vector field. Then* $\mathbf{F}$ *is path independent if and only if there exists a scalar-valued function $f : D \to \mathbb{R}$ such that $\vec{\nabla} f = \mathbf{F}$. If such a function exists, then*

$$\int_C \mathbf{F} \cdot d\mathbf{x} = f(\mathbf{b}) - f(\mathbf{a}),$$

where $\mathbf{a}$ *and* $\mathbf{b}$ *are the endpoints of the path.*

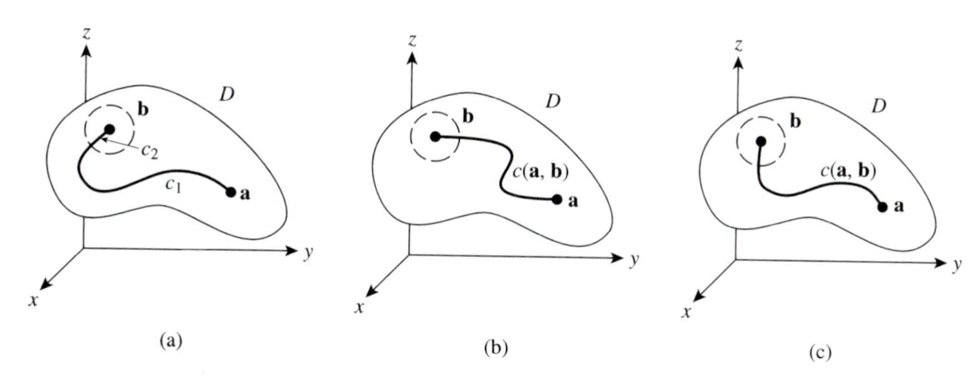

Figure 5.1.1 Paths from $\mathbf{a}$ to $\mathbf{b}$ that are parallel to the x-, y-, and z-axes near $\mathbf{b}$

Proof The backward part of the implication is easy: If there exists f such that $\vec{\nabla} f = \mathbf{F}$, then, by Theorem 5.1.1,

$$\int_C \mathbf{F} \cdot d\mathbf{x} = \int \vec{\nabla} f \cdot dx = f(\mathbf{b}) - f(\mathbf{a}),$$

so that the value of $\int_C \mathbf{F} \cdot d\mathbf{x}$ depends only on the endpoints of C.

To prove the forward part of the implication we must use the path independence to construct a function f whose gradient is $\mathbf{F}$. Choose any point $\mathbf{a}$ in D. For each $\mathbf{x}$ in D, let $C(\mathbf{a}, \mathbf{x})$ be a path in D that joins $\mathbf{a}$ to $\mathbf{x}$. Then define

$$f(\mathbf{x}) = \int_{C(\mathbf{a},\mathbf{x})} \mathbf{F} \cdot d\mathbf{y}.$$

(We are using $\mathbf{y}$ as the "dummy variable" in the path integral.) This function, we will show, has partial derivatives equal to the components of $\mathbf{F}$ (i.e., $\vec{\nabla} f = \mathbf{F}$).

We first fix a value, say $\mathbf{b}$, for $\mathbf{x}$ and a path $C(\mathbf{a}, \mathbf{b})$ that goes from $\mathbf{a}$ to $\mathbf{b}$ so that very close to $\mathbf{b}$, the path coincides with the x_1-axis. See Figure 5.1.1(a). Taking the partial of f with respect to x_1 at $\mathbf{b}$ for this choice of path, we will obtain $\dfrac{\partial f}{\partial x_1}(\mathbf{b}) = F_1(\mathbf{b})$.

Since D is open, there is an open ball $B_r(\mathbf{b})$ of radius r centered at $\mathbf{b}$ such that $B_r(\mathbf{b}) \subset D$. Let C_2 be the line segment from $(b_1+r,\, b_2, b_3, \ldots, b_n)$ to $\mathbf{b} = (b_1, b_2, \ldots, b_n)$, let C_1 be any path in D from $\mathbf{a}$ to $(b_1+r,\, , b_2, \ldots, b_n)$, and let $C(\mathbf{a}, \mathbf{b}) = C_1 \cup C_2$. Again see Figure 5.1.1(a). Then

$$f(b_1 + h,\, b_2, \ldots, b_n) - f(b_1, b_2, \ldots, b_n) = \int_{C_3} \mathbf{F} \cdot d\mathbf{y},$$

where C_3 is the line segment from $\mathbf{b}$ to $(b_1 + h,\, b_2, \ldots, b_n)$. So

$$\begin{aligned}
\frac{\partial f}{\partial x_1}(\mathbf{b}) &= \lim_{h \to 0^+} \frac{f(b_1 + h,\, b_2, \ldots, b_n) - f(b_1, \ldots, b_n)}{h} \\
&= \lim_{h \to 0} \frac{1}{h} \int_{b_1}^{b_1+h} F_1(y,\, b_2, \ldots, b_n)\, dy \\
&= F_1(b_1, b_2, \ldots, b_n) \\
&= F_1(\mathbf{b}).
\end{aligned}$$

By choosing $C(\mathbf{a},\, \mathbf{b})$ to be parallel with the x_2-, x_3-, $\ldots$, x_n-axes near $\mathbf{b}$, we repeat this line of reasoning to see that

$$\frac{\partial f}{\partial x_2}(\mathbf{b}) = F_2(\mathbf{b}), \ldots, \frac{\partial f}{\partial x_n}(\mathbf{b}) = F_n(\mathbf{b}).$$

See Figure 5.1.1(b) and (c). Thus $\vec{\nabla} f(\mathbf{b}) = \mathbf{F}(\mathbf{b})$, and since the choice of $\mathbf{b}$ in D is arbitrary, we have shown that $\vec{\nabla}\mathbf{f} = \mathbf{F}$. We conclude that if $\mathbf{F}$ is path independent, then it is the gradient of some scalar-valued function. ♦

A scalar-valued function f is called a **potential field** for a vector field $\mathbf{F}$ if $\vec{\nabla} f = \mathbf{F}$.

Example 5.1.1 Determine by inspection a potential field for the vector field $\mathbf{F}(x,\, y,\, z) = (yz,\, xz,\, xy)$ on R^3.

Solution The potential field f must satisfy $\partial f/\partial x = yz$, $\partial f/\partial y = xz$, and $\partial f/\partial z = xy$. One such function is $f(x,\, y,\, z) = xyz$, and in fact any other potential field has the form $g(x,\, y,\, z) = xyz + c$ for some constant c. ♦

We can characterize path independence in terms of path integrals around *closed paths* (i.e., those for which the starting and ending points coincide). Suppose that $\mathbf{F}$ is path independent on D and let C be a path in D that begins and ends at a point $\mathbf{a}$. Then, by Theorem 5.1.2,

$$\oint_C \mathbf{F} \cdot d\mathbf{x} = f(\mathbf{a}) - f(\mathbf{a}) = 0,$$

where f is any potential field for $\mathbf{F}$. Thus, if $\mathbf{F}$ is path independent on D, then $\oint_C \mathbf{F} \cdot d\mathbf{x} = 0$ for any closed path C in D. Conversely, suppose that $\oint_C \mathbf{F} \cdot d\mathbf{x} = 0$ for any closed path in D. Now choose $\mathbf{a}$ and $\mathbf{b}$ in D and let C_1 and C_2 be any two paths that join $\mathbf{a}$ to $\mathbf{b}$. See Figure 5.1.2. Let C be

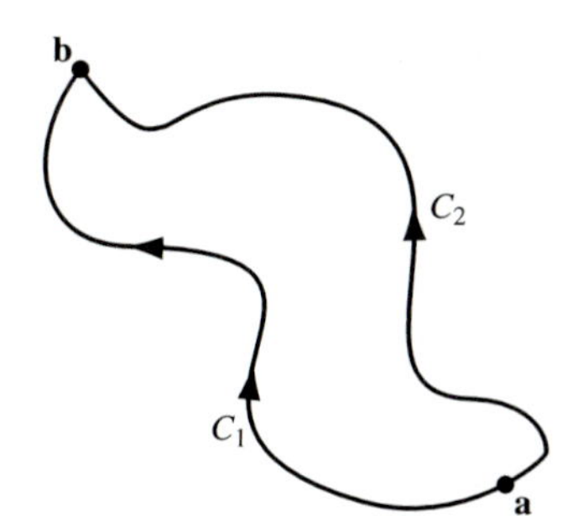

Figure 5.1.2 If $\oint_C \mathbf{F} \cdot d\mathbf{x} = 0$ for all closed paths C, then $\mathbf{F}$ is path independent.

the closed path that starts at $\mathbf{a}$, follows C_1 to $\mathbf{b}$, and then follows $-C_2$ back to $\mathbf{a}$. Then $\oint_C F \cdot d\mathbf{x} = 0$, so

$$\begin{aligned} 0 &= \int_{C_1} \mathbf{F} \cdot d\mathbf{x} + \int_{-C_2} \mathbf{F} \cdot d\mathbf{x} \\ &= \int_{C_1} \mathbf{F} \cdot d\mathbf{x} - \int_{C_2} \mathbf{F} \cdot d\mathbf{x}. \end{aligned}$$

This means that $\int_{C_1} \mathbf{F} \cdot d\mathbf{x} = \int_{C_2} \mathbf{F} \cdot d\mathbf{x}$, so $\mathbf{F}$ is path independent on D. We have proven the following theorem.

Theorem 5.1.3 *Let $D \subset \mathbb{R}^n$ be an open connected set and let $\mathbf{F} : D \to \mathbb{R}^n$ be a continuous vector field. Then $\mathbf{F}$ is path independent on D if and only if $\oint_C \mathbf{F} \cdot d\mathbf{x} = 0$ for all closed paths C that lie in D.*

Theorem 5.1.3 can be used to detect when a vector field is path dependent.

Example 5.1.2 Consider the vector field $\mathbf{F}(x, y) = \left(\dfrac{-y}{x^2 + y^2}, \dfrac{x}{x^2 + y^2} \right)$ on its domain $D = \{(x, y) | x^2 + y^2 \neq 0\}$. Let C be the closed circular path parametrized by $\mathbf{x} = (\cos t, \sin t)$, $0 \leq t \leq 2\pi$. By using the definition of path integral, we see that

$$\begin{aligned} &\oint_C \frac{-y}{x^2 + y^2} dx + \frac{x}{x^2 + y^2} dy \\ &= \int_0^{2\pi} \left(\frac{-\sin t}{\cos^2 t + \sin^2 t} \cdot (-\sin t) + \frac{\cos t}{\cos^2 t + \sin^2 t} \cos t \right) dt \\ &= \int_0^{2\pi} 1 \, dt = 2\pi. \end{aligned}$$

Since this path integral is not zero, by Theorem 5.1.3, the vector field is not path independent. ♦

Is there a more automatic way to determine when a given vector field is path independent? The answer is, "Yes," but it is qualified by a restriction on the domain of the vector field. The domain must possess a characteristic called simple connectedness before we can use our automatic path-independence check.

Definition 5.1.3 A connected set D in $\mathbb{R}^n$ is **simply connected** if any simple closed path lying in D is the boundary of an orientable surface lying entirely in D. In more graphic terms, D is simply connected if it has no holes through it.

Example 5.1.3 The unit disk $D = \{(x, y)|x^2 + y^2 < 1\}$ in $\mathbb{R}^2$ is simply connected since the region enclosed by any closed curve C lies entirely in D.

On the other hand, the annulus $A = \{(x, y)|1 \leq x^2 + y^2 \leq 4\}$ is not simply connected since any curve C that goes around the hole in the center encloses a region with a hole in it.

The set $S = \{(x, y, z)|1 \leq x^2 + y^2 + z^2 \leq 9\}$ in $\mathbb{R}^3$ has a "hole" in it, but it is in fact simply connected: any closed curve that goes around the "bubble" in the center can serve as a "frame" for a piece of fabric that wraps around the "bubble."

The cylindrical set with a hollow middle $H = \{(x, y, z)|1 \leq x^2 + y^2 \leq 4\}$ is not simply connected: The hole goes all the way through the set. ♦

Theorem 5.1.4 *1. Let $D \subset \mathbb{R}^2$ be a simply connected open set, and let $\mathbf{F} : D \to \mathbb{R}^2$ be continuously differentiable. Then $\mathbf{F}$ is a path independent on D if and only if*

$$\frac{\partial F_2}{\partial x_1} - \frac{\partial F_1}{\partial x_2} = 0. \tag{5.1.3}$$

2. Let $D \subset \mathbb{R}^3$ be a simply connected open set and let $\mathbf{F} : D \to \mathbb{R}^3$ be continuously differentiable. Then $\mathbf{F}$ is path independent on D if and only if

$$\operatorname{curl} \mathbf{F} = \mathbf{0}. \tag{5.1.4}$$

This theorem is actually a special case of a more general result that makes remembering (5.1.3) and (5.1.4) easier.

Theorem 5.1.5 *Let $D \subset \mathbb{R}^n$ be a simply connected open set and let $\mathbf{F}: D \to \mathbb{R}^n$ be a vector field that is continuously differentiable. Then $\mathbf{F}$ is path independent if and only if the Jacobian matrix $J\mathbf{F}(\mathbf{x})$ is symmetric for all $\mathbf{x} \in D$.*

Proof Suppose that F is path independent on D. Then, by Theorem 5.1.2, there exists f such that $\vec{\nabla} f = \mathbf{F}$. By definition,

$$J\mathbf{F} = \begin{bmatrix} \partial F_1/\partial x_1 & \partial F_1/\partial x_2 & \cdots & \partial F_1/\partial x_n \\ \partial F_2/\partial x_1 & \partial F_2/\partial x_2 & \cdots & \partial F_2/\partial x_2 \\ \vdots & & & \\ \partial F_n/\partial x_1 & \cdots & \cdots & \partial F_n/\partial x_n \end{bmatrix}.$$

But

$$\frac{\partial F_i}{\partial x_j} = \frac{\partial}{\partial x_j}\left(\frac{\partial f}{\partial x_i}\right) = \frac{\partial^2 f}{\partial x_j\, x_i}$$

for $i = 1, 2, \ldots, n$ and $j = 1, 2, \ldots, n$, so

$$J\mathbf{F} = \begin{bmatrix} \partial^2 f/\partial x_1^2 & \partial^2 f/\partial x_2\, \partial x_1 & \cdots & \partial^2 f/\partial x_n\, \partial x_1 \\ \partial^2 f/\partial x_1\, \partial x_2 & \partial^2 f/\partial x_2^2 & \cdots & \partial^2 f/\partial x_n\, \partial x_2 \\ \vdots & & & \\ \partial^2 f/\partial x_1\, \partial x_n & \cdots & \cdots & \partial^2 f/\partial x_n^2 \end{bmatrix}.$$

To say that $\mathbf{F}$ is continuously differentiable means, in particular, that the mixed partials of f are continuous. By Clairaut's Theorem, the mixed partials are equal, so $J\mathbf{F}$ is symmetric.

A full proof of the converse is beyond the scope of this book. However, later, in Sections 5.2 and 5.4, we will be able to prove the converse in the cases $n = 2$ and $n = 3$. ♦

Example 5.1.4 Determine if $\mathbf{F} = (2xy^3+1,\ 3x^2y^2-2y)$ is path independent on $\mathbb{R}^2$, and if so, find a function f such that $\vec{\nabla} f = \mathbf{F}$ and $f(0, 0) = 0$.

Solution We check that $\partial F_2/\partial x = 6xy^2$ and $\partial F_1/\partial y = 6xy^2$ for every point in $\mathbb{R}^2$, so by Theorem 5.1.5, $\mathbf{F}$ is path independent. Thus there is a function f such that $\vec{\nabla} f = \mathbf{F}$. So

$$\frac{\partial f}{\partial x} = F_1 = 2xy^3 + 1.$$

Integrating with respect to x, we obtain

$$f = x^2y^3 + x + C(y),$$

where C is some unknown function of y. Differentiating with respect to y, we obtain

$$\frac{\partial f}{\partial y} = 3x^2y^2 + C'(y).$$

This must equal $F_2 = 3x^2y^2 - 2y$, so $C'(y) = -2y$, and thus $C(y) = -y^2 + k$ for any constant k. Hence

$$f(x, y) = x^2y^3 + x - y^2 + k.$$

Since f is 0 at the point $(0, 0)$, the value of k is zero, and $f(x, y) = x^2y^3 + x - y^2$. ♦

Example 5.1.5 Evaluate $\oint_C -e^y \sin x\, dx + e^y \cos x\, dy + dz$ where C is the line segment joining $(0, 0, 0)$ to $(\pi, \pi, 1)$.

Solution One way to evaluate the integral is to parametrize the segment and use the definition of the line integral. Another is first to recognize this as the path integral of $\mathbf{F} = (-e^y \sin x,\, e^y \cos x,\, 1)$, then to determine whether $\mathbf{F}$ is path independent by using Theorem 5.1.4, and finally—if the field is path-independent—to construct the potential field f and, by Theorem 5.1.2, evaluate the integral as $f(\pi, \pi, 1) - f(0, 0, 0)$. We take the latter approach.

First,

$$\begin{aligned} \operatorname{curl} \mathbf{F} &= (0 - 0,\, 0 - 0,\, -e^y \sin x - (-e^y \sin x)) \\ &= (0, 0, 0) = \mathbf{0}, \end{aligned}$$

so $\mathbf{F}$ is path-independent. The potential field f then satisfies

$$\frac{\partial f}{\partial x} = -e^y \sin x,$$

so by "partial integration" with respect to x we get

$$f = e^y \cos x + C_1(y, z), \tag{5.1.5}$$

where C_1 is a function of y and z to be determined. Taking the partial of Equation 5.1.5 with respect to y, we obtain

$$\frac{\partial f}{\partial y} = e^y \cos x + \frac{\partial C_1}{\partial y},$$

which must equal the second component of **F**:

$$\begin{aligned} e^y \cos x + \frac{\partial C_1}{\partial y} &= e^y \cos x \\ \frac{\partial C_1}{\partial y} &= 0. \end{aligned} \tag{5.1.6}$$

Partial integration of (5.1.6) with respect to y gives

$$C_1(y,\ z) = C_2(z),$$

where C_2 is a function of z alone. Thus

$$f = e^y \cos x + C_2(z). \tag{5.1.7}$$

Finally, differentiating (5.1.7), we get

$$\frac{\partial f}{\partial z} = C_2'(z),$$

which must equal the third component of **F**:

$$\begin{aligned} C_2'(z) &= 1 \\ C_2(z) &= z + k, \end{aligned}$$

where k is any constant. This yields

$$f(x,\ y,\ z) = e^y \cos x + z + k$$

as the most general potential field. By Theorem 5.1.1,

$$\begin{aligned} \int_c -e^y \sin x\, dx + e^y \cos x\, dx + dz &= f(\pi,\ \pi,\ 1) - f(0,\ 0,\ 0) \\ &= (e^\pi \cos \pi + 1 + k) - (e^\circ \cos 0 + 0 + k) \\ &= -e^\pi + 1 - 1 = -e^\pi. \quad \blacklozenge \end{aligned}$$

Example 5.1.6 Find the work done by the inverse square gravitational field

$$\mathbf{G}(\mathbf{x}) = -\frac{1}{\|\mathbf{x}\|^3}\mathbf{x}$$

on a particle that follows the helical path parametrized by $\mathbf{x} = (\cos t,\ \sin t,\ t)$, $0 \le t \le 2\pi$.

Solution We can easily check, by using differentiation formulas from Section 3.8, that

$$\begin{aligned}
\operatorname{curl}\mathbf{G} &= -\operatorname{curl}\left(\frac{1}{\|\mathbf{x}\|^3}\mathbf{x}\right)\\
&= -\left(\left(\vec{\nabla}\frac{1}{\|\mathbf{x}\|^3}\right)\times\mathbf{x}+\frac{1}{\|\mathbf{x}\|^3}\left(\vec{\nabla}\times\mathbf{x}\right)\right)\\
&= -\left(\left(\frac{3}{\|\mathbf{x}\|^5}\mathbf{x}\right)\times x+\frac{1}{\|\mathbf{x}\|^3}\mathbf{0}\right)\\
&= -\left(-\frac{3}{\|\mathbf{x}\|^5}(\mathbf{x}\times\mathbf{x})+\mathbf{0}\right)=\mathbf{0}.
\end{aligned}$$

Thus $\mathbf{G}$ is path-independent. To obtain a potential field, we again just refer to the differentiation formulas in Table 3.8.1. Since $\vec{\nabla}\left(\|\mathbf{x}\|^{-1}\right)=-\|\mathbf{x}\|^{-3}\mathbf{x}$, the most general potential field is

$$g(\mathbf{x})=\|\mathbf{x}\|^{-1}+c.$$

The endpoints of the given path are $(\cos 0,\,\sin 0,\,0)=(1,\,0,\,0)$ and $(\cos 2\pi,\,\sin 2\pi,\,2\pi)=(1,\,0,\,2\pi)$, so

$$\begin{aligned}
\int_C -\frac{1}{\|\mathbf{x}\|^3}\mathbf{x}\cdot d\mathbf{x} &= g(1,\,0,\,2\pi)-g(1,\,0,\,0)\\
&= \frac{1}{\|(1,\,0,\,2\pi)\|}-\frac{1}{\|(1,\,0,\,0)\|}\\
&= \frac{1}{\sqrt{1+4\pi^2}}-1. \quad \blacklozenge
\end{aligned}$$

A differential form $F_1\,dx_1+F_2\,dx_2+\cdots+F_n\,dx_n$ is said to be **exact** if it is the total differential of some scalar-valued function f. Equivalently, the form is exact if $\mathbf{F}=(F_1,\ldots,F_n)$ is the gradient of some function f. Certain ordinary differential equations are readily solved by rewriting them in terms of exact differential forms. Such equations are called **exact equations**.

Example 5.1.7 Find the equation for the curve in the xy-plane that passes through the point $(5,\,2)$ whose slope at any point $(x,\,y)$ is negative to the slope of the segment from $(0,\,0)$ to $(x,\,y)$.

Solution By inspecting Figure 5.1.3, we see that the points on such a curve satisfy

$$\frac{dy}{dx}=-\frac{y}{x}. \tag{5.1.8}$$

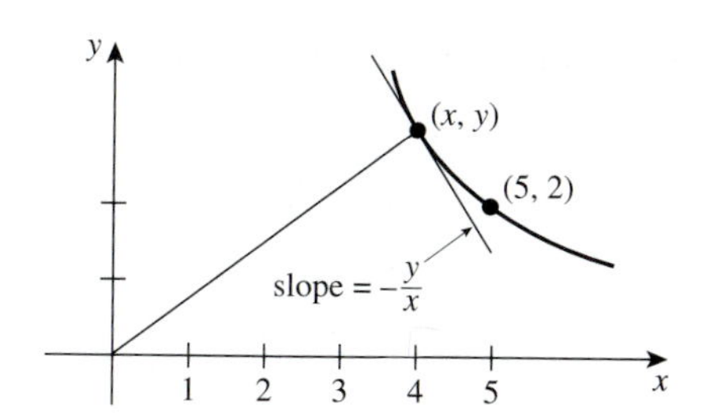

Figure 5.1.3 The curve satisfying the exact differential equation $dy/dx = -y/x$ and the initial condition $y(5) = 2$

Treating the derivative on the left-hand side of (5.1.8) as a quotient of differentials, we can multiply both sides by dx and x and rework this equation to the form

$$y\,dx + x\,dy = 0. \tag{5.1.9}$$

Since $\frac{\partial}{\partial x}x - \frac{\partial}{\partial y}y = 1 - 1 = 0$, by Theorem 5.1.4, the left-hand side of (5.1.9) is an exact differential form. This differential form—you should verify—is the total differential of $f(x,\, y) = xy$. Thus the general solution to (5.1.9) is

$$xy = c$$

for some constant c. Since the curve passes through the point $(5,\, 2)$, it must be that c satisfies $5 \cdot 2 = c$; $c = 10$. The solution curve is the hyperbola $xy = 10$. ♦

Exercises 5.1

Use Theorem 5.1.4 or Theorem 5.1.5 to determine which of the vector fields in Exercises 1 through 7 is path-independent on the given set. For those that are, find a function f whose gradient equals the vector field.

1. $\mathbf{F}(x,\, y) = \left(x \sin y + 1,\, \frac{1}{2}x^2 \cos y\right)$; $D = \mathbb{R}^2$.

2. $\mathbf{F}(x,\, y) = \left(2xy^3 + 1,\, 3x^2y^2 - 1/y^2\right)$; $D = \{(x,\, y) | y > 0\}$.

3. $\mathbf{F}(x,\, y) = (x \sin y,\, -y \sin x)$; $D = \mathbb{R}^2$

4. $\mathbf{F}(x,\, y,\, z) = (x \ln y,\, z,\, x - 1/y)$; $D = \{(x,\, y,\, z) | y > 0\}$

5. $\mathbf{F}(x,\, y,\, z) = \left(\cot y - z \sec^2 x,\, -x \csc^2 y,\, -\tan x\right)$; $D = \{(x,\, y,\, z) | 0 < x < \pi/2,\, 0 < y < \pi/2,\, 0 < z < 1\}$

6. $\mathbf{F}(x_1, x_2, x_3, x_4) = \left(4x_1^4, 3x_2, 2x_3, 1\right)$; $D = \mathbb{R}^4$

7. $\mathbf{F}(x_1, x_2, x_3, x_4) = \left(1, 2x_2x_4^5, 2x_3, 5x_2^2x_4^4\right)$; $D = \mathbb{R}^4$

Determine which of the differential forms in Exercises 8 through 14 are exact. For those that are, find a potential function whose total differential equals the differential form.

8. $(5y + 2x)\,dx + \left(5x - 4y^3\right)dy$

9. $\dfrac{1}{1 + (x + y)^2}dx + \dfrac{1}{1 + (x + y)^2}dy$

10. $e^{-x}\sin y\,dx - e^{2x}\cos y\,dy$

11. $2(xy - x)dx + (x^2 + 2y)dy + 2z\,dz$

12. $x\,dx + x^2dy + x^3dz$

13. $\dfrac{y}{z}\,dx + \dfrac{x}{z}\,dy - \dfrac{xy}{z^2}\,dz,\ z > 0$

14. $x_3x_4\,dx_1 + 3x_2\,dx_2 + x_1x_4\,dx_3 + x_1x_3\,dx_4$

In Exercises 15 through 23, evaluate the line integral over the given path. In each case, check first, by using Theorem 5.1.4 or Theorem 5.1.5, to see if the integrand is path-independent. If it is, use Theorem 5.1.1.

15. $\int_C x\,dx - 3y\,dy$, where C is the line segment from $(0, 0)$ to $(-9, 7)$

16. $\int_C y^{5/3}\,dx + \frac{5}{3}xy^{2/3}\,dy$, where C is the parabolic path $y = 2x^2$ from $(0, 0)$ to $(2, 8)$

17. $\oint_C (x^2 - y)\,dx + (y^2 - x)\,dy$, where C is the circle of radius 5 centered at $\left(-\frac{2}{3}, \frac{4}{3}\right)$

18. $\oint_C y\,dx - x\,dy$, where C is the circle $x^2 + y^2 = 1$

19. $\int_C dx + 2y\,dy + dz$, where C is the line segment joining $(1, 0, 1)$ to $(-6, -5, -3)$.

20. $\int_C ye^x\,dx + e^x\,dy + dz$, where C is the line segment from $(\ln 3, 2, -4)$ to $(0, 1, 2)$

21. $\oint_C 2xy dx + x^2 dy$, where C is the path parametrized by $\mathbf{x} = (2\cos\theta + \cos(\theta/2),\ 2\sin\theta + \cos(\theta/2),\ \sin(\theta/2))$, $0 \le \theta \le 4\pi$.

22. $\oint_C 2xyz\,dx + x^2z\,dy + x^2y\,dz$, where C is the circle in which the sphere $x^2 + y^2 + z^2 = 1$ intersects the plane $z = \frac{1}{2}$.

23. $\int_C x_2\,dx_1 + x_1\,dx_2 + dx_3 + dx_4$, where C is the line segment in $\mathbb{R}^4$ from $(0,\ 0,\ 0,\ 0)$ to $\left(\frac{1}{3},\ -12,\ 16,\ -16\right)$.

24. Identify the following sets as connected or disconnected.

(a) $S = \{(x,\ y) \mid x \ge 0,\ y \ge 0,\ x + y \le 1\}$

(b) $B = \{(x,\ y) \,\big|\, x^2 + y^2 = 81\}$

(c) $T = \{(x,\ y) \mid x < 0 \text{ or } x > 3\}$

(d) $U = \{(x,\ y) \,\big|\, x^2 + y^2 < 1 \text{ or } (x-3)^2 + (y+4)^2 < 1\}$

(e) $A = \{(x,\ y,\ z) \,\big|\, 4 < x^2 + z^2\}$

(f) $K = \{(x,\ y,\ z) \mid |x| \ge 1 \text{ or } |y| \ge 1 \text{ or } |z| \ge 1\}$

(g) $N = \{(x,\ y,\ z) \mid |x| \ge 1\}$

25. Which of the connected sets in Exercise 24 are simply connected?

In Exercises 26 through 29, find the solution to the given exact differential equations that satisfies the given condition.

26. $3x(xy - 2)dx + (x^3 + 2y)dy = 0$; $y(0) = 1$

27. $2xydx + (y^2 + x^2)dy = 0$; $y(1) = 1$

28. $\dfrac{dy}{dx} = \dfrac{3y - 2x}{2y - 3x}$; $y(0) = 0$

29. $\dfrac{dy}{dx} = \dfrac{\cos y}{x \sin y}$; $y(1) = \pi$

30. Explain, using Theorem 5.1.4, why $\mathbf{F} = \left(\dfrac{-y}{x^2 + y^2},\ \dfrac{x}{x^2 + y^2}\right)$ is path independent on $D = \{(x,\ y) \mid |y| > 0 \text{ or } x > 0\}$.

31. Use Theorem 5.1.3 to show that

$$\mathbf{F}(x,\ y) = \left(\frac{2y}{(x-1)^2 + y^2},\ \frac{-2(x-1)}{(x-1)^2 + y^2}\right)$$

is path dependent on $D = \{(x,\ y) \,\big|\, (x-1)^2 + y^2 \ne 0\}$.

32. Find a region D such that the vector field in Exercise 31 is path independent on D.

33. Suppose that an object moves along a path (avoiding the origin) from $(1, 1, -1)$ to $(8, 3, 5)$ under the influence of an inverse first-power gravitational attraction law $\mathbf{G}(\mathbf{x}) = -(1/\|\mathbf{x}\|^2)\mathbf{x}$. Use the method of Example 5.1.6 to find the work done by the gravitational force. [*Hint*: $\vec{\nabla}(\ln \|\mathbf{x}\|) = \mathbf{x}/\|\mathbf{x}\|^2$.]

34. A **central force field**is one in which the gravitational attraction experienced by an object is always directed toward a fixed point (or center) and the magnitude of the force is a function of the distance from the center. In symbols, a central force is one of the form

$$\mathbf{G}(\mathbf{x}) = g(\|\mathbf{x}\|)\mathbf{x},$$

where g is a scalar-valued function of a single real variable defined for all values except possibly 0. (Without loss of generality we have assumed that the center is $\mathbf{0}$.) By using the Chain Rule and properties of the curl, show that any central force $\mathbf{G}$ on $D = \{(x, y, z) \,|\, x^2 + y^2 + z^2 > 0\}$ is path independent.

35. Prove that

$$\int_{C(\mathbf{a},\mathbf{b})} \|\mathbf{x}\|^k \mathbf{x} \cdot d\mathbf{x} = \begin{cases} \frac{1}{k+2}\left(\|\mathbf{b}\|^{k+2} - \|\mathbf{a}\|^{k+2}\right), & k \neq -2 \\ \ln \|\mathbf{b}\| - \ln \|\mathbf{a}\|, & k = -2, \end{cases}$$

where $C(\mathbf{a}, \mathbf{b})$ is any smooth path that joins $\mathbf{a}$ to $\mathbf{b}$. (*Hint*: see Table 3.8.1.)

5.2 Green's Theorem

Green's Theorem relates the line integral around a closed curve of a vector field on $\mathbb{R}^2$ to a certain double integral over the region which the curve bounds. The theorem is readily applied to evaluating line integrals, and it is also important because it provides the key to proving Stokes' Theorem.

Theorem 5.2.1 *(Green's[1] Theorem) Let $D \subset \mathbb{R}^2$ be a connected open set, and let $\mathbf{F} : D \to \mathbb{R}^2$ be a smooth (i.e. $\partial F_1/\partial x$, $\partial F_1/\partial y$, $\partial F_2/\partial x$, and $\partial F_2/\partial y$*

[1]George Green (1828) discovered the theorem, but the result became widely circulated much later, by William Thomson in 1846. The theorem is also known as Ostrogradski's Theorem.

are continuous) vector field. For any region R in D with piecewise smooth counterclockwise-oriented boundary ∂R,

$$\oint_{\partial R} F_1 dx + F_2 dy = \iint_R \left(\frac{\partial F_2}{\partial x} - \frac{\partial F_1}{\partial y} \right) dA. \tag{5.2.1}$$

Proof Partition the portion of $\mathbb{R}^2$ about R into small rectangular regions R_{ij} as shown in Figure 5.2.1. Let P represent the collection of those rectan-

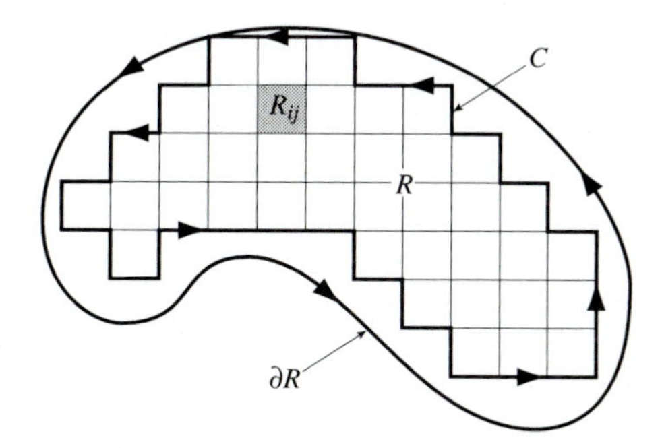

Figure 5.2.1 A partition of a region R

gular regions that lie inside R, and let C be the counterclockwise-oriented boundary of the region that the rectangular regions of P form. For a very fine partition it seems intuitive that C and ∂R are very close and hence that

$$\oint_{\partial R} \mathbf{F} \cdot d\mathbf{x} \approx \oint_C \mathbf{F} \cdot d\mathbf{x}. \tag{5.2.2}$$

This can be shown by arguments that rely on the continuity of $\mathbf{F}$ and the smoothness of ∂R. If we take the path integral of $\mathbf{F}$ around the counterclockwise-oriented boundary of each R_{ij} in P and add these up, we see that the integrals along adjoining edges cancel one another. (See Figure 5.2.2.) Thus, all that remains of this sum is the line integral along the segments that are not common boundaries of the rectangular regions in P. In other words, what remains is the line integral of $\mathbf{F}$ along C. In formal language,

$$\sum_{R_{ij} \in P} \oint_{\partial R_{ij}} \mathbf{F} \cdot d\mathbf{x} = \oint_C \mathbf{F} \cdot d\mathbf{x}. \tag{5.2.3}$$

Combining (5.2.3) with (5.2.2), we see that

$$\oint_{\partial R} \mathbf{F} \cdot d\mathbf{x} \approx \sum_{R_{ij} \in P} \oint_{\partial R_{ij}} \mathbf{F} \cdot d\mathbf{x}, \tag{5.2.4}$$

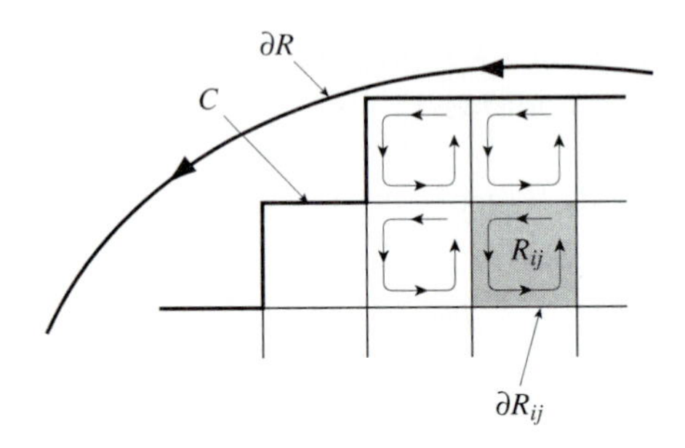

Figure 5.2.2 The path integrals around the boundaries of the rectangles cancel along adjacent edges.

and the approximation is better with a finer partition. The key to completing the proof is to express each path integral in the sum as a double integral over the rectangular region bounded by ∂R_{ij}. Each path integral in the sum in (5.2.3) can be written as the sum of four path integrals $\int_{C_1} + \int_{C_2} + \int_{C_3} + \int_{C_4}$, where C_1, C_2, C_3, and C_4 are shown in Figure 5.2.3. Since $\int_{C_1} \mathbf{F} \cdot d\mathbf{x} = \int_{C_1} F_1\, dx$, $\int_{C_2} \mathbf{F} \cdot d\mathbf{x} = \int_{C_2} F_2\, dy$, $\int_{C_3} \mathbf{F} \cdot d\mathbf{x} = \int_{C_3} F_1\, dx$, and $\int_{C_4} \mathbf{F} \cdot d\mathbf{x} = \int_{C_4} F_2\, dy$, we have

$$
\begin{aligned}
\oint_{\partial R_{ij}} \mathbf{F} \cdot d\mathbf{x} &= \int_{x_{i-1}}^{x_i} F_1(x, y_{j-1})\, dx + \int_{y_{j-1}}^{y_j} F_2(x_i, y)\, dy \\
&\quad + \int_{x_i}^{x_{i-1}} F_1(x, y_j)\, dx + \int_{y_j}^{y_{j-1}} F_2(x_{i-1}, y)\, dy \\
&= \int_{y_{j-1}}^{y_j} [F_2(x_i, y) - F_2(x_{i-1}, y)]\, dy \\
&\quad - \int_{x_{i-1}}^{x_1} [F_1(x, y_j) - F_1(x, y_{j-1})]\, dx.
\end{aligned} \tag{5.2.5}
$$

Since F_1 and F_2 have continuous partial derivatives, we can apply the Fundamental Theorem of Calculus to the integrands in the last expression in (5.2.5):

$$F_1(x,\, y_j) - F_1(x,\, y_{j-1}) = \int_{y_{j-1}}^{y_j} \frac{\partial F_1}{\partial y} dy \tag{5.2.6}$$

and

$$F_2(x_i,\, y) - F_2(x_{i-1},\, y) = \int_{x_{i-1}}^{x_i} \frac{\partial F_2}{\partial x} dx. \tag{5.2.7}$$

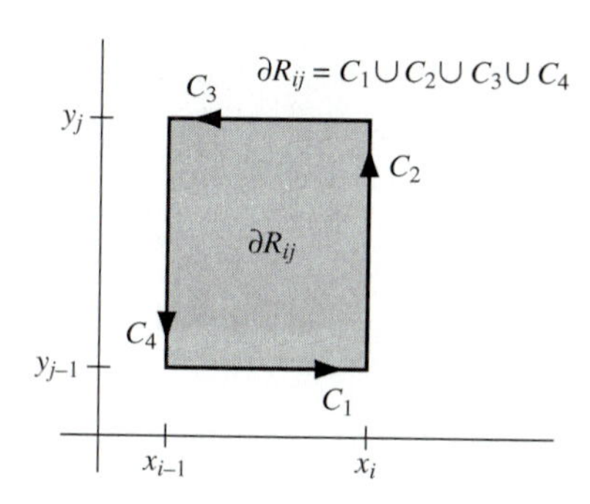

Figure 5.2.3 The path integral around the boundary of the rectangle is the sum of four path integrals.

Using (5.2.6) and(5.2.7), we obtain

$$\begin{aligned}\oint_{\partial R_{ij}} \mathbf{F} \cdot d\mathbf{x} &= -\int_{x_{i-1}}^{x_i} \int_{y_{j-1}}^{y_j} \frac{\partial F_1}{\partial y}\, dy\, dx + \int_{y_{j-1}}^{y_j} \int_{x_{i-1}}^{x_i} \frac{\partial F_2}{\partial x}\, dx\, dy \\ &= \int_{x_{i-1}}^{x_i} \int_{y_{j-1}}^{y_j} \left(\frac{\partial F_2}{\partial x} - \frac{\partial F_1}{\partial y}\right) dy\, dx = \iint\limits_{R_{ij}} \left(\frac{\partial F_2}{\partial x} - \frac{\partial F_1}{\partial y}\right) dA,\end{aligned}$$

By (5.2.4),

$$\oint_{\partial R} \mathbf{F} \cdot d\mathbf{x} \approx \sum_{R_{ij} \in P} \iint\limits_{R_{ij}} \left(\frac{\partial F_2}{\partial x} - \frac{\partial F_1}{\partial y}\right) dA. \tag{5.2.8}$$

As the partition is refined, the approximation becomes better and better, and the right-hand side of (5.2.8) approaches $\iint\limits_{R} (\partial F_2/\partial x - \partial F_1/\partial y)\, dA$. We conclude that

$$\oint_{\partial R} \mathbf{F} \cdot d\mathbf{x} = \iint\limits_{R} \left(\frac{\partial F_2}{\partial x} - \frac{\partial F_1}{\partial y}\right) dA,$$

which is the content of Green's Theorem. ♦

Example 5.2.1 Use Green's Theorem to compute the line integral

$$\oint_C (x + y)\, dx + x^2\, dy,$$

where C is the counterclockwise-oriented triangle with vertices (0, 0), (2, 0), and (0, 1).

Solution The curve is the boundary of the triangular region R given by

$$\begin{aligned} 0 &\leq x \leq 2 \\ 0 &\leq y \leq 1 - x/2. \end{aligned}$$

See Figure 5.2.4. Since the integrand is smooth and the boundary is piecewise

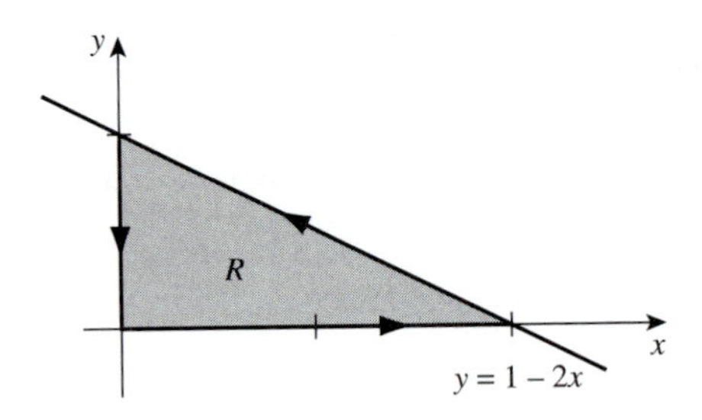

Figure 5.2.4 The region R for Example 5.2.1

smooth, we can apply Green's Theorem:

$$\begin{aligned} \oint (x+y)\,dx + x^2\,dy &= \iint_R \left(\frac{\partial}{\partial x}(x^2) - \frac{\partial}{\partial y}(x+y) \right) dA \\ &= \int_0^2 \int_0^{1-x/2} (2x-1)\,dy\,dx \\ &= \int_0^2 (2x-1)(1-x/2)\,dx \\ &= \int_0^2 \left(-x^2 + \frac{5}{2}x - 1 \right) dx = \frac{1}{3}. \quad \blacklozenge \end{aligned}$$

Example 5.2.2 Can Green's Theorem be applied to evaluate

$$\oint_C \frac{-y}{x^2+y^2}\,dx + \frac{x}{x^2+y^2}\,dy,$$

where C is the counterclockwise-oriented unit circle?

Solution No. The integrand is not smooth on the region bounded by C; in fact, it is not even defined at $(0,\,0)$. Since one of the hypotheses of the theorem is not met, we cannot apply it. ♦

Example 5.2.3 Use Green's Theorem to show that

$$\oint_C \frac{-y}{x^2+y^2}\,dx + \frac{x}{x^2+y^2}\,dy = 0,$$

where C is any piecewise smooth simple curve that does not enclose the origin.

Solution For such a curve as C, the integrand is smooth on the region R that C bounds. Since

$$\frac{\partial}{\partial x}\left(\frac{x}{x^2+y^2}\right) - \frac{\partial}{\partial y}\left(\frac{-y}{x^2+y^2}\right) = \frac{x^2+y^2-2x^2}{(x^2+y^2)^2} + \frac{x^2+y^2-2y^2}{(x^2+y^2)^2} = 0,$$

by Green's Theorem

$$\oint_C \frac{-y}{x^2+y^2}\,dx + \frac{x}{x^2+y^2}\,dy = \iint_R 0\,dA = 0. \quad \blacklozenge$$

If $\mathbf{F}$ is any vector field such that $\partial F_2/\partial x - \partial F_1/\partial y = 1$, then the integral on the right in (5.2.1) is the area of R. In particular, $\mathbf{F} = (-y/2,\, x/2)$, $\mathbf{F} = (-y,\, 0)$, $\mathbf{F} = (0,\, x)$ are such vector fields, so we have a corollary.

Corollary 5.2.1 *If R is a region with a piecewise smooth counterclockwise-oriented boundary ∂R, then the area of R is*

$$A(R) = \frac{1}{2}\oint_{\partial R} -y\,dx + x\,dy = \oint_{\partial R} -y\,dx = \oint_{\partial R} x\,dy. \tag{5.2.9}$$

Example 5.2.4 Find the area of the region bounded by an ellipse $x^2/a^2 + y^2/b^2 = 1$.

Solution A parametrization of the ellipse is $x = a\cos t$, $y = b\sin t$ for $0 \le t < 2\pi$. By the first line of (5.2.9), then

$$\begin{aligned} A(R) &= \frac{1}{2}\int_0^{2\pi} [-(b\sin t)(-a\sin t) + (a\cos t)(b\cos t)]\,dt \\ &= \frac{1}{2}\int_0^{2\pi} ab\ dt = \pi ab. \quad \blacklozenge \end{aligned}$$

Example 5.2.5 Let $(x_1,\, y_1), (x_2,\, y_2), \dots, (x_n,\, y_n)$ be points in $\mathbb{R}^2$ such that when they are joined in succession by line segments [and $(x_n,\, y_n)$ is joined

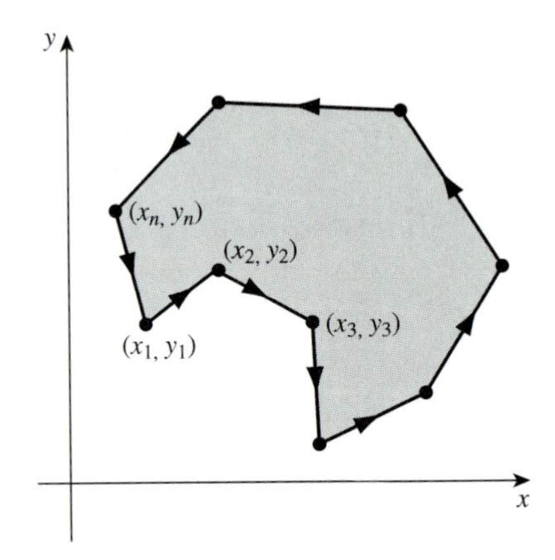

Figure 5.2.5 We find a formula for the area of this region by integrating around its boundary.

to (x_1, y_1)], the segments enclose a polygonal region R. See Figure 5.2.5. Assuming that the polygonal boundary is counterclockwise oriented, find the area of the region in terms of the coordinates of the vertices.

Solution By Corollary 5.2.1, the area is the integral along the boundary of the region of any of three differential forms. We choose $\oint_{\partial R} x\,dy$. This path integral in turn is the sum of path integrals along the line segments C_i in order. A parametrization for the line segment from (x_{i-1}, y_{i-1}) to (x_i, y_i) is

$$\begin{aligned}\mathbf{x} &= (x_{i-1}, y_{i-1}) + t(x_i - x_{i-1}, y_i - y_{i-1}) \\ &= (x_{i-1} + (x_i - x_{i-1})t, y_{i-1} + (y_{i-1} - y_{i-1})t),\end{aligned}$$

where $0 \le t \le 1$. Thus

$$\begin{aligned}\int_{C_i} x\,dy &= \int_0^1 (x_{i-1} + (x_i - x_{i-1})t)\,(y_i - y_{i-1})\,dt \\ &= (y_i - y_{i-1})\left(x_{i-1} + \frac{1}{2}(x_1 - x_{i-1})\right) \\ &= \frac{1}{2}(x_i + x_{i-1})(y_i - y_{i-1}),\end{aligned}$$

for $i = 2, 3, 4, \ldots n$, and (you should check) the integral over the last segment closing up the loop is

$$\int_{(x_n, y_n)}^{(x_1, y_1)} x\,dy = \frac{1}{2}(x_1 + x_n)(y_1 - y_n).$$

So the area is

$$\begin{aligned} A(R) = \oint_{\partial R} x\,dy &= \int_{C_2} x\,dy + \cdots + \int_{C_n} x\,dy + \int_{(x_n,\,y_n)}^{(x_1,\,y_1)} x\,dy \\ &= \frac{1}{2}\left[(x_2 + x_1)(y_2 - y_1) + (x_3 + x_2)(y_3 - y_2) + \cdots \right. \\ &\quad \left. + (x_n + x_{n-1})(y_n - y_{n-1}) + (x_1 + x_n)(y_1 - y_n)\right]. \end{aligned}$$

You are asked in the exercises to show that this can be written as

$$A(R) = \frac{1}{2}\left\{\begin{vmatrix} x_1 & x_2 \\ y_1 & y_2 \end{vmatrix} + \begin{vmatrix} x_2 & x_3 \\ y_2 & y_3 \end{vmatrix} + \cdots + \begin{vmatrix} x_{n-1} & x_n \\ y_{n-1} & y_n \end{vmatrix} + \begin{vmatrix} x_n & x_1 \\ y_n & y_1 \end{vmatrix}\right\}. \quad \blacklozenge \tag{5.2.10}$$

Exercises 5.2

Use Green's Theorem to evaluate the line integrals in Exercises 1 through 7 over the indicated closed path C. Assume in each case that C is oriented counterclockwise.

1. $\oint_C 3xy\,dx + (x - y^2)\,dy$, C is the rectangle with vertices (0, 0), (2, 0), (2, 1), and(0, 1).

2. $\oint_C 2xy\,dx + (y^3 + x^2)\,dy$; C is the quadrilateral with vertices (1, 2), (3, 2), (3, 5), and (0, 7).

3. $\oint_C y\,dx - x\,dy$; C is the triangle with vertices (0, −1) (1, 0), and (0, 1).

4. $\oint_C e^x \sin y\,dx + e^x \cos y\,dy$; C is the unit circle.

5. $\oint_C (xy^2 + x)\,dx + (3x + y^2)\,dy$, where C is the boundary of the region bounded by $y = 0$, $y = 1$, $y = -x$, and $x = y^2$.

6. $\oint_C \frac{1}{2}(y - \sin y \cos y)\,dx + x \sin^2 y\,dy$, where C is the boundary of the region bounded by $x = -1$, $x = 2$, $y = 4 - x^2$, and $y = x - 2$.

7. $\oint_C (x^2 - y)\,dx + (x - y^2)\,dy$, where C is the boundary of the region bounded by the curves $y = x^2$ and $y = 6x + 7$.

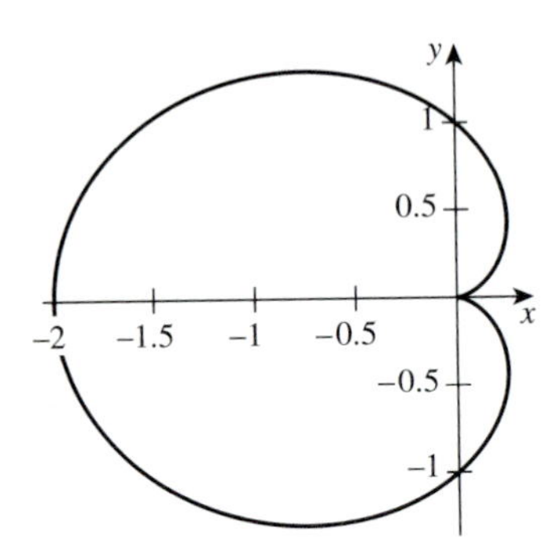

Figure 5.2.6 The cardiod for Exercise 12

8. Can Green's Theorem be applied to evaluate

$$\oint_C \frac{x}{\sqrt{x^2+y^2}}\,dx + \frac{y}{\sqrt{x^2+y^2}}\,dy,$$

where C is a circle of radius 1 centered at the origin? Explain.

9. Can Green's Theorem be applied to evaluate

$$\oint_C \frac{-2x}{(x^2+(y-2)^2)^2}\,dx - \frac{2(y-2)}{(x^2+(y-2)^2)^2}\,dy,$$

where C is the unit circle? Explain.

In Exercises 10 through 15, use the Corollary to Green's Theorem to find the area of the indicated region.

10. R is the region bounded by a circle of radius a.

11. R is the region bounded by the curve $\mathbf{x} = (\sin 2t, \sin t)$, $0 \le t \le \pi$.

12. R is the cardioid bounded by $x = \cos\theta - \cos^2\theta$, $y = \sin\theta - \cos\theta\sin\theta$, $0 \le \theta \le 2\pi$. See Figure 5.2.6.

13. R is the region bounded by the curve $x = t^2 - 1$, $y = t^3 - t$, $-1 \le t \le 1$. See Figure 5.2.7.

14. R is the region bounded by the hypocycloid of four cusps $x^{2/3}+y^{2/3} = 1$.

15. R is the region bounded by the loop of the Folium of Descartes. (See Exercise 24 in Section 1.8.)

16. Use (5.2.10) to find the areas of the polygons shown in Figure 5.2.8.

17. Use (5.2.10) to find a formula for the area of a polygon whose vertices are five equally spaced points on the circumference of a circle of radius r.

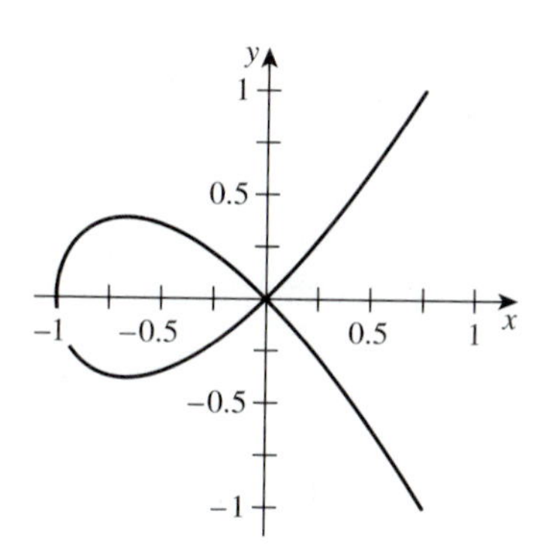

Figure 5.2.7 The region for Exercise 13

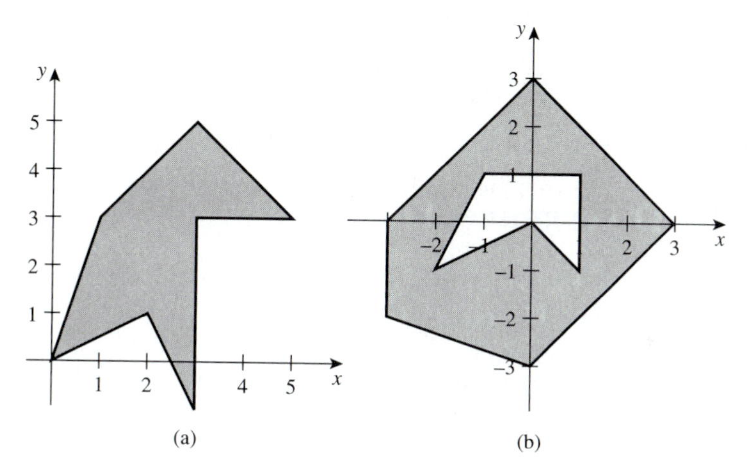

Figure 5.2.8 Polygonal regions for Exercise 16

18. Elizabeth has just inherited from her uncle a tract of land whose boundary is described as follows: "From the maple tree in front of the house, go 1000 yards northeast; from that point, go 1200 yards northwest; from that point, go 800 yards south; from that point, go back to the maple tree." What is the area of Elizabeth's inheritance?

19. There is an alternative way to prove Green's Theorem when the region in question is both x- and y-simple. First, write

$$\iint\limits_R \left(\frac{\partial F_2}{\partial x} - \frac{\partial F_1}{\partial y}\right) dA = \iint\limits_R \frac{\partial F_2}{\partial x}\, dA - \iint\limits_R \frac{\partial F_1}{\partial y}\, dA$$

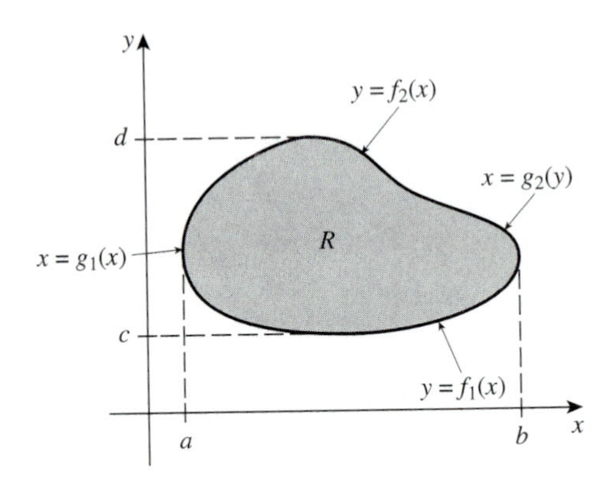

Figure 5.2.9 A region R that is both x-simple and y-simple

and then calculate the separate integrals on the right as iterated integrals using the x-simple description of R for the first and the y-simple description for the second. See Figure 5.2.9.

5.3 The Divergence Theorem

Another form of the Fundamental Theorem of Calculus is the **Divergence Theorem**, often called **Gauss's**[2] **Theorem** by physicists and engineers. It is exactly analogous to Green's Theorem in the plane: The surface integral of a vector field $\mathbf{F}$ over the boundary of a solid region S in three-dimensional space is equal to the triple integral over S of the divergence of $\mathbf{F}$. Moreover, the proof we give here is directly analogous to the one we gave for Green's Theorem.

One word about notation is in order before we state the Divergence Theorem. When we integrate a vector field over a closed surface (i.e., one that encloses a solid region in $\mathbb{R}^3$) we often draw a circle or ellipse through the double integral sign and write $\oiint_S \mathbf{F} \cdot \mathbf{n}\, d\sigma$ simply as a reminder of the special nature of the surface. For example, if S is the unit sphere, we would use this notation, whereas if S is a parallelogram, then we would omit the circle. This is the same practice that we adopted for path integrals along closed paths.

[2]Karl Friedrich Gauss (1777-1855) was one of the greatest mathematicians of modern times. Theorems bearing his name dot the mathematical landscape, testimony to his contributions in such disparate areas as number theory, algebra, differential geometry, and electromagnetic theory.

Theorem 5.3.1 *(Divergence Theorem) Let $D \subset \mathbb{R}^3$ be an open connected region, and let $\mathbf{F} : D \to \mathbb{R}^3$ be a smooth vector field. For any solid region S contained in D whose boundary ∂S is piecewise smooth,*

$$\oiint_{\partial S} \mathbf{F} \cdot \mathbf{n} \, d\sigma = \iiint_{S} \operatorname{div} \mathbf{F} \, dV. \tag{5.3.1}$$

In differential form, (5.3.1) is written

$$\oiint_{\partial S} F_1 \, dy \wedge dz + F_2 \, dz \wedge dx + F_3 \, dx \wedge dy = \iiint_{S} \left(\frac{\partial F_1}{\partial x} + \frac{\partial F_2}{\partial y} + \frac{\partial F_3}{\partial z} \right) dx \, dy \, dz. \tag{5.3.2}$$

Proof The intuition behind this proof is just like that for Green's Theorem.

Partition the portion of $\mathbb{R}^3$ about S into small rectangular boxes B_{ijk} with boundaries ∂B_{ijk} that are parallel to the coordinate planes. The union of those boxes that lie inside S forms a solid U whose boundary ∂U consists of tiny rectangular faces which, together, approximate the boundary surface ∂S. See Figure 5.3.1. Thus it is plausible (and it can be proved using the

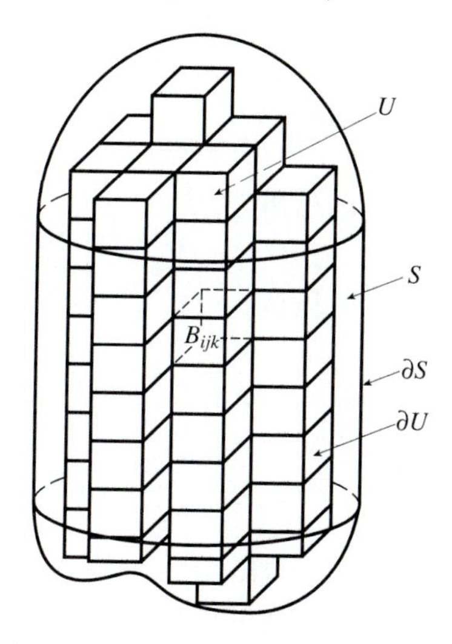

Figure 5.3.1 The sum of the surface integrals over each box B_{ijk} in the partition approximates the surface integral over the boundary of the solid S.

continuity of $\mathbf{F}$ and the piecewise smoothness of the boundary) that

$$\oiint_{\partial S} \mathbf{F} \cdot \mathbf{n} \, d\sigma \approx \oiint_{\partial U} \mathbf{F} \cdot \mathbf{n} \, d\sigma. \tag{5.3.3}$$

Since the outward normals to adjacent boxes are directed opposite one another, when we add up the surface integrals of $\mathbf{F}$ over the box boundaries, the portions over common faces cancel, leaving only surface integrals over the faces not common with those of other boxes in the partition. That is, the sum of the surface integrals of $\mathbf{F}$ over ∂B_{ijk} is the same as the surface integral of $\mathbf{F}$ over ∂U:

$$\oiint_{\partial U} \mathbf{F} \cdot \mathbf{n}\, d\sigma = \sum_{B_{ijk} \subset S} \oiint_{\partial B_{ijk}} \mathbf{F} \cdot \mathbf{n}\, d\sigma. \tag{5.3.4}$$

Each surface integral in the sum on the right hand side of (5.3.4) is the sum of six surface integrals, one over each face of the box. The outward normals are just $\mathbf{i}$, $-\mathbf{i}$, $\mathbf{j}$, $-\mathbf{j}$, $\mathbf{k}$, and $-\mathbf{k}$. See Figure 5.3.2. Consequently, two of the three terms in each surface integral are zero, and the surface integral over ∂B_{ijk} can be written

$$\begin{aligned}
&\oiint_{\partial B_{ijk}} \mathbf{F} \cdot \mathbf{n}\, d\sigma \\
&= \int_{z_{k-1}}^{z_k} \int_{y_{j-1}}^{y_j} \mathbf{F}(x_i, y, z) \cdot \mathbf{i}\, dy\, dz + \int_{z_{k-1}}^{z_k} \int_{y_{j-1}}^{y_j} \mathbf{F}(x_{i-1}, y, z) \cdot (-\mathbf{i})\, dy\, dz \\
&+ \int_{z_{k-1}}^{z_k} \int_{x_{i-1}}^{x_i} \mathbf{F}(x, y_j, z) \cdot \mathbf{j}\, dx\, dz + \int_{z_{k-1}}^{z_k} \int_{x_{i-1}}^{x_i} \mathbf{F}(x, y_{j-1}, z) \cdot (-\mathbf{j})\, dx\, dz \\
&+ \int_{y_{j-1}}^{y_j} \int_{x_{i-1}}^{x_i} \mathbf{F}(x, y, z_k) \cdot \mathbf{k}\, dx\, dy + \int_{y_{j-1}}^{y_j} \int_{x_{i-1}}^{x_i} \mathbf{F}(x, y, z_{k-1}) \cdot (-\mathbf{k})\, dx\, dy \\
&= \int_{z_{k-1}}^{z_k} \int_{y_{j-1}}^{y_j} [F_1(x_i, y, z) - F_1(x_{i-1}, y, z)]\, dy\, dz \\
&+ \int_{z_{k-1}}^{z_k} \int_{x_{i-1}}^{x_i} [F_2(x, y_j, z) - F_2(x, y_{j-1}, z)]\, dx\, dz \\
&+ \int_{y_{j-1}}^{y_j} \int_{x_{i-1}}^{x_i} [F(x, y, z_k) - F(x, y, z_{k-1})]\, dx\, dy
\end{aligned} \tag{5.3.5}$$

Consider the first of the last three double integrals in (5.3.5). Since $\partial F_1/\partial x$ is continuous, by the Fundamental Theorem of Calculus applied to the integrand, we have

$$F_1(x_i, y, z) - F_1(x_{i-1}, y, z) = \int_{x_{i-1}}^{x_i} \frac{\partial F_1}{\partial x}(x, y, z)\, dx$$

for each choice of y and z. Similarly, the integrands of the remaining two integrals can be written

$$F_2(x, y_j, z) - F_2(x, y_{j-1}, z) = \int_{y_{j-1}}^{y_j} \frac{\partial F_2}{\partial y}(x, y, z)\, dy$$

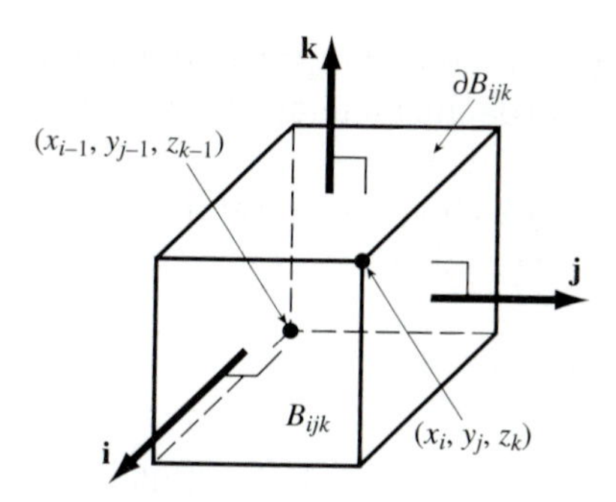

Figure 5.3.2 The surface integral is the sum of six surface integrals, where **n** takes the values **i**, **−i**, **j**, **−j**, **k**, and **−k**.

and

$$F_3(x, y, z_k) - F_3(x, y, z_{k-1}) = \int_{z_{k-1}}^{z_k} \frac{\partial F_3}{\partial z}(x, y, z)\,dz.$$

Substituting these integrals for the differences in (5.3.5), we obtain

$$\begin{aligned}
\oiint_{\partial B_{ijk}} \mathbf{F}\cdot\mathbf{n}\,d\sigma &= \int_{z_{k-1}}^{z_k}\int_{y_{j-1}}^{y_j}\int_{x_{i-1}}^{x_i} \frac{\partial F_1}{\partial x}\,dx\,dy\,dz \\
&+ \int_{z_{k-1}}^{z_k}\int_{x_{i-1}}^{x_i}\int_{y_{j-1}}^{y_j} \frac{\partial F_2}{\partial y}\,dy\,dx\,dz \\
&+ \int_{y_{j-1}}^{y_j}\int_{x_{i-1}}^{x_i}\int_{z_{k-1}}^{z_k} \frac{\partial F_3}{\partial z}\,dz\,dx\,dy \\
&= \int_{z_{k-1}}^{z_k}\int_{y_{j-1}}^{y_j}\int_{x_{i-1}}^{x_i} \left(\frac{\partial F_1}{\partial x} + \frac{\partial F_2}{\partial y} + \frac{\partial F_3}{\partial z}\right) dx\,dy\,dz \\
&= \iiint_{B_{ijk}} \operatorname{div}\mathbf{F}\,dV.
\end{aligned}$$

Then, by (5.3.4),

$$\oiint_{\partial U} \mathbf{F}\cdot\mathbf{n}\,d\sigma = \sum_{B_{ijk}\subset S} \oiint_{\partial B_{ijk}} \mathbf{F}\cdot\mathbf{n}\,d\sigma = \sum_{B_{ijk}\subset S} \iiint_{B_{ijk}} \operatorname{div}\mathbf{f}\,dV = \iiint_U \operatorname{div}\mathbf{F}\,dV. \tag{5.3.6}$$

Combining (5.3.6) and (5.3.3), we have

$$\oiint_{\partial S} \mathbf{F}\cdot\mathbf{n}\,d\sigma \approx \iiint_U \operatorname{div}\mathbf{F}\,dV.$$

For increasingly fine partitions, the approximation gets better, and also the triple integral over U approaches the triple integral over S. Therefore,

$$\oiint_{\partial S} \mathbf{F}\cdot\mathbf{n}\,d\sigma = \iiint_S \operatorname{div}\mathbf{F}\,dV,$$

which is the formula we wanted to prove.

Example 5.3.1 Find the flux out of the unit box $S=[0,1]\times[0,1]\times[0,1]$ of the vector field $\mathbf{F}=10(y-x)\mathbf{i}+(28x-xz)\mathbf{j}+\left(xy-\frac{8}{3}z\right)\mathbf{k}$.

Solution The flux is just the surface integral of $\mathbf{F}$ over the boundary ∂S of the box. Since $\mathbf{F}$ is smooth and the boundary of the box is piecewise smooth, we can use the Divergence Theorem. First,

$$\begin{aligned}\operatorname{div}\mathbf{F} &= \frac{\partial}{\partial x}(10(y-x))+\frac{\partial}{\partial y}(28x-xz)+\frac{\partial}{\partial z}\left(xy-\frac{8}{3}z\right)\\ &= -10+0-\frac{8}{3}=-\frac{38}{3}.\end{aligned}$$

Thus

$$\oiint_{\partial S} \mathbf{F}\cdot\mathbf{n}\,d\sigma \iiint_S \operatorname{div}\mathbf{F}\,dv = \int_0^1\int_0^1\int_0^1 -\frac{38}{3}\,dx\,dy\,dz = -\frac{38}{3}.$$

What does a negative value mean? The net flow is *inward.* Note, moreover, that if the surface had been different, then we would still integrate $-\frac{38}{3}$; the flux would be this number times the volume enclosed by the surface. ♦

Example 5.3.2 Suppose that $\operatorname{div}\mathbf{F}>0$ at all points inside a solid S. What can be said about the flux across ∂S? Simply that the flux is positive also: $\operatorname{div}\mathbf{F}>0$ implies that $\iiint_S \operatorname{div}\mathbf{F}\,dV>0$, so by the Divergence Theorem and part 6 of Theorem 4.4.3,

$$\oiint_{\partial S} \mathbf{F}\cdot\mathbf{n}\,d\sigma = \iiint_S \operatorname{div}\mathbf{f}\,dV>0. \quad ♦$$

The Divergence Theorem can give us added insight into the physical meaning of the divergence of a vector field. Suppose that $\mathbf{F}$ is smooth and represents the velocity field of some fluid. We've already remarked that, in some sense, $\operatorname{div}\mathbf{F}(\mathbf{a})$ is a measure of the per-unit-volume expansion rate of the volume occupied by an infinitesimal bit of fluid at the point $\mathbf{a}$. In slightly different terms, we can say that this is the instantaneous per-unit-volume flow rate through the boundary of an infinitesimal bit of fluid at the point $\mathbf{a}$. We can make the latter statement more precise by using the Divergence Theorem.

Let $\mathbf{a}$ be a point in space and let S represent a "small" solid region centered at $\mathbf{a}$. Since $\operatorname{div}\mathbf{F}$ is continuous, by the Mean Value Theorem for Triple Integrals, there is a point $\mathbf{b} \in S$ such that

$$\iiint_S \operatorname{div}\mathbf{F}\,dV = \text{Volume}\,(S)\cdot\operatorname{div}\mathbf{F}(\mathbf{b}).$$

Thus

$$\begin{aligned}\lim_{\operatorname{diam}(S)\to 0}\frac{\iiint_S \operatorname{div}\mathbf{F}\,dV}{\text{Volume}\,(S)} &= \lim_{\operatorname{diam}(S)\to 0}\frac{\text{Volume}\,(S)\operatorname{div}\mathbf{F}(\mathbf{b})}{\text{Volume}\,(S)}\\ &= \lim_{\operatorname{diam}(S)\to 0}\operatorname{div}\mathbf{F}(\mathbf{b}) = \operatorname{div}\mathbf{F}(\mathbf{a}). \quad (5.3.7)\end{aligned}$$

(Note that the last step follows by the continuity of $\operatorname{div}\mathbf{F}$.) On the other hand, by the Divergence Theorem, we can replace the triple integral of $\operatorname{div}\mathbf{F}$ in (5.3.7) with the flux of $\mathbf{F}$ through ∂S, so

$$\lim_{\operatorname{diam}(S)\to 0}\frac{\oiint_{\partial S}\mathbf{F}\cdot\mathbf{n}\,d\sigma}{\text{Volume}\,(S)} = \operatorname{div}\mathbf{F}(\mathbf{a}). \qquad (5.3.8)$$

Equation (5.3.8) says that *the divergence of* $\mathbf{F}$ *at* $\mathbf{a}$ *is the limit as* S *shrinks to* $\mathbf{a}$ *of the flux of* $\mathbf{F}$ *through* ∂S *per unit volume.*

If the divergence is negative, then there is actually a flow inward at the point: molecules of the fluid in question are being compressed in tighter and tighter. When $\operatorname{div}\mathbf{F} = 0$ in a region, the flow is said to be **incompressible** (or **divergence-free**), for obvious reasons. Water is an essentially incompressible fluid, so for instance the velocity field of water flowing through some region would be incompressible. In electromagnetic theory, the magnetic induction vector is divergence-free. This corresponds to the fact that, in Maxwell's theory, **magnetic monopoles** do not exist.

Example 5.3.3 Evaluate

$$\oiint_{\partial S} \ln x\,dy\wedge dz + \ln y\,dz\wedge dx + \ln z\,dx\wedge dy,$$

where S is the box $[1, 2]\times[2, 3]\times[3, 4]$.

Solution Here the surface integral is written in differential notation. We just use, then, the differential form of the Divergence Theorem. The integral equals

$$\iiint_S \left(\frac{\partial}{\partial x} \ln x + \frac{\partial}{\partial y} \ln y + \frac{\partial}{\partial z} \ln z \right) dV = \int_1^2 \int_2^3 \int_3^4 \left(\frac{1}{x} + \frac{1}{y} + \frac{1}{z} \right) dz\, dy\, dx.$$

You can easily check that the triple integral evaluates to $\ln 4$. ♦

Example 5.3.4 Evaluate

$$\oiint_A xy^2\, dy \wedge dz + yz^2\, dz \wedge dx + x^2 z\, dx \wedge dy,$$

where A is the sphere of radius 3 centered at the origin.

Solution The Divergence Theorem applies since the vector field and the boundary surface are smooth: This surface integral equals

$$\iiint_B \left(\frac{\partial}{\partial x} \left(xy^2\right) + \frac{\partial}{\partial y} \left(yz^2\right) + \frac{\partial}{\partial z} \left(x^2 z\right) \right) dV = \iiint_B (y^2 + z^2 + x^2)\, dV, \tag{5.3.9}$$

where B is the ball of radius 3 centered at the origin. We can evaluate this integral by converting it to spherical coordinates. Indeed, the ball is described by

$$\begin{array}{ccccc} 0 & \leq & \rho & \leq & 3 \\ 0 & \leq & \theta & \leq & 2\pi \\ 0 & \leq & \phi & \leq & \pi. \end{array}$$

So the triple integral in (5.3.9), and hence the surface integral, is

$$\begin{aligned} \int_0^3 \int_0^{2\pi} \int_0^{\pi} \rho^2 \cdot \rho^2 \sin\phi\, d\phi\, d\theta\, d\rho &= 2 \int_0^3 \int_0^{2\pi} \rho^4\, d\theta\, d\rho \\ &= 4\pi \int_0^3 \rho^4\, d\rho = \frac{972\pi}{5}. \end{aligned}$$ ♦

Example 5.3.5 Show that if f and g are continuously differentiable scalar-valued functions on an open set D in $\mathbb{R}^3$, then for any solid region S contained in D,

$$\iiint_S \vec{\nabla} f \cdot \vec{\nabla} g\, dV = \oiint_{\partial S} f\, \vec{\nabla} g \cdot \mathbf{n}\, d\sigma - \iiint_S f\, \vec{\nabla}^2 g\, dV. \tag{5.3.10}$$

(This is sometimes called **Green's identity** or the **integration-by-parts formula**.)

Solution Recalling that the ordinary integration-by-parts formula is just the product rule of differentiation stated in terms of integrals, we first consider an analogous product rule for taking the divergence of a product. By formula 13 in Table 3.8.1,

$$\begin{aligned}\operatorname{div}(f\vec{\nabla}g) &= \vec{\nabla}f\cdot\vec{\nabla}g + f\operatorname{div}(\vec{\nabla}g)\\ &= \nabla f\cdot\vec{\nabla}g + f\,\vec{\nabla}^2 g.\end{aligned}$$

Thus, solving for $\vec{\nabla}f\cdot\vec{\nabla}g$ and integrating over S, we obtain

$$\begin{aligned}\iiint_S \vec{\nabla}f\cdot\vec{\nabla}g\,dV &= \iiint_S \left(\operatorname{div}(f\,\vec{\nabla}g) - f\,\vec{\nabla}^2 g\right)dV\\ &= \iiint_S \operatorname{div}(f\,\vec{\nabla}g)\,dV - \iiint_S f\vec{\nabla}^2 g\,dV. \qquad (5.3.11)\end{aligned}$$

We can now apply the Divergence Theorem to the first integral on the right-hand side of (5.3.11):

$$\iiint_S \operatorname{div}(f\,\vec{\nabla}g)\,dV = \oiint_{\partial S} f\,\vec{\nabla}g\cdot\mathbf{n}\,d\sigma.$$

Substituting this surface integral into (5.3.11), we obtain (5.3.10). ♦

Problems in applied mathematics are often formulated as **boundary-value problems**: An unknown function u satisfies a partial differential equation for (x, y, z) in some solid region S of space and its values (or other conditions) are specified on the boundary ∂S of the region. For example, suppose that a cylindrical metal tank contains water, and the metal has been kept at the constant temperature 0°C for a very long time. On an intuitive level, we would certainly say that the temperature at each point in the water is also zero. But we can also demonstrate by purely mathematical means that a mathematical model for the temperature predicts the same conclusion. This demonstration relies critically on the Divergence Theorem.

Example 5.3.6 Suppose that the time-independent distribution of temperature u in a solid region S of space is described by the partial differential equation (Laplace's Equation)

$$\frac{\partial^2 u}{\partial x^2} + \frac{\partial^2 u}{\partial y^2} + \frac{\partial^2 u}{\partial z^2} = 0 \qquad (5.3.12)$$

throughout S and that $u(x, y, z) = 0$ whenever (x, y, z) lies on the boundary ∂S of S. Show that $u(x, y, z) = 0$ for (x, y, z) anywhere in S.

Solution Note first that (5.3.12) is equivalent to $\operatorname{div}(\vec{\nabla} u) = 0$. Now, by taking both f and g in (5.3.10) to be u, we obtain

$$\begin{aligned}
\iiint_S \vec{\nabla} u \cdot \vec{\nabla} u \, dV &= \oiint_{\partial S} u \vec{\nabla} u \cdot \mathbf{n} \, d\sigma - \iiint_S u \operatorname{div}(\vec{\nabla} u) \, dV \\
&= \oiint_{\partial S} u \vec{\nabla} u \cdot \mathbf{n} \, d\sigma - \iiint_S 0 \, dV \\
&= \oiint_{\partial S} u \vec{\nabla} u \cdot \mathbf{n} \, d\sigma. \qquad (5.3.13)
\end{aligned}$$

Since the values of u are zero on ∂S, the integrand in the surface integral is zero, so (5.3.13) reduces to

$$\iiint_S \|\vec{\nabla} u\|^2 \, dV = 0. \qquad (5.3.14)$$

Since the integrand in (5.3.14) is continuous and nonnegative, it must be that $\vec{\nabla} u = \mathbf{0}$ throughout S, which means that u is a constant function on S. Since the value of u on the boundary is zero, that constant must be $0 : u(x\, y, z) = 0$ for $(x, y, z) \in S$. ♦

Exercises 5.3

Use the Divergence Theorem to calculate the flux integrals in Exercises 1 through 13.

1. $\displaystyle\oiint_{\partial S} \mathbf{F} \cdot \mathbf{n} \, d\sigma$; $\mathbf{F}(x, y, z) = (z, y^2, -x^3)$, $S = [-1, 1] \times [1, 4] \times [-3, 5]$.

2. $\displaystyle\oiint_{\partial S} \mathbf{F} \cdot \mathbf{n} \, d\sigma$; $\mathbf{F}(x, y, z) = (x \sin^2 y, \cos z, z \cos^2 y)$, $S = [0, \pi] \times [0, \pi/2] \times [-\pi, 0]$.

3. $\displaystyle\oiint_{\partial S} \mathbf{F} \cdot \mathbf{n} \, d\sigma$; $\mathbf{F}(x, y, z) = (y - 3xz^2, x^2 + z^3, xy + z^3)$, S is the solid bounded above by the sphere $x^2 + y^2 + z^2 = 1$ and below by the xy-plane.

4. $\displaystyle\oint\!\!\oint_{\partial S} \mathbf{x}\cdot\mathbf{n}\,d\sigma$; S is the tetrahedral solid in the first octant bounded by the coordinate planes and the plane $6x + 3y + 2z = 6$.

5. $\displaystyle\oint\!\!\oint_{\partial S} \|\mathbf{x}\|^{-3}\mathbf{x}\cdot\mathbf{n}\,d\sigma$; S is any solid region not containing the origin.

6. $\displaystyle\oint\!\!\oint_{\partial S} (\mathbf{a}\times\mathbf{x})\cdot\mathbf{n}\,d\sigma$; $\mathbf{a}$ is a constant vector and S is any solid region.

7. $\displaystyle\oint\!\!\oint_{\partial S} 2x\,dy\wedge dz + 3xy\,dz\wedge dx + 4xyz\,dx\wedge dy$; S is the box $[0, 1]\times[-1, 1]\times[-1, 0]$.

8. $\displaystyle\oint\!\!\oint_{\partial S} y\,dy\wedge dz + xyz\,dz\wedge dx + x\,dx\wedge dy$; S is the solid bounded by the coordinate planes and the plane $x + 2y + 4z = 8$.

9. $\displaystyle\oint\!\!\oint_{\partial S} (x+z)\,dy\wedge dz + (y-x)\,dz\wedge dx + (z-y)\,dx\wedge dy$; S is the solid lying above the region in the xy-plane bounded by $y = x - x^2$ and $y = -x$, between the surfaces $z = x$ and $z = 1 + x^2 + y^2$.

10. $\displaystyle\oint\!\!\oint_{\partial S} y^2\,dy\wedge dz + z\ln(x^2+y^2+1)\,dx\wedge dy$; S is the cylindrical solid bounded by $z = 1$, $z = 4$, and $x^2 + y^2 = \frac{1}{4}$.

11. $\displaystyle\oint\!\!\oint_{\partial S} (x^2 + y^2 + z^2)^{3/2}\,dz\wedge dx$; S is the unit ball.

12. $\displaystyle\oint\!\!\oint_{\partial S} z^2\,dx\wedge dy$; S is the solid region between the spheres $x^2 + y^2 + z^2 = 1$ and $x^2 + y^2 + z^2 = 9$.

13. $\displaystyle\oint\!\!\oint_{\partial S} 2x\,y\,dy\wedge dz - y^2\,dz\wedge dx + z\,dx\wedge dy$; S is the solid portion of the ball $x^2 + y^2 + z^2 \le 4$ that remains after the portion inside the cylindrical solid $x^2 + y^2 \le 1$ is removed.

14. Show that if M is a piecewise smooth surface and $\mathbf{a}$ is any constant vector, then $\displaystyle\oint\!\!\oint_{M} \mathbf{a}\cdot\mathbf{n}\,d\sigma = 0$.

15. Let $\mathbf{F}$ be such that $\operatorname{curl}\mathbf{F}$ is continuously differentiable on an open connected set D in $\mathbb{R}^3$. Show that

$$\oiint_{\partial S} \operatorname{curl}\mathbf{F}\cdot\mathbf{n}\,d\sigma = 0$$

for any solid region S in D.

16. Let $D \subset \mathbb{R}^3$ be an open connected set and let $f : D \to \mathbb{R}$ have continuous second order partials on D. Show that

$$\oiint_{\partial S} \vec{\nabla} f\cdot\mathbf{n}\,d\sigma = \iiint_{S} \vec{\nabla}^2 f\,dV$$

17. Let S be a solid region in $\mathbb{R}^3$ with outward unit normal $\mathbf{n}$, and let u be a function that satisfies Laplace's equation on S and the boundary condition

$$\vec{\nabla} u(x,\,y,\,z)\cdot\mathbf{n} = 0$$

for $(x,\,y,\,z) \in \partial S$. Show that u is a constant function.

18. Use Exercise 17 to show that if u and v are solutions to Laplace's equation on a solid S that satisfy $\vec{\nabla} u\cdot\mathbf{n} = \vec{\nabla} v\cdot\mathbf{n}$ on ∂S, then $u = v + c$ for some constant c.

19. Let S be a solid region in $\mathbb{R}^3$ and let V denote its volume. Use the Divergence Theorem and (4.4.6) to show that

$$\begin{aligned} V &= \oiint_{\partial S} x\,dy\wedge dz = \oiint_{\partial S} y\,dz\wedge dx = \oiint_{\partial S} z\,dx\wedge dy \\ &= \frac{1}{3}\oiint_{\partial S} x\,dy\wedge dz + y\,dz\wedge dx + z\,dx\wedge dy. \end{aligned} \tag{5.3.15}$$

[Note that this is directly analogous to the area formula (5.2.9) that came out of Green's Theorem.]

20. By calculating the surface integral directly, verify that at least one of the formulas in (5.3.15) works when S is

(a) the ball $x^2 + y^2 + z^2 \le r^2$

(b) the cylindrical solid $x^2 + y^2 \le r^2$, $0 \le z \le h$.

(c) the rectangular solid $[0,\,a] \times [0,\,b] \times [0,\,c]$.

21. Let S be a solid with boundary ∂S, and let $\mathbf{x}_0$ be a fixed point in space. See Figure 5.3.3. Show that the volume of S is given by

$$V(S) = \frac{1}{3} \oiint_{\partial S} (\mathbf{x} - \mathbf{x}_0) \cdot \mathbf{n}\, d\sigma,$$

where $\mathbf{n}$ is the outward unit normal to ∂S.

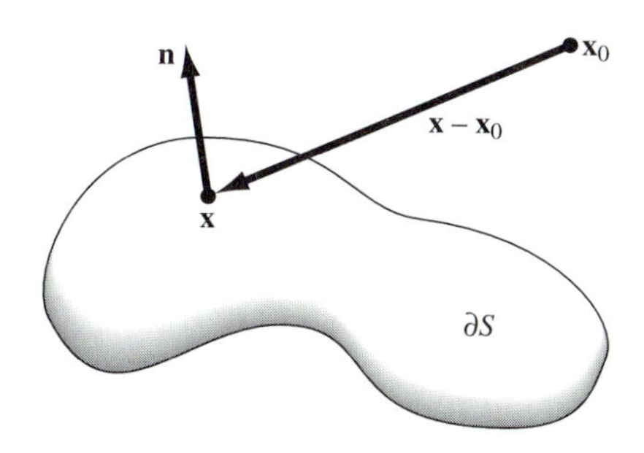

Figure 5.3.3 The geometry of Exercise 21

22. Let C be a solid cone generated by a point $\mathbf{x}_0$ and a surface M that lies in a plane $\mathbf{m} \cdot (\mathbf{x} - \mathbf{x}_1) = 0$. See Figure 5.3.4. Show that

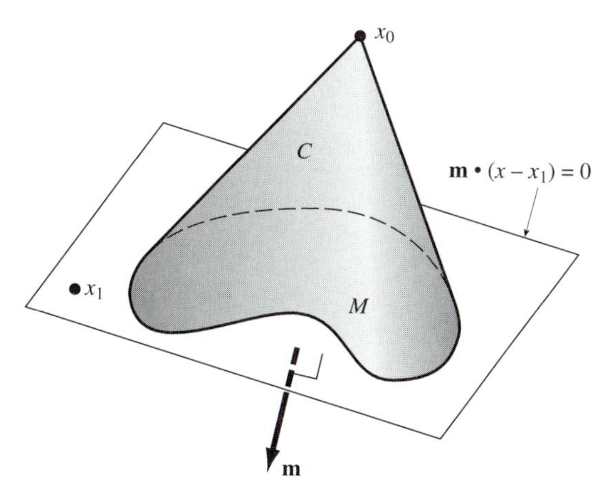

Figure 5.3.4 The volume of C is one-third the area of M times the distance between $\mathbf{x}_0$ and the plane $\mathbf{m} \cdot (\mathbf{x} - \mathbf{x}_0) = 0$.

$$V(C) = \frac{1}{3} \frac{|(\mathbf{x}_1 - \mathbf{x}_0) \cdot \mathbf{m}|}{\|\mathbf{m}\|} \sigma(M). \tag{5.3.16}$$

(This proves the familiar geometric formula for the volume of a cone: If the cone has base of area A and altitude h, then its volume is $V = \frac{1}{3}Ah$.)

23. Use (5.3.16) to show that the volume of a tetrahedron with vertices $\mathbf{a}$, $\mathbf{b}$, $\mathbf{c}$, and $\mathbf{d}$ is

$$V = \frac{1}{6}\left|(\mathbf{d}-\mathbf{a})\cdot((\mathbf{b}-\mathbf{a})\times(\mathbf{c}-\mathbf{a}))\right|. \tag{5.3.17}$$

24. **(a)** Use (5.3.17) to calculate the volume of the tetrahedron with vertices (0, 0, 0), (0, 1, 0), (1, 0, 0), and (0, 0, 1).

(b) Calculate the volume of the tetrahedron with vertices (1, 1, 1), (0, 3, 4), (5, 0, 2), and (2, 2, 0).

5.4 Stokes' Theorem

Stokes'[3] Theorem is a generalization of Green's Theorem that gives a relationship between the line integral of a vector field $\mathbf{F}$ around a simple closed path C in $\mathbb{R}^3$ and the surface integral of $\operatorname{curl}\mathbf{F}$ over any surface for which C is the boundary. For instance, if $\mathbf{F}$ is a force field, Stokes' Theorem says

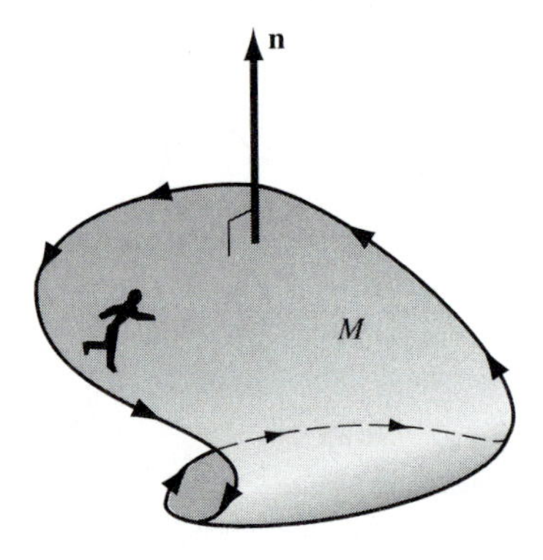

Figure 5.4.1 A positively-oriented boundary goes counterclockwise as viewed from the positive side of the surface.

that the sum of the component of $\operatorname{curl}\mathbf{F}$ in the direction of the unit normal

[3]The attribution of this theorem to the British mathematician George Gabriel Stokes(1819-1903) is at least partly a result of his having posed it as an examination question in 1854. However, according to [1], the identity had been discovered earlier by William Thomson (Lord Kelvin) and, in 1850, communicated to Stokes in a letter.

to a surface M is the same as the sum of the tangential component of $\mathbf{F}$ along the boundary of M.

Before we state the theorem, we need to discuss one bit of terminology. Let M be an oriented surface with a boundary[4] (i.e., a curve that forms the "edge" of the surface). A path along the boundary curve is said to be **positively oriented** with respect to the surface if a person walking along this path on the positive side of the surface has the surface to his or her left. In other words, the boundary path goes counterclockwise as viewed from the positive side of the surface. See Figure 5.4.1.

Theorem 5.4.1 *(Stokes' Theorem) Let $D \subset \mathbb{R}^3$ be an open connected set, and let $\mathbf{F} : D \to \mathbb{R}^3$ be a vector field that is continuously differentiable. Let M be any piecewise smooth, simple, oriented surface lying in D, let $\mathbf{n}$ be its unit normal, and let ∂M be a positively oriented boundary path. Then*

$$\oint_{\partial M} \mathbf{F} \cdot d\mathbf{x} = \iint_M \operatorname{curl} \mathbf{F} \cdot \mathbf{n} \, d\sigma. \tag{5.4.1}$$

In differential notation, this is stated

$$\oint_{\partial M} F_1\,dx + F_2\,dy + F_3\,dz = \iint_M \left(\frac{\partial F_3}{\partial y} - \frac{\partial F_2}{\partial z}\right) dy \wedge dz \tag{5.4.2}$$
$$+ \left(\frac{\partial F_1}{\partial z} - \frac{\partial F_3}{\partial x}\right) dz \wedge dx + \left(\frac{\partial F_2}{\partial x} - \frac{\partial F_1}{\partial y}\right) dx \wedge dy.$$

Proof We assume that M and its boundary are smooth. (In the case that M is only piecewise smooth, we can add the Stokes' formulas for each smooth piece. Since the path integrals over the edges where subpieces adjoin cancel out, the formula in (5.4.1) will hold when M is piecewise smooth. See Figure 5.4.2 for an informal description of this reasoning.) Let $\mathbf{f} : R \subset \mathbb{R}^2 \to \mathbb{R}^3$ be a smooth function such that $\mathbf{x} = \mathbf{f}(s, t)$, for $(s, t) \in R$, is a parametrization of M. Then $\mathbf{f}(\partial R) = \partial M$. See Figure 5.4.3. We proceed in three steps. First, we "pull back" from $\mathbb{R}^3$ to $\mathbb{R}^2$ by way of $\mathbf{f}$ so that $\oint_{\partial M} \mathbf{F} \cdot d\mathbf{x}$ can be represented as a line integral around ∂R. Second, we apply Green's Theorem to the resulting line integral. Finally, we "push forward" by recognizing the resulting double integral as a surface integral over M.

---

[4]This use of the term *boundary* is different from that of Section 3.3.

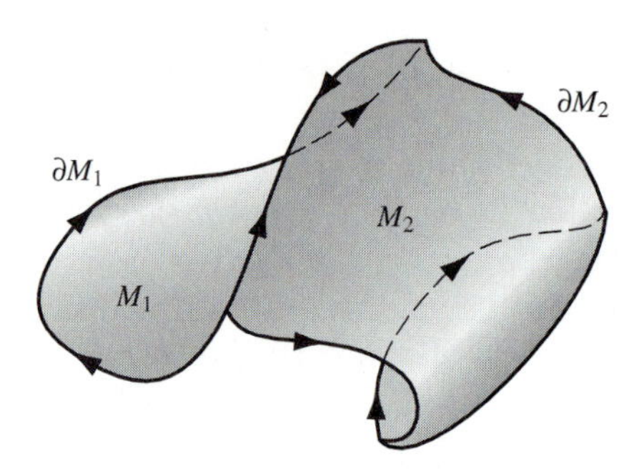

Figure 5.4.2 If M consists of the union of the two smooth surfaces M_1 and M_2, then Stokes' formula applied to each gives $\oint_{\partial M_1} \mathbf{F} \cdot d\mathbf{x} = \iint_{M_1} \operatorname{curl} \mathbf{F} \cdot \mathbf{n}\, d\sigma$ and $\oint_{\partial M_2} \mathbf{F} \cdot d\mathbf{x} = \iint_{M_2} \operatorname{curl} \mathbf{F} \cdot \mathbf{n}\, d\sigma$. Adding these equations, we obtain $\oint_{\partial M} \mathbf{F} \cdot d\mathbf{x} = \iint_{M} \operatorname{curl} \mathbf{F} \cdot \mathbf{n}\, d\sigma$ because the line integrals over the portion of boundary where M_1 and M_2 are adjacent cancel.

By regarding our vectors as column matrices, we can write $\mathbf{F} \cdot d\mathbf{x} = \mathbf{F}^T d\mathbf{x}$. Since $\mathbf{x} = \mathbf{f}(s, t)$, by the Chain Rule we have

$$d\mathbf{x} = J\mathbf{f}(s, t) d(s, t),$$

where $d(s, t) = \begin{bmatrix} ds \\ dt \end{bmatrix}$. So

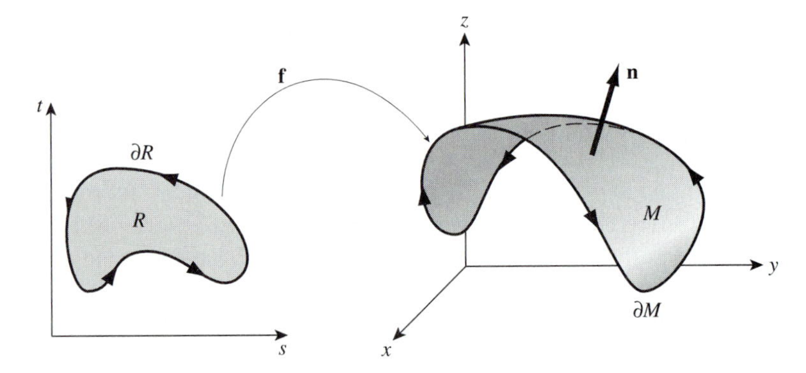

Figure 5.4.3 The first step in our proof of Stokes' Theorem is to "pull back" from $\mathbb{R}^3$ to $\mathbb{R}^2$ by regarding the surface M over which we are integrating as the image under $\mathbf{f}$ of a region R in $\mathbb{R}^2$.

$$\oint_{\partial M} \mathbf{F} \cdot d\mathbf{x} = \oint_{\partial M} \mathbf{F}^T d\mathbf{x} = \oint_{\partial R} \mathbf{F}(\mathbf{f}(s, t))^T J\mathbf{f}(s, t) d(s, t)$$

$$= \oint_{\partial R} \left(\mathbf{F}(\mathbf{f}(s,\, t))^T J\mathbf{f}(s,\, t) \right) \cdot (ds,\, dt). \tag{5.4.3}$$

Now let $\mathbf{G}(s,\, t) = \mathbf{F}(\mathbf{f}(s,\, t))^T J\mathbf{f}(s,\, t)$, and let $G_1(s,\, t)$ and $G_2(s,\, t)$ be the components of $\mathbf{G}$. Then the last integral in (5.4.3) is

$$\oint_{\partial R} G_1\, ds + G_2\, dt. \tag{5.4.4}$$

By Green's Theorem, the expression in (5.4.4) is equal to

$$\iint_R \left(\frac{\partial G_2}{\partial s} - \frac{\partial G_1}{\partial t} \right) ds\, dt. \tag{5.4.5}$$

By applying the Product Rule and Chain Rule, and by doing some rearranging of terms, we calculate the integrand in (5.4.5) in terms of the component functions of $\mathbf{F}$ and $\mathbf{f}$:

$$\begin{aligned} \frac{\partial G_2}{\partial s} - \frac{\partial G_1}{\partial t} &= \left(\frac{\partial F_3}{\partial y} - \frac{\partial F_2}{\partial z} \right) \left(\frac{\partial f_2}{\partial s}\frac{\partial f_3}{\partial t} - \frac{\partial f_2}{\partial t}\frac{\partial f_3}{\partial s} \right) \\ &+ \left(\frac{\partial F_1}{\partial z} - \frac{\partial F_3}{\partial x} \right) \left(\frac{\partial f_3}{\partial s}\frac{\partial f_1}{\partial t} - \frac{\partial f_3}{\partial t}\frac{\partial f_1}{\partial s} \right) \\ &+ \left(\frac{\partial F_2}{\partial x} - \frac{\partial F_1}{\partial y} \right) \left(\frac{\partial f_1}{\partial s}\frac{\partial f_2}{\partial t} - \frac{\partial f_1}{\partial t}\frac{\partial f_2}{\partial s} \right) \\ &= \operatorname{curl} \mathbf{F}\,(\mathbf{f}(s,\, t)) \cdot \left(\frac{\partial \mathbf{f}}{\partial s} \times \frac{\partial \mathbf{f}}{\partial t} \right). \end{aligned}$$

Thus the integral in (5.4.5) equals

$$\iint_R \operatorname{curl} \mathbf{F}\,(\mathbf{f}(s,\, t)) \cdot \left(\frac{\partial \mathbf{f}}{\partial s} \times \frac{\partial \mathbf{f}}{\partial t} \right) ds\, dt = \iint_M \operatorname{curl} \mathbf{F} \cdot \mathbf{n}\, d\sigma. \tag{5.4.6}$$

Following the steps from (5.4.6) back to (5.4.3), we see that

$$\oint_{\partial M} \mathbf{F} \cdot d\mathbf{x} = \iint_M \operatorname{curl} \mathbf{F} \cdot \mathbf{n}\, d\sigma. \quad \blacklozenge$$

Example 5.4.1 Use Stokes' Theorem to evaluate

$$\oint_C (x^2 + z^2)\, dx + y\, dy + z\, dz,$$

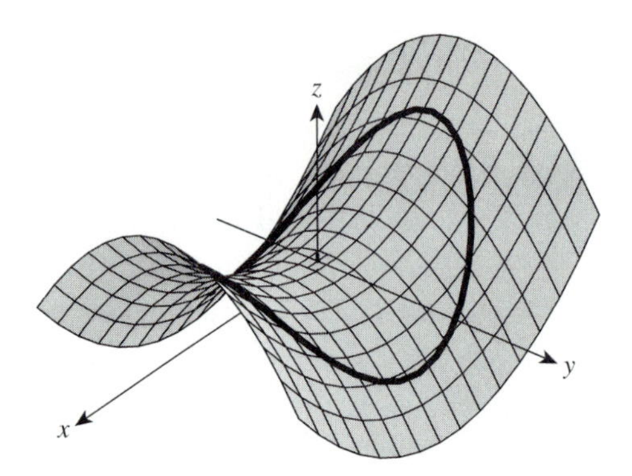

Figure 5.4.4 The curve parametrized by $\mathbf{x} = (\cos t, \sin t, \cos^2 t - \sin^2 t)$, $0 \le t \le 2\pi$ forms the boundary of a surface that lies in the graph of $z = x^2 - y^2$.

where C is the curve parametrized by

$$\mathbf{x} = (\cos t, \sin t, \cos^2 t - \sin^2 t), \qquad 0 \le t \le 2\pi.$$

Solution By inspection, we can see that the curve lies in the surface $z = x^2 - y^2$ and that it is the boundary of the portion of the surface for which $x^2 + y^2 \le 1$; call that portion M. See Figure 5.4.4. The curve is counterclockwise oriented as viewed from the positive z-axis, so we choose the upward normal. By the differential form of Stokes' Theorem, (5.4.2), the integral equals

$$\begin{aligned} \iint_M \left(\frac{\partial}{\partial y} z - \frac{\partial}{\partial z} y\right) dy \wedge dz \; &+ \; \left(\frac{\partial}{\partial z}(x^2 + z^2) - \frac{\partial}{\partial x} z\right) dz \wedge dx \\ &+ \; \left(\frac{\partial}{\partial x} y - \frac{\partial}{\partial y}(x^2 + z^2)\right) dx \wedge dy \\ &= \iint_M 0\, dy \wedge dz + 2z\, dz \wedge dx + 0\, dx \wedge dy \\ &= \iint_M 2z\, dz \wedge dx. \end{aligned}$$

On M, $z = h(x, y) = x^2 - y^2$, for $-1 \le x \le 1$ and $-\sqrt{1 - x^2} \le y \le \sqrt{1 - x^2}$, so we have

$$\begin{aligned} dz \wedge dx &= \begin{vmatrix} \partial z/\partial x & \partial z/\partial y \\ \partial x/\partial x & \partial x/\partial y \end{vmatrix} dy\, dx \\ &= \begin{vmatrix} 2x & -2y \\ 1 & 0 \end{vmatrix} = 2y\, dy\, dx, \end{aligned}$$

so

$$\begin{aligned}\iint_M 2z\,dz \wedge dx &= \int_{-1}^{1}\int_{-\sqrt{1-x^2}}^{\sqrt{1-x^2}} 2(x^2-y^2)\cdot 2y\,dy\,dx \\ &= \int_{-1}^{1}\int_{-\sqrt{1-x^2}}^{\sqrt{1-x^2}} 4(x^2y-y^3)\,dy\,dx.\end{aligned}$$

Since the integrand is an odd function of y and the y-limits are symmetric about 0, this double integral is zero. Thus

$$\oint_C (x^2+z^2)\,dx + y\,dy + z\,dz = 0.$$

Does this mean that $\mathbf{F} = (x^2+y^2,\, y,\, z)$ is path independent? ♦

Example 5.4.2 Evaluate $\oint_C \mathbf{F}\cdot\mathbf{n}\,d\sigma$, where C is the triangle with vertices $(1,\, 0,\, 0)$, $(0,\, 2,\, 0)$, and $(0,\, 0,\, 3)$ oriented by the order in which the points are given and $\mathbf{F} = (yz,\, -xz,\, xy)$.

Solution To apply Stokes' Theorem, we must identify a surface M for which the triangle is the boundary. One such choice is the triangular region determined by the three points. You should check that a normal to this plane consistent with the orientation of the triangle is

$$(-1,\, 2,\, 0)\times(-1,\, 0,\, 3) = (6,\, 3,\, 2),$$

so an equation for the plane is

$$\begin{aligned}6(x-1)+3y+2z &= 0 \\ z &= \frac{6-6x-3y}{2}.\end{aligned}$$

We want M to be the portion corresponding to x and y in the triangle

$$\begin{aligned}0 \le\ & y \le 2-2x \\ 0 \le\ & x \le 1.\end{aligned}$$

See Figure 5.4.5.

Since $\operatorname{curl}\mathbf{F} = (x+x,\, y-y,\, -z-z) = (2x,\, 0,\, -2z)$ and a unit normal to M is $\mathbf{n} = (6,\, 3,\, 2)/7$, by Stokes' Theorem,

$$\oint_C \mathbf{F}\cdot\mathbf{n}\,d\sigma = \iint_M \operatorname{curl}\mathbf{F}\cdot\mathbf{n}\,d\sigma$$

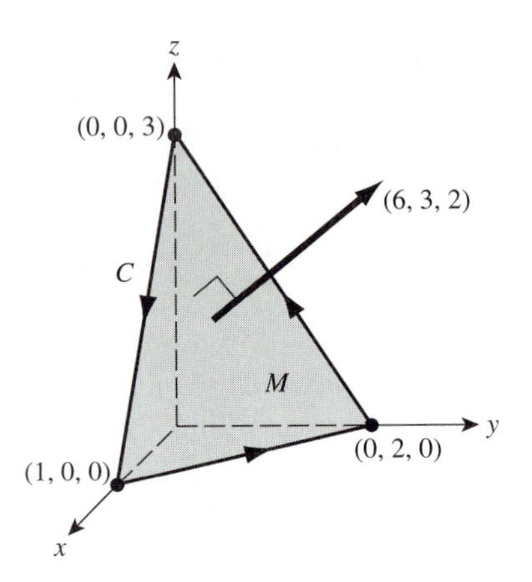

Figure 5.4.5 The surface for Example 5.4.2

$$
\begin{aligned}
&= \iint_M (2x,\, 0,\, -2z) \cdot \frac{(6,\, 3,\, 2)}{7}\, d\sigma \\
&= \int_0^1 \int_0^{2-2x} \frac{1}{7}\left(12x - 4\left(\frac{6-6x-3y}{2}\right)\right) \cdot \sqrt{3^2 + \left(\frac{3}{2}\right)^2 + 1}\, dy\, dx \\
&= \int_0^1 \int_0^{2-2x} (12x + 3y - 6)\, dy\, dx \\
&= \int_0^1 \left(-6 + 24x - 18x^2\right) dx \\
&= \left[-6x + 12x^2 - 6x^3\right]\Big|_0^1 \\
&= 0. \quad \blacklozenge
\end{aligned}
$$

Recall that in Section 5.1 we gave only half the proof of Theorem 5.1.5. We are now in a position to prove the other half in the setting of $\mathbb{R}^3$. That is, we will demonstrate that if $\mathbf{F}$ is continuously differentiable on a simply connected open set $D \subset \mathbb{R}^3$ and $\operatorname{curl} \mathbf{F} = \mathbf{0}$ on D, then $\mathbf{F}$ is path independent on D.

Let C be any simple closed path that lies in D. Since D is simply connected, we can find a surface M that lies in D whose boundary is C. By Stokes' Theorem,

$$\oint_C \mathbf{F} \cdot d\mathbf{x} = \iint_M \operatorname{curl} \mathbf{F} \cdot \mathbf{n}\, d\sigma = \iint_M \mathbf{0} \cdot \mathbf{n}\, d\sigma = 0.$$

This says, by Theorem 5.1.3, that $\mathbf{F}$ is path independent on D.

Stokes' Theorem also provides insight into the physical meaning of the curl of a vector field. Suppose that $\mathbf{F}$ is the velocity field for a fluid. Fix a point $\mathbf{a}$ and a unit vector $\mathbf{n}$, and consider the plane $\Pi_{\mathbf{n}}$ through $\mathbf{a}$ perpendicular to $\mathbf{n}$. Now let C be a simple closed curve in $\Pi_{\mathbf{n}}$ that encircles $\mathbf{a}$ and is oriented positively with respect to $\mathbf{n}$. Let M be the portion of the plane inside C. See Figure 5.4.6. By our discussion at the end of Section 4.2,

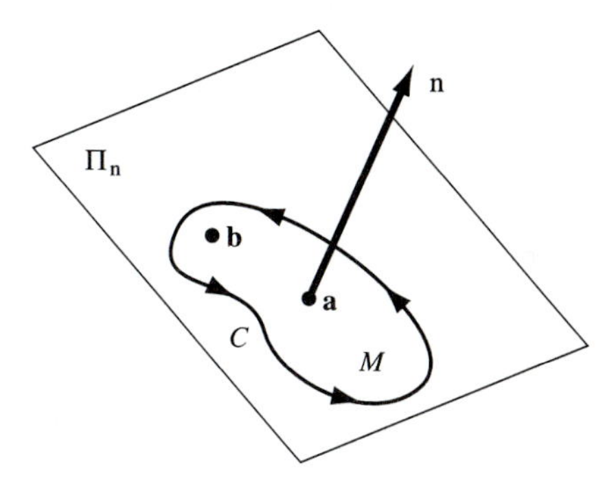

Figure 5.4.6 For a given point $\mathbf{a}$, a unit vector $\mathbf{n}$, and a "small" closed path C in the plane $\Pi_{\mathbf{n}}$, $\operatorname{curl}\mathbf{F}\cdot\mathbf{n}$ approximates the per-unit-area circulation of $\mathbf{F}$ around C.

$\oint_C \mathbf{F}\cdot d\mathbf{x}$ is the instantaneous rate of circulation of fluid particles about C. Therefore, the quotient

$$\frac{1}{\sigma(M)}\oint_C \mathbf{F}\cdot d\mathbf{x} \tag{5.4.7}$$

is the per-unit-area circulation of particles around the boundary of M. What happens to (5.4.7) as we shrink C down to the point $\mathbf{a}$? By Stokes' Theorem, $\oint_C \mathbf{F}\cdot d\mathbf{x} = \iint_M \operatorname{curl}\mathbf{F}\cdot\mathbf{n}\,d\sigma$, and by the Mean Value Theorem for double integrals, there exists a point $\mathbf{b}$ in M such that

$$\iint_M \operatorname{curl}\mathbf{F}\cdot\mathbf{n}\,d\sigma = \operatorname{curl}\mathbf{F}(\mathbf{b})\sigma(M).$$

Substituting this last expression for the integral in (5.4.7) and letting C shrink to $\mathbf{a}$, we have (abusing the limit notation slightly)

$$\begin{aligned}\lim_{C\to\mathbf{a}}\frac{1}{\sigma(M)}\oint_C \mathbf{F}\cdot d\mathbf{x} &= \lim_{C\to\mathbf{a}}\frac{1}{\sigma(M)}\cdot\sigma(M)\operatorname{curl}\mathbf{F}(\mathbf{b})\\ &= \lim_{C\to\mathbf{a}}\operatorname{curl}\mathbf{F}(\mathbf{b}) = \operatorname{curl}\mathbf{F}(\mathbf{a}).\end{aligned}$$

This says that $\operatorname{curl}\mathbf{F}(\mathbf{a})\cdot\mathbf{n}$ is the per-unit-area circulation of $\mathbf{F}$ at the point $\mathbf{a}$ about the axis $\mathbf{n}$. If $\operatorname{curl}\mathbf{F}(\mathbf{a})\neq\mathbf{0}$, then the per-unit-area circulation of $\mathbf{F}$ at $\mathbf{a}$ is maximized when $\mathbf{n}=\operatorname{curl}\mathbf{F}(\mathbf{a})/\|\operatorname{curl}\mathbf{F}(\mathbf{a})\|$, and its value in this direction is $\|\operatorname{curl}\mathbf{F}(\mathbf{a})\|$. Thus $\operatorname{curl}\mathbf{F}(\mathbf{a})$ is the axis about which the rotational tendency in $\mathbf{F}$ at $\mathbf{a}$ is greatest. If $\operatorname{curl}\mathbf{F}(\mathbf{a})=\mathbf{0}$, then $\operatorname{curl}\mathbf{F}\cdot\mathbf{n}=0$ for all choices of $\mathbf{n}$; there is no rotational tendency at $\mathbf{a}$ about any axis. For this reason, a velocity field $\mathbf{F}$ that satisfies $\operatorname{curl}\mathbf{F}(\mathbf{x})=\mathbf{0}$ for all $\mathbf{x}$ in the domain of $\mathbf{F}$ is called **irrotational**.

Exercises 5.4

In Exercises 1 through 4, use Stokes' Theorem to evaluate the path integral.

1. $\oint_C (x+y)\,dx-(x+y+2z)\,dy+(5x-8z)\,dz$; C is parametrized by $\mathbf{x}=(\cos t,\,\sin t,\,4)$, $0\le t\le 2\pi$.

2. $\oint_C 5yz\,dx+(x^2-y)\,dy+yz\,dz$; C is parametrized by $\mathbf{x}=(\cos t,\,\sin t,\,\cos t-\sin t)$, $0\le t\le 2\pi$.

3. $\oint_C z^2\,dx+x^2\,dy+y^2\,dz$; C is parametrized by $\mathbf{x}=(\sin 2t,\,\cos t,\,\sin t)$, $0\le t\le 2\pi$.

4. $\oint_C x^2\,dx+y^2\,dy+z^2\,dz$; C is parametrized by

$$\mathbf{x}=\begin{cases}(6t,\,0,\,5t), & 0\le t\le 1\\ (12-6t,\,4t-4,\,3t+2), & 1\le t\le 2\\ (0,\,6-4t,\,9-3t), & 2\le t\le 3.\end{cases}$$

In Exercises 5 through 8, evaluate the line integrals by using Stokes' Theorem.

5. $\oint_C \mathbf{F}\cdot d\mathbf{x}$; C is the square with vertices $(0,0,4)$, $(3,0,4)$, $(3,2,4)$, and $(0,2,4)$ oriented by this ordering of the points; $\mathbf{F}=(y/z,\,x^2y,\,x+z)$.

6. $\oint_C \mathbf{F}\cdot d\mathbf{x}$; C is the triangle with vertices $(0,1,0)$, $(0,1,5)$, and $(3,1,0)$ oriented by this ordering of the points; $\mathbf{F}=(-xy,\,-xz,\,-yz)$.

7. $\oint_C \mathbf{F}\cdot d\mathbf{x}$; C is the curve in which the surface $z=x^2-y^2$ intersects the cylinder $x^2+y^2=1$, oriented counterclockwise as viewed from the positive z-axis; $\mathbf{F}=(-y/2+z,\,x/2+3z/2,\,-x-3y/2)$.

8. $\oint_C \mathbf{F} \cdot d\mathbf{x}$; C is the curve in which the plane $x = 2$ intersects the sphere $x^2 + y^2 + z^2 = 16$, oriented counterclockwise as viewed from the positive x-axis; $\mathbf{F} = (z - x^2,\, x - y^2,\, y - z^2)$.

9. Use Stokes' Theorem to show that if f has continuous second-order partial derivatives in a simply connected set D, then

$$\oint_C \vec{\nabla} f \cdot d\mathbf{x} = 0$$

for any simple piecewise smooth closed curve C lying in D.

Bibliography

[1] D. Bressoud, *Second Year Calculus*, Springer-Verlag, New York, 1991.

[2] B. Kolman and C. G. Denlinger, *Applied Calculus*, Harcourt Brace Jovanovich, San Diego, 1989.

[3] J. E. Marsden and A. J. Tromba, *Vector Calculus*, W. H. Freeman and Co., San Francisco, 1976.

[4] John Stillwell, *Mathematics and its History,* Springer-Verlag, New York, 1989.

[5] Angus E. Taylor and W. Robert Mann, *Advanced Calculus, 2nd ed.*, John Wiley & Sons, New York, 1972.

Answers to Selected Exercises

Section 1.1

1.

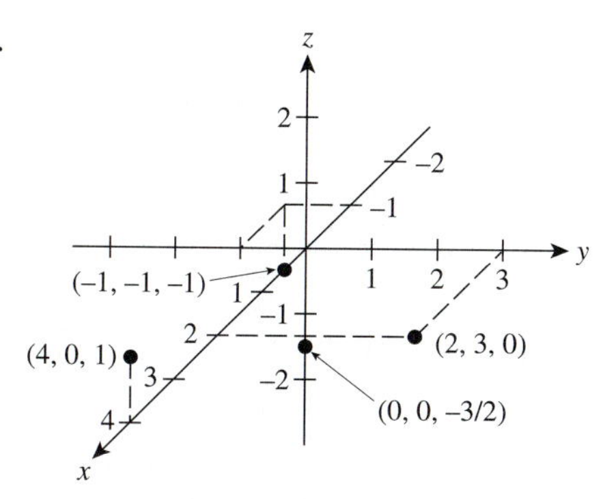

3.

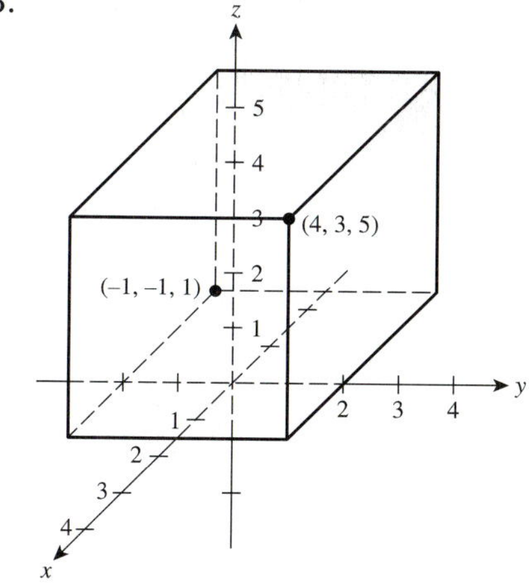

4. (a) $\sqrt{306}$, $(2, -11/2, 13/2)$
   (c) $\sqrt{19.53}$, $(11/2, 0, 121/20)$
   (e) $\sqrt{37}$, $(1, 0, 1/2)$

5. $z = 3$ describes the plane that is parallel to the xy-plane and passes through the point $(0, 0, -3)$.

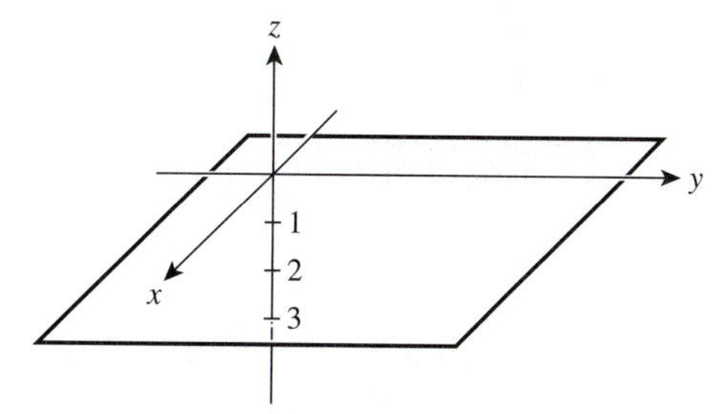

7. $2x+z = 2$ is the plane that passes through the points $(0, 0, 2)$ and $(1, 0, 0)$ and is parallel to the y-axis.

9. $x^2+y^2+z^2+4x-10y-2z+26 = 0$ is the sphere with center at point $(-2, 5, 1)$ and radius 2.

11. $(y-1)^2 + (z+3)^2 = 0$ is the line that passes through the point (0,1, -3) and is parallel to the x-axis.

13. $-1 \leq z \leq 1$ describes the space between and including the planes $z = -1$ and $z = 1$.

15. $(x-1)^2 + (y+1)^2 + (z+3)^2 > 81$ describes the space that surrounds, but does not include, the sphere $(x-1)^2 + (y+1)^2 + (z+3)^2 = 81$.

17. 607 ft

19. $(x-2)^2+(y+1)^2+(z-7)^2=1$

21. 33

23. cylinder with the line in question as its axis; $x^2+z^2=25$

26. (a) $2(x+1)-7(y-7/2)-3(z-3/2)=0$
    (b) $(x-5/2)-4(y+1)-2(z-3)=0$

27. Let the midpoint be $(\overline{x},\overline{y},\overline{z})$ and let the two endpoints be (a_1, a_2, a_3) and (b_1,b_2,b_3). Then $(b_1, b_2, b_3)=(2\overline{x}-a_1, 2\overline{y}-a_2, 2\overline{z}-a_3)$.

Section 1.2

3.

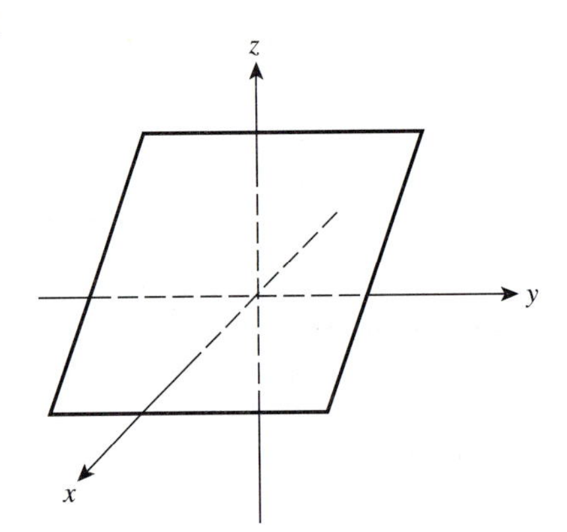

5.

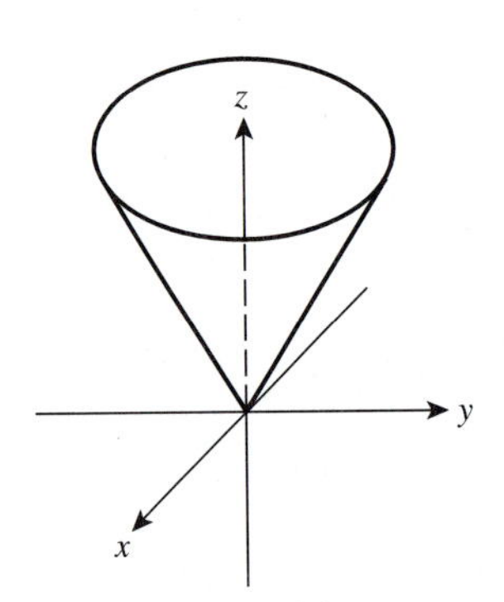

7. (a) 5; (b) 3; (c) 1; (d) 6; (e) 4; (f) 2

9. $f(x,y)=-\sqrt{x^2+y^2}$

11. $f(x,y)=\begin{cases}1-|y|, & -x<y<x\\ 1-|x|, & -y<x<y\end{cases}$

13. sphere with center $(1,-2,-5)$ and radius 3

15. hyperbolic paraboloid

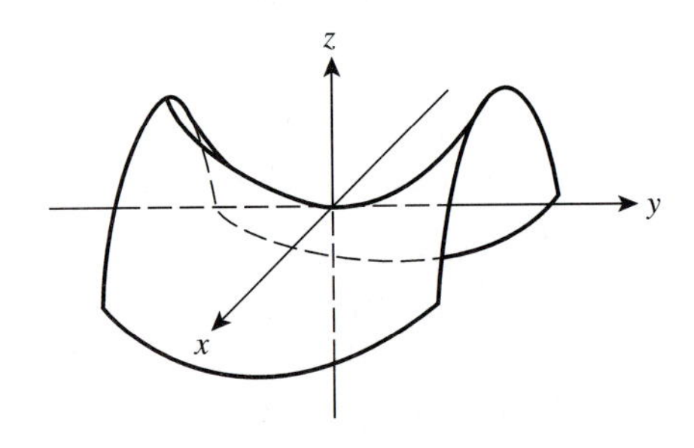

17. elliptic paraboloid

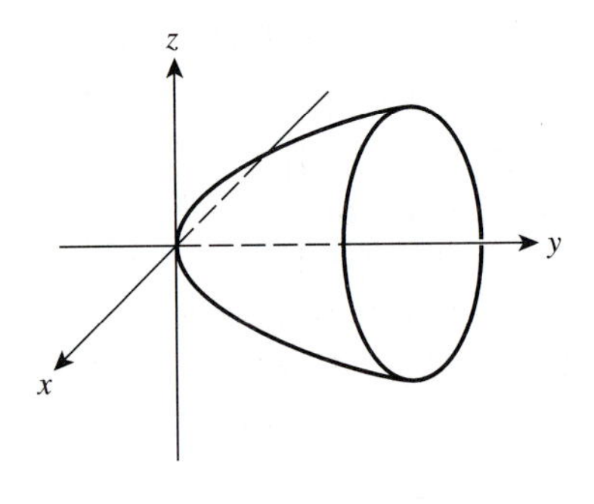

19. ellipsoid with center $(1,-2,0)$

21.

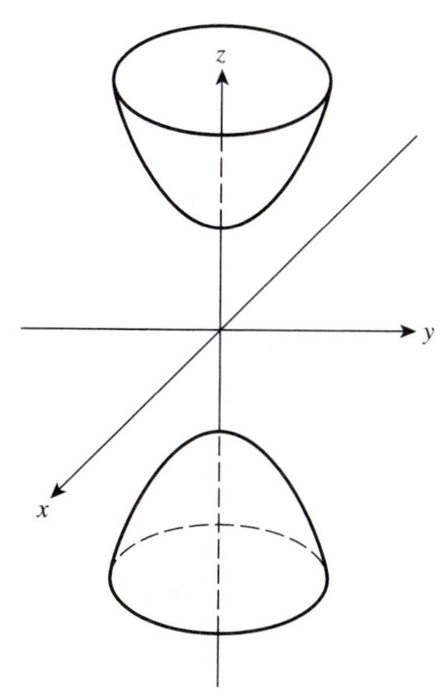

23. hyperboloid of one sheet
25. hyperboloid of two sheets
27. The elliptic paraboloid $z = -2x^2 - 2y^2$ is the first one ($z = x^2 + y^2$) turned upside-down and stretched vertically by a factor of 2. Both share the same vertex at the origin.
29. The surface $z = e^x/4 - 2$ is obtained from $z = e^x$ by scaling vertically by a factor of 1/4 and then translating vertically by -2.
31. $25x^2 + y^2 + 4(z-1)^2 = 100$ is obtained from the ellipsoid $x^2 + 4y^2 + 25z^2 = 100$ by reflection in the plane $x = y$, reflection in the plane $y = z$, and translation 1 unit in the z-direction.

Section 1.3

3. (a) $(3/\sqrt{2}, 3/\sqrt{2}, 2)$
   (b) $(1, 0, -4)$
   (c) $(-1, \sqrt{3}, 1)$
   (d) $(1/\sqrt{2}, -1/\sqrt{2}, 0)$
5. (a) $(\sqrt{2}, \pi/4, 1)$; $(\sqrt{3}, \pi/4, \arccos 1/\sqrt{3})$
   (b) $(\sqrt{2}, 3\pi/4, -1)$; $(\sqrt{3}, 3\pi/4, \arccos 1/\sqrt{3})$
   (c) $(2\sqrt{2}, 2\pi/3, 2\sqrt{2})$; $(4, 2\pi/3, \pi/4)$
   (d) $(\sqrt{3}\pi/2, 3\pi/2, \pi/2)$; $(\pi, 3\pi/2, \pi/3)$
7. (b) $r = 2\cos\theta$

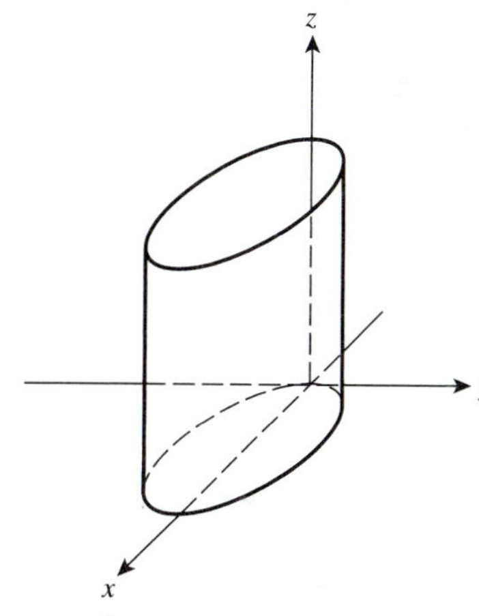

   (d) $z = \theta$

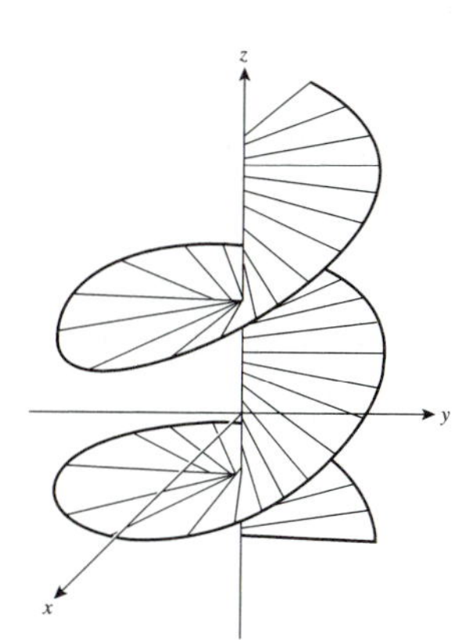

9. (a) all the space between and including the planes $z = 1$ and

$z = 2$

(c) all the space between and including the planes $y = x$ and $x = 0$ $(0 \leq x; 0 \leq y)$ and between and including the cylinders $x^2 + y^2 = 1$ and $x^2 + y^2 = 4$

10. (a) the semicircle formed by the intersection of the sphere $x^2 + y^2 + z^2 = 4$ and the plane $y = x$ $(0 \leq y)$

(c) the space between the spheres $x^2 + y^2 + z^2 = 4$ and $x^2 + y^2 + z^2 = 9$, the half planes $y = 0$ $(0 \leq x)$ and $y = x$ $(0 \leq x)$, and the half cones $z = (3x^2 + 3y^2)^{1/2}$ and $z = ((x^2 + y^2)/3)^{1/2}$

11. All points on the z-axis could have any and every possible θ-value

12. (a) $7\pi/4$; (b) $5\pi/3$; (c) $\pi/4$

Section 1.4

1. (a) $5\mathbf{i} - 6\mathbf{j} + \mathbf{k}$
   (b) $31\mathbf{i} - 11\mathbf{j} + 6\mathbf{k}$
   (c) not defined
   (d) $-24\mathbf{i} + 13\mathbf{j} - 8\mathbf{k}$
   (e) not defined
   (f) $\sqrt{365}$
   (g) $1/\sqrt{26}(5\mathbf{i} - \mathbf{j})$
   (h) $-3.68\mathbf{i} - 7.42\mathbf{j} + 10.2\mathbf{k}$

3. (1.124, 2.396, 1.413)

7. (a) yes; (b) no; (c) yes; (d) no

9. 1.19 miles east, 4.67 miles north, 0.66 miles up

11. $c(-0.012\mathbf{i} - 0.0203\mathbf{j})$

12. $(\mathbf{a} - \mathbf{x})/||\mathbf{a} - \mathbf{x}||^3 - (\mathbf{b} - \mathbf{x})/||\mathbf{b} - \mathbf{x}||^3$

15. 69.6° with x-axis; 35.5° with y-axis; 62.3° with negative z-axis

19. (a) $(-23/16, 1/4, -3/4)$; (b) $(0, 0, 1/5)$

20. $m_1 = \sqrt{6} - \sqrt{2}$, $m_2 = 2\sqrt{2}$, $m_3 = 2$

Section 1.5

1. (a) 9; (b) 15; (c) -12; (d) $-2/\sqrt{87}$

3. (a) yes; (b) no; (c) yes; (d) no; (e) yes; (f) yes

5. (a) $\mathbf{i} + \mathbf{j} + 2\mathbf{k}$; (b) $\mathbf{k}$; (c) $\mathbf{j}$; (d) $(1, \sqrt{2}, 1)$, $(2, 2\sqrt{2}, 2)$

7. $69.8°, 60.3°, 49.9°$

9. 73

19. -28.8 N·m

21. 109.5°

22. $\phi = (r_1 r_2 \cos(\theta_1 - \theta_2) + z_1 z_2) \cdot [(r_1^2 + z_1^2)(r_2^2 + z_2^2)]^{-1/2}$

25. $6{,}370 \arccos[\cos(\Phi_1 - \Phi_2) \cdot \cos^2((\Theta_1 - \Theta_2)/2) + \cos(\Phi_1 + \Phi_2) \cdot \sin^2((\Theta_1 - \Theta_2)/2)]$

Section 1.6

1. (a) $(3, -3, 2)$
   (b) $(-18, 0, 0)$
   (c) $(13, -17, -3)$
   (d) $(0, 0, 0)$

14. (a) -22; (c) 48; (e) -62; (g) -729

15. 33

17. 48

19. 17

21. $19/\sqrt{2}$

22. (b) 10, 4; (c) $2\sqrt{30}$

33. 36

35. 90

36. (a) -10; (b) -4

Section 1.7

1. (a) $6x + y - z + 3 = 0$
   (b) $(-1/2)x + (1/3)y - (1/4)z = 0$
   (c) $-x + 7y + 16z = 0$
   (d) $7x + 3y + z - 30 = 0$
   (e) $7x - 6y - 56 = 0$

2. (a) $x = -t$, $y = 4t + 5$, $z = t - 1$
   (c) $x = -8$, $y = 6t+1$, $z = 6t+4$

3. $(-2/3, 8/9, -77/9)$

5. $(2.34, -0.552, 2.24)$, $(-0.521, -1.507, -2.535)$

7. $11x + 67y + 12z = 0$

9. $-9x - 7y + 10z = 200$

11. $(1/2)y - (1/3)z = 0$

13. $x = -t, y = t, z = t$

17. $27/\sqrt{74}$

22. $\mathbf{x} = (-6 + 3s + t,\ 4 + s - 5t,\ 7 - s + 2t)$

23. $\mathbf{x} = (-s-4t,\ 8-6s-15t,\ s-11t)$

25. $\mathbf{x} = (s,\ -13 + 12s + 17t,\ t)$

27. $40x - 8y - 15z = 25$

Section 1.8

1. $t \geq 1$

3. $2 < t < 4$

7.

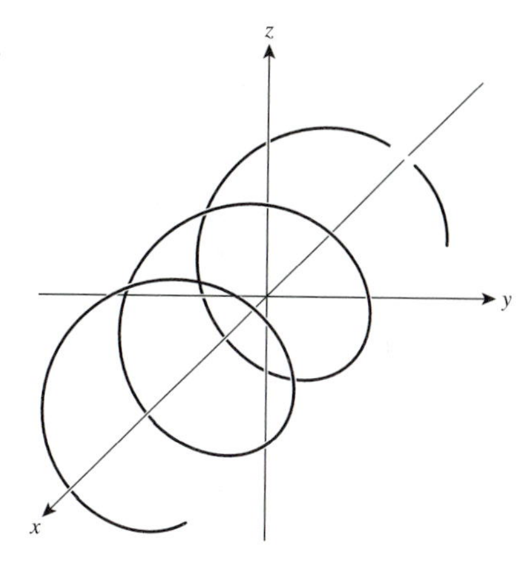

9.

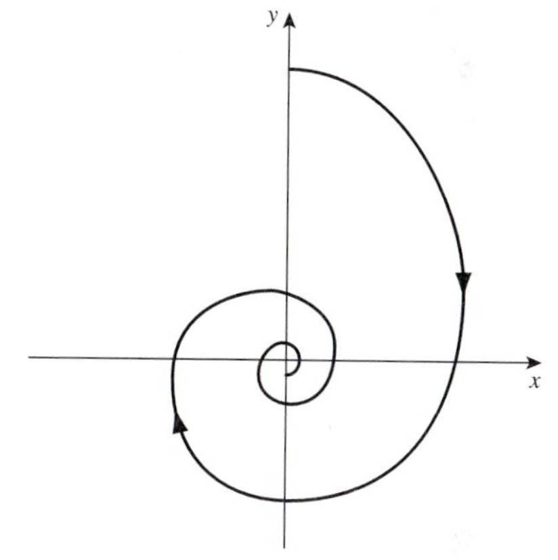

10. (a) 6; (b) 3; (c) 2; (d) 1; (e) 4; (f) 5

11. $\mathbf{x} = t(1, 2, 10) + (1, 5, 8)$, $-1 \leq t \leq 0$

13. $\mathbf{x} = (\cos t,\ t \sin t,\ t)$, $0 \leq t \leq 4\pi$

15. $\mathbf{x} = (1, 2\cos t, (5/2)\sin t)$, $0 \leq t \leq 2\pi$

16. $\mathbf{x} = 8(0, t - \sin t, 1 - \cos t)$, $t \geq 0$

21. $\mathbf{x}(t) = a(\sin\theta\cos\theta + \cot\theta, \sin^2\theta + 1)$

24. (b) $\mathbf{x}(t) = 3at/(1 + t^3)\mathbf{i} + 3at^2/(1 + t^3)\mathbf{j}$

(c) asymptote: $y = -x - a$

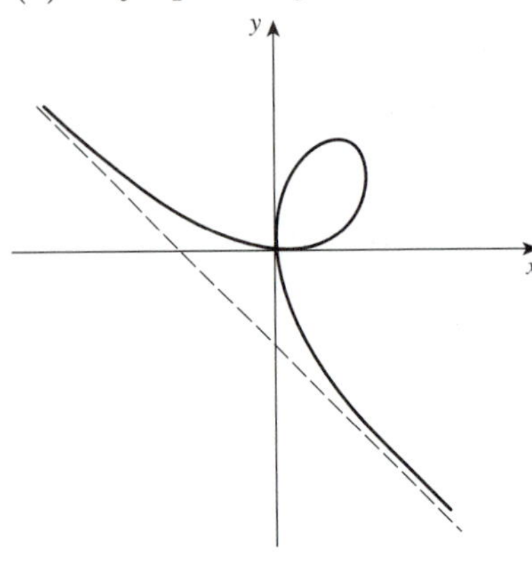

Section 1.9

1. $\mathbf{i} + \mathbf{j} + \mathbf{k}$
3. $\mathbf{i}$
5. $\mathbf{A}'(t) = (3t^2 - 1)\mathbf{i} + (2t)\mathbf{j} + \mathbf{k}$
7. $\mathbf{C}'(t) = (2t/t^2 + 1)\mathbf{i} + 4\mathbf{k}$
9. $\mathbf{y} = (t+1)\mathbf{i} + t\mathbf{j} + (1-t)\mathbf{k}$
11. $\mathbf{y}(\pi) = (-\pi, 3\pi, -3\pi/2)$; $\mathbf{y}''(\pi) = (0,0,0)$
13. $\mathbf{v}(t) = (3t^2 + 3)\mathbf{i} + (8t)\mathbf{j} + 6\mathbf{k}$; $\mathbf{a}(t) = (6t)\mathbf{i} + 8\mathbf{j}$; $\mathbf{f}(0) = -\mathbf{j}$; $\mathbf{v}(0) = 3\mathbf{i} + 6\mathbf{k}$; $\mathbf{a}(0) = 8\mathbf{j}$
23. $(4t + c_1)\mathbf{i} - (3t + c_2)\mathbf{j} + (2t + c_3)\mathbf{k}$
25. $((1/2)\ln(t^2+1) + c_1)\mathbf{i} + (\arctan t + c_2)\mathbf{j} + (t^2/2 + \ln t + c_3)\mathbf{k}$
27. $((1/2)\sin^2 t + c_1)\mathbf{i} + (\sin t + (1/3)\sin^3 t + c_2)\mathbf{j} + ((\cos 2t - 1)/2 + c_3)\mathbf{k}$
29. $\mathbf{r}(t) = (2 - \cos t)\mathbf{i} + ((1/2)\sin 2t + 1)\mathbf{j} + (4/3 - (1/3)\cos 3t)\mathbf{k}$
31. $\mathbf{r}(t) = ((t+1)(\ln(t+1) - \ln 2 - 1) + 2)\mathbf{i} + (t^5/20 + t^3/6 - 3t/4 + 8/15)\mathbf{j} + (t^4/12 + t^2/2 - 4t/3 + 3/4)\mathbf{k}$
36. (a) $\mathbf{s}(t) = (1/3)(t^2+4)^{3/2} - 8/3$
    (c) $\mathbf{s}(t) = t\sqrt{a^2 + b^2 + c^2}$
37. (a) $\mathbf{x} = (1/2)((3s+8)^{2/3} - 4)^{3/2}\mathbf{i} + ((3s+8)^{2/3} - 4)\mathbf{j}$
    (b) $\mathbf{x} = \cos(s/\sqrt{2} + a)\mathbf{i} + \sin(s/\sqrt{2} + a)\mathbf{j} + (s/\sqrt{2} + a)\mathbf{k}$
    (c) $\mathbf{x} = (as, bs, cs)/\sqrt{a^2 + b^2 + c^2}$

Section 2.1

1. (a) $(3, 1/2, 51/10, 3/2, 31/8, 3)$; $(3, -1/2, 49/10, -3/2, 33/8, 1)$; 2
   (b) $(0, -1/\sqrt{2}, 1/\sqrt{2}, 0)$; $(\sqrt{2}, -1/\sqrt{2}, 1/\sqrt{2}, \sqrt{2})$; -1
   (c) $(3/2, 7/6, 13/12, \ldots, 111/110)$; $(1/2, -1/6, -5/12, \ldots, -89/110)$; $55, 991/27, 726$
   (d) $(2, 3, 4, \ldots, n, n+1)$; $(0, 1, 2, \ldots, n-2, n-1)$; $n(n+1)/2$
2. (a) $3\sqrt{6}$; $\sqrt{5,641}/40$; $81.66°$
   (c) 1.24; 2.55; 50.4°
3. (a) $(11, 11, 5, 8, 10)$
   (b) $(13/2, 4, 5/2, -3/2, -1)$
   (c) does not make sense
   (d) $(8, 14, 0, 22, 20)$
   (e) $(8/\sqrt{286}, 14/\sqrt{286}, 0, 22/\sqrt{286}, 20/\sqrt{286})$
   (f) 66
   (g) does not make sense
   (h) $(12, 8, 0, 0, -4)$
4. (a) $(\sqrt{5}, 0, 0, 0, 0, 0, 0, 0, 0, 0)$
   (c) $(1, 2, 3, 4)$, $(-1, 0, -3, 0)$, $(0, -2, 0, -4)$
14. (a) $(-1/3, 0, 1/3, 0, -1/3)$, $-1/\sqrt{3}$

(c) $(12.75, -12.75, 12.75, \ldots, 12.75, -12.75)$, 127.5

20. $49/(4\sqrt{165})$

22. $d_M(\mathbf{x}_1, \mathbf{x}_2) = 2$;
$d_M(\mathbf{x}_1, \mathbf{x}_3) = 0.9999985$;
$d_M(\mathbf{x}_3, \mathbf{x}_2) = 0.9999985$.
In Euclidean space, the distance is 5.2×10^5.

Section 2.2

1. $\begin{bmatrix} 25 \\ -5 \end{bmatrix}$

3. $\begin{bmatrix} 12 & 26 & 8 \\ -6 & -7 & -1 \end{bmatrix}$

4. undefined

5. $\begin{bmatrix} 60 \\ 74 \\ 12 \end{bmatrix}$

7. $\begin{bmatrix} -14 & 54 \\ 1 & 6 \end{bmatrix}$

9. $\begin{bmatrix} 3 \\ -2 \\ 0 \end{bmatrix}, \begin{bmatrix} -1 \\ 3 \\ 7/3 \end{bmatrix}, \begin{bmatrix} 1 \\ 2 \\ -3 \end{bmatrix}$

11. $\begin{bmatrix} -2 & 3 \\ 1 & -1 \end{bmatrix}$

13. $\begin{bmatrix} 3/4 & 9/4 & 3/2 \\ 9/4 & -9/4 & -3/2 \\ 3/2 & -3/2 & 0 \end{bmatrix}$

14. not invertible

16. $\begin{bmatrix} 0 & 0 & 1 & 0 \\ 1 & 0 & 0 & 0 \\ 0 & 0 & 0 & 1 \\ 0 & 1 & 0 & 0 \end{bmatrix}$

17. not invertible

18. $x = \frac{16}{3}$, $y = \frac{2}{3}$

19. $x = \frac{3}{7}$, $y = \frac{5}{7}$, $z = \frac{3}{7}$

33. $\mathbf{a} = (-a_{23}, a_{13}, -a_{12})$

38. (a) $3\sqrt{7}$;
(c) $\sqrt{30}$;

40. (a) 1; (c) 10

Section 2.3

1. $\begin{bmatrix} 3 & 5 \\ 0 & 1 \\ -1 & 4 \end{bmatrix}$

3. $\begin{bmatrix} 4 & 1/2 & 1/3 \\ 1/2 & 3 & -1/4 \\ 1/3 & 1/4 & 2 \end{bmatrix}$

7. $(5x_1/9 - x_2/9,\ -x_1/9 + 2x_2/9)$

9. $(21x_1/4 - 9x_2/4 + 3x_3/2, -9x_1/4 + 9x_2/4 - 3x_3/2, 3x_1/2 + 3x_2/2)$

11. $(x_5, x_4, x_3, x_2, x_1)$

14. (a) $(-x_1 + \frac{11}{2}x_2, -\frac{7}{5}x_1 + \frac{1}{2}x_2)$
(c) $(-x_1 + 7x_2 + 13x_3 - x_4, 14x_1 - 7x_2 + 14x_4)$

15. (a) $(\frac{1}{2}x_1 + \frac{11}{5}x_2, -\frac{7}{2}x_1 - x_2)$
(c) undefined

16. (a) $\mathbf{T}^i = \begin{cases} (x_1, x_2) & i = 4k \\ (x_2, -x_1) & i = 4k-3 \\ (-x_1, -x_2) & i = 4k-2 \\ (-x_2, x_1) & i = 4k-1 \end{cases}$
for some $k = 0, 1, 2, 3, \ldots$
(c) $\mathbf{T}^i = ((7/8)^i x_1, (-2/3)^i x_2, (1/4)^i x_3)$

Section 2.4

1. $\{\mathbf{x} \mid \mathbf{x} = s(-1,-1) + t(40,51), 0 \le s \le 1,\ 0 \le t \le 1\}$; area = 11

3. $\{\mathbf{x} \mid \mathbf{x} \in s(3,-4,10)+t(6,-4,5)+u(3,-8,5)+(3,-4,5),\ 0 \le s \le 1, 0 \le t \le 1, 0 \le u \le 1\}$; volume $= 240$

8. 9.5, 19

10. $\mathbf{T}(\mathbf{x}) = \begin{bmatrix} 6 & 0 & 0 \\ 0 & 6 & 0 \\ 0 & 0 & 6 \end{bmatrix} \mathbf{x}$

12. $\mathbf{T}(\mathbf{x}) = \begin{bmatrix} 1/2 & -\sqrt{3}/2 & 0 \\ \sqrt{3}/2 & 1/2 & 0 \\ 0 & 0 & 1 \end{bmatrix} \mathbf{x}$

13. $\mathbf{T}(\mathbf{x}) = \begin{bmatrix} \sqrt{2}/2 & \sqrt{2}/2 & 0 \\ -\sqrt{2}/2 & \sqrt{2}/2 & 0 \\ 0 & 0 & 1 \end{bmatrix} \mathbf{x}$

15. $\mathbf{T}(\mathbf{x}) = \begin{bmatrix} 0 & 0 & 0 \\ 0 & 1 & 0 \\ 0 & 0 & 1 \end{bmatrix} \mathbf{x}$

17. $\mathbf{T}(\mathbf{x}) = \begin{bmatrix} 25/34 & 0 & 15/34 \\ 0 & 1 & 0 \\ 15/34 & 0 & 9/34 \end{bmatrix} \mathbf{x}$

19. $\mathbf{T}(\mathbf{x}) = \begin{bmatrix} \sqrt{2}/2 & 0 & -\sqrt{2}/2 \\ 0 & 0 & 0 \\ \sqrt{2}/2 & 0 & \sqrt{2}/2 \end{bmatrix} \mathbf{x}$

20. $\mathbf{T}(\mathbf{x}) = \begin{bmatrix} 1 & 0 & 0 \\ 0 & 1 & k\tau \\ 0 & 0 & 1 \end{bmatrix} \mathbf{x}$

21. The image is a rectangle with base 3 parallel to the x-axis and height 2/3 parallel to the y-axis.

23. The image is the ellipsoid $x^2/16 + 4y^2 + 9z^2 = 1$.

27. a counterclockwise rotation about the z-axis by an angle of $60°$, followed by a coordinate rescaling, followed by a clockwise rotation about the z-axis by an angle of $60°$

Section 2.5

1. $\begin{bmatrix} -2/3 & 2/3 \\ 2/3 & 5/3 \end{bmatrix}$

3. $\begin{bmatrix} 0 & 0 & 0 & 1/2 \\ 0 & 0 & -1/2 & 0 \\ 0 & -1/2 & 1 & 0 \\ 1/2 & 0 & 0 & 0 \end{bmatrix}$

4. I

5. $\begin{bmatrix} 6 & 1/2 & 4 & -1 \\ 1/2 & 5 & 1/2 & 9/2 \\ 4 & 1/2 & -2 & 0 \\ -1 & 9/2 & 0 & 0 \end{bmatrix}$

7. negative definite

9. positive definite

11. indefinite

13. indefinite

18. $4b^2 + 3a < 0$

Section 3.1

1. $\{(x,y) | x > -y\}$

3. $\{(x,y) | x \neq 0, y \neq 0\}$

5. $\{(x,y,z) | x^2 + y^2 + z^2 \le 25\}$

7. $\{(x_1,x_2,x_3,x_4) | x_2 \neq -6x_3\}$

9. 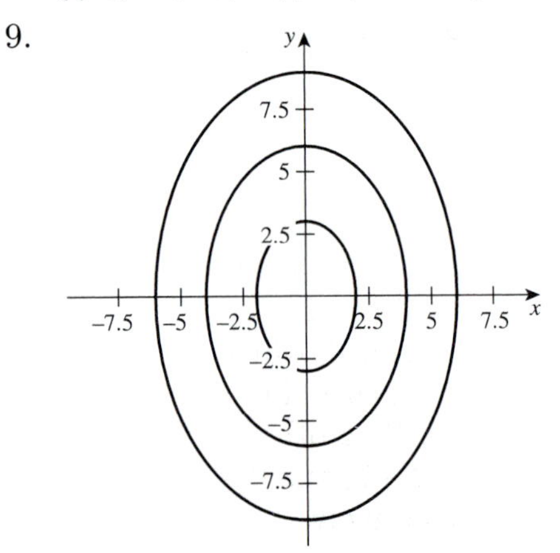

11.

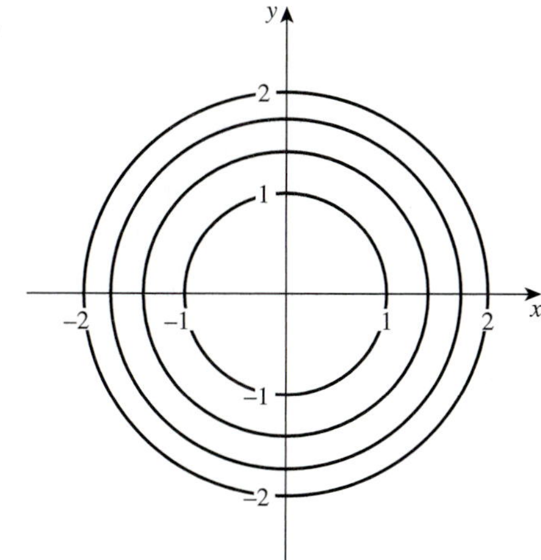

13.

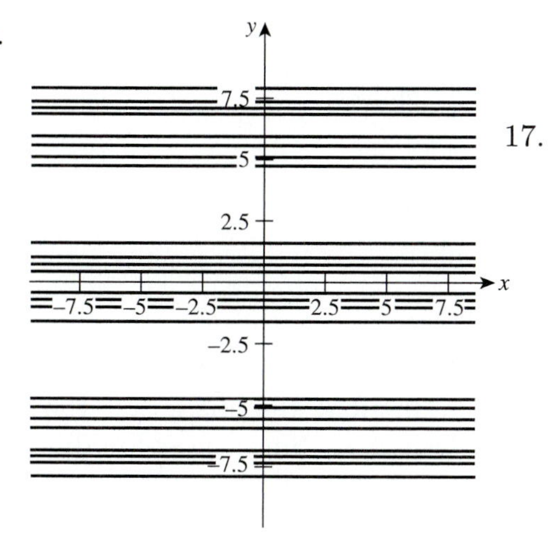

17.

15. The level surfaces are all planes with normal vector $(1, 1, 1)$. The first one lies below the rest and has -3 as an intercept on all three axes. The others have intercepts of -2,-1, and 0 on all three axes, respectively.

19. the unit disk in the xy-plane

21.

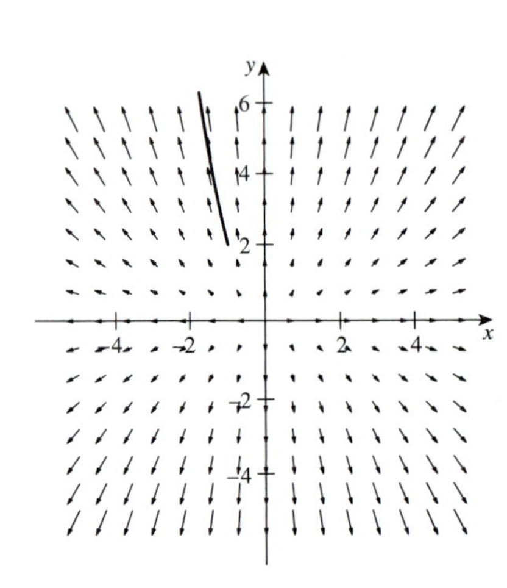

23.

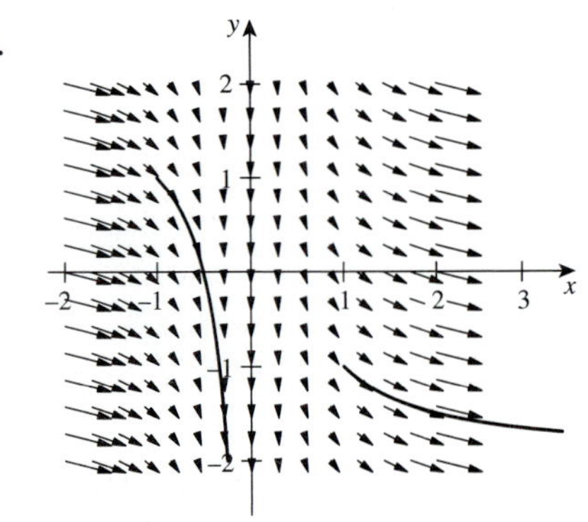

25.

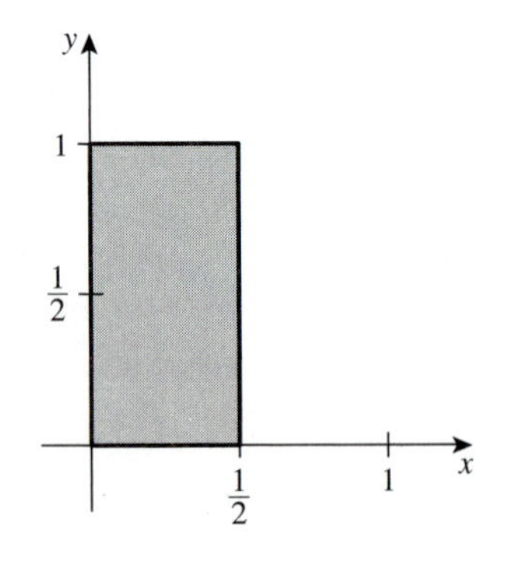

27.

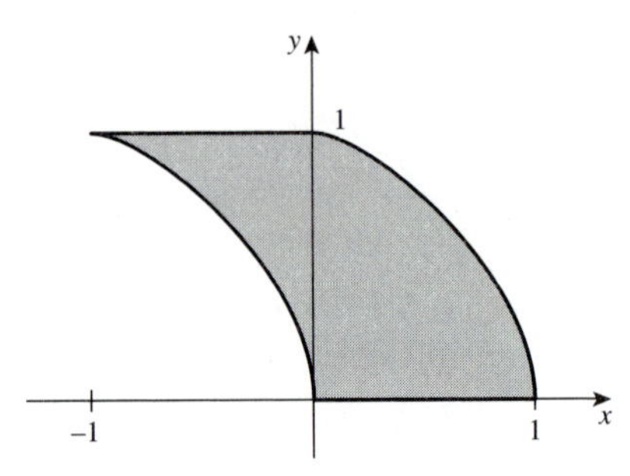

29. If two level curves of $f(x, y)$ intersected, then there would be two different c values for one point (x, y), which contradicts the fact that $f(x, y)$ is a function.

Section 3.2

1. 3
3. 26.5°
5. 1
7. 0
9. 1
11. 1
13. Along the line $x = 0$, the limit is 0; along the line $y = x$, the limit is $\pm\infty$. Since these are different, the limit does not exist.
15. Along the line $y = 0$, the limit is 0; along the curve $y = x^3$, the limit is $1/2$. Since these are different, the limit does not exist.
17. Note that $\lim\limits_{(x,y)\to(0,1^+)} h(x, y) = 0$ and $\lim\limits_{(x,y)\to(0,1^-)} h(x, y) = 1$.
19. $(0, 0)$; removable
21. $(1, 2, 3, 4)$

23. $B(x, y) = h(x, y) + g(f(x, y))$, where $h(x, y) = x - y$, $g(u) = \ln u$, and $f(x, y) = xy$. Observe that f is continuous at $(1, e)$, g is continuous at $f(1, e) = e$, and h is continuous at $(1, e)$. By Theorems 3.2.6 and 3.2.2, B is continuous at $(1, e)$.

25. $\mathbf{D}(x, y) = (D_1(x, y), D_2(x, y))$, where $D_1(x, y) = e^x/y$ and $D_2(x, y) = xe^{-y}$. By Theorem 3.2.2, D_1 and D_2 are continuous at $(1, -1)$; by Theorem 3.2.5, $\mathbf{D}$ is continuous at $(1, -1)$.

Section 3.3

1. neither open nor closed; $\partial A = \{(x, y)|x^2 + y^2 = 1$ or $x^2 + y^2 = 4\}$; $\mathbb{R}^2 \backslash A = \{(x, y)|x^2 + y^2 \leq 1$ or $x^2 + y^2 > 4\}$
3. open; $\partial C = \{(x, y)||x| = 5\}$; $\mathbb{R}^2 \backslash C = \{(x, y)||x| \geq 5\}$
5. $\mathbb{R}^{10} \backslash A = \{(x_1, x_2, \ldots, x_{10})|x_3 < 0$ or $x_3 \geq 1$ or $x_6 \leq 0$ or $x_6 > 2\}$ int $(A) = \{(x_1, x_2, \ldots, x_{10})|0 < x_3 < 1, 0 < x_6 < 2\}$; $\partial A = \{(x_1, x_2, \ldots, x_{10})|x_3 = 0$ or $x_3 = 1$ or $x_6 = 0$ or $x_6 = 2\}$
7. $\mathbb{R}^3 \backslash C = \{\mathbf{x} \in \mathbb{R}^3 | x_2 \leq -1\}$; $\text{int}(C) = C$; $\partial C = \{\mathbf{x} \in \mathbb{R}^3 | x_2 = -1\}$
9. closed
11. closed
13. min $= -9\sqrt{17}$, max $= 9\sqrt{17}$
15. min $= -5$, max $= 5$
16. $\max_{x \in S} = 1 = g(\pi/2)$; $\min_{x \in S} g(x) = -1 = g(3\pi/2)$; $\max_{x \in T} = 1 = g(\pi/2)$; $\min_{x \in T} g(x) = -1/\sqrt{2} = g(-\pi/4)$
17. $\max_{(x,y) \in S} h(x, y) = 2$ attained on $x^2 + y^2 = 4$; $\min_{(x,y) \in S} h(x, y) = 0 = h(0, 0)$
19. $\max k(x, y) = 1 = k(0, 0)$; k has no minimum
21. max $= 11$; min $= -1$

Section 3.4

1. $\partial g/\partial x = 2xy^3 - 3x^2y^2$, $\partial g/\partial y = 3x^2y^2 - 2x^3y$
3. $\partial f/\partial x = ye^{xy} - ye^x$, $\partial f/\partial y = xe^{xy} - e^x$
5. $\partial g/\partial x = \sin(2y + z)$, $\partial g/\partial y = 2x \cos(2y + z)$, $\partial g/\partial z = x \cos(2y + z)$
7. $\partial B/\partial \rho = \sin\phi \cos\theta$, $\partial B/\partial \theta = -\rho \sin\phi \sin\theta$, $\partial B/\partial \rho = \rho \cos\phi \cos\theta$
9. $\partial F/\partial x_1 = 1/(x_3 + x_4)$, $\partial F/\partial x_2 = -1/(x_3 + x_4)$, $\partial F/\partial x_3 = (x_2 - x_1)/(x_3 + x_4)^2$, $\partial F/\partial x_4 = (x_2 - x_1)/(x_3 + x_4)^2$
11. $\partial f/\partial x_1 = \partial f/\partial x_2 = 1/(x_1 + x_2)$, $\partial f/\partial x_3 = 1/(x_4 - x_3)$, $\partial f/\partial x_4 = 1/(x_3 - x_4)$
13. $\partial M/\partial x_i = a_i$
15. uphill, 1 vertical unit per 1000 feet; downhill 1 vertical unit per 1000 feet
17. A small increase in temperature will have a bigger effect on pressure.
19. (a) 0.8; (b) −1.25
21. $\partial g/\partial x = e^y$, $\partial g/\partial y = xe^y$

23. $\partial F/\partial x = 0$, $\partial F/\partial y = 0$

25. $\partial^2 u/\partial x^2 = 24xy^4 + 2y^5$, $\partial^2 u/\partial y\partial x = \partial^2 u/\partial x\partial y = 48x^2y^3 + 10xy^4 + 1$, $\partial^2 u/\partial y^2 = 48x^3y^2 + 20x^2y^3$

27. $\partial^2 u/\partial x^2 = y^2e^{xy}$, $\partial^2 u/\partial y\partial x = \partial^2 u/\partial x\partial y = xye^{xy} + e^{xy}$, $\partial^2 u/\partial y^2 = x^2e^{xy}$

29. $\partial^2 r/\partial v^2 = (2\sin 2\theta)/g$, $\partial^2 r/\partial\theta\partial v = \partial^2 r/\partial v\partial\theta = (4v\cos 2\theta)/g$, $\partial^2 r/\partial\theta^2 = -(4r^2\sin 2\theta)/g$

42. $c_1 + c_2 = -k$

Section 3.5

1. $[8 \quad -7]$, $De(-4,-5)(x,y) = 8x - 7y$

3. $[3/4 \quad -1/4]$, $Dg(1/2,3/2)(x,y) = 3x/4 - y/4$

5. $[\sqrt{6}/4 \quad 0 \quad 0]$, $Dm(0,\pi/4,\pi/3)(x,y,z) = \sqrt{6}x/4$

7. $[-1/15 \quad \cdots \quad -1/15]$, $Dq(-1,1,\ldots,1)\mathbf{x} = -\frac{1}{15}(x_1 - x_2 + x_3 - \cdots + x_{30})$

9. $\begin{bmatrix} 6 & -2 \\ 2 & 6 \end{bmatrix}$; $\mathbf{DF}(3,1)(x,y) = \begin{bmatrix} 6x - 2y \\ 2x + 6y \end{bmatrix}$

11. $J = \begin{bmatrix} 0 & \sqrt{2} \\ -1 & 0 \end{bmatrix}$; $\mathbf{DH}(\sqrt{2},3\pi/2)(\mathbf{x}) = J\mathbf{x}$

13. $J = \begin{bmatrix} 0 & -2/3 & 1/3 \\ 1 & 0 & -1/3 \\ -1 & 2/3 & 0 \end{bmatrix}$; $\mathbf{DK}(1/2,1/3,1/6)(\mathbf{x}) = J\mathbf{x}$

15. $J = \begin{bmatrix} 1/6 & 0 & 0 & 0 & 0 \\ 0 & 0 & 1/2 & -2 & 0 \\ 11 & 0 & 0 & 0 & 1 \end{bmatrix}$, $\mathbf{DN}(0,-\ln 6,4,\ln 2,11)(\mathbf{x}) = J\mathbf{x}$

17. $\begin{bmatrix} 2.73 \\ 6.10 \end{bmatrix}$

19. $\begin{bmatrix} 0.867 \\ -0.465 \end{bmatrix}$

21. $df = dx/2 + dy/4$

23. $df = e^x/(2(e^x - y))\,dx - 1/(2(e^x - y))\,dy$

25. $df = x_2x_3/(x_4x_5)\,dx_1 + x_1x_3/(x_4x_5)\,dx_2 + x_1x_2/(x_4x_5)\,dx_3 - x_1x_2x_3/(x_4^2x_5)\,dx_4 - x_1x_2x_3/(x_4{x_5}^2)\,dx_5$

27. 1.3×10^{-5} m^3

29. 1.34×10^{-3} cm^3

30. 50.9 km^3

32. $[5 \quad -8]$

33. $[4 \quad 1]$

35. $\begin{bmatrix} 0 & 0 & 0 \\ 1 & 0 & -1 \\ 0 & 0 & 2 \end{bmatrix}$

Section 3.6

1. $du/dt = -4x^3\sin t - 8x^3t\cos t^2$

3. $du/dt = (3xt - y)/(2(y^2 + x^2)\sqrt{t})$

5. $du/dt = -2yf'(t)/(x-y)^2 + 2xg'(t)/(x-y)^2$

7. $du/dt = 3\partial g/\partial x + 4\partial g/\partial y - \partial g/\partial z$

9. $du/dt = 4{x_1}^3f_1'(t) + 3{x_2}^2f_2'(t) - 2x_3f_3'(t) + f_4'(t)$

11. $\partial u/\partial s = 1$, $\partial u/\partial t = 2$

13. $\partial u/\partial t_1 = 1/x_1+1/x_2-1/(x_3(t_1-t_2)^2)$, $\partial u/\partial t_2 = 1/x_1 - 3/x_2 + 1/(x_3(t_1-t_2)^2)$

15. $\partial u/\partial s = 2s\partial g/\partial x + 2t\partial g/\partial y$, $\partial u/\partial t = -2t\partial g/\partial x + 2s\partial s/\partial y$

17. $\partial u/\partial s = t(\partial f/\partial x(s+t, 2s+3t) + 2\partial f/\partial y(s+t, 2s+3t)) - s(2\partial f/\partial y(2s+3t, s+t) + \partial f/\partial y(2s+3t, s+t)) - f(2s+3t,\ s+t)$

19. $\partial u/\partial t = r\partial f/\partial x + s\partial f/\partial y$, $\partial u/\partial r = t\partial f/\partial x$, $\partial u/\partial s = t\partial f/\partial y$

21. $\partial u/\partial x = -\partial f/\partial x$, $\partial u/\partial y = -\partial f/\partial x$

23. $J(\mathbf{g} \circ \mathbf{f})(3,2) = \begin{bmatrix} 108 & 108 \\ -6 & -3 \end{bmatrix}$

25. $J(\mathbf{g} \circ \mathbf{f})(\mathbf{a}) = \begin{bmatrix} 0 & 1 & 0 \\ \sqrt{3}/2 & \sqrt{3}/2 & \sqrt{3}/2 \\ 1 & 0 & 0 \end{bmatrix}$

27. $-5/6$

29. $0.012R$

31. increasing at a rate of 0.0367

35. $\partial z/\partial x = (z^3 \sin(x/y) - yz^2 \sin(y/z))(yz^2 \cos(x/y) - xy^2 \cos(y/z))^{-1}$

37. $\partial x/\partial z = (-2z - e^{x+2y})/(ze^{x+2y} - 1)$; $\partial x/\partial y = (1 - 2ze^{x+2y})/(ze^{x+2y} - 1)$

Section 3.7

1. $\vec{\nabla} f = (\frac{3}{2}xy^4 - \frac{5}{4}y^6, 3x^2y^3 - \frac{15}{2}xy^5)$

3. $\vec{\nabla} h = (e^{yz}, xze^{yz}, xye^{yz})$

5. $\vec{\nabla} G = -2((x_1-x_4)^2+(x_2-x_5)^2+(x_3-x_6)^2)^{-2} \cdot (x_1-x_4,\ x_2-x_5,\ x_3-x_6,\ x_4-x_1,\ x_5-x_2,\ x_6-x_3)$

7. $\vec{\nabla} P = 2\exp(\sum_{i=1}^n x_i{}^2)(x_1, \ldots, x_n)$

9. $\vec{\nabla} f = \frac{2}{||\mathbf{x}||^4}(x_1, x_2, \ldots, x_n)$

11. $8\sqrt{5}/3$

13. $1/\sqrt{e^2+1}$

15. $16/\sqrt{5}$

17. $(3+\sqrt{3})/(4\sqrt{5})$

19. $\frac{1}{3}(12 - 3e + 2e^3)(35e^2 - 12e^3 + 2e^4 + 1)^{-1/2}$

21. $3\pi e^{12}/\sqrt{432+\pi^2}$

23. Increasing most rapidly in direction of $\left(\frac{-5}{\sqrt{374}}, \frac{18}{\sqrt{374}}, \frac{-5}{\sqrt{374}}\right)$; decreasing most rapidly in direction of $\left(\frac{5}{\sqrt{374}}, \frac{-18}{\sqrt{374}}, \frac{5}{\sqrt{374}}\right)$

28. $x = -5$

29. $3x + y = 4$

31. $3y - x + \pi - 3 = 0$

32. $-6x - 4 + 3y - 1 = 0$

33. $2y - 2x + 2 = 0$

35. $27(x-1) + 7(y-3) - (z+5) = 0$

37. $3x - 9y + z - 4 = 0$

39. $2x + 10y - 2z - 54 = 0$

41. $2x + 4y - 2z - 12 = 0$

43. $((1/3)^{1/4}, (1/3)^{1/4}, (1/3)^{1/4})$, $(-(1/3)^{1/4}, -(1/3)^{1/4}, -(1/3)^{1/4})$

45. $((9-\sqrt{7})/4, (3-\sqrt{7})/4, (26-3\sqrt{7})/4)$, $((9+\sqrt{7})/4, (3+\sqrt{7})/4, (13+3\sqrt{7})/2)$

51. $(\frac{1}{\sqrt{3}}, \frac{1}{\sqrt{3}}, \frac{1}{\sqrt{3}})$, $(\frac{1}{\sqrt{3}}, -\frac{1}{\sqrt{3}}, \frac{1}{\sqrt{3}})$, $(\frac{1}{\sqrt{3}}, \frac{1}{\sqrt{3}}, -\frac{1}{\sqrt{3}})$, $(\frac{1}{\sqrt{3}}, -\frac{1}{\sqrt{3}}, -\frac{1}{\sqrt{3}})$, $(-\frac{1}{\sqrt{3}}, \frac{1}{\sqrt{3}}, \frac{1}{\sqrt{3}})$, $(-\frac{1}{\sqrt{3}}, -\frac{1}{\sqrt{3}}, \frac{1}{\sqrt{3}})$, $(-\frac{1}{\sqrt{3}}, \frac{1}{\sqrt{3}}, -\frac{1}{\sqrt{3}})$, $(-\frac{1}{\sqrt{3}}, -\frac{1}{\sqrt{3}}, -\frac{1}{\sqrt{3}})$

52. $f(x, y, z) = (3x + 5z, 8x - y + 2z, 4y + 6z)$

Section 3.8

1. $\operatorname{div} \mathbf{B} = \frac{4}{3}$, $\operatorname{curl} \mathbf{B} = (0, 0, 0)$
3. $\operatorname{div} \mathbf{E} = 1$, $\operatorname{curl} \mathbf{E} = (5, -8, 6)$
5. $\operatorname{div} \mathbf{G} = 1/y + 1/z + 1/x$, $\operatorname{curl} \mathbf{G} = (y/z^2, z/x^2, x/y^2)$
7. $\operatorname{div} \mathbf{J} = yz \cos(xy) - x^2 \sin(xy)$; $\operatorname{curl} \mathbf{J} = (xy \sec^2(xy) + \tan(xy), \sin(xy) - y^2 \sec^2(xy), -xy \sin(xy) + (1 - xz) \cos(xy))$
19. $\operatorname{curl}(\mathbf{F}/f) = (f \operatorname{curl} \mathbf{F} - \vec{\nabla} f \times \mathbf{F})/f^2$
27. (a) $\operatorname{div}(\mathbf{v}) = 3$; the volume is increasing at a rate of 3 units of volume per unit time with expansion at a rate of 1 in all directions
    (c) $\operatorname{div}(\mathbf{v}) = 0$; volume remains the same, but expanding at a rate of 3 units per unit time in the x-direction, contracting at a rate of 1 unit per unit time in the y-direction and 2 in the z-direction
    (e) $\operatorname{div}(\mathbf{v}) = -3$; the volume is contracting at a rate of 3, with contraction in all directions at a rate of 1
28. (a) They move parallel to the y-axis at a speed proportional to their height (z-coordinate).
    (c) $\operatorname{curl} \mathbf{F} = (-k, 0, 0)$; there is a clockwise rotation on an axis parallel to the x-axis.

Section 3.9

1. $Hf(1, 4) = \begin{bmatrix} 4 & -3 \\ -3 & 28 \end{bmatrix}$
3. $Hf(0, 3, 1) = \begin{bmatrix} 0 & 0 & 0 \\ 0 & 0 & 1 \\ 0 & 1 & 0 \end{bmatrix}$
5. $Hf(1, 1, 1, 1) = \begin{bmatrix} 0 & 0 & -1/4 & -1/4 \\ 0 & 0 & -1/4 & -1/4 \\ -1/4 & -1/4 & 1/2 & 1/2 \\ -1/4 & -1/4 & 1/2 & 1/2 \end{bmatrix}$
7. $Hf = 2I$
8. $p_2(x, y) = 7 + 8(x - 1) - 6(y + 1) + 4(x - 1)^2 + 3(y + 1)^2$
9. $p_2(x, y) = 2(x - 1)^2 + 3(y - 7)^2 + 84(y - 7) + 584$
11. $p_2(x, y) = x^2 + y^2$
13. $p_2(x, y, z) = -2\sqrt{2} + (1/\sqrt{2})(x + 4) + 2\sqrt{2}(y - \pi/4) + (\pi/\sqrt{2})(z - 1) - (1/\sqrt{2})(x + 4)(y - \pi/4) - (\pi/4\sqrt{2})(x + 4)(z - 1) + (4 + \pi/\sqrt{2})(y - \pi/4)(z - 1) + \sqrt{2}(y - \pi/4)^2 + (\sqrt{2}\pi^2/16)(z - 1)^2$
15. $p_2(x, y, z, w) = 3(x - 1) + 2(y + 1) + w - 41$
21. $p_4(x, y) = 1 + (x + y) + (1/2!)(x + y)^2 + (1/3!)(x + y)^3 + (1/4!)(x + y)^4$
23. $p_4(x, y) = (x - y) + (1/4)(x - y)^2 + (1/32)(x - y)^3 + (1/384)(x - y)^4$
25. $p_2(x, y) = 1 - (x + y)^2/72$
27. $p_6(x, y) = 1 + (x - y)^2 + (x - y)^4 + (x - y)^6$

Section 3.10

1. $(0, 0)$
3. (x, y) such that $y = (5/2)x$

5. all points (x, y) such that $y = 0$
7. $(k\pi, (2m+1)\pi/2)$, $((2k+1)\pi/2, m\pi)$ where $k = 0, \pm 1, \pm 2, \ldots$ and $m = 0, \pm 1, \pm 2, \ldots$
9. no critical points
11. $(0, 0, 0)$
13. $f(2, 0)$ is a local maximum.
14. local maximum at $(-1/\sqrt{3}, 0)$; saddle at $(-1/\sqrt{3}, 0)$
15. Second Derivative Test is inconclusive; saddle at $(0, 0)$.
17. $f(0, 0) = 0$ is a local minimum; f has a saddle at the point (0, 2).
19. local minimum at $(0, 0, 0)$
21. local minimum at $(1, 1, 1, 1/2)$
22. $(\pm 1, 0, 0)$
23. Box is $\frac{a}{3} \times \frac{b}{3} \times \frac{c}{3}$.
26. $y = -1.098x + 9.527$
29. $p(t) = 1.40812 - 0.47665t$
31. $p(t) = 0.77404t$
33. $p(t) = 0.996294 + 1.10364t + 0.53672t^2$
35. no local extrema; $2x^2$; 2; 0

Section 4.1

1. $\mathbf{f}'(0) = \mathbf{0}$
3. Both parametrize the circle $x^2 + y^2 = 1$.
8. $\mathbf{f}(t) = (1+t, 3+t, 2-(3/2)t)$, $0 \le t \le 4$; $\mathbf{g}(t) = (5-t, 7-t, -4+(3/2)t)$, $0 \le t \le 4$
9. $\mathbf{f}(t) = (\cos t, 2\sin t)$, $0 \le t \le 6\pi$; $\mathbf{g}(t) = (-4/t^2 + 1, -4/t)$, $-\infty \le t \le \infty$
10. $((40)^{3/2} - (13)^{3/2})/27$
11. $2\sqrt{195}$
13. 6
15. $8a$
17. $s = 4a(1 - \cos(t/2))$
19. $\mathbf{f}(s) = (a\sin(s/a), a\cos(s/a))$, $0 \le s \le 2a\pi$

Section 4.2

1. $40\sqrt{10}/3$
3. $(-17/18)\sqrt{613}$
5. $a \ln 2$
7. $6\sqrt{5}$
9. $8/3$
11. -34
13. $-161/2$
15. $16/3$
17. $(-5e^2 + 10e + 1)/e^2$
19. 4
21. (a) $-gc$; (b) $-gc$
24. $GmM/(R_1 + R_e)$
25. As $R \longrightarrow \infty$, work $\longrightarrow -\infty$ (i.e., it takes infinitely much energy to send the ship up!).
26. 18π
27. 5
29. 0

Section 4.3

1. 21
3. 467/64
5. 4.85834
7. 256

8. $(\pi^2 + 4)/2$
9. 2.99
11. $-1/2 \ln(1/\sqrt{2}) + \ln(\cos(\pi/8))$
12. 126
13. $(9/16)\ln(17) - 1$
14. $(23/50)e^{25} + (27/50)$
15. $(1/2)(1 - \ln 2)$
16. 121/108
17. 1/2
19. 8
21. 1
23. $(2/3)A$
25. $\int_0^1 \int_{\sqrt{1-y^2}}^{\sqrt{9-y^2}} \frac{k}{x^2+y^2}\, dx\, dy + \int_1^3 \int_0^{\sqrt{9-y^2}} \frac{k}{x^2+y^2}\, dx\, dy$
26. $\int_0^2 \int_0^y f(x,y)\, dx\, dy$
27. $\int_1^e \int_{\ln x}^1 f(x,y)\, dy\, dx$
29. $\int_{-4}^4 \int_{-\sqrt{16-x^2}}^{\sqrt{16-x^2}} f(x,y)\, dy\, dx$
31. $\int_{-4}^{-2} \int_{-1}^{x+3} f(x,y)\, dy\, dx + \int_{-2}^0 \int_{-1}^1 f(x,y)\, dy\, dx + \int_0^1 \int_{-1}^{\sqrt{x}} f(x,y)\, dy\, dx + \int_0^1 \int_{\sqrt{x}}^1 f(x,y)\, dy\, dx$
33.

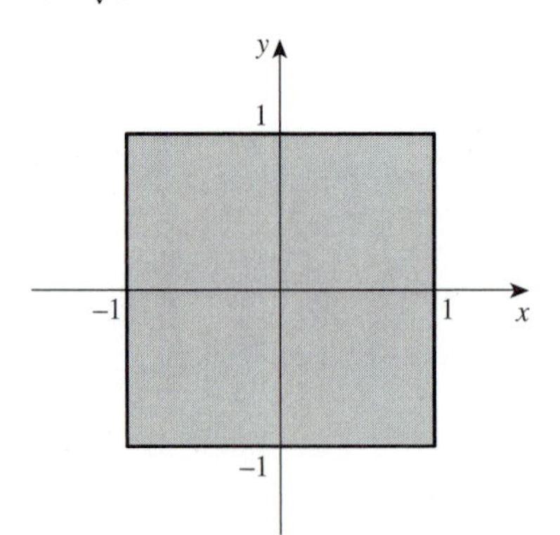

positive

35.

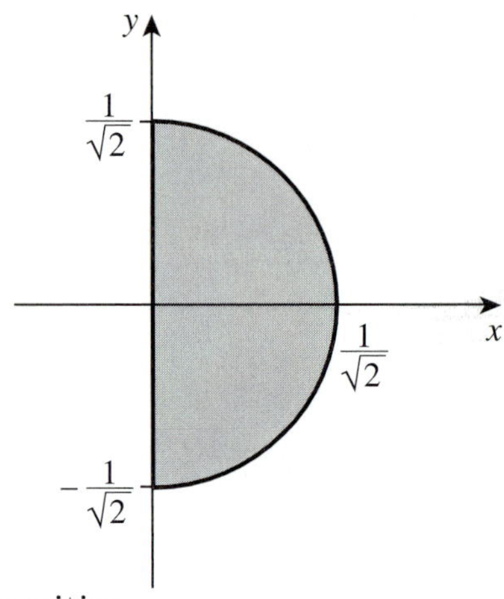

positive

37. 0
39. 250
41. 9/2
42. $\int_0^1 \int_0^2 x^2 y^3\, dx\, dy$; R is the rectangle with vertices $(0,0)$, $(1,0)$, $(1,2)$, and $(0,2)$.

Section 4.4

1. 1.501
3. 0.227
4. $((5 \ln 2)/2)(e - 1)$; R is the cube with width, length, and height 1.
5. $-(20/3) - 5 \ln 3$; R is the cube with length 2, width 5, and height 1.
7. $9/2$; R is the tetrahedron in the first octant bounded by the planes $x = 0, y = 0, z = 0$, and $6x + 12y + 18z = 36$.
9. 1; R is the solid formed by the planes $y = 0, y = x, x = 1$, and $z = 1 + x + y$.
11. 72.6

13. $256/15$

15. $1/6$

17. $16a^3/3$

19. $1/720$

21. $\int_1^2 \int_0^1 \int_{-1}^0 f(x,y,z)\,dx\,dz\,dy$; $\int_{-1}^0 \int_1^2 \int_0^1 f(x,y,z)\,dz\,dy\,dx$

23. $\int_0^1 \int_0^{3\sqrt{1-x^2}} \int_0^{2\sqrt{1-x^2-y^2/9}} f\,dz\,dy\,dx$; $\int_0^3 \int_0^{2\sqrt{1-y^2/9}} \int_0^{\sqrt{1-y^2/9-z^2/4}} f\,dx\,dz\,dy$

25. zero

27. zero

28. negative

Section 4.5

1. $\mathbf{x} = (\sin\phi\cos\theta, \sin\phi\sin\theta, \cos\phi)$, $\pi/2 \le \phi \le \pi$, $0 \le \theta \le 2\pi$

3. $\mathbf{x} = (w, \cos\theta, \sin\theta)$, $0 \le \theta \le 2\pi$, $0 \le w \le 5$

5. $\mathbf{x} = (x, y, (8 - x + 2y)/4)$, $-1 \le x \le 0$, $3 \le y \le 7$

7. $\mathbf{x} = (\theta\cos\theta, \theta\sin\theta, w)$, $0 \le \theta \le 6\pi$, $-2 \le w \le 2$

9. $\mathbf{x} = ((t-s)/3, (t+2s)/3, 1-(2t-s)/3)$, $0 \le s \le 1$, $0 \le t \le 1$

11. $3x - 2y - 3z = 0$

13. $y = x^2/4 - z^2$

19. $\int_{-1}^1 \int_0^1 \sqrt{30}\,ds\,dt = 2\sqrt{30}$

21. $\sqrt{2}\int_0^1 \frac{s^2\sqrt{s^4+s^2+1}}{s^2+1}\,ds$

23. $\pi(5^{3/2} - 1)/6$

24. $32\sqrt{10}$

25. $(16\sqrt{3} - 8)/(3\sqrt{2})$

27. $\pi\sqrt{2}$

28. (a) $(f_1(s)\cos t, f_2(s), f_1(s)\sin t)$

30. (a) $(r\cos\theta, (r\sin\theta - R)\cos\phi, (r\sin\theta - R)\sin\phi)$, $0 \le \theta \le 2\pi$, $0 \le \phi \le 2\pi$

31. area $= 2\pi\int_c^d |f(y)|\sqrt{1 + f'(y)^2}\,dy$

33. $\int_0^1 \int_a^b ||(\mathbf{g}_2(t) - \mathbf{g}_1(t)) \times (s\mathbf{g}_2'(t) + (1-s)\mathbf{g}_1'(t))||\,dt\,ds$

35. $9\pi/2$

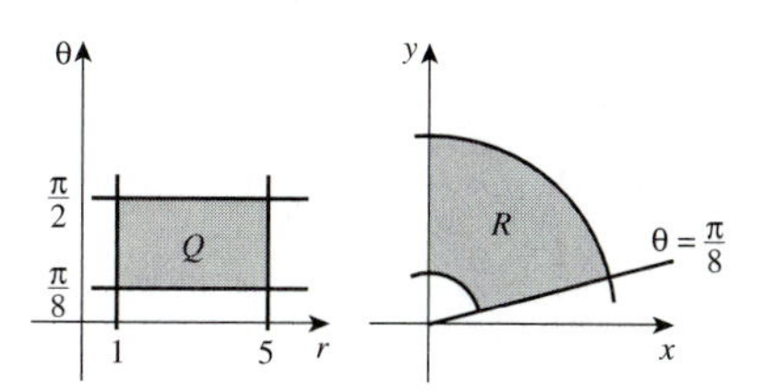

37. $\pi/8$ 

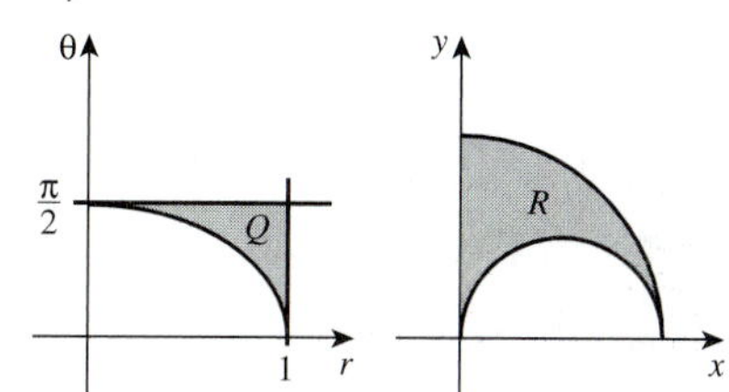

39. 1

40. $5\ln 2$

42. $3\sqrt{2,490}$

Section 4.6

1. $\sqrt{3}/4$

3. $\pi^3/24$

5. 0.10851

7. 1,039

9. π

10. $5\sqrt{2}/3$

11. 0
13. $\pi/2 - 1$
15. 39
17. $\iint_M x \sin y \, dy \wedge dz + z \sin y \, dz \wedge dx + x \cos z \, dx \wedge dy$
19. $\iint_M x \, dy \wedge dz + y \, dz \wedge dx + z \, dx \wedge dy$
20. 3
21. $(40/2) \ln 2 - 5$
23. $-96/35$
25. 0
27. (a) $64K/3$; (b) $-16\pi K r^3/3$; (c) 0

Section 4.7

1. $(23/2) \ln \sqrt{2}$
3. 2π
5. $15\pi^2/2, 304$
7. $\pi(1 - e^{-4})$
9. $3 - \ln(\sqrt{6} - \sqrt{3})$
11. 2π
14. $-(1/9)$
15. $95/648$
17. $35/48$
19. π
21. 2

Section 4.8

1. $\pi/4$
3. $4\pi/3$
5. $17\pi/240$
6. $81\pi/8$
7. 0
9. $128\pi/3$
11. $81/5$
13. $2/3$
15. 2
16. $(h^2 + a^2 R^4 - 2ahR^2)\pi/(2a)$
17. $(4\pi/3)(R^2 - r^2)^{3/2}$
19. 8π

Section 5.1

1. path independent; $f = (x^2/2) \sin y + x + k$
2. path independent; $f = x^2 y^3 + x + (1/y) + k$
3. not path independent
5. path independent; $f = x \cot y - z \tan x + k$
7. path independent; $f = x_1 + x_2^2 + x_4^5 + x_3^2 + k$
8. exact; $f = 5yx + x^2 - y^4$
9. exact; $f = \arctan(x + y)$
11. exact; $f = x^2 y - x^2 + y^2 + z^2$
12. not exact
13. exact; $f = xy/z$
15. -33
17. 0
19. 14
21. 0
23. -4
24. (a) connected
(b) connected
(c) disconnected
(d) disconnected
(e) connected
(f) connected

(g) disconnected

25. a and f

26. $x^3y - 3x^2 + y^2 = 1$

27. $x^2y + y^3/3 = 4/3$

29. $x \cos y = -1$

33. $-\ln(7\sqrt{6}/3)$

Section 5.2

1. -4

3. -2

5. $31/12$

7. $512/3$

9. Yes. The integrand is smooth on the region bounded by C; the point of discontinuity (0, 2) is not contained in this region.

10. πa^2

11. $4/3$

13. $8/15$

15. $3a^2/2$

17. $(5/2)\sin(2\pi/5)r^2 \approx 2.38r^2$

Section 5.3

1. 240

3. 0

5. 0

7. 7

9. $256/35$

11. 0

13. $-4\pi\sqrt{3}$

24. (a) $1/6$; (b) $25/6$

Section 5.4

1. -2π

3. $-\pi$

5. $33/2$

7. π

Index